PHYSICS AS A CALLING

CORNELL HISTORY OF SCIENCE SERIES

Editor: L. Pearce Williams

Mersenne and the Learning of the Schools
by Peter Dear

The Scientific Reinterpretation of Form
by Norma E. Emerton

Geology in the Nineteenth Century:
Changing Views of a Changing World
by Mott T. Greene

Physics as a Calling: Discipline and Practice in the Königsberg
Seminar for Physics
by Kathryn M. Olesko

PHYSICS AS A CALLING

DISCIPLINE AND PRACTICE IN THE KÖNIGSBERG SEMINAR FOR PHYSICS

KATHRYN M. OLESKO

CORNELL UNIVERSITY PRESS

Ithaca and London

Cornell University Press and L. Pearce Williams, general editor of the Cornell History of Science series, gratefully acknowledge several grants that have aided in bringing this book to publication. We thank the Herbert and Roseline Gussman Foundation, Ronald P. Lynch, Emmett W. MacCorkle, Jeffrey W. MacCorkle, and M. K. Whyte.

First published 1991 by Cornell University Press.

International Standard Book Number 0-8014-2248-5
Library of Congress Catalog Card Number 90-55717
Printed in the United States of America
Librarians: Library of Congress cataloging information
appears on the last page of the book.

⊗ The paper in this book meets the minimum requirements
of the American National Standard for Information Sciences—
Permanence of Paper for Printed Library Materials, ANSI Z39.48-1984.

FOR WAYNE

Foreword

HISTORIANS OF SCIENCE, for some time, have examined, analyzed and theorized about the contributions of philosophy, religion, mathematics, theory, and experiment to the rise of modern science. Sociologists of science have only recently, and rather stridently, called attention to the social dimensions of science, which involve the structure of social institutions, the institutionalizing of science itself, and that old familiar deus ex machina, the Zeitgeist.

Curiously, no one before Kathryn Olesko has taken a serious and detailed look at one of the places where the essentially intellectual currents examined by the historians run into the social currents of a given society, namely in the education of scientists. This book is about the creation of a new profession. The very word "physicist" was introduced into the English language only in 1840, according to the Oxford English Dictionary. Before 1800, "physics" generally meant the qualitative study of all physical bodies in the world, animal, vegetable, and mineral. In 1900, it meant the precise and mathematical presentation of laws applying to things such as fields and bodies in motion and hypothesized molecular and atomic particles. This profound change was the result of complex interactions, among which we may here cite the growth of French mathematical physics in the early nineteenth century, the rapid development of precise instrumentation throughout the nineteenth century, the discovery of important new mathematical and experimental tools for the analysis of Nature, and, by no means least important, the creation of institutions for the education of the young to carry on and push forward these new approaches.

Dr. Olesko has made a long and careful examination of one of these institutions, the Königsberg seminar founded by Franz Neumann in

1834, which continued under his direction until his retirement in 1876. Neumann was not one of the giants of nineteenth-century physics; indeed, it could be argued that his work was seriously out of date by the end of the century, superseded both institutionally and intellectually by Maxwell's field theory and the new, great research laboratories at Cambridge, Berlin, and Paris. But that would be to miss a major point. What Neumann achieved was a synthesis of French mathematical physics and the older empiricism based upon careful experiments monitored by increasingly precise instruments. Thus Neumann underlined the emphasis of his teaching; mathematical physics had to feed upon instruments, experimental techniques, and critical techniques, such as the method of least squares, if it were not to be led astray either by enthusiastic empiricists unable to see the logical consequences of observations because of lack of mathematical sophistication or by mathematical enthusiasts ever eager to generalize beyond the empirical evidence. The major concern of the Königsberg seminar was to prepare teachers of physics on both the secondary and higher levels of science education to pass on Neumann's vision of physics.

It was in the nineteenth century that Germany evolved from a region that only occasionally produced a natural philosopher of the first rank to a nation that dominated the scientific world. To make this possible, talent had to be discovered, nurtured, and trained. It is this process that Dr. Olesko has studied, in microcosm, in this work.

L. PEARCE WILLIAMS

Ithaca, New York

Contents

Figures and Tables   xi
Preface   xiii
Note on Citations   xvi
Abbreviations   xvii
Introduction   1

PART I. DISCIPLINE

1. Educational Reform and the Natural Sciences   21
2. Quantification in Physics Teaching and Research   61
3. Shaping a Pedagogical Physics   99
4. Mechanics and the Besselian Experiment   128
5. Successes and Realizations   172
6. Error Analysis and Measurement   205
7. Interpolation and Certainty   235
8. Error over Truth   266

PART II. PRACTICE

9. The Workaday World of *Physiklehrer*   317
10. The Ethos of Exactitude   366
    Epilogue   451

APPENDIXES
1. Provisional Statutes of the Seminar for Mathematics and
   Physics at the University of Königsberg   461
2. Neumann's Cycle of Lecture Courses   464
3. Directory of Identified Seminar Students   469
   Glossary of German Terms   473
   Index   477

Figures and Tables

FIGURES

1. Friedrich Wilhelm Bessel   27
2. Bessel's observatory at Königsberg, 1815   28
3. Floor plan of Bessel's observatory   29
4. Bessel's seconds pendulum apparatus   71
5. Carl Gustav Jacob Jacobi   101
6. Franz Ernst Neumann   102
7. Page from Paul Peters's seminar notes on theoretical
   physics, 1870/71   152
8. Portion of title page from Kirchhoff's first investigation   181
9. Kirchhoff's drawing of the curves of equal tension   181
10. Schematic drawing of Wild's polarimeter   225
11. Wild's polarimeter   227
12. Wild's graph of optical intensities as a function of angular
    distance from the sun   229
13. Hagen's graph representing absorption coefficients as
    a function of wavelength   240
14. Quincke's graph depicting equipotential lines of an electric
    current on a square lead plate   244
15. Quincke's graph depicting equipotential lines of an electric
    current on a circular plate   245
16. Quincke's schematic drawing of a mercury drop   249
17. Quincke's apparatus for determining capillary constants   249
18. Meyer's graph depicting the viscosity constant of salt
    solutions as a function of concentration   258
19. Müttrich's graphs of optical constants as a function of
    temperature   259

20. Page from Paul Peters's seminar notes on capillary theory,
    winter semester 1872/73   277
21. Neumann's apparatus for measuring elasticity constants   291
22. Voigt's "curve tables"   293
23. Voigt's representation of the coefficients of dilation and
    torsion   296
24. Neumann's apparatus for measuring the thermal
    conductivity of metals   307
25. Neumann in his study   309
26. Baumgarten's graph depicting elasticity as a function of
    weight   342
27. A Königsberg geothermometer   351
28. Dorn's method of interpolation for thermometer calibration   352
29. Mischpeter's apparatus for measuring thermal expansion   353
30. Ground floor plan of the Königsberg Physical Institute   440

TABLES

1. Bonn University natural sciences seminar curriculum   48
2. Professions of Neumann's seminar students   318
3. Secondary school teachers from Neumann's seminar   320
4. Secondary school teachers identifying with physics   323
5. The practice of exact experiment in physics   369

Preface

Physics as a Calling is a history of the Königsberg physical seminar and its meaning in the scientific practices of the students who attended it. Established in 1834, the seminar played an important role in integrating methods of quantification, including methods of error and data analysis, into physics instruction in Germany. We know little about how that was accomplished and what problems were encountered along the way. The director of the seminar, Franz Ernst Neumann, developed a scientific pedagogy that not only engaged students directly in mathematical and measuring exercises but also inculcated rigorous standards for evaluating empirical data and theoretical results. In striving for a deeper understanding of the relation between theoretically calculated and experimentally observed values, his students helped to extend the method of least squares; to refine the analysis of constant, or systematic, errors; and to raise questions about certain quantitative techniques of generalization, such as interpolation and the graphical analysis of data. In the process, they espoused a conception of scientific truth tied to the type of experiment practiced by the astronomer Friedrich Wilhelm Bessel, who had strongly influenced the practical element in Neumann's science pedagogy. The relative weights they assigned to these techniques marked their distinctive investigative and pedagogical styles, which helped to shape not only what they called theoretical physics but also the exact experimental physics of nineteenth-century Germany.

While this book went through several versions as I rethought the meaning of the seminar, I received research assistance and support from individuals and institutions, to whom I extend my heartfelt thanks. The holdings of three archives and libraries were particularly crucial for the development of my themes. My understanding of the evolution of Neumann's teaching during the early years of the seminar is drawn principally from the official annual seminar reports he submitted to the Prussian Kultusministerium, now located in the Zentrales Staatsarchiv, Abteilung II, Merseburg. The Merseburg archive also contains much of the material on other science seminars in Prussia, with which I compared Neumann's, and on Bessel's observatory. The Franz Neumann Nachlaß, housed in the Handschriftenabteilung of the Niedersächsische Staats- und Universitätsbibliothek, Göttingen, contains the lecture notes, correspondence, and research notes essential for understanding the content of Neumann's teaching in the early years of the seminar and the effect of his teaching upon the subsequent scientific practices of his students. Draft seminar reports, also in that collection, rounded out and confirmed the developments I had charted from material found in Merseburg. The directors of the Handschriftenabteilung, Dr. Klaus Haenel and his successor Dr. H. Rohlfing, have graciously and promptly responded to my requests for material and extended their expertise, friendship, and hospitality when I traveled to Göttingen. The Staatsbibliothek Preußischer Kulturbesitz in Berlin contains a good portion of the *Schulprogramme* I needed for the secondary school activity of Neumann's former seminar students. Dr. Regina Mahlke helped me find these in the stacks of the library and offered advice on other useful sources. The Handschriftenabteilung of the Staatsbibliothek in Berlin, as every student of German science knows, contains a rich collection of letters. I thank the staff of that division for helping me to track down the correspondence of students who had attended the Königsberg physical seminar.

My research is also based on material in the university archives of Berlin, Bonn, Erlangen, Freiburg, Giessen, Göttingen, Heidelberg, Marburg, Munich, Tübingen, and Würzburg; and in the Staatsarchiv des Kantons Basel-Stadt, the Geheimes Staatsarchiv Preußischer Kulturbesitz (including the Staatsarchiv Königsberg), the Zentrales Akademie-Archiv of the Akademie der Wissenschaften in Berlin, the Nordrhein-Westfälisches Hauptstaatsarchiv in Düsseldorf, the Handschriftenabteilungen of the Universitätsbibliothek Giessen and Universitätsbibliothek Heidelberg, the Niedersächsisches Hauptstaatsarchiv in Hannover, the Generallandesarchiv Karlsruhe, the Firmenarchiv of the Siemens-Museum in Munich, the Sondersammlungen of the Deutsches Museum in Munich, the Bayerisches Hauptstaatsarchiv in Munich, the Landesarchiv Schleswig-Holstein in Kiel, and the Hessisches Staatsarchiv in Marburg. For access to collections and for permission to quote from them, I thank the officials of all these institutions.

Some archival material relating to nineteenth-century German science is still in private hands. For the opportunity to discuss the family letters of Woldemar Voigt, I thank Voigt's niece, Frau Maria Voigt, and Prof. Dr. Friedrich Hund, both of Göttingen.

Thomas Barrett and Keyko Lüders provided research assistance on this project. Unless otherwise noted, David Hagen of the Georgetown University Photographic Services prepared the figures for publication. Fellow historians of science Kenneth L. Caneva and Gert Schubring graciously supplied copies of articles I could not find in the Washington, D.C., area. The Washington community of historians of science offered intellectual stimulation, moral support, and gentle prodding. It was my pleasure to use the collections of the Library of Congress, one of the finest libraries in the world.

Parts of my research and much-needed release time from teaching were supported by the National Science Foundation and the National Endowment for the Humanities. Additional financial support for summer grants and research supplies came from the Graduate School, the School of Foreign Service, and the Provost of Georgetown University.

L. Pearce Williams unselfishly and patiently extended guidance and encouragement while a project he considered finished a long time ago continued to metamorphose. At the later stages of writing, Frederic L. Holmes watched this book evolve, read most versions of it, and gracefully offered criticisms and suggestions.

As Bessel found, errors are a part of truth; I am responsible for any that remain. All translations are my own.

Wayne and John became best friends while this book was being written. This book is lovingly dedicated to Wayne and to all he stands for.

KATHRYN M. OLESKO

Springfield, Virginia

Note on Citations

FULL BIBLIOGRAPHIC CITATIONS are supplied on the first instance in each chapter; subsequent references generally use only the author's last name and an abbreviated title. Citations of archival material vary according to the standards requested by each archive. I have, however, included more information than required for citations from acts contained in the Zentrales Staatsarchiv in Merseburg because these documents are so central to my arguments.

Schulprogramme often present cataloging problems to libraries in the United States and Germany; they also present citation problems for scholars who use more than a few. Their official titles are lengthy, cumbersome, and inconsistent. Because major depositories arrange them alphabetically by city, and within each city by school and year, I have used only this information in citing them. (This system is similar to the way a university *Vorlesungsverzeichniss* or *Chronik* is customarily cited.) Most of the *Schulprogramme* I have used are found in the Staatsbibliothek Preußischer Kulturbesitz in Berlin and the Universitätsbibliothek in Giessen.

Abbreviations

AFWB: Abhandlungen von Friedrich Wilhelm Bessel, ed. Rudolf Engelmann, 3 vols. (Leipzig: W. Engelmann, 1875–76)
AHES: Archive for History of Exact Sciences
AN: Astronomische Nachrichten
AP: Annalen der Physik und Chemie (to 1900); *Annalen der Physik* (after 1900)
BJHS: British Journal for the History of Science
Carl's Repertorium: Repertorium für physikalische Technik, für mathematische und astronomische Instrumentenkunde
Darms. Samml.: Darmstaedter Sammlung, SBPK-Hs
DM: Deutsches Museum, Sondersammlungen, Munich
EZ: Elektrotechnische Zeitschrift
FNN: Franz Neumann Nachlaß, UBG-Hs
GLA: Generallandesarchiv Karlsruhe
GN: C. F. Gauss Nachlaß, UBG-Hs
GSPK: Geheimes Staatsarchiv Preußischer Kulturbesitz, Berlin
GW: Franz Neumanns Gesammelte Werke, ed. by his students, 3 vols. (Leipzig: B. G. Teubner, 1906–28)
HS: History of Science
HSPS: Historical Studies in the Physical and Biological Sciences
JDMV: Jahresberichte der Deutschen Mathematiker-Vereinigung
JRAM: Journal für die reine und angewandte Mathematik
LN: Luise Neumann, *Franz Neumann: Erinnerungsblätter von seiner Tochter* (Leipzig: J. C. B. Mohr, 1904)
MA: Mathematische Annalen
MZ: Meteorologische Zeitschrift
NWH: Nordrhein-Westfälisches Hauptstaatsarchiv, Düsseldorf
PZ: Physikalische Zeitschrift
SAKBSt: Staatsarchiv des Kantons Basel-Stadt

SBPK-Hs: Staatsbibliothek Preußischer Kulturbesitz, Handschriftenabteilung, Berlin
SPGK: Schriften der physikalisch-ökonomischen Gesellschaft zu Königsberg
UAB: Universitätsarchiv Bonn
UAE: Universitätsarchiv Erlangen
UAF: Universitätsarchiv Freiburg
UAG: Universitätsarchiv Göttingen
UAGi: Universitätsarchiv Giessen
UAH: Universitätsarchiv Heidelberg
UAM: Universitätsarchiv Munich
UAT: Universitätsarchiv Tübingen
UAW: Universitätsarchiv Würzburg
UBG-Hs: Niedersächsische Staats- und Universitätsbibliothek Göttingen, Handschriftenabteilung
UBGi-Hs: Universitätsbibliothek Giessen, Handschriftenabteilung
WVN: Woldemar Voigt Nachlaß, UBG-Hs
ZAA: Zentrales Akademie-Archiv, Akademie der Wissenschaften, Berlin
ZMNU: Zeitschrift für mathematischen und naturwissenschaftlichen Unterricht
ZPCU: Zeitschrift für den physikalischen und chemischen Unterricht
ZStA-M: Zentrales Staatsarchiv, Abteilung II, Merseburg

PHYSICS AS A CALLING

Introduction

Sᴇᴍɪɴᴀʀꜱ ᴛʜᴀᴛ ᴄᴏɴᴄᴇɴᴛʀᴀᴛᴇᴅ on research methods appeared in the German universities of the nineteenth century. By then, seminars had been in existence for about two centuries. Initially they were institutes for training preachers, but by the late eighteenth century they became settings for educating secondary school teachers as well. Concurrent with the emergence of the modern academic disciplines and the enhancement of the research function of the professoriate at the end of the eighteenth and the beginning of the nineteenth centuries, many seminars assumed, in addition to their professional functions, a scholarly one: rigorous instruction in the academic disciplines.[1]

Seminars of the past would be unfamiliar to us. Having evolved from informal meetings of professors and students, nineteenth-century seminars would appear closer in structure and function to what are now known as recitation or discussion sections. Their contributions to the transformation of university pedagogy were nonetheless profound. Seminar instruction helped shape the intellectual contours of the disciplines. Seminars replaced the monologues of lecture courses with dialogues between professors and students, thereby not only helping to transform the nature of teaching and of learning but also raising the suspicions of government authorities, who sometimes viewed the free discussion in seminars as politically dangerous. In principle the scholarly purpose of seminar instruction was to engage students in independent investiga-

---

1. There is no comprehensive history of the German seminar system, despite its strategic importance in the history of higher education. Only one other German seminar has received a book-length treatment, the Bonn history seminar: Paul Egon Hübinger, *Das Historische Seminar der Rheinischen Friedrich-Wilhelms-Universität zu Bonn: Vorläufer-Gründung-Entwicklung* (Bonn: Röhrscheid, 1963).

[1]

tions. In practice, however, seminars often became settings for the creation of curricula and other pedagogical strategies, such as exercises and problem sets, which were intended to guide students systematically in learning both exact procedures of investigation and critical standards for evaluating sources and conclusions. Such pedagogical strategies were necessary for making research methodologies accessible to students, and in them we can readily recognize the precedents for modern methods of instruction. Although introductory and advanced exercises were created in seminars for the humanities, they were more prominent in seminars for the natural sciences, especially for physics. Decades before the establishment of the laboratory-based physical institutes that secured hegemony in physics instruction and research for the German universities, directors of physical seminars were creating exercises—chiefly mathematical and measuring—that prepared students for advanced training and independent investigations in physics.

In the nineteenth century, fifteen seminars in one or more branches of the natural sciences were established at universities within the boundaries of what became Imperial Germany after 1871. General seminars for the natural sciences were founded at Bonn (1825), Königsberg (1834), Halle (1839), and Freiburg (1846); the latter two also incorporated instruction in mathematics. More specialized seminars, exclusively for mathematics and physics, were inaugurated at Königsberg (1834), Munich (1856), Breslau (1863), Heidelberg (1869), Tübingen (1869), Erlangen (1874), and Rostock (1880). Despite its name, the Göttingen mathematico-physical seminar (established in 1850) included instruction in the descriptive natural sciences. A seminar for physics alone was in operation at Giessen between 1862 and roughly 1880. In addition, there were two pedagogical seminars for secondary school teachers of mathematics and physics: one in Berlin (established in 1855 outside the university) and another at Giessen (1888).

Not one of these seminars survived beyond the first decades of the twentieth century. All general seminars for the natural sciences had disappeared by the 1890s; all had experienced operational problems almost from the beginning. Giessen's physical seminar deteriorated after 1878 when its first director, Heinrich Buff, died. By the 1930s, all the eight mathematico-physical seminars had either split into separate mathematical and physical seminars or become seminars for mathematics alone. Physical seminars had by then either faded in importance or been absorbed into the much larger and more amply endowed physical institutes. Science seminars were thus, at best, transitional institutional forms.

Among these fifteen seminars, one has traditionally been considered pivotal: the Königsberg mathematico-physical seminar. In contrast to the two general science seminars established earlier at Bonn and Königsberg, this seminar had conspicuously novel features. It was the first state-approved seminar to incorporate mathematical instruction. It was the earliest formal institutional setting for the combined teaching of mathematics and physics, and hence for mathematical physics. Moreover, its statutes did not explicitly address the traditional function of seminars, teacher training, and they seemed to encourage the pursuit of original investigations by students more strongly. Purportedly, several, but by no means all, of the mathematico-physical seminars established after Königsberg's found their inspiration in that seminar or drew upon its practice as a model for seminar exercises, especially in physics.[2]

The mathematical division of the Königsberg seminar, directed by Carl Gustav Jacobi from 1834 to 1844 and then by his student Friedrich Julius Richelot from 1844 to 1875, is generally more well known than the physical division; for its students' achievements contributed to the formation of a school. Jacobi's high profile in the mathematical community enhanced its reputation, as did Richelot's teaching, his contributions to the reform of Prussia's examination for secondary school teachers in 1866, and the central role of mathematics in the neohumanist curriculum in general. The literature on the mathematical division, although modest, has portrayed the seminar primarily as a research institute. Wilhelm Lorey, especially, in his work on mathematical seminars, emphasized the dominance of research in seminar instruction, a feature he called "the Königsberg principle."[3] Upon closer inspection, however, one finds that the mathematical division did not always operate at an advanced level. During Jacobi's tenure, continuity in instruction was sometimes lost because he was away from Königsberg so often. Moreover, the viability of the division would have been seriously impaired without a periodic return to fundamentals. How a balance between introductory instruction and advanced projects was achieved in the mature seminar may never be known, though, because the records of the mathematical division after 1860 are lost.[4]

The richer and more complete records of the physical division offer, by

---

2. Although the claims made by Wilhelm Lorey in *Das Studium der Mathematik an den deutschen Universitäten seit Anfang des 19. Jahrhunderts* (Leipzig: B. G. Teubner, 1916), pp. 71–80, 117–20, and in "Die Physik an der Universität Giessen im 19. Jahrhundert," *Nachrichten der Giessener Hochschulgesellschaft* 15 (1941): 80–132, on 92, and by Albert Wangerin in *Franz Neumann und sein Wirken als Forscher und Lehrer* (Braunschweig: F. Vieweg & Sohn, 1907), pp. 21, 152, are exaggerated, proposals for the mathematico-physical seminars at, for example, Göttingen and Breslau did invoke the Königsberg model, especially for combining instruction in mathematics and physics. See M. Stern to Minister Dr. Braun, 10 October 1849, and W. Weber, J. B. Listing et al. to Göttingen University Curator, 19/31 January 1850, UAG 4/Vh/20; M. Frankenheim to Franz Neumann, 15 November 1849, FNN 53.IA: Briefe von Collegen; M. Frankenheim to Breslau University Curator, 26 March 1863, ZStA-M, Die Errichtung eines mathematisch-physikalischen Seminars bei der Königlichen Universität zu Breslau, Rep. 76Va, Sekt. 4, Tit. X, Nr. 56, Bd. I: 1862–1934, fol. 4a. On other seminars see Chap. 10.

3. Lorey, *Studium der Mathematik*, p. 101.

4. Records of the mathematical division are found only in ZStA-M, Acta betreffend das mathematisch-physikalische Seminar an der Universität zu Königsberg, Rep. 76Va, Sekt. 11, Tit. X, Nr. 25, Bd. I: 1834–61.

contrast, a rare view not only of the evolution of physics pedagogy—
especially of the introduction of quantitative techniques into instruction—but also of the relation between pedagogy and practice in physics.
Along with Gustav Magnus's experimental practicum and colloquium at
Berlin and Wilhelm Weber's exercises and mathematico-physical seminar
at Göttingen, the physical seminar at Königsberg has generally been
considered a center of innovative physics teaching in Germany during the
middle decades of the nineteenth century. Directed by the physicist and
mineralogist Franz Ernst Neumann from its founding in 1834 until his
retirement in 1876, the Königsberg physical seminar had the longest
continuity in leadership of the three institutes. Before the 1830s, physics
instruction had not yet completely incorporated mathematical topics,
including mechanics; offered few opportunities for performing practical
measuring exercises; and was largely based on textbooks and lecture
courses. Self-instruction was not uncommon. In the 1870s, by contrast,
physics instruction was more formal and standardized. It routinely included precision laboratory exercises and recitation sections for the review of problem sets, as well as the customary lecture courses. Learning
theoretical and experimental techniques of quantification was considered
essential. Mechanics, formerly a part of applied mathematics, had become a fundamental component of the lecture-course cycle in physics.
Neumann's teaching, especially in the seminar, had helped shape these
pedagogical changes.

Neumann's students remembered the seminar thus: Once a week they
sat at a round table in the university's main building and discussed the
mathematical methods of physics. They watched and listened as Neumann, chalk in hand, went back and forth between the table and the
blackboard, where he derived the equations he had used in his lectures.
At home in the evening, they began the problems he assigned. They
found his homework difficult because it sometimes involved working with
elliptic integrals and transcendental functions. Usually students had to
integrate equations to fit specific conditions, identifying the constants and
coefficients that could—and should—be determined through measurement. Later in the week, they regrouped in Neumann's teaching laboratory, an extension of his seminar located in his home, where they conducted introductory and advanced measuring exercises, sometimes with
instruments of their own design. With the seminar's lathe, oven, glass-
works, and other equipment, they tried their hand at constructing instruments. They analyzed their instruments theoretically, determining
their limitations. Afterward, measurements in hand, they returned to the
mathematical methods of physics, especially to the methods of error
analysis. They no doubt hoped, as Neumann certainly did, that their
seminar exercises would evolve into publishable original investigations;
but only a few articles ever appeared. At year's end, they submitted their
results to Neumann, who included them with his annual seminar report
to the Prussian educational ministry in Berlin.

These student memories are confirmed by Neumann's official seminar reports after 1850. What neither indicate, however, is how and why Neumann had shaped physics instruction as he did. When he opened his seminar, he had not anticipated that the route to original investigations by students would be through assigned homework problems and routine measuring exercises executed in common. His initial vision had been shaped by his goal of extending mathematical methods to new areas of physics; the reality, however, was determined in part by his students' needs and available resources. The international reputation of his institute, acknowledged by the 1840s, rested largely upon the research productivity of its students, especially Gustav Kirchhoff. Its daily operation, however, was dominated by more basic matters. Looking to Neumann's seminar as a model, directors of other seminars often cited what his seminar was supposed to do, not what it actually did. In the beginning Neumann had hoped to draw more students to the study, and especially to the practice, of physics; in the end he had educated but a handful of university physicists. The contrasts between vision and reality, reputation and operation, and hopes and results led to the writing of this book.

We have come a long way in understanding how scientists do research. We know less about how they teach, even though teaching is a time-consuming career activity that forms the character and quality of the next generation of investigators. Finding out what goes on in the classroom, though, is no easy task, a fact that prompted Konrad H. Jarausch to identify classroom experiences as "research lacunae" in the history of education.[5] To understand both the origins of science curricula and the functions of exercises and problem sets, however, we must view teaching as a creative enterprise, not as one inferior to research and depending for its continued vitality on research's results. It is in the classroom, after all, that what Erwin Panofsky called "mental habits" or what Lucien Febvre called "mental equipment" are shaped, creating styles of thinking and practice.[6] It is especially important to uncover German classroom experiences in the sciences for the period between 1820 and 1850, when the pedagogical form of several scientific disciplines took shape and rigorous professional training for scientists was ascendant, although not yet firmly in place. But how can classroom experiences be investigated in a way that will contribute to an *Alltagsgeschichte* of science instruction?

One obvious record of science instruction is the lecture course. Primarily because of his cycle of lecture courses, Neumann's students and contemporaries credited him with having established mathematical or theoretical physics as an independent discipline in Germany. In 1838 one

---

5. Konrad H. Jarausch, "The Old 'New History of Education': A German Reconsideration," *History of Education Quarterly* 26 (1986): 225–41, on 235.

6. Erwin Panofsky, *Gothic Architecture and Scholasticism* (Latrobe, Pa.: Archabbey, 1951), pp. 20–21, 27; Lucien Febvre, *The Problem of Unbelief in the Sixteenth Century: The Religion of Rabelais*, trans. Beatrice Gottlieb (Cambridge, Mass.: Harvard University Press, 1982), p. 150.

of his students, Carl Senff, wrote that through Neumann's teaching he found "a new life in science" through "a new discipline, mathematical physics."[7] On the occasion of the fiftieth anniversary of his doctorate in 1876, the Academy of Sciences in Berlin honored Neumann in part for his teaching: "You were the first to hold a continuous series of lectures on mathematical physics at a German university," the academy's congratulations read. "This discipline, which is of highest importance not just for mathematics and physics, but also for the entire structure of human knowledge, is now cultivated in many higher schools of our fatherland. This is chiefly your work."[8] Statements linking Neumann's teaching to the establishment of mathematical physics in Germany were especially prominent after his death in 1895, the strongest written by two of his doctoral students, Albert Wangerin and Woldemar Voigt.[9] In an earlier and much broader study, I examined how Neumann's teaching, research, and students contributed to the emergence of theoretical physics in Germany.[10] Neumann and several of his seminar students figure prominently in histories of German theoretical physics, and his seminar has also been linked to the inauguration of physics as a modern, independent discipline.[11]

To the extent that discipline formation depends upon teaching, Neumann made a significant contribution to the institutionalization of mathematical physics after 1830.[12] He did so partly through lecture courses,

7. Carl Senff to Neumann, 19/31 January 1838, FNN 53.IIB: Briefe anderer Schüler.

8. Quoted in LN, p. 393.

9. Woldemar Voigt, "Gedächtnissrede auf Franz Neumann" [1895], in GW 1:1–19; Albert Wangerin, "F. E. Neumann," JDMV 4 (1894–95): 54–68, on 56; idem, "F. E. Neumann," Leopoldina 32 (1896): 51–54, on 53; idem, "Franz Neumann als Mathematiker," PZ 11 (1910): 1066–72, on 1067; idem, Forscher und Lehrer.

10. Kathryn M. Olesko, "The Emergence of Theoretical Physics in Germany: Franz Neumann and the Königsberg School of Physics, 1830–1890" (Ph.D. diss., Cornell University, 1980). Some of the principal features of Neumann's seminar were first sketched out in this earlier study, which dealt only in part with the seminar; see pp. 300–322, 343–53, 366–70, 372–74, 388–94, 411–15.

11. On German theoretical physics see, e.g., Christa Jungnickel and Russell McCormmach, Intellectual Mastery of Nature: Theoretical Physics from Ohm to Einstein, 2 vols. (Chicago: University of Chicago Press, 1986). On Neumann's seminar see esp. 1:84–89, 97–101; 2:113, 148; cf. Russell McCormmach, "Editor's Foreword," HSPS 3 (1971): ix–xxiv, esp. xi–xiii. On the seminar and physics as a discipline see Lewis Pyenson, "Mathematics, Education, and the Göttingen Approach to Physical Reality, 1890–1914," Europa 2 (1979): 91–127, on 103; and David Cahan, "The Institutional Revolution in German Physics, 1865–1914," HSPS 15 (1985): 1–66, on 5. On German theoretical physics see, in addition, Rudolf Stichweh, Zur Entstehung des modernen Systems wissenschaftlicher Disziplinen: Physik in Deutschland, 1740–1890 (Frankfurt a.M.: Suhrkamp. 1984); Kenneth L. Caneva, "From Galvanism to Electrodynamics: The Transformation of German Physics and Its Social Context," HSPS 9 (1978): 63–159; and idem, "Conceptual and Generational Change in German Physics: The Case of Electricity, 1800–1846" (Ph.D. diss., Princeton University, 1974).

12. The central role of teaching in discipline building has been argued by Owen Hannaway, The Chemists and the Word: The Didactic Origins of Chemistry (Baltimore: Johns Hopkins University Press, 1975); William Coleman, "The Cognitive Basis of the Discipline: Claude Bernard on Physiology," Isis 76 (1985): 49–70; and Stichweh, Physik in Deutschland.

which he created in three distinct periods. The first was 1830–31, when he offered courses on theoretical optics (most likely based on the work of Augustin Jean Fresnel) and on the analytical theory of heat, in which he followed Joseph Fourier. Between 1838 and 1840 he introduced a second group of courses related to mechanics: one, from 1838, was a survey; the other two were on elasticity and capillarity. The final cluster of courses appeared in 1843–44 and dealt with magnetism and electrodynamics. Only the first and third periods coincided with Neumann's involvement in major research projects, and he seems to have used his teaching to organize material and express ideas on which he was working. He did in fact introduce several of his research results into the classroom, although they cannot always be identified with certainty. Courses created during the second period, by contrast, had less to do with his research than with problems in his teaching; these courses enabled him to operate his seminar with a stability and purpose that had previously eluded him. After 1844, Neumann offered only four additional "new" courses, and all treated special topics derived from earlier courses: his course on hydrodynamics (1847–48) was a specialized version of his course on mechanics; and his courses on the theory of the electric current (1849–50), potential theory (1852–53), and the induced electric current (1861–62) grew out of his courses on electrodynamics and magnetism. No set of lecture courses purporting to teach a mathematical physics before 1850 were as continuous, comprehensive, or thorough as his. One common objective that linked his lecture courses was his ceaseless pursuit of the highest degree of certainty and generality, achieved by eradicating hypothetical ideas; by using analogies to extend mathematical methods from one area of physics to another; and by subjecting theory to the rigorous tests of an exact experimental physics.[13]

Although valuable for an understanding of mathematical physics in Germany, a study of Neumann's lecture courses alone would provide a limited view of what he taught, how he viewed the practice of mathematical physics, and especially how his teaching contributed to its acceptance in Germany. Some of his lecture notes and some taken by his students are still extant, but the earliest date from the 1850s; most date from the 1860s and 1870s, when Neumann's cycle of lecture courses was firmly estab-

13. Carl Neumann, Neumann's son and former student, commented on Neumann's lectures in *GW* 2:313–16. Neumann's lecture courses on theoretical physics (mechanics), capillarity, the electric current, magnetism, theoretical optics, elasticity and the optical ether, and potential theory were edited by several of his former students and published in the series Franz Neumann, *Vorlesungen über mathematische Physik gehalten an der Universität Königsberg von Dr. F. Neumann*, 7 vols. (Leipzig, B. G. Teubner, 1881–94). His lectures on heat, which were to have been edited by Carl Neumann, were never published in full; parts appear as Franz Neumann, "Ein Kapitel aus den Vorlesungen von Franz Neumann über mechanische Wärmetheorie: Königsberg, 1854/55," *Abhandlungen der Bayerischen Akademie der Wissenschaften*, Math.-naturw. Klasse 59 (1950): 7–27; see also FNN 5: Carl Neumann's Manuscript über mech. Wärmetheorie.

lished. Unfortunately, student notebooks do not permit one to follow the evolution of his courses in fine detail, nor do they provide direct access to what went on in the classroom, especially during the early crucial years when Neumann was shaping his seminar pedagogy. When his students published his lectures during the 1880s, they edited his courses (despite their statements to the contrary) and emphasized theoretical components over practical ones. They combined several years' notes on the same topic, thereby creating a composite course that not only had never been offered but that also was more coherent and more conceptually oriented than any single course had been.

There is an additional difficulty that arises when Neumann's teaching is viewed primarily through his lecture courses: how to identify and treat the students in them. The only existing attendance list is incomplete, includes only some of the students who attended his private courses, and is arranged by the semester they matriculated at Königsberg rather than by the course they attended.[14] Hence, accurate identification of the lecture courses' student clientele is severely hampered. Yet even with accurate identification, one would be left with the problem of explaining the significance of having attended Neumann's lectures. What, precisely, did students take away from a *lecture* course? Some commentators believed that Neumann's teaching so shaped the intellectual orientation of several (the number varies) of his students that one could speak of a "Königsberg school" of mathematical or theoretical physics,[15] the members of which "followed through the *ideas* developed by [Neumann]."[16] Upon closer inspection, however, the definition of a research school as developed by Jerome Ravetz, Gerald L. Geison, and others seems not to apply.[17] Unlike Geison's "Cambridge school of physiology," which "descended from and crystallized around the problem of the heartbeat,"[18] the investigations of Neumann and his students were not as strongly linked by any one problem or family of problems. That does not mean, of course, that Neumann's contemporaries were misguided in calling some group of his students a "school"; but only that the current historiographic definition of

14. Paul Volkmann, *Franz Neumann: Ein Beitrag zur Geschichte Deutscher Wissenschaft* (Leipzig: B. G. Teubner, 1896), pp. 59–68; idem, "Erinnerungen an Franz Neumann (Nachtrag.)," *SPGK* 40 (1899): 41–51, on 46–51.

15. Wangerin, "Neumann als Mathematiker," p. 1067; Voigt, "Gedächtnissrede," p. 17; H. Wild, ["Franz Neumann,"] *Bulletin de l'Académie impériale des sciences de St.-Pétersbourg* 3 (1895): ix–x, on x; Joh. Wislicenus, ["Franz Neumann,"] *Verhandlungen der Gesellschaft Deutscher Naturforscher und Aerzte* 67.1 (1895): 27–28, on 27; 67.2 (1895): 5–6.

16. C. Voit, "Franz Ernst Neumann," *Sitzungsberichte der math.-physikal. Classe der k. b. Akademie der Wissenschaften zu München* 26 (1896): 338–43, on 339 (emphasis added).

17. Jerome R. Ravetz, *Scientific Knowledge and Its Social Problems* (New York: Oxford University Press, 1971), and Gerald L. Geison, "Scientific Change, Emerging Specialties, and Research Schools," *HS* 19 (1981): 20–40.

18. Gerald L. Geison, *Michael Foster and the Cambridge School of Physiology: The Scientific Enterprise in Late Victorian Society* (Princeton, N. J.: Princeton University Press, 1978), p. xv.

a school, one that emphasizes primarily the existence of a family of related problems or a coherent research program, inadequately describes the linkages between Neumann and his seminar students.

Just as research embraces more than what is found in published investigations, teaching involves more than what is said in lecture courses or written in textbooks. Just as others have used laboratory notebooks for reconstructing the investigative pathways of individual scientists, I have worked from Neumann's annual seminar reports to retrace the steps he took in shaping his pedagogical physics, especially during the formative years from 1834 to 1849. His seminar reports allow us to recreate his teaching semester by semester, and sometimes month by month. I have followed them as closely as the historical record will allow, covering not only the seminar's successes but also its routine operation. This strategy proved fruitful.

For the most part I have used official Prussian ministerial files that contain Neumann's seminar reports from the winter semester of 1834/35 to the academic year 1859/60.[19] The ministerial files also provide surrounding documents that elaborate further upon the operation of the seminar, including documents related to its founding; reports for the mathematical division; ministerial evaluations and recommendations; budgetary requests and awards; and correspondence concerning problems that arose in the course of its operation. An equally rich set of files for the period after 1860 was lost or destroyed during World War II. Fortunately, other records are available. Neumann's draft reports for the academic years from 1860/61 to 1875/76 are preserved in his papers.[20] Although not always as complete as his official seminar reports, his draft reports proved useful as guides to the operation of the seminar in its last years. They span the years from 1847 to 1875 with some years missing. The draft reports from 1847 to 1860 differ from his official reports in that sometimes they are clearly working documents (often failing to mention student investigations), and occasionally the student lists are variant. Lists recommending his students for the seminar's prizes and a few other small items accompany the draft reports.

Altogether, twelve semesters are missing from this combined set of reports: the period from the winter semester of 1836/37 to the winter semester of 1838/39, when Neumann held informal exercises but did not submit formal reports; the winter semester of 1839/40 and the summer semesters of 1868, 1872, and 1874, when the seminar was not held; and the winter semester of 1872/73 and the academic year 1875/76, when the seminar was held but no reports are extant. Passages from his seminar reports—most from after 1849, when its operation had stabilized—are

reprinted or discussed in biographies by Albert Wangerin, Paul Volkmann, and Neumann's daughter Luise Neumann.[21] (All three relied
upon Königsberg University curatorial files destroyed during World War
II.) Collectively these printed passages are but a small portion of the
documentary record and mostly concern the operation of the mature
institute after 1850.[22]

There are several advantages offered by Neumann's seminar reports.
First, 150 students (about three-quarters of the total number) and the
semesters they attended the seminar can be identified. That list is necessarily incomplete because only after 1852, when he was chastised by the
ministry, did Neumann accurately record all active members of his seminar, whom he defined as students who submitted written work. He never
mentioned the names of auditors, who were excluded until the 1860s and
who attended in increasing numbers after then. If we exclude auditors
and interpolate the number of students who attended during the years
for which reports are missing or incomplete, then approximately 208
students attended Neumann's seminar over forty-two years, a modest
number compared with the enrollment in physical seminars elsewhere.[23]

The second advantage of the seminar reports is that they are dense
enough to permit reconstructing the evolution of Neumann's teaching.
They provide as detailed an *Alltagsgeschichte* of the classroom as we are
likely to get. Especially from documents covering the period from 1834 to
1849, the factors shaping Neumann's pedagogical program become apparent. Although during those fifteen years he was very productive in
mathematical physics and had integrated some of his own work into his
seminar curriculum, his pedagogical program was not one constructed
exclusively by distilling the results of research for classroom use. The
survival of his institute depended upon another pedagogical strategy.
Early on, he had viewed teaching as a creative enterprise in its own right
and sought to use it as a vehicle for extending mathematical methods into
new areas in physics. But Neumann was not so driven by that goal, or so
committed to introducing research into the classroom, that he ignored his
students' untrained abilities. As I argue below, he decided to base his
lecture cycle and his seminar exercises on a course on "physical" mechanics—one oriented not toward learning analytic equations of motion but
rather toward training in quantitative methods, including the analysis of
constant and accidental errors and the principles of instrumentation—
partially to meet his student's needs. The involved engagement in learn-

---

21. Volkmann, *Ein Beitrag*, pp. 50–54; Wangerin, *Forscher und Lehrer*, pp. 152–75; LN, pp. 368–69, 371–72, 443–44.

22. Hence most studies that discuss Neumann's seminar rely primarily upon documents describing the mature institute; see, e.g., Jungnickel and McCormmach, *Intellectual Mastery* 1:97–101. Although Wangerin points out that Neumann modified his plans for the seminar and that Neumann's exercises became elementary, he does not explain how or why they did (*Forscher und Lehrer*, pp. 151, 154).

23. See App. 3 for the students in Neumann's seminar.

ing that the seminar format promoted in fact made it necessary for him to take this or a similar pedagogical step in order to introduce his students systematically to the tools of research.

The third and most important advantage of using Neumann's seminar reports is that they permit, in ways his lecture courses do not, a view of the *activity* of learning. The significant feature of Neumann's seminar instruction was less the ideas and theories that he taught than the mathematical and measuring methods in the exercises he assigned. Voigt described Neumann's seminar exercises as "problems which for the most part were more than simply exemplars [*Uebungsbeispiele*] and had great scientific interest."[24] Likewise Wangerin linked the seminar exercises to independent investigations.[25] So important did Paul Volkmann, who attended Neumann's lectures in the mid-1870s, consider the exercises that he thought it desirable to publish a collection of them, along with Neumann's solutions, so that future generations could see "how Neumann taught, worked, and trained investigators."[26] Neumann's seminar exercises—which, contrary to Voigt and Wangerin, were not independent investigations but smaller problems—are therefore crucial indexes to how he trained his students. Unfortunately the completed exercises are no longer available. Neumann had sent bound sets of problems completed by his students with his annual seminar reports to the ministry, but the ministry returned them to Neumann, who either gave them back to his students or kept them on file. Problems still in Neumann's possession when he retired were displayed in the museum of the Königsberg physical laboratory, where they were destroyed in World War II.

But even without the exercises, the ways in which his students deployed the techniques cultivated in them are not entirely inaccessible. I considered it reasonable to assume that early in their careers his students used methods recently learned. I therefore adopted the strategy of analyzing, in conjunction with Neumann's seminar reports, the first publications of his seminar students when those publications derived from seminar exercises, as they frequently did. I used these first publications as additional windows on the activity of the seminar. Letters written by Neumann's students confirmed that my strategy was justified. The rich archive of their correspondence offers a rare view not only of their relationship to Neumann but also of the problems and difficulties they encountered in executing their fledgling investigations. Their letters especially reveal the frustrations they encountered when they wanted to do things "Neumann's way" but found they could not.[27]

Thus the operation of the seminar, viewed through official reports and

24. Voigt, "Gedächtnissrede," p. 10.
25. Wangerin, "Neumann," *Leopoldina*, p. 53; idem, "Neumann," *JDMV*, p. 56.
26. Volkmann, *Ein Beitrag*, pp. 25–26.
27. The richest such archive of correspondence is FNN 53.IIA: Briefe von Schülern and FNN 53.IIB: Briefe anderer Schüler.

student investigations, demonstrates how students acquired what Lud-
wick Fleck called the professional habits needed to become a "trained person."[28] Although Fleck also claimed that the "technical skills required for
any scientific investigation" cannot be "formulated in terms of logic,"[29] we
do know Neumann's seminar students conducted their exercises and
investigations according to explicit and rigorous quantitative techniques.
These techniques turned out to be less the mathematical methods taught
in Neumann's lecture courses and elaborated in the seminar than the
quantitative methods of error analysis incorporated into seminar exercises, which enabled the students to evaluate the precision and reliability of their experimental data. Especially important in shaping their
professional habits was a particular set of quantitative techniques associated with exact experiment: the analytic determination of constant (or
systematic) errors; the computation of accidental (or random) errors by
the method of least squares; and other methods for processing data, such
as graphs (which remained persistently problematic for them) and approximation techniques. Minor themes in Neumann's research thus became dominant tools in his science pedagogy.

That the historical significance of a seminar singled out in its own time
for its role in institutionalizing mathematical physics should be found in
its cultivation of the techniques of an exact experimental physics is not so
surprising when one considers German views on physical theory in the
late 1820s and early 1830s, when the seminar was founded. The eventual
acceptance of Georg Simon Ohm's pathbreaking mathematical theory of
the galvanic circuit, published in 1827, as well as of French theories in
mathematical physics, relied on much more than the dissemination and
acceptance of the mathematical methods used in them. It also relied on
developing the means for judging the accuracy of those theories through
a rigorous quantitative assessment of their empirical foundations. Hence
the new mathematical techniques appearing in German physical investigations in the 1830s included not only partial differential equations, series
expansions, and the like but also methods for determining constant and
accidental errors. Error analysis helped establish the reliability of theory
and make mathematical methods legitimate in physics. The stronger
experimental tradition then extant among German physicists—a tradition that was even defining the subject matter of physics in terms of what
could be treated with instruments—thus appears to have shaped in significant ways the context within which mathematical physics took hold and
matured in some German institutional settings.

Concerning the function of measurement in modern physical science,
Thomas S. Kuhn remarked that the agreement between theoretically

---

28. Ludwik Fleck, *Genesis and Development of a Scientific Fact*, trans. Fred Bradley and
Thaddeus J. Trenn (Chicago: University of Chicago Press, 1979), pp. 89, 90.
29. Ibid., p. 35.

calculated and experimentally observed data is more than a numerical agreement; it also involves evaluating experimental procedures and understanding limits of accuracy.[30] In Neumann's seminar the quantification of error was thus a means to hone the skills and to inculcate the values used in computing and judging the degree of epistemological certainty believed essential for theory formulation or confirmation. Yet there were problems in this way of teaching physics. What Neumann's instruction seemed not always to provide was a *sense* of when the further pursuit of accuracy and precision should end. His students used rigorous standards for theory confirmation and hence for scientific truth. They were at times overly cautious in formulating mathematical laws. They were skeptical about certain quantitative techniques, such as the graphical analysis of data. Only by examining the operation of the seminar and the investigations that began in it can we understand why Neumann's students had such high standards of reasonable agreement as part of what Ravetz called the "craft knowledge" that enables one "to decide which sort of functional relation is represented by the discrete set of points obtained from [one's] readings."[31] The quantitative techniques and values of an exact experimental physics strongly shaped the fledgling investigations of seminar students. It was probably with these techniques in mind that the British scientist Arthur Schuster, impressed by what he heard about Neumann's teaching, concluded that "there was never, probably, a school of original research conducted in so systematic a manner as this seminar, in which Neumann was the leading spirit."[32]

I have drawn the title of this book, *Physics as a Calling*, from Neumann's remarks in 1876 on the necessity of a physical laboratory.[33] In them he claimed that instruction in theoretical physics—by which he then meant his entire program of instruction in mathematical physics—was designed for teachers of physics and for *Physiker von Beruf*. I have translated *Beruf* as "calling," its older meaning, rather than as "profession," as it came to be understood. Until the late 1840s there were noticeable ambiguities in how Neumann viewed his teaching. In public he upheld idealistic notions of *Wissenschaft* and especially of *Bildung*, the tenets of which maintained that instruction, particularly participation in independent exercises and the

30. Thomas S. Kuhn, "The Function of Measurement in Modern Physical Science," in idem, *The Essential Tension: Selected Studies in Scientific Tradition and Change* (Chicago: University of Chicago Press, 1977), pp. 178–224.

31. Ravetz, *Scientific Knowledge*, p. 84.

32. Arthur Schuster, "Franz Ernst Neumann," *Proceedings of the Royal Society of London* 60 (1896–97): viii–xi, on xi. Stichweh has noted the importance of the "reduction of error" and the "calculation of error" in the mathematization of physics and the professional training of physicists in the nineteenth century (*Physik in Deutschland*, pp. 232–38, 241).

33. Neumann to [Königsberg University Curator], c. April–May 1876 (draft), FNN 61.7: Kampf um das Laboratorium. An altered version appears in LN, pp. 455–56, and presumably what is the final version, in Wangerin, *Forscher und Lehrer*, pp. 183–85.

process of discovery, shaped character and drew out natural talents and hence a student's "calling." But Neumann did not always practice what he preached. For even as he spoke in the 1840s of idealistic notions of *Bildung*, he was developing a program of systematic instruction (*Ausbildung*) in physics that in principle shaped skills more than it drew out natural talents. Traces of this ambiguity remained until 1876 when he retired from teaching and stepped down as director of the seminar. That is not to say that Neumann's notion of *Beruf* was entirely devoid of elements that now define the term *profession*. Above all else, his mode of instruction cultivated the kinds of skill-based activities and critical perspectives that are central to professional practices. But to have translated *Physiker von Beruf* as "professional physicists" would have meant ignoring the ambiguities present in his own mind as well as to have invited a sociological definition of the term *profession* that I believe would have masked aspects of his teaching.

The two themes of this book are discipline and practice. The term *discipline* is especially important. In her provocative analysis of Michel Foucault's *Discipline and Punish*, Jan Goldstein argues that one of the connotations of Foucault's use of the word *discipline* is the "rigorous 'disciplined' training to which the professional has himself submitted . . . and through which he has gained mastery over a body of knowledge and has come to view his possession of this knowledge as entailing a serious commitment or higher calling."[34] From 1780 to 1840 several different forms of disciplining students were taking shape in Prussia: the state reformed its examinations for the professions; in certain fields, sets or hierarchies of courses constituting curricula were emerging; and exercises that systematically organized skills of investigation were created. Reforms such as these rationalized instruction. In contrast to England, where a "curriculum-as-fact" dominated science instruction at first,[35] in early-nineteenth-century Prussia a "curriculum-as-practice" grew alongside, challenged, and eventually transformed the idealistic conception of

34. Jan Goldstein, "Foucault among the Sociologists: The 'Disciplines' and the History of the Professions," *History and Theory* 23 (1984): 170–92, on 179. Goldstein's observation is more than a sociological nicety. Anthony LaVopa has given a penetrating commentary on the late-eighteenth-century transition from religious connotations to professional ideology in the German meaning of *Beruf* or calling; see his *Grace, Talent, and Merit: Poor Students, Clerical Careers, and Professional Ideology in Eighteenth Century Germany* (New York: Cambridge University Press, 1988).

35. Keith Hoskin, "Examinations and the Schooling of Science," in *Days of Judgement: Science, Examinations, and the Organization of Knowledge in Late Victorian England*, ed. Roy MacLeod (Driffield, U.K.: Nafferton, 1982), pp. 213–36, on p. 230. In British physics the introduction of practical exercises in precision measurement began in William Thomson's Glasgow laboratory in the 1850s and 1860s and spread to other British teaching laboratories by the late 1860s and early 1870s; see Graeme Gooday, "Precision Measurement and the Genesis of Physics Teaching Laboratories in Victorian Britain," *BJHS* 23 (1990): 26–51; Crosbie Smith and M. Norton Wise, *Energy and Empire: A Biographical Study of Lord Kelvin* (Cambridge: Cambridge University Press, 1989), pp. 117–35.

education as *Bildung*. Following Goldstein, one can view these pedagogical changes not only as foundations for professional scientific training but also as agents that guarantee genuine commitments to scientific practice. In the sciences, students became especially involved in instruction through skill-based exercises—either written or manual—that tested not only their ability to master and perform but also to persevere as the intellectual tasks became more difficult. I use "discipline" to mean training the mind to follow certain rules of investigative protocol and rigorous techniques of investigation. In the physical seminar at Königsberg, Neumann disciplined his students in the mathematical and measuring methods of physics.

The second theme is that of "practice." Almost all Neumann's former seminar students—from archaeologists, physicians, and chemists to instrument makers, school teachers, and physicists—found his rigorous instruction in physics worthwhile. But the meaning of the seminar is found less in the professions represented by his former students than it is in how they actually worked in science, especially during the early years of their careers. By "practice" I do not mean "research program." Studying scientific practice entails de-emphasizing the products of science, ideas and theories, in favor of exploring the labor of science, especially its mental and material tools. The investigative strategies cultivated by Neumann's seminar exercises placed a premium on certain skills, techniques, and even values, thus limiting the range of tools available for scientific practice.

Why Neumann chose and continued to cultivate a certain way of teaching physics is a complicated issue that touches upon several factors: his own intellectual development; the state of physics; local institutional conditions at Königsberg; ministerial policies that shaped the examinations for the professions and set levels of financial support for the sciences; political opinions that supported the breakdown of more traditional forms of educational and social discipline (and that later reacted against newer ones); a social climate that supported the pursuit of abstract knowledge by certain "classes"; a commercial economy that found accurate measurements useful before physicists considered precision indispensable in their own work; and finally, an emerging industrial economy that slowly but irreversibly absorbed individuals with highly specialized training in physics. Not all factors impinged upon the history of the seminar to the same degree. The evolution of the career patterns of Neumann's student clientele, for example, followed approximately the expansion and contraction of school and university teaching positions and, to a lesser extent, the emergence of positions in precision concerns; but in contrast to career patterns in chemistry before 1876, those in physics were not significantly affected by industrial growth. Although the relative importance of each factor differs, all are important in some way. Nevertheless, it proved impossible to pursue all equally. Educational institutes are, to be sure, the points where social, economic, and political forces confront the structure and function of abstract knowledge most directly. But in order to offer a balanced assessment of the role of larger forces characteristic of certain regions such as Prussia, we need to know first not just the fine structure of the operation of one institute, but of many. This study is accordingly a local one.

Part I of this book relates in fine detail the evolution of instruction in the seminar. Chapters 1 to 3 cover the first, or founding, phase of the seminar from its formal establishment in 1834 to its reopening in 1839 after five semesters of unofficial operation. These were troublesome years. Neumann was not only exploring various pedagogical tactics to make his institute viable; he was also becoming more strongly drawn to the investigative techniques of an exact experimental physics. The astronomer Friedrich Wilhelm Bessel figured prominently in shaping Neumann's investigative and pedagogical styles. Neumann's decision to integrate quantitative techniques into physics teaching set him apart from most of his science colleagues at Königsberg.

The second phase of the seminar, from 1839 to 1849, is covered in Chapters 4 and 5. Chapter 4 focuses on a new course on mechanics that Neumann introduced in 1838 and its strategic role in his pedagogical physics. The particular way he fashioned this course—it was not simply on motion—enabled him to teach the principles of Bessel's exacting method of experiment so that they could be applied in seminar exercises to problems of theoretical concern. The decade of the 1840s was in many respects an exciting one for Neumann. His seminar began to achieve stability, largely in consequence of his course on mechanics; its curriculum was all but complete; and the investigations of his students, especially those of Gustav Kirchhoff, brought it national and international recognition. Neumann completed ground-breaking investigations of his own on electrodynamic induction, and he achieved a position of prominence in the university. A more favorable political environment and the university's tricentennial celebration in 1844 seemed to portend greater state support for the sciences, including a physical laboratory at Königsberg, but that support never materialized. The ministry's generally favorable view of Neumann's requests for a physical laboratory was eclipsed by its renewed concern for liberal political events, including those on university campuses, but more generally by the onset of fiscal and other forms of conservatism. Chapter 5 closes on a note of resignation. Throughout the decade, Neumann had pursued both his original goal of promoting student research and the unexpected task of constructing an introductory course for beginning students in the seminar. But by 1849, when finally he was able to provide the laboratory he and his students so needed, he acknowledged that the function of the seminar was to provide systematic instruction in physics.

Chapters 6 to 8 deal with the third, or mature, phase of the seminar from 1849 to 1876, when Neumann retired. His teaching did not evolve conceptually to any great extent during those years. Instead, he deepened his commitment to a program of hierarchical learning and to the kinds of technical exercises he had developed earlier. What we begin to see emerging against the background of an *Alltagsgeschichte* of the seminar are patterns in student investigations. So much did the perfection of technique dominate seminar exercises that students began to consider the rational execution of technique an important subject of investigation in itself. Hence their publications reflected less the expansion of theory or the discovery of something new than the problems they encountered in practicing physics, especially in processing data. The themes of these three chapters—error analysis and measurement, interpolation and certainty, and error over truth—are drawn from their investigations.

Part II focuses on the relation between education and practice. The literature on and writings of Neumann's seminar students—particularly those who achieved positions of prominence in the physics community— are vast, and they reveal much about career opportunities for physicists, the institutional history of physics in Germany, and changes in theoretical physics. A study of the social origins and career patterns of former seminar students (to the extent that both variables can be identified) would be an important contribution to the history of the German seminar system. But what distinguished Neumann's seminar most strongly from other physical seminars and institutes was a distinctive set of investigative techniques: an emphasis on both constant and accidental errors; a sensitivity to problems arising from graphical analysis and interpolation; and the epistemological and technical concern for certainty that at times bordered on obsession. The final two chapters are devoted to a thematic examination of the meaning of those techniques and the values associated with them in the scientific and pedagogical practices of Neumann's seminar students. Chapter 9 explores how secondary school teachers—the most popular profession among former seminar students—selectively deployed these techniques to create for themselves a professional identity consonant with their training. Chapter 10 examines how the teaching and investigative practices of other former students reinforced or transformed the seminar's ethos of exactitude, especially its experimental style based on the quantification of error.

For more than a decade after Neumann's death in 1895, several of his students defended the position they believed he had held in the evolution of mathematical physics in Germany. Voigt concluded that it was "understandable" that the generation that had once surrounded Neumann had forgotten him, so widely dispersed was his influence.[36] Historiographic traditions that valued original scientific contributions, not teaching programs, have also been responsible since then for Neumann's relatively

36. Voigt, "Gedächtnissrede," p. 4.

unexceptional historical reputation. But what probably contributed most to his peripheral status by the end of the century was that the particular combination of mathematical and measuring methods he taught as a unitary set of tools for investigation were no longer regarded as so important in experimental physics, where a new experimental style was taking shape, and no longer characterized the investigative enterprise in either mathematical or theoretical physics.[37]

Yet memories of an exceptional teacher remained. One of his obituaries described Neumann as a "giving and selfless teacher as there will be no more."[38] Understanding the role of teachers in science, as in other fields, has never been easy. Reflecting on the intellectual indebtedness of Albert Einstein to his teachers, Lewis Pyenson asked "whether the affinities between his thought and several of his early teachers indicate causal relationships or contingent ones."[39] Pyenson concluded that although one could not reduce Einstein to his textbooks or his teachers, both had to be taken into consideration when explaining Einstein's conceptual orientation. Similar historiographic issues arose in Gerald L. Geison's study of Michael Foster. Geison pointed out that the history of science education cannot be written without "a more systematic examination of . . . 'great teachers.' " He added, quite appropriately, that "we need not be excessively skeptical about the possibility of doing meaningful influence studies" that demonstrate what kind of linkage existed between teachers and students.[40] The studies of Pyenson, Geison, and others, as well as mine on the Königsberg seminar, have convinced me that Fleck was right when he wrote that "initiation into science is based on special methods of teaching."[41] Part of my purpose in writing this book has been to understand the science teachers, as well as the science teaching, of the past.

Viewing classroom activities that took place in the historical past brought with it certain pleasures. Among them was the playful experience of thinking again like a student when I tried to understand why Neumann taught this way and not that. It also brought disappointment. There is an intangible element in learning that depends upon the rapport between teacher and student. I could not hear the dialogues between Neumann and his students in the seminar, and only vicariously, through the written word, could I experience the bond between them. What I sensed of his students' feelings convinced me that attending the seminar meant more to them than merely learning to use exact techniques of investigation.

37. For instance Max Planck, considered by Jungnickel and McCormmach to be the first truly theoretical physicist (*Intellectual Mastery* 1:xix), did not practice the techniques of an exact experimental physics. At the same time, a leading experimental physicist at the end of the century, August Kundt, eschewed precision measurement; see Chap. 10.

38. J. Hann, "Nekrolog [Franz Neumann]," *Almanach der Kaiserlichen Akademie der Wissenschaften zu Wien* 64 (1896): 271–80, on 280.

39. Lewis Pyenson, "Einstein's Education: Mathematics and the Laws of Nature," *Isis* 71 (1980): 399–425, on 424.

40. Geison, *Michael Foster*, pp. xiii, xiv; see also Ravetz, *Scientific Knowledge*, p. 105.

41. Fleck, *Scientific Fact*, p. 112.

PART I

DISCIPLINE

CHAPTER ONE

Educational Reform and the Natural Sciences

In 1834 when the Prussian educational ministry approved a seminar for mathematics and physics at its provincial university in Königsberg, most of the reforms that shaped the state's university system were already in place. From the 1790s to 1818, Prussian universities were dissolved, amalgamated, re-created, or born anew, so that by 1818 there were six: Berlin, Bonn, Breslau, Greifswald, Halle, and Königsberg. Over the first half of the nineteenth century, institutional and intellectual reforms heightened faculty sensitivity to pedagogical issues; accentuated the importance of faculty research; reorganized the structure of knowledge, so as to establish new courses, curricula, and study plans, especially in the philosophical faculties; and began to enhance the material resources, organization, and administration of university institutes that supported and sustained novel forms of teaching and learning, trained students, and facilitated research. Especially after Napoleon's victory over the Prussian army at Jena in 1806, Prussian university reform allegedly took on enhanced social and political dimensions, serving in part as a psychological compensation for military defeat.

Lacking a detailed study of how these changes were rooted in Prussian culture, economics, politics, and society of the second half of the eighteenth century, historians have customarily assigned a prominent, but not necessarily causal, role to the reforms of Wilhelm von Humboldt, educational minister in 1809–10. His reforms have been viewed as accelerating and refining changes at all levels of the state's educational system. Humboldt's vision of a hierarchical educational system culminating in university instruction and his philosophy of the unity of teaching and research in the university especially helped give further shape to what are traditionally considered the two principal ideological pillars of Prussian pedagogy: *Bildung*, a self-cultivation believed best achieved through a neo-humanist curriculum based on the classics and leading to the refinement of one's personality and the development of one's talents; and *Wissenschaft*, the pursuit of pure knowledge through research and independent study. *Bildung* was achieved through *Wissenschaft*. Eventually both ideals were believed best (although not exclusively) exemplified by the philosophical faculty of the Prussian university.

Far reaching as these changes were, neither the gross structural features of Prussian university reform nor the ideologies of *Bildung* and *Wissenschaft* addressed in detail their significance for actual classroom practice. In principle the Humboldtian reforms did define a kind of learning: school instruction, for example, was supposed to include empirical exercises that would guide the student to the point where he could teach himself. University instruction in turn was supposed to enhance the skills of independent learning; the professor was not the teacher and the student the learner, but the student was to be his own teacher, guided along by the greater wisdom of the professor.[1] Yet without specifying the precise means whereby students could attain this goal, the reforms and ideologies shaping the Prussian university system did not firmly establish a path to their objectives. At best they suggested fluid boundaries within which classroom practices were supposed to take shape.

Certainly the research ethos, which became a prominent part of the Prussian university in the first half of the nineteenth century, placed new and different pedagogical demands on a professoriate now compelled to develop curricula that could in some way accommodate methodological concerns as well as changes in the content and structure of their disciplines. By the 1860s, university instruction had changed permanently, thoroughly, and more decisively than ever in the past. But the transformation cannot be explained simply by presuming that research methods were introduced intact into the classroom. Undoubtedly the professoriate felt a tension between its deeper intellectual mission to promote *Wissenschaft* and its daily participation in teaching, but it is not at all clear that the ideals of *Wissenschaft*, or of *Bildung* for that matter, can account for the fine details of the changes in the classroom. To be sure, student research and independent study were accomplished; but the reform of classroom practices before midcentury, especially in the natural sciences, was dominated more by exercise sessions supplementing lecture courses; examinations, homework problems, or other ways of checking student performance; the arrangement of lecture courses of a single discipline into a curriculum based on special introductory courses; laboratory exercises wherein students practiced the investigative techniques of a science on a

---

1. Wilhelm von Humboldt, "Der Königsberger Schulplan" [1809], in *Zur Geschichte der höheren Schule*, Bd. II, *19. and 20. Jahrhundert*, ed. A. Reble (Bad Heilbrunn: J. Klinkhardt, 1975), pp. 7–15.

small scale; and the creation of new instructional materials and facilities. Original student research was not always the intention or goal of these reforms.

Changes in science classes of the philosophical and medical faculties were in two respects sometimes contingent upon reforms that concerned the state's secondary schools. First, state reform of the gymnasium curriculum assigned priority to the classics, the humanities, and mathematics, making it clear that the natural sciences, including physics, were minor subjects, of little importance to the gymnasium's educational mission. The debates that had first emerged in the 1780s over whether the classical and semiclassical schools were adequate for dealing with contemporary social needs eventually led to the formal establishment of the classical gymnasium in 1812, when a gymnasium leaving-examination, the *Abitur*, was instituted. Over the next two decades the natural sciences found a more hospitable home in the state's various *Realschulen,* the semiclassical schools that taught Latin but not Greek and that emphasized science and mathematics more than did gymnasiums. Unlike the gymnasium, formed from above by the state's interests, *Realschulen* were shaped from below by social groups with their own objectives. *Realschulen* proved so difficult to fit into the state's educational system that, especially in the 1820s, their role and purpose were pressing educational concerns. Ernst Gottfried Fischer—*Physiklehrer* at the new Cöllnisches Realgymnasium, author of an influential physics textbook, and *außerordentlicher* professor at Berlin University—argued, for example, that because economic growth (which in his view Prussia sorely needed) demanded new ways of thinking— rational, practical, and technical—the natural sciences, not the classics, should be central to the curriculum. But Fischer's suggestion that educational reform went hand-in-hand with economic reform went largely unheeded. Shortly after *Realschule* graduates acquired the right to enter certain civil service positions in the early 1830s, the gymnasium's role in the state's educational system became even more strategic: in 1834, passing the *Abitur* became a prerequisite for university study. Although by 1859 *Realschulen* came more firmly under the formal jurisdiction of the state's educational ministry, it was not until 1870 that *Realschule* graduates could enter the university, and then only in certain fields.[2] Humboldt had believed that a strong educational system would provide a smooth transition from the school to the university. For most of the century, however, that transition in the natural sciences was determined by the weaker of the two secondary school curricula. What knowledge of the sciences a student did gain at the gymnasium therefore determined in part what could be taught in introductory courses at the university.

Second, social changes begun at the end of the eighteenth century

2. Kathryn M. Olesko, "Physics Instruction in Prussian Secondary Schools before 1859," *Osiris* 5 (1989): 94–120, esp. 95–103.

included measures designed to redefine social position from status inherited at birth to objective qualifications usually achieved through education. The state itself dramatically shaped and articulated that transformation in its civil service examinations, which emphasized "functional expertise" as the distinguishing feature of the professions of law, theology, and medicine.[3] Obviously, to display the skills necessary for certain professions, students would have to be trained in them. After 1810, when Prussian authorities established a state examination for gymnasium teachers (thereby conferring civil service status upon the profession and so upgrading it), teacher training became the utilitarian goal of the philosophical faculty. The state's teaching examination was in principle a scholarly one, testing in particular the candidate's philological, historical, and mathematical knowledge rather than teaching abilities. The relative unimportance of didactics was also underscored by the fact that candidates with a doctorate were not initially obligated to take the examination. By 1827 prospective *Realschullehrer* also had to take the examination, a requirement that became a formal part of the examination statutes in the 1830s, when the 1810 examination was revised.[4]

The 1831 teaching examination was considerably more specialized but still heavily oriented toward classical subjects, defining the natural sciences only in very broad terms. One concession to candidates who wished to teach mathematics or the natural sciences in *Realschulen* was that they were not obligated to be tested in Greek and Latin. Given that the state examination system as a whole worked to define the professions of law, theology, medicine, and secondary school teaching in terms of functional expertise, it seems likely that learning that cultivated skills and methods (as well as abstract knowledge) would probably be favored over learning that cultivated abstract knowledge alone. Although not enough is yet known about what kinds of questions in the natural sciences were asked on the examination, the broad outlines of what that functional expertise meant are nonetheless discernible in its regulations. In principle, functional expertise for prospective physics teachers was defined partly in terms of the ability to express physical concepts mathematically (to the extent that they then could be).[5] Yet despite the central importance of the state's examination in setting teaching qualifications, there is little evi-

---

3. R. Steven Turner, "The *Bildungsbürgertum* and the Learned Professions in Prussia: The Origins of a Class," *Histoire Sociale–Social History* 13 (1980): 150–35; idem, "The Prussian Professoriate and the Research Imperative, 1790–1840," in *Epistemological and Social Problems of the Sciences of the Early Nineteenth Century*, ed. H. N. Jahnke and M. Otte (Dordrecht: Reidel, 1981), pp. 109–21, esp. pp. 113, 117.

4. *Das Reglement vom 20. April 1831 für die Prüfungen der Candidaten des höheren Schulamts in Preußen mit den späteren Erläuterungen, Abänderungen, und Erweiterungen* (Berlin: Enslin, 1865), pp. 1–4; Otto Schmeding, *Die Entwicklung des realistischen höheren Schulwesens in Preußen bis zum Jahre 1933 mit besonderer Berücksichtigung der Tätigkeit des deutschen Realschulmännervereins* (Cologne: Balduin Pick, 1956), pp. 27–28.

5. *Reglement vom 20. April 1831*, pp. 24–35, 63.

dence that the examination *determined* to any great extent the content of university instruction in the natural sciences, including physics, at a local level. Instead, it seems to have reinforced certain existing practices over others, making it more expedient and profitable, for example, to study a mathematically oriented physics than a nonmathematical one.

What neither of these two state-implemented reforms for upgrading secondary schools suggest in detail, however, is why natural science instruction came to include practical exercises; what the constitution of those exercises were; and how, subsequently, they shaped scientific practice.

In the midst of these changes, Königsberg University was favored by the state for a while but was never as special as Berlin or Bonn, nor as ignored as Greifswald. Shortly after the Prussian defeat at Jena in 1806 and the establishment of an educational reform commission by Friedrich Wilhelm III in 1807, the royal family took up residence in Königsberg, as did Wilhelm von Humboldt months later. During the Stein reform era of 1807–08, Königsberg was a "laboratory for experimenting with new forms of municipal structure" that sought to strengthen citizenship and modernize the bureaucracy while retaining older, entrenched commercial-class interests.[6] The university that had cradled both the chief exponent of the German Enlightenment, Immanuel Kant, and the popularizer of Adam Smith's economic theories, Christian Jakob Kraus, found its fortune enhanced even as the city entered a period of economic decline after the blockades that diminished the trade handled by its port. Reforms intended to improve Prussia's economic, social, and political profile vis-à-vis other nations also impinged upon local concerns, including the city's university. In 1809 the university's budget doubled; later, new institutes were created and new appointments made. A prominent part of the reforms at Königsberg involved the natural sciences.

The focal point for the evolution of the natural sciences at Königsberg in the early nineteenth century was the chair held by Karl Gottfried Hagen. Hagen, who had taught at Königsberg University since 1783 and was responsible for courses in zoology, pharmacy, botany, chemistry, and mineralogy in the medical faculty, was later transferred to the philosophical faculty, where his teaching responsibilities expanded to include physics when the professor of philosophy, poetry, and physics retired. Hagen used his practical expertise as town pharmacist to create simple laboratory exercises in chemistry and pharmacy before they were common at most other German universities; he held them in his own shop. He promoted an interest in the natural sciences among a broad-based stu-

6. Marion W. Grey, *Prussia in Transition: Society and Politics under the Stein Reform Ministry of 1808*, Transactions of the American Philosophical Society, 76, pt. 1 (Philadelphia: American Philosophical Society, 1986), pp. 105–7, 154–55, on p. 105.

dent clientele and townspeople alike; even the royal family had attended his lectures.[7] Despite his popularity, by the time of his death in 1829, others had changed natural science instruction at Königsberg considerably.

By then the astronomer Friedrich Wilhelm Bessel (1784–1846), who had come to the natural sciences by way of commerce and trade, had enhanced the university's program. Bessel's own educational background was unusual. Loathing his Latin classes but drawn to mathematical calculation, he left the Minden Gymnasium in *Tertia* to take up commercial pursuits. In 1799 he joined the Kuhlenkamp trade firm in Bremen for a seven-year apprenticeship and became immersed in the world of commerce. When the prospect of joining a trade expedition to the French and Spanish colonies and to China appeared, Bessel reviewed the commercial transactions of these regions, learned English and Spanish, and then studied navigation, especially the techniques used to calculate a ship's position. Finding that captains of German trade ships, in contrast to their English counterparts, used few astronomical techniques, he set out to master them. About the same time, he was drawn into the scientific culture of Bremen, experiencing science as it impinged upon the world of commerce and trade. By 1806, after he had performed exemplary calculations on the path of the comet of 1607, he decided to leave commerce to take up an inspector's position at a private observatory in Lilienthal. The change in his career plans did not diminish his commitment to a commercial outlook: until the end of his life he believed that commerce and trade had the potential to produce the kinds of social and economic change necessary for improving general welfare. And although Bessel exchanged bookkeeping for astronomy, he retained a business mentality by accounting for every value in a transaction, even in his measurements of the heavens. In Lilienthal, while working with apparatus of less than desirable quality, he developed ways of identifying and computing instrumental errors and their effect upon astronomical observations.[8]

Bessel's skill as an observer was well enough known by 1809 that when Friedrich Wilhelm III approved eight thousand taler for the construction of an observatory as part of the reform of Königsberg, Humboldt recommended that Bessel be appointed its director. Bessel (Figure 1) arrived at Königsberg in May 1810, without having completed the gymnasium,

7. On Hagen see L. Friedländer, "Aus Königsberger Gelehrtenkreisen," *Deutsche Rundschau* 88 (1896): 41–62, 224–39, on 48–49; Hans Prutz, *Die Königliche Albertus-Universität zu Königsberg i. Pr. im neunzehnten Jahrhundert* (Königsberg: Hartung, 1894), p. 147; Götz von Selle, *Geschichte der Albertus-Universität zu Königsberg in Preußen*, 2d ed. (Würzburg: Holzner, 1956), pp. 221, 280.

8. F. W. Bessel, "Lebensabriss: Kurze Erinnerungen an Momente meines Lebens," in *AFWB* 1:xi–xxxi; idem, "Uebervölkerung," in *AFWB* 3:483–86, esp. 486; A. Peters, "Uebersicht der Leistungen Bessel's in der Stellar-Astronomie und in der Theorie der astronomischen Instrumente," *Königsberger Naturwissenschaftliche Unterhaltungen*, vol. 2 (Königsberg: Bornträger, 1851), pp. 101–40.

Figure 1. Friedrich Wilhelm Bessel. From *AFWB*,
courtesy of Yale University Photographic Services.

without university training, and without a doctorate. The Königsberg
faculty approved his courses for the university's lecture catalog only after
Göttingen's astronomer, Carl Friedrich Gauss, awarded him an honorary
doctorate in 1811. Until his observatory was completed in 1813, Bessel
used a room in Hagen's residence for teaching and for astronomical
observations.[9]

Nothing was spared in the construction of his observatory, Bessel later
wrote. Constructed "entirely" according to his wishes, it was, in his words,
"one of the most magnificient observatories in the world," a "new temple
of science" (Figure 2). No other observatory at the time enjoyed such
unobstructed views of the horizon. Living quarters for his family were on

9. On the doctorate problem see F. W. Bessel to C. F. Gauss, 20 August 1810, 19 October
1810, 10 March 1811, GN.

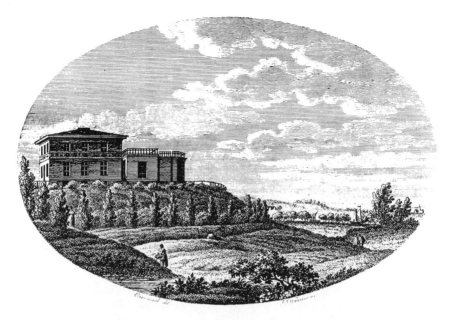

Figure 2. Bessel's observatory at Königsberg, 1815. From Bessel, *Astronomische Beobachtungen*, I Abt.

the top floor, but the much larger ground floor was for science alone (Figure 3). The ground-floor plan of Bessel's observatory was cruciform church-like. Those who worked in service of the observatory (by 1829 a staff of four conducted and reduced observations and maintained the institute) or who came seeking enlightenment and inspiration in its lecture hall (*Hörsaal*) could be found in its nave, which also housed Bessel's office (*Arbeits-Zimmer*), an entrance room (*Eintritts-Zimmer*), and a small sleeping room. Bessel conducted observations in the transept and apse sections (*Stern-Warte*); significantly his major instruments were located in the apse, a scientific altar to the heavens. The space of the altar was truly sacred: Bessel did not allow his students to climb the steps from the lecture hall to the transept until they had thoroughly learned the quantitative and technical foundations of the art of observation. Over the apse, a dome was erected in 1829, completing the likeness of the structure to a church.[10]

Although the number of students trained at the Königsberg observatory is at present unknown, the educational functions of Bessel's institute

10. F. W. Bessel, "Kurze Beschreibung der Königsberger Sternwarte," in *AFWB* 3:296–97; idem, *Astronomische Beobachtungen auf der Königlichen Universitäts-Sternwarte in Königsberg, I Abt., Vom 12. November 1813 bis 13. Dezember 1814* (Königsberg: Nicolovius, 1815), pp. i–iii, on p. i; Karl Faber, *Die Haupt- und Residenz-Stadt Königsberg i. Pr.* (Wiesbaden: Sändig, 1840; rpt. 1971), pp. 289–90.

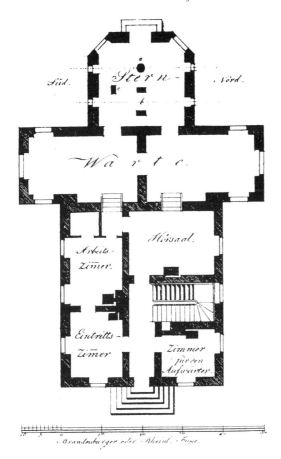

Figure 3. Floor plan of Bessel's observatory. From Bessel, *Astronomische Beobachtungen*, I Abt.

were as important to him as the observations he conducted there. At first doubting his ability to lecture, Bessel quickly found that he did so "rather well, and always before a full auditorium." He developed a keen interest in educational matters concerning mathematics and the natural sciences and worked to improve instruction in mathematics and to increase student and public interest in the exact sciences. He served on the local board for administering the state's teaching examination, for which he tested teaching candidates in mathematics. Along with Hagen, he had founded in 1812 the *Königsberger Archiv für Naturwissenschaft und Mathematik*, a publication intended to communicate news about the natural sciences and mathematics to a broad audience. But the real focus of his efforts was the improvement of instruction in mathematics. Buoyed by his successes in training students in his observatory, he began to promote mathematical education more vigorously. In 1825, just before several appointments in the exact sciences were made at Königsberg, Bessel reported that he was

happy "to hear from Altenstein's own mouth that around here the resources for mathematical studies would be *generally* renewed," adding
that "if this actually takes place at the state's secondary schools as well as its
universities, then not only we will attain the first *general* preparation of the
true foundation for arts and trade, but we will also be protected from the
many undesirable elements that have accompanied the *Zeitgeist*," namely,
the relative neglect of science and mathematics in Prussia's neohumanist
curriculum based on classical studies.[11]

In Bessel's view, the curricular hegemony of the classics could be overcome if the public's perception of mathematics changed. He condemned
the contemporary image of the mathematician "as a caricature . . . incapable of making contact with the world" and hoped to counteract that
image by reforming mathematical instruction. "We cannot hope to make
mathematical education commonplace," he pointed out, "if we only emphasize *pure theory*." As his own experience had taught him, students were
not incapable of learning mathematics, but they resisted doing so, viewing
the subject as a rigidly constructed "accumulated mass of theorems."
Bessel taught applied mathematics in part as an antidote to this image. He
found his own field, astronomy, particularly well suited for this task; for in
his view, "no science is as rich in the application of mathematics as astronomy, and if mathematics is to be learned, then *precisely this science* must be
taught."[12]

Bessel knew, however, that the success of his efforts depended not only
on what he did but on the quality of his student clientele. A few years after
he arrived at Königsberg, he told Gauss that "only seldom do I have the
luck of having students who are very capable and active" but that it would
be "inappropriate" for him "to complain loudly" about the others.[13] After
having trained several assistants in his observatory and after years of
teaching mathematics, his success improved. He wrote to Alexander von
Humboldt in 1828 that he regarded it a "considerable accomplishment"
to have spread "mathematical knowledge." He explained that he had
"tried for eighteen years to convey some mathematics to students, always
through its application, especially in astronomy. In the beginning, I met
with little success. But soon quite a number of students, who had an
inclination to study *modern* theorems, appeared." To ensure a steady
stream of interested university students, though, Bessel believed Prussia's
secondary school curriculum had to change. "The dominance of the
classical languages must cease if the best direction of the mind is not to go
unused," he reiterated.[14] Although he suggested that "many existing

11. Bessel to Gauss, 20 August 1810, 27 December 1810, 10 March 1811, GN; F. W.
Bessel to Ober-Regierungsrath, 21 November 1825, Darms. Samml. Sig. J 1844.

12. Bessel to Ober-Regierungsrath, 21 November 1825, Darms. Samml. Sig. J 1844.

13. F. W. Bessel to C. F. Gauss, 12 January 1812, *Briefwechsel zwischen Gauss und Bessel*, ed.
A. Auwers (Leipzig: W. Engelmann, 1880), p. 166.

14. F. W. Bessel to Alexander von Humboldt, 25 December 1828, A. v. Humboldt
Nachlaß, SBPK-Hs.

regulations will have to be altered" in order to enhance the role of the natural sciences in the curriculum, Bessel never attempted to change the ordinances that governed instruction.[15] Instead he shaped his own curriculum so as to compensate for what was not taught in gymnasiums and to promote a mathematical orientation to the study of the exact sciences. By offering elementary lectures in mathematics, he was able in principle to open his introductory astronomy courses to more students and to conduct other astronomy courses at more advanced levels.

But he was not always so accommodating. One student from the 1820s reported that Bessel often said, "If you don't understand me, then you shouldn't be in my course."[16] Although Bessel's lectures were heavily attended at the beginning of the semester, attendance allegedly decreased considerably thereafter because he assumed a quick mastery of fundamentals and moved on rapidly to more difficult topics. Moreover, the intensity of Bessel's courses often forced students, including those familiar with his teaching methods, to ignore all other course work. Yet those students who remained in his courses found themselves well prepared to study higher mathematics and to enter professions requiring mathematical expertise.[17] Bessel assumed that ordinary students could learn the mathematical techniques needed for professional practices, such as those of astronomy, if they were taught systematically. His pedagogy worked. By the 1830s and 1840s he had the pleasure of writing to Gauss about the exceptional students who came his way; among them were Carl Eduard Senff and Carl Wilhelm Borchardt; both had also studied physics at Königsberg.[18] By drawing students to the study of mathematics, Bessel helped strengthen the exact sciences at Königsberg.

Before 1820 the ministry appointed three other natural scientists to Königsberg: August Friedrich Schweigger in botany and Karl Friedrich Burdach and his prosector Karl Ernst von Baer in anatomy and physiology. Burdach and von Baer reformed instruction in anatomy and physiology. Von Baer's interest in the content and format of natural science instruction dated from his secondary school days. He valued independent study (although not necessarily independent research) and abhorred the mere transmission of knowledge in lectures. During his university years at Dorpat (under Burdach) and Würzburg, he came to appreciate the role of exercises in cultivating critical modes of thinking, which he believed to be the highest goal of education. As for the "means suitable for exercising the mind," he advocated, near the end of his career, instruction in mathematics and the exact sciences because they were based upon "measure-

15. Ibid.

16. Quoted in *Erinnerungen aus dem Leben von Hans Victor von Unruh (1806–1886)*, ed. H. von Poschinger (Stuttgart: Deutsche Verlags-Anstalt, 1895), p. 29.

17. As did Hans Victor von Unruh, ibid., p. 30. Several of Bessel's students worked on, for instance, triangulation projects throughout Europe.

18. Bessel to Gauss, 17 November 1832, 11 October 1843, *Briefwechsel zwischen Gauss und Bessel*, pp. 509, 559.

ment and numbers." Earlier, he had directed his pedagogical efforts elsewhere. Shortly after his appointment in 1817 he and Burdach implemented reforms that methodically and systematically reorganized lecture courses in anatomy and physiology and introduced demonstrations into them. Von Baer distanced himself from earlier university pedagogy by creating lectures that were "not designed to be copied word for word" but that were intended "to provide the audience with an idea." He also supplemented his lecture courses with special review sessions on Saturdays—he called them "refresher meetings"—at which "those present were requested at random to demonstrate their recently acquired knowledge."[19]

What Bessel, Burdach, and von Baer began to offer was a type of science instruction that involved students in more critical ways of thinking than the older eighteenth-century methods of rote learning and passive listening had done. Not enough is known about what went on in their classrooms to say with certainty how they organized anatomical, physiological, and astronomical knowledge in their lectures, or how their student clienteles helped to shape their courses. Yet their efforts undeniably reoriented science instruction at Königsberg, integrating practical exercises and skills into theoretical instruction and making it clear that science instruction was best organized systematically. Their courses reshaped the contours of the scientific disciplines they represented and as their students later understood them.

Other appointments made at Königsberg in 1826 decentralized natural science instruction even more and made it possible to specialize and enhance science teaching still further. Ernst Heinrich Friedrich Meyer replaced Schweigger, and four *Privatdozenten* arrived: Carl Gustav Jacob Jacobi (1804–51) in mathematics, Friedrich Philipp Dulk in chemistry, Heinrich Wilhelm Dove in physics, and Franz Ernst Neumann (1798–1895) in mineralogy. The polymath approach to science teaching exemplified by Hagen had ended. Jacobi, gifted in mathematics since his gymnasium years and favored candidate of the Prussian ministry, had requested to go to Königsberg after lecturing for one semester at Berlin. Dulk eventually built upon the exercises in chemistry and pharmacy that Hagen had introduced, but not without difficulty and considerable delay. Although the ministry brought Dove and Neumann to Königsberg to teach different subjects, it must have known their interests overlapped; Neumann in fact had hoped to teach physics in addition to mineralogy.

Called to teach physics, Heinrich Dove had just completed a thesis on climatology at Berlin. His research interests included meteorology, geomagnetism, optics (especially crystalline optics), and several problems in experimental physics, later including electrical induction. He was among

19. Karl Ernst von Baer, *Autobiography of Karl Ernst von Baer*, ed. and trans. Jane Oppenheimer (Canton, Mass.: Science History Publications, 1986), pp. 42–48, 60–72, 120–31, 157–64, 181; on pp. 68, 72, 157, 164.

the first to find a system in weather changes. Generally comprehensive, his work relied upon an inductive methodology that stressed observation but had little room for speculative theory and mathematical expressions. He viewed theory merely as the accumulation of results, arguing that "one must always stop short in physics, because in the science of nature, things are as in nature itself."[20] While at Königsberg, Dove lectured on experimental physics to a diverse student clientele. In 1829 he returned to Berlin, where he did not become a full professor until 1844. In Berlin he held auxiliary positions at the Friedrich Werder Gymnasium, the Friedrich Wilhelm Gymnasium, and the Allgemeine Kriegsschule. He became part of the Berlin physics group—a group that by 1834 included Johann Poggendorff, Gustav Magnus, and August Seebeck—whose accomplished members worked to promote a strong following in physics at Berlin educational institutes but whose collective approach to physics, manifest in both pedagogy and practice, was less quantitative than what emerged elsewhere, especially at Göttingen and Königsberg.[21]

Dove's departure from Königsberg was in part occasioned by the evolution of Neumann's career. Neumann's interest in mathematics and the physical sciences in general had its origins in a culture quite different from Bessel's. Classically educated, possessing a university degree, earlier drawn to popular intellectual movements in the university and to Prussian causes in the Wars of Liberation (in which he served), Neumann's interest in the physical sciences had not been stimulated by their practical applications but, rather, by ideal intellectual concerns. Schooled in Berlin, he began his university studies in theology there in 1817 and then migrated to Jena for 1818/19, where he abandoned theology and was drawn politically to the *Burschenschaften*, student organizations unsettling enough to authorities to be banned on university campuses in the 1820s, and intellectually to *Naturphilosophie*. Although the latter interest appears to have been brief, it did surface in one early publication.[22] Owing to the excesses of the *Burschenschaften*, Neumann returned to Berlin in 1819 and continued his studies under the mineralogist Christian Weiss, whose research centered on the geometric structure of crystals and who may have stimulated Neumann's further interest in *Naturphilosophie*. Neumann adopted Weiss's geometric approach to crystallography and became deeply attached to Weiss, later remarking he "was the only person from whom I learned something."[23] Although his tutelage under Weiss was interrupted from 1821 to 1823, when Neumann was called home to supervise his

---

20. H. W. Dove, *Ueber Elektricität* (Berlin: Reimer, 1848), p. 38.

21. Hans Neumann, *Heinrich Wilhelm Dove* (Liegnitz: Krumblaar, 1925), pp. 6–7, 11–15, 28–31.

22. In his *Beiträge zur Krystallonomie* (Berlin: Mittler, 1823), Neumann suggested that directionality in crystal structure was the spatial representation of forces at work in the crystal. This statement was deleted from the *Beiträge* in *GW* (1:175–322, on 318).

23. Quoted in LN, p. 10.

mother's modest estate, Weiss nonetheless asked Neumann upon his return to Berlin to present a series of lectures on crystallography and mineralogy for a select audience that included Alexander von Humboldt and the geologist Leopold von Buch. These lectures proved decisive for Neumann's career, not only because he became known within Berlin scientific and educational circles but also because when he expressed an interest in the physical properties of crystals and minerals, someone in the audience recommended he take a look at Joseph Fourier's recently published *Théorie analytique de la chaleur*, a work that later shaped part of Neumann's investigative style.[24]

Being known in Berlin probably also helped him obtain a university teaching position as *Privatdozent* under exceptional circumstances: with a modest salary and without fulfilling the requirement of the *Habilitation*. When in 1825 Neumann first asked to become *Privatdozent* at a university with a mineralogical collection, the ministry did not respond favorably but instead encouraged him to continue working in crystallography, awarding him a 150 taler stipend. Because of the exceptional quality of his doctoral dissertation and the publication of some of his results, he was excused from the written requirement of the *Habilitation*, the examination and lecture exercises for *Privatdozenten*. Finally, on 6 May 1826, the ministry invited Neumann to become *Privatdozent* at Königsberg, where he collected an annual salary of 200 taler.[25]

Despite the unusual award of a salary (*Privatdozenten* were normally expected to draw their income from lecture fees alone), Neumann was not altogether pleased with the offer to go to Königsberg. He had wanted to stay with Weiss, where he could continue working on a mineralogical collection both familiar and essential for his investigations. But his appointment at Berlin was precluded by that of Gustav Rose, who in 1823 had joined Weiss as Berlin's second mineralogist. Though Königsberg's mineralogical collection was of acceptable quality, Hagen guarded it closely, so Neumann tried, unsuccessfully, to convince the ministry that he should become *Privatdozent* at Breslau. He apparently did not know that Ernst Friedrich Glocker, a gymnasium teacher, had already taken that post when Heinrich Steffens went on leave during 1824/25.[26] Without offering to clarify Neumann's relationship to Königsberg's mineralogical collection, the ministry pressed him to make a decision on the offer by approving an additional 50 taler for mineralogical models and 50 taler for

24. LN, pp. 89–146; Joseph Fourier, *Théorie analytique de la chaleur* (Paris: Firmin Didot, 1822).

25. Kultusminister Karl Altenstein to Neumann, 6 May 1826, rpt. in LN, p. 230.

26. Neumann to Kultusminister Altenstein, 8 July 1826, rpt. in LN, pp. 232–34. On Glocker see *Festschrift zur Feier des hundertjährigen Bestehens der Universität Breslau*, Teil 2, *Geschichte der Fächer, Institute, und Aemter der Universität Breslau, 1811–1911*, ed. G. Kaufmann (Breslau: F. Hirt, 1911), pp. 467–68.

his journey. Neumann accepted but wrote to his mother that he hoped his stay would not last more than a year.[27]

At the time of his appointment, Neumann's interests were not confined to mineralogy. Over a year earlier he had decided to explore the application of mathematics to physical problems. On 2 January 1825 when he made his first attempt to secure a position in a Prussian university, he wrote to the ministry that he had indeed occupied himself "with the physico-mathematical parts of mineralogy, but I do not wish to see myself restricted to these subjects in teaching. I would like to be permitted to include in my teaching in general those sections of physics which have received a higher mathematical treatment *or those which are now capable of doing so*."[28] Neumann's scientific work before 1825 gives us few clues to what inspired his initial goals in science pedagogy. A mathematical orientation was consistent with his approach to mineralogy, and beyond that, as we shall see, his reading of Fourier and other French works was stimulating and suggestive. But the route he would take to achieve that "higher mathematical treatment" was as yet undefined.

"Mathematical physics" was by 1825 a term in common use. In the German-speaking areas of Europe, it meant principally the quantitative determination of physical constants. Sometimes the term also embraced the more abstract style of the French, including Poisson and Fourier; both French and German practitioners classified mathematical physics institutionally as mathematics or applied mathematics. Yet "mathematical physics" was a term hitherto used principally to define research styles rather than a kind of physics instruction. To be sure, courses in mathematical physics had existed since early in the eighteenth century, but the general belief was that pedagogically the difference between mathematical and experimental physics was artificial.[29] Combining the two proved difficult. A few individuals, such as the Berlin gymnasium teacher Ernst Gottfried Fischer, tried to teach physics from a mathematical perspective but only on rare occasions were able to go beyond simple methods of quantifying physical variables.[30] Institutional and other circumstances prevented Fischer and others from pursuing a mathematical physics in the university classroom more deeply, along the lines of French physical research. Calculus was not yet regularly taught in Prussian gymnasiums, nor did students who studied the natural sciences at Prussian universities necessarily take calculus or other forms of higher mathematics.

27. Kultusminister Altenstein to Neumann, 6 May 1826, 12 August 1826, rpt. in LN, pp. 230, 234; Neumann to Gräfin Wilhilmine, n.d. [September 1826], rpt. in LN, pp. 234–35.

28. Neumann to Kultusminister Altenstein, 2 January 1825, rpt. in LN, pp. 225–27; on p. 226 (emphasis added).

29. Johann Karl Fischer, *Physikalisches Wörterbuch*, 10 vols. (Göttingen: Dieterich, 1798–1827), 3:887–903 ("Physik"), on 890.

30. Ernst Gottfried Fischer, *Lehrbuch der mechanischen Naturlehre*, 3d ed., 2 vols. (Berlin: Nauck, 1826–27). The first edition of Fischer's text appeared in 1805; the second, 1819.

In an institutional environment where research and teaching were being drawn closer together, it was not unusual for professors to incorporate research results into courses. Drawing into physics teaching recently mathematized areas of physics was Neumann's way of doing just that. Yet Neumann's statement of January 1825 seems to indicate that he intended to do more. Through exploratory teaching, Neumann hoped to enhance the use of mathematics in physics and to make the classroom a creative disciplinary setting. Because at this time he gave no indication of how this might be done, his statement was rash. He was aware that because this kind of teaching helped to locate the intellectual and methodological boundaries of a field, it was best accomplished in an environment where competition for the same "intellectual territory" was kept to a minimum. So when he learned in July 1826 that Dove had been called to Königsberg to teach physics, he wrote to the ministry, requesting he be sent instead to Breslau, where, he thought, he would find "a *free field* for lectures . . . not only in the mathematical and physical parts of mineralogy, but also in mathematical physics in general."[31] Sent to Königsberg after all, Neumann did not at first offer physics courses that allowed him to develop his mathematical interests, nor did he even teach courses in physics proper. Between 1826 and 1829, it was Dove who taught experimental physics, optics, electricity, acoustics, and the theory of heat, while Neumann offered courses in mineralogy, crystallography, the physics of the earth, and the physical characteristics of minerals.[32]

But in Neumann's unpublished research between 1826 and 1828, quantitative techniques were prominent enough so that when Bessel recommended him for *außerordentlicher* professor of mineralogy on 7 October 1828, he stressed the mathematical quality of Neumann's work, arguing that his "mastery of knowledge and the caution with which he conducts his scientific investigations along with the diligence and perseverance which he brings to them are so great, that I safely predict that he will soon attain one of the first places among mathematical physicists." By then Bessel was also impressed by Neumann's teaching abilities; for he ended his letter of recommendation with the remark that "everyone who enters university teaching should display a similar maturity and decisive preference for his scholarship."[33] Less than a year later, on 17 May 1829 after Hagen had died, the ministry appointed Neumann the *Ordinarius* for physics and mineralogy. Dove, passed over in favor of someone whose pedagogical intentions and intellectual inclinations were more in line with Bessel's, remained bitter.

Bessel's support of Neumann's research style and his endorsement of

31. Neumann to Kultusminister Altenstein, 8 July 1826, rpt. in LN, pp. 232–34; on p. 233 (emphasis added).

32. H. Neumann, *Dove*, p. 13; below, App. 2.

33. F. W. Bessel to Kultusminister Karl Altenstein, 7 October 1828, rpt. in LN, pp. 254–55.

Neumann's goals for physics pedagogy can be viewed as part of his own agenda to reform instruction in mathematics and the natural sciences. Bessel's efforts reached outside his own classes and beyond his support of curricular reform at the secondary school level. His views were progressive enough to assure him a strategic role as informal advisor to the liberal chief provincial administrator of East Prussia, Theodor von Schön, through whom the educational ministry in Berlin administered Königsberg University. Bessel shared this role with Carl Jacobi and the Pestalozzian pedagogue and philosopher Johann Friedrich Herbart, who was Kant's successor and who had arrived at Königsberg when Bessel did. All three had similar enough views on curricular reform in the exact sciences so that when Schön wanted to plan a polytechnic school in the vicinity of Königsberg, for instance, he consulted them together.[34] Bessel even assisted in Herbart's pedagogical seminar, established in 1810 to train primarily elementary and middle school teachers. Like Johann Heinrich Pestalozzi, Herbart incorporated a strong practical component into instruction, and he tried in his seminar to provide teaching experience for his students and to tie practice teaching to the production of original and, he hoped, publishable pedagogical essays. Although he was often unable to bring his seminar to the desired level (his students, still attending lectures, were unable to perform the critical thinking necessary for original work), Herbart never abandoned his goal of improving instruction at all levels and in several subjects. With Bessel's help, he offered his seminar students instruction in mathematics and several of the natural sciences: Bessel taught mathematics and elementary mechanics, while Herbart lectured on differential and integral calculus, conic sections, elementary astronomy, statics, and higher mechanics. Herbart's emphasis on didactics contrasted sharply with the customary training of secondary school teachers, whose education omitted didactics despite the inauguration in 1826 of a required, unpaid probationary year of teaching. Without being expressly designed as an institute for training mathematics teachers, Herbert's seminar was nonetheless the first institute in Prussia for mathematics and science pedagogy.[35]

Bessel came to Königsberg not only without having attended a university but also without any formal experience in teaching. His close association with Herbart during his first years at Königsberg helped him to form his scientific pedagogy. Bessel adapted Pestalozzian principles to science instruction in the 1810s and used them to shape the practical exercises accompanying his lectures. They also helped him to view curricular reform within a larger context: he came to consider Pestalozzian methods

34. W. Ahrens, *C. G. J. Jacobi als Politiker* (Leipzig: B. G. Teubner, 1907), pp. 6, 21n.
35. K. Kehrbach, "Das pädagogische Seminar J. F. Herbarts in Königsberg," *Zeitschrift für Philosophie und Pädagogik* 1 (1894): 31–47; Prutz, *Albertus-Universität zu Königsberg*, pp. 153–56; Wilhelm Lorey, *Staatsprüfung und praktische Ausbildung der Mathematiker* ((Leipzig: B. G. Teubner, 1911), pp. 102–3.

responsible for exciting the "great hope" that educational reform was
capable of changing the outlook of students and providing them with the
mental apparatus for dealing imaginatively with contemporary social,
political, and economic problems.[36]

Bessel's teaching had two important practical components. The first
was exercises to accompany his courses. He pioneered in viewing homework problems as essential to science learning; his students, in turn,
brought his pedagogical methods to other institutions, especially secondary schools.[37] Every student in Bessel's classes completed problem sets; he
did not reserve active learning for an elite. The second component was
more difficult to realize: training scientific assistants. His observatory was
the first natural science institute at Königsberg to unite science instruction with professional training in workaday experiences. Although the
proximity of the lecture hall to the observatory symbolized the union of
theoretical and practical learning, Bessel allowed students to work by his
side only after they had been adequately trained in mathematics. His own
success in integrating professional training into science teaching was due
generally to his view that science and mathematics could only be learned
by uniting "theory to practice." More specifically it was due to what his
students remembered—notably in terms of an economic metaphor—as
his ability to create a "division of labor" among his young assistants, which
was possible largely because of the strategic role of error analysis in the
measuring operations of his observatory. He was especially pleased when
his students extended mathematical methods to other sciences because he
found it so seldom that "the natural sciences were comprehended in this
way." By the early 1820s he reported to the ministry on the progress of the
young men who had studied with him. Although he was obviously proud
that he had trained a few astronomers, he was nonetheless disappointed
that few other careers were available to students who, out of economic
necessity, had to abandon the exact sciences or—what to Bessel was
tantamount to the same—to teach secondary school, where he believed
advanced academic work had little chance of maturing. That even a few
of his students were able to enter an academic career was a source of
pleasure to him.[38]

Bessel and Herbart were not alone in trying to reform mathematics and
natural science instruction at Königsberg, but few others before the

36. F. W. Bessel, "Uebervölkerung," in *AFWB* 3:485.
37. H. F. Scherk brought them to Halle; C. T. Anger, to secondary schools. Both had
assisted Bessel's teaching at Königsberg. F. W. Bessel to Johannes Schulze, 21 May 1825, rpt.
in E. Schoenberg and A. Perlick, "Unbekannte Briefe von C. F. Gauss und F. W. Bessel,"
Abhandlungen der Bayerischen Akademie der Wissenschaften, Math.-naturw. Klasse 71 (1955):
25–58, on 33–34; C. T. Anger, "Grundzüge der neueren astronomischen Beobachtungskunst," *Programm*, Städtisches Gymnasium zu Danzig, 1846/47.
38. F. W. Bessel to Kultusministerium, 10 December 1821, 26 December 1822, ZStA-M,
Acta betr. die Sternwarte bei der Universität zu Königsberg, Rep. 76Va, Sekt. 11, Tit. X, Nr.
16, Bd. II: 1814–25, fols. 106–7, 143–44; Anger, "Grundzüge," pp. 16–17, 31.

mid-1820s were able to integrate practical exercises into instruction with their success. Some of the new university institutes that appeared in the first few decades of the century were designed in part to accommodate similar novel means of instruction. When a clinical hospital was established at Königsberg in 1809, for instance, its director, Wilhelm Remer, singled out clinical experience as "among the first necessities of medical instruction" because that experience gave "young physicians certainty in the evaluation of the sick" by exposing them beforehand to what they would encounter when they assumed their duties as physicians. Remer also considered clinical instruction part of a refined, hierarchical curriculum. Employing a pedagogical tactic similar to one recently introduced in humanistic seminars, Remer divided students into two sections based on level of achievement and required that "everyone who wanted to take part in clinical instruction [present] me with a certified list of courses hitherto taken so that I might be in the position to judge at which level he could enter and whether or not he possessed the necessary prerequisites [*Vorkenntnisse*]."[39]

Several other facilities in the natural sciences were in principle available to students at Königsberg. A botanical garden established in 1809 and expanded in 1828 was the smallest of all such gardens in Prussia, but its holdings were diverse enough to require six separate greenhouses representing six different climatic regions. In 1817, Burdach supervised the reconstruction of the university's eighteenth-century anatomical theater. Added to it was a mechanical table that moved cadavers and other dissecting specimens from the cellar to the middle of the auditorium, around which were arranged sixty seats from which students could view the dissection. Within this renovated anatomical institute, von Baer established in 1819 a zoological museum primarily for classroom instruction; in 1821 he made it a separate institute. A mineralogical collection began in 1812 with a donation of specimens to the university; it was enhanced in 1822 and again in 1835. Directed by Neumann, the collection was housed in four rooms of the main university building, two of them very large. The physical cabinet, a gift of the crown in 1803, had less spacious facilities—only a single storage room—but in 1817 it absorbed Hagen's physical instruments, most of them built in England. The physical collection at first housed chemical instruments as well; these were weeded out sometime in the 1820s in order to establish a separate collection.[40]

These museums, institutes, and collections constituted the material resources for the natural sciences at Königsberg in the late 1820s, after the new appointments were made. Not all of them offered students the

---

39. Wilhelm Remer, "Die klinische Lehranstalt der Universität zu Königsberg, von ihrer Eröffnung am ersten November 1809 bis zum Anfang des Augusts 1811," *Königsberger Archiv für Naturwissenschaft und Mathematik* 1 (1812): 192–205, on 199.
40. Faber, *Die Haupt- und Residenz-Stadt Königsberg*, pp. 288–89, 290–92, 294–98.

opportunity to engage in practical exercises on a regular basis. Occasionally Neumann used the mineralogical collection and the rooms housing it
either for his own investigations or for practical exercises for student
experiments, but demonstration remained the primary purpose. The
botanical gardens and the physical and chemical collections were also
used primarily for demonstrations and secondarily for research; little is
known about what kind of student exercises and experiments, if any, were
conducted in them before the mid-1830s. In the renovated anatomical
institute, by contrast, Burdach was able to supply his students with about
forty cadavers a year for classroom dissections. For almost a decade the
curricular reforms of von Baer and Burdach focused on improved and
refined demonstrations, including the assembly and viewing of microscopic preparations, as the principal pedagogical innovation. Not until
von Baer took over the institute in 1827 were regular scientific exercises
other than the dissection of cadavers incorporated into the curriculum for
anatomy and physiology. Thereafter, von Baer conducted zootomical
exercises in the summer semester, while the prosector of the institute
supervised the dissection of cadavers in the winter semester. With the
exception of the anatomical dissections, regular practical scientific exercises were available before 1827 only in Bessel's observatory and Herbart's seminar. It is worth noting that before 1830 or so, Bessel, von Baer,
Remer, and Herbart created curricula of hierarchically ordered courses
either before or concurrently with practical exercises.[41] Sustained practical exercises in the natural sciences were closely tied to more general
curricular reforms.

   That this curricular reform took place at all has sometimes been attributed to faculty discontent over student apathy and concern over student enrollments. Reform proposals during the late eighteenth century,
for instance, included criticisms of *Brotstudien* and the cavalier attitude of
students who sought during their university years only what was directly
relevant to their professional careers afterward. Even during the first
decades of the nineteenth century, student apathy seemed to stimulate
the reform effort. At a time when a student subculture became more
prominent and campuses were politicized by the *Burschenschaften*, the
aftermath of the Wars of Liberation, and the restrictions on student
activities imposed by the Karlsbad Decrees, the excitement and tensions
outside the classroom stood in marked contrast to the doldrums within.
Or so it appreared to Königsberg faculty members annoyed by their
students' lack of interest and diligence.[42]

   41. Karl Ernst von Baer, *Berichte von der königlichen anatomischen Anstalt zu Königsberg*, VII
Bericht, *Mit dem Be[s]chl[ü]sse der Uebersicht von parasitischen und gedoppelten Menschenkörpern*
(Leipzig: Dyk, 1824), pp. 3–7; Karl Burdach, *Berichte von der königlichen anatomischen Anstalt
zu Königsberg*, VIII Bericht, *Mit Bemerkungen über die ernährenden Gefässe der Puls- und Blut-
Adern* (Königsberg: Bornträger, 1835), pp. 3–11.
   42. Prutz, *Albertus-Universität zu Königsberg*, pp. 1–120.

Powerful and encompassing curricular changes have also been attributed to student demographics. Lacking the faculty of some of its more prestigious counterparts, for instance, Königsberg seemed to suffer more in an environment of professionally minded students: university enrollments were low at the beginning of the nineteenth century, and even the medical faculty, with four or five professors, at times had only twice as many students. Ostensibly to increase and stabilize the student population of what would always remain a provincial university, the administration proposed to reorganize its faculty to make it more representative of current fields of knowledge; offered higher salaries so professors could concentrate more on scholarship; and promised to establish anew or to reorganize university institutes to create environments more appealing to students and conducive to effective instruction. To these general proposals were later added more specific plans for several institutes, principally for the natural sciences. Among them were not only Bessel's observatory, the botanical garden, and the clinical institutes of the medical faculty but also institutes for physics and chemistry.[43]

Upon closer scrutiny, however, the causal connection between student apathy and demographics, on the one hand, and curricular reform, on the other, proves weak, especially when material resources such as institutes are invoked as linkages between the two. Faculty members who sought better students and higher enrollments through material improvements were often disappointed. Funding was scarce and improvements were made inconsistently, often because perspicacious faculty members, aware of better conditions elsewhere, argued their cases effectively and persuasively. Material resources that did exist were not always deployed for active student learning, as evidenced by the time lag between the material improvement of the anatomical institute and zoological collection and the inauguration of von Baer's zootomical exercises. Moreover, by the first decades of the nineteenth century, student apathy seemed less important than student activism, which made the attitudes and performance levels of students a disciplinary matter separate from curricular reform. It was the *extra*curricular activities of students that were at times regarded with disapproval or suspicion between 1824 and 1834, when restrictions imposed by the Karlsbad Decrees were strengthened, participation in banned student organizations became punishable, and seminar membership was withheld from students belonging to secret student societies.[44]

The curricular reform that permanently transformed science instruction at the beginning of the nineteenth century was instead in large part the continuation of changes in classroom practices from the late eighteenth century, when systematic lectures, more fluid and accommodating

43. Ibid., pp. 146–73.
44. Ibid., pp. 99–120, 151.

of new material, began to replace the reading of canonical texts. About the same time (and perhaps because of the more demanding nature of lectures), exercise sessions, private academic societies, reading clubs, and other small meetings of students with professors appeared as forums for understanding material presented in lectures, not only in the philosophical faculty but also in the faculties of law, theology, and medicine. These forums were designed for learning through practical application what had been conveyed theoretically in lectures; exercises, in turn, helped shape and further define disciplinary knowledge by highlighting important methods and central topics. Often held privately, such forums were sometimes the forerunners of state-supported seminars. Although at present we know little about the causes and extent of these eighteenth-century changes, we do know that they were instigated from below, by professors and students, and not from above, by the state. We also know that they not only predated the broader structural university reforms of the early nineteenth century but also persisted through them.[45]

Sporadic in the eighteenth century, these reform efforts were enhanced considerably in the nineteenth through the efforts of Bessel, von Baer, and other science teachers. There was undeniably a strong intellectual component to their classroom reforms. In 1829, for instance, Dove responded to a special inquest into classroom reform at Königsberg by claiming that a major impediment to effective instruction in the natural sciences was the inability of students to read scientific papers in English or French, modern languages not taught in the gymnasium.[46] His complaint gave clear expression to the expectations of more intellectually aggressive faculty members who considered exposure to recent literature a necessary part of their teaching. But it also was intended to undermine the prominent role of classical studies in the gymnasium curriculum and so to weaken the position of those served by it.

Yet the pedagogical reform at Königsberg before the mid-1820s was for the most part anything but the result of pursuing *Wissenschaft* for its own sake. The exercises introduced into science instruction were designed primarily to aid the understanding of introductory-level material. They incorporated training in research methodologies and skills only to the extent that they became more complicated and rigorous in advanced courses. Even so, exercises did not necessarily introduce students to the research interests of the professor. Nor did it matter. The practical exercises of Remer, Bessel, and Herbart assumed an expertise in the handling of abstract knowledge, but their objective was to mimic the professional activity of the physician, the astronomer, and the teacher. What

45. Kathryn M. Olesko, "On Institutes, Investigations, and Scientific Training," in *The Investigative Enterprise: Experimental Physiology in Nineteenth Century Medicine*, ed. W. Coleman and F. L. Holmes (Berkeley and Los Angeles: University of California Press, 1988), pp. 295–332, esp. pp. 299–305.

46. Prutz, *Albertus-Universität zu Königsberg*, p. 114.

strengthened their educational value was their professional utility. To be sure, exercises had secondary benefits. Bessel eventually turned this pedagogical reform to his personal advantage by using advanced students as his scientific assistants, which further enhanced their own professional training.[47] And at least Bessel and Herbart saw in such exercises the possibility of changing a student's way of thinking so that he could deal with contemporary issues. But the principal value of exercises was in their ability to unite theory and practice in science instruction.

By the mid-1820s, science instruction at Königsberg no longer consisted solely of lecture courses in which students remained passive recipients of knowledge. Yet lecture courses still dominated science instruction. Specialization had begun, but with little coordination between subjects. Bessel, for instance, did not send his students to the mathematics professor to learn mathematics; he constructed his own mathematics courses or sent his students to Gauss at Göttingen University. Practical exercises were most likely to be found where they could be related to the practical issue of professional development. Finally, despite their complaints about secondary school science instruction, Königsberg natural scientists had not yet come to believe it was within their power to improve secondary school teaching, even though training teachers was by then the professional charge of the philosophical faculty.

In 1827, Königsberg natural scientists—with the notable exception of the two most mathematically minded among them, Bessel and Jacobi—discussed going beyond individual efforts at reform. They began to think in terms of a larger cooperative enterprise that would pool their expertise and resources: a natural sciences seminar for training secondary school teachers. It is significant that they were led in this effort by von Baer, who had just become director of the anatomical institute and inaugurated his zootomical exercises. On Christmas Day 1827, von Baer wrote to the ministry about the interest he and his science colleagues had in establishing a natural sciences seminar along the lines of one that had opened at Bonn University two years earlier. Although von Baer did not yet have a copy of the Bonn seminar statutes, he told the ministry that Königsberg natural scientists intended to follow the Bonn model of offering practical exercises for teaching candidates. What additional plans and motives von Baer and his colleagues may have had in mind were not as yet made known; they provided the ministry with only an abstract of their proposed statutes.[48]

47. Bessel to Schulze, 21 May 1825, rpt. in Schoenberg and Perlick, "Unbekannte Briefe," pp. 33–34.

48. Karl Ernst von Baer to Kultusminister Karl Altenstein, 25 December 1827, ZStA-M, Acta betr. eines Seminars für die naturwissenschaftlichen Studien auf der Universität zu Königsberg, Rep. 76Va, Sekt. 11, Tit. X, Abt. X, Nr. 21, [1827–76], fol. 1; copy in Darms. Samml. Sig. 3 K 1827.

The Königsberg seminar proposal seems on the surface to have been an adaptation to the natural sciences of an institutional form already highly developed for training teachers, running exercises, and promoting scholarship: the humanistic seminar, especially the philology seminar. The first philology seminar for training teachers had opened at Göttingen University in 1737. Under its second director, Christian Gottlob Heyne, the seminar turned in the 1760s to promoting more scholarly goals, including defining philology more firmly as an academic discipline. Heyne tightened the operation of the seminar by restricting its membership, by instituting regular meeting times, and by introducing more scholarly and intellectual exercises. This "Göttingen model" was adopted at several other German universities. By 1812 there were twelve philology seminars, the strongest of which had been Friedrich August Wolf's seminar at Halle University, established in 1788.[49]

Before the Humboltian reforms, secondary school teaching in Prussia matured in the context of these seminars and became strongly linked to rigorous training in the disciplines taught in them. Humanistic seminars traditionally have been viewed as elite institutes for exceptional students drawn to participate by an *innerer Beruf*, an inner calling, that expressed not the students' professional interests but rather a personal quality that defined the seriousness and diligence with which they took up a field of study. Seminars have also been viewed as institutes dedicated to the ideals of *Bildung* and *Wissenschaft*, as places where *Wissenschaft* could be "preserved, propagated and expanded."[50]

But because most seminars, including those with a decided preference for research, originated in exercise sessions or in some effort to teach in novel ways, the ideals of *Bildung*, *Wissenschaft*, and *innerer Beruf* were rarely achieved in practice. Seminars were thus less the result of the research ethos and neohumanist ideals of the Prussian universities than of a more fundamental desire to change didactic practices. Just as in the exercise sessions, students could not promote *Wissenschaft* until they had acquired more basic knowledge. Creating meaningful exercises and workable problems in seminars helped to institute the systematic learning of disciplinary knowledge, thereby fostering training more than self-cultivation. The establishment of prerequisites, entrance examinations, and introductory divisions in seminars not only promoted training even further but also helped to bridge the gap between school and university learning. Even the notion of *innerer Beruf* disappeared by the 1820s. Students had never been "called" to the seminar; they often had to be lured by free meals, financial prizes, and other material and academic privileges. The ideology of seminar learning was thus not only one that in practice failed

49. Wilhelm Erben, "Die Entstehung des Universitäts-Seminars," *Internationale Monatsschrift für Wissenschaft, Kunst, und Technik* 7 (1913): cols. 1247–64, 1335–48; *Geschichte der Friedrichs-Universität zu Halle*, ed. W. Schrader, 2 vols. (Berlin: Dümmler, 1894), 2:455–60.
50. Olesko, "On Institutes," pp. 300, 301.

to promote Prussian pedagogical ideals consistently but also one that eventually challenged the principles of *Lehrfreiheit* and *Lernfreiheit*.[51]

When Königsberg natural scientists proposed a seminar, the university already had some very strong societies and seminars for training gymnasium teachers for school districts in and around East Prussia. As in many other seminars, utilitarian concerns often outweighed the promotion of *Wissenschaft*, shaping seminar exercises more toward practical goals. The Königsberg historian Friedrich Wilhelm Schubert began in 1821 to hold regular meetings of history students to discuss recent literature, review student essays, and train teaching candidates. His students learned how to teach history, to construct courses, and to prepare lessons. Although Schubert instructed in historical methodologies, he did not encourage his students to develop or even to apply them. Instead, he considered methodology to be only one part of the overview of historical learning prospective history teachers should possess. In 1832, Schubert's historical exercises became the foundation of Prussia's first history seminar.[52] Other humanistic seminars at Königsberg had equally utilitarian orientations. The Lithuanian and Polish language seminars of the theological faculty offered teaching candidates the opportunity to develop effective classroom communication in those languages; more sophisticated independent projects as well as more elementary language exercises were specifically excluded.[53] Königsberg's philology seminar, which had existed informally since the 1810s, was designed in principle for the promotion and transmission of scholarly methods. But when its statutes were written in 1822, the ministry required that the director's annual reports make note of topics of instruction that had proven valuable in placing seminar graduates in teaching positions.[54]

To focus closely on the humanistic seminar, and specifically on the philology seminar, as the model for educational reform in the natural sciences at Königsberg (and elsewhere) would distort both the precedents and the motivations for that reform. It would mean both overlooking the common origins of humanistic seminars and exercises in the sciences in eighteenth-century reforms and diminishing the impact of more recent changes in science instruction at Königsberg. In 1827, Königsberg natural scientists already had compelling and far more appropriate models for transforming instruction in the natural sciences: the classroom practices of Bessel, Herbart, and most recently, von Baer. Just as their exercises

---

51. Ibid., pp. 300, 301, 303–7, 308–9.
52. Prutz, *Albertus-Universität zu Königsberg*, pp. 183–89; *Die preußischen Universitäten: Eine Sammlung der Verordnungen*, ed. J. W. F. Koch, 2 vols. (Berlin: Mittler: 1839–40), 2.2:855–58; Walther Hubatsch, "Die Albertus-Universität zu Königsberg i. Pr. in der deutschen Geistesgeschichte, 1544–1944," in *Die Deutschen Universitäten und Hochschulen im Osten*, ed. W. Hubatsch et al. (Cologne: Westdeutscher Verlag, 1964), pp. 9–39, esp. pp. 32–35.
53. *Die preußischen Universitäten* 2.2:847–50.
54. Ibid., 2.2:854.

were tied to professional goals, so were those of the newly proposed natural sciences seminar. It is not at all surprising, then, that more of the natural sciences at the university incorporated exercises into instruction only when their utility—in this case, to prepare students for secondary school teaching—appeared greater. The Königsberg natural sciences seminar continued and enhanced the reform of university instruction through exercises quietly begun in the eighteenth century. What the humanistic seminar offered was something other educational reforms did not: a model for gaining official state approval and financial support, and hence the means to an operational stability that would transcend staffing changes. Still, the Königsberg natural sciences seminar was not entirely of a piece with recent reforms. In at least one important respect, it diverged from the direction in which natural science instruction at Königsberg had been heading. Designed as a seminar for *several* natural sciences, it did not cultivate the intellectual specialization that the establishment of several faculty positions and exercise sessions in individual sciences had fostered.

On 15 September 1828, von Baer, Meyer, Dulk, and Neumann submitted to the ministry a revised proposal in which they elaborated upon their seminar's function. Although they thought Prussian secondary schools generally offered "good, solid instruction," they noted that "only in the natural sciences does instruction frequently remain superficial, if not neglected entirely." Because there were few school teachers competent to teach the natural sciences, they reasoned, another seminar besides Bonn's seemed necessary. Arguing that the natural sciences should occupy more of the gymnasium curriculum, they requested substantial state funding on the grounds that the seminar served broader needs than just the university's. But von Baer and his colleagues turned out to have something more in mind than just offering practical exercises suitable for training secondary school teachers. The plans they had for the thirteen hundred taler initial appropriation and the eight hundred taler annual budget they requested seemed to suggest other goals. They wanted a small lending library "in order to maintain a lively interest [in the sciences] among students." They immediately wanted to establish a small physical and mineralogical collection. They wanted funding for four seminar stipends and two substantial awards for prize questions. These expenditures suggested that they were interested in promoting scientific excellence and presumably scientific research in addition to enhancing the qualifications of teaching candidates through the routine execution of seminar exercises.[55]

Although the ministry awarded preliminary approval in 1829, it took

55. Karl Ernst von Baer, Franz Neumann, Ernst Meyer, and Heinrich Dove to Kultusministerium, 15 September 1828, ZStA-M, Naturw. Seminar Königsberg, fols. 7–8, 13–14 (budget proposal), 15–19 (budget revision).

no further steps to support their decision until 1833, when Herbart, professionally and personally discontent with Königsberg, left for Göttingen and the funds for his pedagogical seminar became available for other university institutes. Königsberg natural scientists again resubmitted their proposal. The ministry approved the seminar's statutes on 17 March 1834, but the first meeting was not held until a year later. When the seminar opened, its operation differed substantially from its sister institute, the natural sciences seminar at Bonn.[56]

The Bonn natural sciences seminar, upon which Königsberg scientists wished to model their own, had originated in a student society for the natural sciences. Established in June 1821, the society was disbanded a year later by the Bonn University curator on the grounds that it, like other similar student gatherings, could promote the activism associated with the *Burschenschaften*. When the society was resurrected as a seminar in 1825, largely at the request of the ministry, which was interested in establishing an institute for training natural science teachers, it had the most rigorous requirements of all Prussian seminars. In a three-year course of study, students moved through various subjects in the natural sciences, initially as auditors and then as first- and second-class ordinary members who not only participated in exercises but also later taught the auditors and introductory students below them (Table 1).[57] Students could in principle begin attending the seminar upon matriculation, immediately after graduating from the gymnasium. The comprehensive course of study promoted the unity of *Wissenschaften* that neohumanist reformers wanted and that, as Gert Schubring has argued, also issued from *Naturphilosophie*.[58]

After their first year in the seminar, students were given a set of exercises that tested both their knowledge and their lecturing skills. Members of the lower division were required to present a lecture in one of the natural sciences to demonstrate an overview of the field and comprehension of particular theories within it. They were also required to give a general summary of the ideas, principles, and main theories of one of the sciences without recourse to experimental demonstration, assisted only

56. Gert Schubring, "The Rise and Decline of the Bonn Natural Sciences Seminar," *Osiris* 5 (1989): 57–93, on 61. On the Bonn seminar see, in addition to the primary sources listed below, Hans-Jürgen Apel, "Gymnasiallehrerbildung zwischen Wissenschaft und fachdidaktischer Orientierung: Konzeption und Praxis der Lehrerbildung am Seminar für die gesamten Naturwissenschaften der Universität Bonn zwischen 1825 und 1848," *Vierteljahrschrift für wissenschaftliche Pädagogik* 62 (1986): 289–319. My discussion of the Bonn seminar, like Schubring's and Apel's, focuses on the tensions between teacher training and scholarly instruction. Unlike their studies, which are broadly conceived and, in the case of Schubring's, comprehensive in scope, I focus only on issues raised in the seminar's reports (some not mentioned by Schubring and Apel) that are most relevant for a comparison with later developments at Königsberg.

57. "Reglement für das Seminarium für die gesammten Naturwissenschaften, vom 3. Mai 1825 [Bonn Universität]," in *Die preußischen Universitäten* 2.2:624–31; on 626–28.

58. Schubring, "Rise and Decline of the Bonn Seminar," p. 66.

Table 1. Bonn University natural sciences seminar curriculum

Semester	Lecture	Seminar exercises		
		Auditors	Ordinary members, first class	Ordinary members, second class
First WS	Chemistry, physics	1 hr. chemistry, 1 hr. physics		
First SS	Mineralogy, botany, analytic chemistry	1 hr. mineralogy, 1 hr. botany	1 hr. chemistry, 1 hr. physics	
Second WS	Geognosy, cryptogam, natural history	1 hr. general natural history and zoology	1 hr. mineralogy, 1 hr. botany	1 hr. chemistry, 1 hr. physics
Second SS	Zoology and zootomy; topics in chemistry, physics, botany, and mineralogy	Preparing objects for storage	1 hr. zoology and natural history	1 hr. chemistry and physics, 1 hr. mineralogy, 1 hr. botany
Third WS	Higher mathematics, astronomy, human anatomy			1 hr. chemistry, physics, and mineralogy; exercise and repetition in the museum and laboratory
Third SS	Higher mathematics, astronomy, human anatomy, physiology			1 hr. for all subjects combined

Source: "Reglement für das Seminarium für die gesammten Naturwissenschaften," in *Die preußischen Universitäten* 2.2:628.
Note: WS, winter semester; SS, summer semester.

by notes and perhaps an illustrated chart. Upon successful completion of these exercises, the students could enter the upper division of the seminar, where they were expected to engage in more independent exercises, such as

(a) scholarly and literary demonstrations or systematic representations of theories of experiments in an area of natural history;
(b) the display of observations and trials;
(c) the derivation of new consequences out of known observations or trials;
(d) a suitable critique of individual theories; and finally,
(e) the evaluation, refutation, or elaboration of the work of another member.

In addition, the seminar had each week a sixth hourly meeting designed "to give unity to a divided institute" by providing students with the opportunity to consider connections between different parts of the natural sciences by giving lectures on them.[59]

The purpose of the Bonn seminar exercises was in principle the coordination of the collective knowledge of the natural sciences and an understanding of what was presumed to be their methodological similarity. They also cultivated certain skills. They offered the opportunity to develop a valuable asset in teaching: the well-ordered and unambiguous presentation of knowledge. The practical laboratory exercises could also have trained students in research skills, but they seem to have been less important than the seminar's teacher-training function, as the seminar's first few years of operation indicate. Training teachers of the natural sciences proved more problematic than the departmental directors had at first anticipated. In the seminar's first year, 1825/26, a few students registered for the seminar, most intending to become natural science teachers; but they bypassed the formal seminar curriculum and planned instead to study only the subjects most needed for the state's teaching examination, especially physics. In the next year the directors found themselves handling a more diverse clientele, including one student who wanted to be a secondary school teacher and another, Theodor Bischoff, who merely wished to augment his medical studies but who had difficulty balancing his medical and scientific interests. In both years a half dozen or so students with various other interests also attended the seminar as auditors. Although the directors were satisfied that their wide-ranging curriculum appeared to work, they also admitted that students were unevenly prepared to study certain subjects not covered in the gymnasium, such as natural history and chemistry.[60] Overall during its first

59. "Reglement für das Seminarium für die gesammten Naturwissenschaften [Bonn]," pp. 624, 629.
60. Seminar reports for 1825/26 (20 September 1826), 1826/27 (9 November 1827), ZStA-M, Acta betr. die Einrichtung eines Seminars für die naturwissenschaftlichen Studien auf der Universität Bonn, Rep. 76Va, Sekt. 3, Tit. X, Nr. 4, Bd. I: 1823–31, fols. 86–98, 134–41.

three years the Bonn seminar was not as popular as its directors had hoped.

Over the next few years the directors, backed by the university, tried to promote the study of the natural sciences and to establish their relevance to didactics and education. "The secondary school is the home of the natural sciences," they argued, for in the school "all parts of the sciences can be cultivated separately and set in relation to one another." Their own seminar, they believed, sharpened and deepened the collective study of the natural sciences. As proof they detailed to the ministry the studies undertaken by their students, mostly in preparation for teaching the natural sciences.[61] Between 1826 and 1831 the directors worked to improve science teaching by using seminar exercises as a forum for developing lecturing skills, by arranging for seminar students to hold guest lectures at local gymnasiums, and by assisting their students in getting teaching positions.

But in the Bonn seminar there were some studies that constituted original investigations, facilitated in part by the close relationship between certain divisions of the seminar and the nascent natural science institutes of the university and in part by the directors' own efforts to acquire instruments for the seminar. For example, the directors purchased several microscopes, including one achromatic microscope equipped with a micrometric apparatus, because, they believed, "many of the most recent physiological discoveries require not only the highest possible magnification offered by, for example, the Amici microscope, but also the most accurate micrometric measurements."[62] Although such instruments were intended for training teachers, they (along with other intellectual advantages the seminar offered) were considered by some students to be reasons in themselves for attending the seminar. Among the two students in 1827/28 who did not want to become a teacher was Bischoff, who attended to acquire the natural scientific knowledge he believed he needed for his medical education. His education was, according to the directors, "an exception to the usual studies of our contemporary physicians." Bischoff's pursuit of natural science instruction, the directors believed, brought him "honor." It also brought recognition to the seminar; for by 1830, Bischoff had taken a doctoral degree in the natural sciences before returning to the study of medicine.[63] In the seminar's early years, specializations of other sorts detracted from its curriculum and from teacher training as originally envisioned. Although the directors seemed to sense a tension between teacher training and advanced, specialized study of the natural sciences, they did not at first admit that the curriculum was in any way problematic.

61. Bonn University officials to Kultusminister Karl Altenstein [memorandum on the promotion of the study of the natural sciences], 11 January 1827, ibid., fols. 104–9, on fol. 106; seminar report for 1826/27, fols. 137–38.

62. Seminar report for 1826/27, ibid., fol. 138.

63. Seminar report for 1828/29 (8 September 1829), ibid., fols. 173–74; on fol. 174.

Yet both the operation and statutes of the Bonn seminar demonstrated how difficult it was not only to construct a natural science curriculum that prepared a student well for secondary school teaching but also to make the natural sciences relevant to other professions. In principle the seminar was designed to prevent specialization. To be eligible for membership, for instance, students only had to "intend to occupy themselves with the natural sciences." The sheer multiplicity of the science courses students were supposed to take meant that their time was so divided that they could not specialize and thus left the seminar ill-equipped to practice science. Even though a favorite subject (*Lieblingsfach*) was allowed, the possibility of identifying *one* natural science as a *Lebensberuf* was remote because the goal of the seminar exercises was "to comprehend all fields of the natural sciences in their totality."[64] Hence the student seeking to utilize his seminar instruction for professional ends could look forward to little more than secondary school teaching in the middle forms of the gymnasium, where only a few classes in natural history were taught and for which there were very few positions.

Although seminar membership grew by the early 1830s, problems continued to plague the Bonn seminar curriculum. Some seminar students concentrated their efforts upon learning physics only, so that they could teach the upper forms of the gymnasium, where more science was taught and at a higher level. The comprehensive coverage of the natural sciences that the directors had planned—which was designed not for penetrating science more deeply but rather for cultivating "overviews and clarity"—was not always, the directors soon realized, the best way to become a science teacher. Those who did follow the seminar's curriculum could be at a disadvantage on the examination; for the examiners, who were often not natural scientists themselves, sometimes failed to elicit from students their training in the natural sciences. What students really had to study in order to pass the examination was philology.[65]

Especially for students who wanted to teach the upper forms of the gymnasium, instruction in the Bonn seminar and the requirements of the state teaching examination remained misaligned because mathematics, a central part of the examination and of the gymnasium curriculum, was not taught in the Bonn seminar. "Therefore it is obvious why," the directors wrote on 10 January 1831, "our pupils turn their efforts chiefly toward the mathematical sciences after studying the natural sciences." They considered it impossible for their students to "study the higher parts of these [mathematical] disciplines" while in the seminar because "no time would be left over" for studying anything else.[66] Although several direc-

64. "Reglement für das Seminarium für die gesammten Naturwissenschaften [Bonn]," pp. 625, 629.

65. Ibid., p. 631; seminar reports for 1827/28 (24 November 1828), 1828/29 (8 September 1829), 1829/30 [n.d.], ZStA-M, Naturw. Seminar Bonn, fols. 151, 153–63, 170–72, 173–74, 193–99; *Reglement vom 20. April 1831*, pp. 24–45, esp. pp. 24–28.

66. Directors of the Bonn natural sciences seminar to Kultusministerium, 10 January 1831, ZStA-M, Naturw. Seminar Bonn, fols. 197–98; on fol. 198.

tors independently acknowledged the importance of science teachers learning mathematics for natural science instruction, especially physics, they persisted in excluding comprehensive mathematical instruction from the seminar itself. Even after the requirements for the state teaching examination were rewritten in 1831 to make the combined areas of mathematics and the natural sciences a major field on the examination, Bonn seminar directors still did not incorporate mathematics into their curriculum. To be sure, the statutory requirements of the examination were not always rigidly enforced, but the mathematical component could not go unheeded forever by those who trained natural science teachers, even though the revised regulations (which were not official until 1834) seemed to assign the natural sciences parity with mathematics and to open the way for creating a *Lehrerberuf* in the natural sciences as a whole. Changes in the examination during the early 1830s could in fact support both critical questions about science curricula not tied to mathematical instruction and arguments in favor of establishing training centers for teachers of mathematics.[67]

By 1831 the Bonn directors viewed the state's examination as a "certain hostility against our institute" and acknowledged that it was extremely difficult to train natural science teachers: "Because the scope of the natural sciences is so vast, all disciplines cannot be comprehended by one scholar at once." Nevertheless, they sought neither to modify their curriculum nor to propose changes in the state examination. Instead, disregarding the limited number of teaching positions in science then available in Prussian secondary schools, the directors, fearing a decline in seminar enrollment and dissatisfied with their inability to attract students outside Westphalian and Rhine regions, asked for and received a special dispensation from the natural sciences section of the teaching examination for their students who attended for three years and obtained the seminar's certificate. They also asked that gymnasium directors give their institute special attention by directing to it students with a "natural calling in the natural sciences." So rather than try to resolve the tension between professional teacher training and scholarly curricular concerns, the Bonn directors merely reaffirmed their commitment to the broad training program they believed appropriate, and the ministry, from all appearances, supported them.[68]

In 1828 when von Baer and his science colleagues at Königsberg had submitted their first formal proposal for a seminar, they had not exploited the nuances of the state's teaching examination as a tactic to win approval.

---

67. *Reglement vom 20. April 1831*, pp. 28–29.
68. Directors of the Bonn natural sciences seminar to Kultusministerium, 10 January 1831, 24 January 1831, ZStA-M, Naturw. Seminar Bonn, fols. 197, 191; *Reglement vom 20. April 1831*, pp. 49–50; "Verfügung an die Königl. Provinzial-Schulkollegien der östlichen Provinzen, wegen Benutzung des Seminars für die gesammten Naturwissenschaften: Vom 20. Mai 1831," *Die preußischen Universitäten* 2.2:631.

They had embarked instead upon what was perhaps a less effective strategy. Whereas the directors of the Bonn seminar had sought "to secure for the natural sciences . . . their proper role in the general *wissenschaftliche Bildung* of the students," the representatives of the natural sciences at Königsberg, anxious to see the natural sciences achieve parity with classical studies, argued that "the natural sciences are at least as important an element of general popular education [*allgemeine Volksbildung*] as historical and grammatical studies. Especially the Prussian state, which has done so much for instruction in the natural sciences, is called to the task of introducing them into popular education in general."[69]

Arguing for the parity of the natural sciences, as opposed to their "proper role," was neither a popular nor a promising position in a state that had in many ways made clear its preferences for classical instruction and all that it entailed. It was, however, a position that found favor locally in Theodor von Schön (who had to approve the request for the seminar before it went on to the ministry) and Bessel. Although Schön's liberal views were often constrained by qualifications that acknowledged the practical difficulties in implementing them, he was an enthusiastic supporter of science and technology who had studied with Kant and Kraus at Königsberg, who admired the English economic system, and who filled his travel notebooks with sketches of English machinery.[70] Bessel's criticism of the neohumanist educational ideology in Prussia was undoubtedly positively received by Schön, who sought similar liberal reforms. Bessel argued that the claim of the philologists that the "*Bildung des Geistes*" could only be accomplished through instruction in Greek and Latin "is not proven and can be strongly doubted." He stressed the progressive nature of the natural sciences over the traditionalism of the classical languages, quipping that "the Greeks could learn . . . a hundred times more from us then we could from them. I mean in the great kingdom of truth, mathematics, and in the equally great kingdom of observation, nature." He wanted the natural sciences to be integrated into daily life, and he considered "false views of nature to be just as significant a sign of deficient *Bildung* as an erroneous [grammatical] case."[71] Like Bessel, von Baer and other natural scientists hoped to transform negative public opinion about the natural sciences into support for their cause, especially by arguing that the low regard in which most of the public held the natural sciences was the consequence of the dearth of school teachers well trained in them.

69. "Reglement für das Seminarium für die gesammten Naturwissenschaften [Bonn]," p. 624; von Baer, Neumann, Meyer, and Dove to Kultusministerium, 15 September 1828, ZStA-M, Naturw. Seminar Königsberg, fol. 7.

70. Grey, *Prussia in Transition*, pp. 52–53, 89, 141, 153; Theodor von Schön Nachlaß, Nrs. 57–58, GSPK.

71. F. W. Bessel to Theodor von Schön, 25 September 1828, Theodor von Schön Nachlaß, Nr. 12, GSPK; Bessel quoted in von Selle, *Geschichte der Albertus-Universität*, p. 282.

Despite their earlier intention to model their institute upon Bonn's, Königsberg natural scientists did not adopt the most problematic feature of the Bonn seminar: its rigid curriculum. In the Königsberg seminar, the natural sciences were represented by the professors of physics, chemistry, mineralogy, botany, and zoology, all codirectors of the seminar. Open to matriculating students and to practicing teachers, and requiring no special prerequisites, the seminar was designed to accommodate students for a three-year course of study, although that period could be adjusted at the student's discretion and with the approval of one director. As at Bonn, the seminar was based on lecture courses; but in contrast to Bonn, Königsberg students were expected to enroll for courses in physics, chemistry, geology, mineralogy, botany, zoology, descriptive anthropology, comparative anatomy, and related applied topics only "as much as their time allows," with no specific sequence of courses or specific relationship between the courses and the seminar.[72] Like the Bonn seminar, though, the Königsberg seminar separated students by level of accomplishment. Students could join the upper section only after passing an examination in the natural sciences and after demonstrating the skills in ancient and modern languages and in mathematics needed for passing the state's teaching examination. So although mathematical instruction was not formally incorporated in the Königsberg seminar, mathematics seems not to have been as tangential to it as in Bonn. As did the Bonn seminar statutes, those at Königsberg did not create a curriculum that promoted the systematic study of individual natural sciences.

A significant difference between the Bonn and Königsberg seminars was in their sixth hour. Although independent research was allowed at Bonn, it was not especially promoted; for the sixth hour was designed principally as an opportunity to synthesize material from various sciences. At Königsberg, by contrast, the sixth hour was intended as a forum for discussing the results of original student investigations; teacher training thus seemed to be more closely linked to performing an original investigation. According to its statutes, the seminar was designed "especially to educate teachers of the natural sciences of gymnasiums and *Bürgerschulen* who are capable *not only* of propagating these sciences, *but also* of expanding them." Furthermore, the section of the statutes written by von Baer argued that because a student "is only capable of penetrating deeper into the natural sciences through [his] own observations, a principal task of the seminar is to introduce students to their own investigations, to give them the necessary introductory [knowledge and techniques], *and to invite them gradually to undertake larger investigations and to support these.*" The support the Königsberg seminar directors planned to offer for research was considerable and surpassed that available at Bonn. Although the seminars at

72. "Reglement des Seminariums für die Naturwissenschaft zu Königsberg: Vom 17. März 1834," *Die preußischen Universitäten* 2.2:860–62; on p. 861.

Bonn and Königsberg each had four premiums to award for exceptional performance, at Königsberg there were three additional premiums for the best answers to annual prize questions. Finally, the Königsberg directors had at their disposal twenty taler to support student investigations.[73]

The original student investigations that were in principle promoted in the Königsberg seminar could have prepared students for practicing the natural sciences. In practice, though, students were not especially encouraged to move in that direction because the directors of the seminar consistently viewed these "larger investigations" as contributions to education itself. The directors offered publication subvention for student investigations, but they intended that published investigations would be sent to the ministry with their annual seminar reports or be submitted as doctoral dissertations, not to scholarly journals. So in both the natural science seminars, instruction and student investigations were mainly directed at one professional goal: school teaching.

Shortly after the Königsberg seminar began to hold meetings, a secondary school teacher at the Friedrichs Collegium in Königsberg observed that although philology students had had seminars for some time, "future teachers of mathematics and the natural sciences lacked the advantage offered by a seminar, even though in the natural sciences practical exercises, the judgment and handling of instruments, the art of observation, and so on, demand the guiding hand and intelligence of a master" no less than exercises in philological interpretation did. He considered it a "true enrichment" of his time that seminars for the natural sciences had been established at Bonn and Königsberg; for now teachers could be trained not only theoretically but also practically and technically.[74] But upon close inspection, it is not at all clear that the careful guidance he thought both natural science institutes offered was actually there. The directors of both seminars tried in different ways to create a curriculum for the natural sciences, but so many subjects were covered in them that it was difficult for students to grasp the conceptual connections between the sciences and to learn the methodological foundations of each. In their effort to achieve recognition and respect for the natural sciences within Prussia's educational system, the directors of both seminars blurred distinctions between the natural sciences and overemphasized what they viewed as their methodological similarity and pedagogical unity. The successful operation of both seminars rested upon the assumption, however unwarranted, that the "collective natural sciences" constituted a legitimate body of knowledge that could be effectively conveyed in instruction designed for training secondary school teachers.

For personal and intellectual reasons, the professional unity among

73. Ibid., pp. 861 (emphasis added), 862.

74. C. F. Lentz, "An- und Aussichten die Mathematik und Physik an den Gymnasien betreffend," *Programm*, Friedrichs Collegium zu Königsberg, 1836/37, pp. 4–5.

seminar directors necessary for the seminar to work was difficult at first to achieve at Königsberg. By 1834, circumstances had so changed that the intellectual vitality and cooperation they had initially envisioned was threatened. Von Baer, who had done so much to promote natural science instruction in the philosophical and medical faculties, left the academic position for which he had longed at an earlier time. Overworked, disappointed that he had not received sufficient financial support for his work, and pressured by family obligations to obtain a higher salary, he accepted an appointment to the Academy in St. Petersburg.[75] His departure exposed other weaknesses.

Bessel thereafter feared the complete disintegration of the natural sciences at Königsberg. For years he had been close to von Baer, supporting his efforts to construct a natural history museum at Königsberg and seeing in him a like-minded individual who knew there was more to life than "eating, drinking, and breathing."[76] Unable to contain his disappointment over the departure of his colleague, he wrote to state officials on 28 June 1834 that he was "disconsolate over this great loss"; for he realized that his efforts of twenty-five years would come to nought unless something were done to keep outstanding faculty members at the university. Hoping to salvage what he could, he explained that in the natural sciences,

> I find *three* individuals upon whom, so far as I know, all Europe has bestowed a distinguished reputation. One is von Baer, whom we have now lost. Another is Jacobi, whom we still have because until now he has only insignificant needs that could be satisfied by his salary, which is funded through the university treasury. . . . The third is Neumann, who would not have been able until now to support his small family had he not drawn additional income from his own assets.—What bond ties Neumann and Jacobi to Königsberg?—Will they not use their distinction to obtain a better position elsewhere?

The "distinguished" philosophical faculty at Königsberg, Bessel pointed out, "will be strewn about in a short time just because it is distinguished." Bessel found it "deplorable" that nothing could be done to prevent that from happening because under present conditions one had to "leave

75. Von Baer had initially wanted to teach at a Prussian university because he thought research was promoted at them. In 1819 he requested a transfer from Königsberg to Bonn because Bonn was closer to Paris and because he considered the natural resources of the Rhine more abundant than those of the Pregel (K. E. von Baer to Kultusminister K. Altenstein, 25 January 1819, Darms. Samml. Sig. 3 K 1827). He discussed his call to St. Petersburg and complained about his Königsberg salary in other letters to Altenstein (26 January 1829 and 20 July 1830, ibid.). On von Baer see also Bessel to Kultusministerium, 29 October 1829, ZStA-M, Acta betr. die Sternwarte, Bd. III: 1825–38, fols. 82–83.

76. F. W. Bessel to K. E. von Baer, 4 April 1825, 21 September 1830 (quote), von Baer Nachlaß, UBGi-Hs.

everything to fate," which could "quickly reduce us to mediocrity." He urged the state to prevent "similar losses, like the one we have now suffered, while there is still time."[77]

With good reason, Bessel expressed such grave concern. It was not only von Baer's work with Burdach in reforming instruction in anatomy and physiology that made him a valuable asset to Königsberg. He had also worked very hard to reorient the activity of the Königsberg Physico-economic Society, one of the oldest Prussian scientific organizations engaged in creating ties between science and commercial-class interests. While serving as the society's president in the early 1830s, he used its lecture series to encourage social support for Königsberg's growing scientific culture, inviting university scientists to talk to a broad audience about their work and about science in general. Bessel did so and as a result had come to view the society as an instrument of political and economic change because of its potential "to expand the views" of persons whose customary activities confined them to a "narrow circle of ideas."[78] Von Baer's departure sensitized Bessel to signs of dissatisfaction in other Königsberg natural scientists. Jacobi, who had antagonized almost everyone but Bessel when he arrived in 1826, often entertained the idea of leaving. Having begun a comprehensive intellectual and institutional reform of mathematics instruction, he yearned to do still more to promote mathematics—and himself. He had already begun to look elsewhere. To Bessel, Jacobi's ties to Königsberg looked tenuous in 1834. He seemed to have believed Neumann's were equally uncertain because, months before his letter to the ministry about von Baer's departure, he had watched representatives of the natural sciences bare their methodological and pedagogical assumptions in ways that divided them.

Neumann's situation was different. His 1829 appointment to full professor had proved liberating, allowing him to turn more decisively to physics instruction. Despite his earlier statement that he had hoped to teach the mathematical parts of physics, he had not by then included either physics or techniques of quantification in his courses, with the exception of his course on the physics of the earth. In it he discussed principally the thermal, magnetic, and gravitational properties of the earth with a view toward determining its shape, size, and density. He acknowledged that this kind of physics could not rely entirely upon rigorously designed experiments but had to be content mainly with simple observational results. Principally through geodetic examples he demonstrated how the reliability and accuracy of observations (of the length of the simple seconds pendulum, for example) could be determined by

---

77. Bessel to Ober-Regierungsrath, 28 June 1834, Darms. Samml. Sig. J 1844.

78. "Historische Einleitung und Mitgleider Verzeichnis," *SPGK* 1 (1860): i–xvi; K. Stieda, "Zur Geschichte der physikalisch-ökonomischen Gesellschaft," *SPGK* 31 (1890): 38–84; Bessel, "Uebervölkerung," 3:485.

comparing readings taken by different observers. Some of the instruments used in geodetic observations he explained in his extensive coverage of the earth's gravity, a section of the course that could have served as an introduction to mechanics. His mathematics, however, did not go beyond geometry. Only in one or two places did he illustrate by concrete examples—what later became known as textbook problems—the theoretical principles he had developed.[79] Because Neumann had thusly shown an interest in exposing students to the principles of physical investigations, he stood to profit by participating in the natural sciences seminar. The seminar's proposed physical and mineralogical collection could have aided his teaching, and the promotion of student investigations could have helped to sharpen his own pedagogical goals by forcing him to create exercises that would bring the investigative techniques he thought most important into prominence.

But between 1829 when he became *ordentlicher* professor and 1833 when the seminar came under consideration again, Neumann's courses changed considerably as he turned more decisively to physics proper. In the winter semester of 1829/30 he taught a course on experimental physics. How he attempted to approach the subject is not known. The fact that he abandoned the course after teaching it once, however, suggests that it was probably a service course for students less interested in physics. Moreover, it was probably offered to replace what Dove, who had by then left and for whom no replacement had been chosen, had taught. In the summer semester of 1830, Neumann for the first time offered a specialized course in physics, on theoretical optics. His transition to physics, though, was not as complete as it might at first seem because he concentrated on the optical properties of crystals. Although he later incorporated several contemporary mathematical treatments of optical phenomena (Fresnel, Arago, Biot, Brewster, Young, and Fizeau), it is likely that initially he followed primarily Augustin Jean Fresnel, with minor treatments of Isaac Newton's theory and especially Christian Huygens's. His discussion of the largely French-constructed undulatory theory was neither as abstract nor as theoretical as it might have been. He did not then begin from the principles of elasticity theory, and he even doubted the applicability of mechanical principles to all areas of optics.[80]

The following summer he used Joseph Fourier's *Théorie analytique de la chaleur* as the foundation of a course on the propagation of heat in minerals. In the winter semester of 1831/32, he repeated his theory of light course, which he offered again in the summer of 1833. That he intended to expand his course offerings in a mathematically based physics, but may not have known precisely what subjects to cover, was evident

79. FNN 32: Physik der Erde (lecture notes from the winter semesters of 1827/28 and 1830/31).

80. Below, App. 2; F. E. Neumann, *Vorlesungen über theoretische Optik*, ed. E. Dorn (Leipzig: B. G. Teubner, 1885), pp. iii–iv.

by the summer semester of 1832, when he offered for the first time a course merely called topics in mathematical physics. This course presumably gave him the opportunity to explore areas of physics that had not been mathematized, perhaps even to include in a single semester several issues treated from similar mathematical perspectives. He repeated this course (perhaps with different topics) in the winter semester of 1832/33. Then in the summer semester of 1833 he offered a course on general physics. What it included and for whom it was intended are not known, but it was probably a service course; as with his course on experimental physics, he never offered it again.[81]

Neumann's teaching had evolved over these few years. By 1833/34 he could have asked for the responsibility of handling the physical division of the natural sciences seminar, even though Dove's former position had by then been filled. He did not, preferring to retain responsibility for its division for mineralogy. Instead Neumann joined Jacobi and the *Privatdozent* of mathematics, Ludwig Adolf Sohncke, in 1833 in proposing a second science seminar, this one dedicated exclusively to mathematics and physics. It was this proposal that provoked discussion about the direction of reform in natural science instruction at Königsberg.

Until 1833, educational reform in the natural sciences at Königsberg had been manifest for the most part in the creation of curricula that organized courses and of practical exercises that accompanied them. There were two broad patterns to these changes in classroom teaching. The first was exemplified most strongly by Bessel, who drew upon the pedagogical precepts of Pestalozzi in shaping his practical exercises and who valued these exercises for their sheer educational value; their utility in training astronomers; their ability to train assistants for his own research; and finally, their potential to forge critical ways of thinking that students could apply to larger social, political, and economic problems. The second pattern was exemplified by the directors of the natural sciences seminar, who viewed the value of exercises less broadly, as useful primarily in education and in training teachers. Only secondarily did they view exercises as a means of dismantling Prussia's neohumanist curriculum. While they never explicitly identified social change as among their goals, it was implied nonetheless in their curricular reform. Both patterns drew upon similar eighteenth-century pedagogical changes and expressed a common discontent with a curriculum based on the classics that ignored the natural sciences. Both also served intellectual and disciplinary ends without explicitly drawing upon the ideals of *Wissenschaft* and *Bildung*. But in contrast to the natural science seminars of Bonn and Königsberg, where the vehicle for introducing practical exercises into science instruction had been teacher training, and even to the kind of exercises introduced by Bessel, the proposed mathematico-physical semi-

81. Below, App. 2.

nar at Königsberg, especially its physical division, seemed to the remaining directors of the Königsberg natural sciences seminar to be tied less to practical exercises, utilitarian goals, and past traditions in educational reform than to the unadulterated pursuit of pure *Wissenschaft*.

CHAPTER TWO

Quantification in Physics Teaching and Research

JUST AS NEUMANN's lecture courses changed after 1829, so did his investigative trail. Between 1829 and 1834 his research interests shifted decisively from the geometry to the physics of crystals and minerals. In a letter of 1829 written to the Berlin mineralogist Christian Weiss, his former teacher, Neumann discussed his research plans. He explained that he had decided to go to Königsberg in part because he believed that there he would have at his disposal the material aids to execute his experimental investigations into the physical properties of crystals and minerals. He had been disappointed, though, with what he had found. He despairingly likened Königsberg to a place without a mechanic or an instrument maker. Undaunted, he had nevertheless pressed on, constructing his own instruments. Uppermost in his mind now was the perfection of experimental protocol and techniques. He intended "to carry out *accurate quantitative measurements* of the various physical characteristics of inorganic nature." In a draft of this letter, Neumann mentioned a special concern for the accuracy of his measurements, acknowledging that such investigations would have to be methodically executed. A special problem was then on his mind: the relation of specific heats to crystalline forms. The exact relation was still unclear because he did not have minerals of sufficient purity and variety to be able to calculate how accurate his measurements were.[1]

Neumann had begun to look beyond the geometric properties of crys-

---

1. Neumann to Christian Samuel Weiss, 20 April 1829, Darms. Samml. Sig. Flc(2) 1841 (emphasis added). Drafts located in FNN 53.IA: Briefe von Collegen.

tals to their physical and mathematical characteristics before his arrival at Königsberg. The prospect of investigating the experimental areas of physics—light, heat, electricity, and magnetism—had captured his imagination in 1821, while he was away from Berlin. When he later resumed his studies under Weiss, he occupied himself with refining standards and procedures for crystallographic classification but tried to augment Weiss's geometric approach with one that treated the physical properties of crystals as differentiating features. In his *Beiträge zur Krystallonomie*, completed in 1822 and published in 1823, he thought about the process of crystallization in physical terms, suggesting that crystalline forms were "only the spatial representations of forces" that acted along major crystalline axes, which themselves were "the geometric expression of these forces, which have their law in measure and number." So convinced was he of his interpretation of crystalline axes that he criticized systems of crystal classification that relied on differentiating parameters not accessible to measurement and that employed mathematical expressions that were "abstractions . . . not found in nature."[2]

What he meant by "measure and number" was not yet clearly expressed, and Neumann does not appear to have been deeply concerned to elaborate his convictions. He was still learning the mathematical tools useful in scientific practice, as well as examining the relations among experimental data, mathematical expressions, and physical reality. He explored three different quantitative techniques in a reading notebook dating principally from the mid-1820s,[3] which he had begun after the publication of his *Beiträge zur Krystallonomie*. The first was the method of least squares, used by astronomers to determine accidental errors (what Gauss called *"zufällige Fehler"*[4]) in their observations. This method had been used almost exclusively in astronomy and geodesy and was not yet considered essential for studying observational data in other sciences. Neumann did not identify the source of the eight pages of notes he had taken on least squares; but what he wrote suggests that he was interested in the method as a tool for analyzing data.

The second technique involved the use of partial differential equations in physics and was drawn from Joseph Fourier's *Théorie analytique de la chaleur*, which Neumann began to study in 1823, a year after the work had been published in full.[5] Neumann copied Fourier's lengthy essay almost

2. "Franz Neumann's *Beiträge zur Krystallonomie* aus den Jahren 1823 und 1826. Ein Versuch, den wesentlichen Inhalt dieser vor fast hundert Jahre erschienenen fundamentalen Schriften in übersichtlicher und lückenloser Weise darzustellen," *Abhandlungen der kgl. Sächsischen Gesellschaft der Wissenschaften* 23 (1916): 379; *GW* 1:179. The statement on forces does not appear in *GW* (1:318).

3. FNN 50: Auszüge F.N.'s aus wichtigeren Arbeiten.

4. [Report on "Theoria combinationis observationum erroribus minimus obnoxiae, pars prior," 26 February 1821], in *Carl Friedrich Gauss Werke*, ed. Kgl. Gesellschaft der Wissenschaften zu Göttingen, vol. 4 (Göttingen: Dieterich, 1880), pp. 95–100, on p. 95.

5. Joseph Fourier, *Théorie analytique de la chaleur* (Paris: Firmin Didot, 1822). Neumann appears not to have paid much attention to Fourier's adaptation of the error curve in this study.

completely, paying special attention to its mathematical techniques. His notes depict a student still learning, not yet ready to comment upon his source. He seemed especially interested in the application of Fourier's methods to the motion of heat in the earth's interior. Only with a firm foundation in higher calculus could Neumann tackle the novel symbolism, the separation of variables in linear partial differential equations, the infinite series solutions, and the special techniques Fourier used to incorporate boundary conditions. Neumann also showed passing interest in the graphical representation of Fourier's periodic sine and cosine functions, the basis of what became known as Fourier series. Presumably he learned from Fourier's work a wide variety of mathematical techniques applicable to the natural sciences, thereby expanding his repertoire to include the abstract language of analysis as well as the more concrete geometric expressions he had previously employed. Although there is no indication that at this time he intended to apply Fourier's quantitative methods to other areas of physics, merely learning them meant that he now had a way to describe physical processes and to express quantifiable variables, especially physical constants, that could not be accomplished with geometry alone. Moreover, Fourier's methods supplied Neumann with a particular model for mathematization, one that de-emphasized underlying physical causes.

The third technique also involved partial differential equations, but it had different physical implications and suggested a slightly different relation between theory and experiment. Passages from Siméon-Denis Poisson's thermal studies and his integration of the wave equation appear last in Neumann's reading notebook.[6] He seems to have studied Poisson's works for over a decade, until just after 1835, when Poisson's *Théorie mathématique de la chaleur* was published. Whereas Fourier dealt only with observable or measurable quantities characteristic of the thermal properties of matter and avoided hypotheses on the nature of heat, Poisson's physics was a continuation of the Laplacian program, wherein molecules or corpuscles were presumed to exist; wherein physical processes were assumed to originate from the action of intermolecular forces; and wherein a Newtonian mechanics provided the principal conceptual framework. Fourier's physics was phenomenological, focused on macroscopic, observable processes, with no strong role for underlying micro-

---

6. Siméon-Denis Poisson, "Extrait d'un mémoire sur la distribution de la chaleur dans les corps solides," *Journal de physique* 80 (1815): 434–41; idem, "Mémoire sur la distribution de la chaleur dans les corps solides," *Journal de L' École royale polytechnique*, Tome 12, Cahier 19 (1823): 1–162; idem, *Théorie mathématique de la chaleur* (Paris: Bachelier, 1835). See also D. H. Arnold, "The Méchanique Physique of Siméon-Denis Poisson: The Evolution and Isolation in France of His Approach to Physical Theory (1800–1840), pt. 4, Disquiet with Respect to Fourier's Treatment of Heat," *AHES* 28 (1983): 299–320, esp. p. 309, where Arnold cites Poisson's intent to preserve "mathematical certainty" in the analytic expression of physical theory. Kenneth L. Caneva has noted that Fourier, Ampère, and Poisson exemplify "the desire . . . for absolute, incontestable, objective, immortal certainty"; see "Conceptual and Generational Change in German Physics: The Case of Electricity, 1800–1846" (Ph.D. diss., Princeton University, 1974), p. 255.

scopic models. Poisson's was reductionist, hypothetical, and concerned with microscopic events.

An important point about these exemplars and quantitative techniques was what they suggested about how one could compare theory and experiment, an issue of increasing concern to Neumann over the years that he added entries to his notebook. By providing a way to treat *data* in quantitative terms, the method of least squares altered the exact experimental physics of the late eighteenth century, which had taken shape largely through the quantification of constant errors and the creation of empirical formulas. The method of least squares could also be used with profit in comparing theoretical calculations more precisely with experimental results because it gave a *range* within which those results were valid. Although the method entailed probabilistic considerations, it was not yet entirely clear to investigators how that was so. Fourier's and Poisson's studies also provided models for the comparison of theory and experiment, but in slightly different ways. Both Fourier and Poisson considered observable thermal properties and measurable thermal constants, but in Fourier's theory, mathematical form had a slight priority over modifications suggested by experimental results. Poisson's mathematical physics, on the other hand, linked as it was to hypothetical microscopic entities and processes, insisted on a quantitative description on a slightly more refined scale. Hence up to a point Poisson used experimental results to modify what may at first have been only an ideal mathematical form, deploying approximation techniques in the process.

It is difficult to say which of these three approaches or combinations of them appealed most to Neumann. On the basis of his comments in *Beiträge zur Krystallonomie*, one could conclude that he preferred mathematical expressions grounded in the appearances of reality and devoid of hypothetical physical constructions (such as atoms). To jump to the conclusion, however, that he rejected Poisson and was immediately drawn to Fourier (as several of his students and contemporary commentators later suggested[7]) would be to overlook the appeal of certain elements of Poisson's method—especially the somewhat more detailed consideration of observational data—in the context of Neumann's interest in the method of least squares, a method that allowed for an even deeper examination of the relation between theoretical calculations and experimental results than either Fourier or Poisson offered, despite their own familiarity with the method. What techniques Neumann would use and what strategy he would take in quantifying aspects of crystallography, mineralogy, or physics was as yet unclear. More than one pathway was open to him.

7. Commenting on Neumann's indebtedness to Fourier were, e.g., Carl Neumann (Paul Volkmann, *Franz Neumann: Ein Beitrag zur Geschichte Deutscher Wissenschaft* [Leipzig: B. G. Teubner, 1896], p. 7) and Woldemar Voigt ("Gedächtnissrede auf Franz Neumann" [1895], in *GW* 1:1–19, on 7).

In his published work up to 1829—his last investigation had been completed in 1825—Neumann had already dealt with problems of experimental precision and accuracy. He believed that mathematical expressions had to be bound to reality, but having acknowledged that measured and calculated values had to agree and that hypotheses had to be used with caution, he did not explain how numerical agreement was achieved and how close it should be. Although he felt free to criticize the unsatisfactory agreement between theory and experiment in the work of his colleagues, before 1829 he achieved "accuracy" in his own work largely through increasingly sophisticated geometric representations of crystalline forms. Despite his earlier exposure to other quantitative and measuring techniques, including the method of least squares, precision and accuracy were at this time more a matter of intellectual principle than of experimental practice.

Only from time to time did he describe his experimental protocol and how his instruments, protocol, and methods of data analysis contributed to accuracy and precision. Measurements of angles were crucial to his crystallographic investigations, but at first he tended to attribute differences between his measurements and those of others to the particular samples used. He noted in his "Ueber das Krystallsystem des Axinits" (1825) the care with which he set up and used an ordinary reflection goniometer so as to minimize the effect of parallax on instrumental readings and to achieve measurements that enabled him to criticize Haüy's notion of the crystallographic primitive form. But he still took and evaluated his readings in simple ways. Only by estimating could he express errors. He believed that by conducting "repeated measurements . . . one immediately after the other" he could guarantee that "external conditions" had "only a small influence" on the final readings. By comparing the degree of agreement between his unanalyzed measurements and his theoretically calculated values, not only could he judge the "accuracy" of his own work, but he could also use that "accuracy" as a measure of how valid his theory was in comparison to others.[8] Notably absent in his early work was any special sensitivity to the exact numerical computations of errors which could have provided a quantitative assessment of the precision of his results. Now in his 1829 letter to Weiss he expressed an interest in these.

Because he had already taught a course on the propagation of heat in minerals in which he adopted Fourier's approach, Neumann probably followed Fourier from the start in his investigation of specific heats. But as his letter to Weiss indicated, the evolution of his investigation was now also guided by the dictates of an exact experimental physics. His adoption of that investigative strategy over others by decade's end coincided with

---

8. *GW* 1:361–73; on 368; see also "Wegen Haidingers Aufsatz über axotomen Bleibaryt," in *GW* 1:353–60, on 360.

his appointment to full professor; with the increasing emphasis on geo-detic issues in his course on the physics of the earth; and with his immersion in his working environment at Königsberg, especially his growing ties to Bessel. Although his students' retrospective accounts have suggested that Bessel offered Neumann a compelling practical model for attaining precise results, they did not always explain when and how Neumann found Bessel's work appealing.[9] Nor have they suggested why it continued to be so important when so little of Neumann's research turned out to be in an exact experimental physics.

By the time Bessel was appointed to Königsberg in 1810, he had developed a distinctive and rigorous approach to the analysis of astronomical data that fundamentally transformed assumptions about instrumentation in astronomical observations. Drawn at first to the purely calculational aspects of astronomy, Bessel found the differences between sets of readings taken by several observers a challenging feature of astronomical practice in need of more rigorous, quantitative explanations. Before 1809 he often referred to such irregularities as "disharmonies." As he learned or developed techniques of data reduction that accounted analytically for statistical and especially material disturbances that marred data, he used the term less frequently. Part of Bessel's matured understanding of the nature of data was due to his growing friendship with Carl Friedrich Gauss, who provided Bessel with "theoretical" assistance in the calculation and reduction of data and appears to have acquainted him with the method of least squares. Significantly when Bessel first mentioned the method in a letter to Gauss of February 1809, he also spoke for the first time of the "harmony" of data whose errors had been reduced. By August 1810 Bessel believed that he had "made the first completely consistent application of [Gauss's] method of least squares" to the computation of the paths of comets. He called least squares "a magnificient method because it banishes arbitrariness where one does not want to have it."[10]

Important as the method of least squares was in the computation of the statistical errors of data, it was in the analysis of constant experimental errors that Bessel at first made more significant contributions to astro-

9. E.g., Voigt, "Gedächtnissrede," pp. 11–12; Paul Volkmann, "Franz Neumann als Experimentator," *Verhandlungen der Deutschen Physikalischen Gesellschaft* 12 (1910): 776–87; on 777–78. In portraying Neumann largely as a technician in the construction of more accurate instruments, Volkmann situates Neumann more firmly than is justified in the context of late nineteenth-century experimental physics, when precision instrumentation was more common. Intriguing as Volkmann's discussion of Neumann's technical expertise is, it is often difficult to corroborate from other sources; Neumann himself left only sketchy evidence of his instrumental modifications. Volkmann's approach leads him to conclude that the geothermal station at Königsberg was not of a piece with Neumann's other experimental work (p. 787), whereas I view it as part and parcel of a particular exact experimental tradition that took shape at Königsberg; see below, Chap. 9.

10. F. W. Bessel to C. F. Gauss, 16 March 1805, 4 December 1805, 9 February 1809, 20 August 1810, GN.

nomical practice. By May 1806, Bessel considered it essential to look in more detail at the "practical difficulties" in making observations as well as at their "theoretical explanation." He meant much more than a simple reconsideration of the well-known disturbances afflicting astronomical data. Bessel began to develop the theory of astronomical and other auxiliary instruments in order to compute quantitatively the constant errors an instrument contributed to observational data. The turning point in his consideration of errors occurred in 1807, when he began a fixed-star catalog based on observations taken by James Bradley between 1750 and 1762. Concerned that the "limit of accuracy of this catalog" was inadequate, Bessel believed that, good as Bradley's observations were, they nevertheless exhibited "disharmony." One should be able to undertake "a test of the instruments" to rid the observations of errors. Whereas most investigators before him who were concerned about constant errors had merely estimated them, Bessel sought an exact quantitative determination by deriving their values from theoretical expressions confirmed by experiments.[11]

The correction that at first concerned Bessel most was for the refraction of light by the atmosphere. He found Pierre Simon Laplace's theory of refraction inadequate for accounting for the errors in Bradley's data, but he did not yet have sufficient data of his own to form an alternative theory that could eliminate what he considered to be Laplace's "arbitrary" hypothesis about the constant density of the atmosphere. He chose in the end not to formulate an alternative theory of atmospheric density but rather to examine theoretically the two instruments used to gather the data for computing the refractive index, the thermometer and the barometer, to see if his results would account for problems in the data. The refractions, he then found, not only were directly proportional to the density of the air as given by thermometric and barometric readings but also depended on the corrections for constant errors calculated for each instrument. His analysis of errors led to a closer examination of the pattern in accidental errors in these observations. Although his results provided empirical proof for Gauss's theory of error distribution, they also suggested that other distributions of error were possible.[12]

After the opening of his observatory at Königsberg in November 1813, Bessel continued his theoretical examination of astronomical instruments, especially of the errors arising from material imperfections in the construction of scales and from human inaccuracies in reading them. By the 1820s he had differentiated kinds of errors, developed methods for quantifying them, and perfected techniques for identifying and quantitatively expressing constant errors. During the 1820s he extended error

11. Ibid., 28 May 1806, 18 March 1808, GN.

12. Ibid., 18 March 1808, 20 July 1808, 18 September 1808, GN; Stephen M. Stigler, *The History of Statistics: The Measurement of Uncertainty before 1900* (Cambridge, Mass.: Harvard University Press, 1986), p. 203.

analysis from astronomical instruments to several used in physics, including the thermometer and the barometer, which he subjected to more refined analyses. In 1826, the year Neumann arrived at Königsberg, Bessel examined the theory of thermometric calibration and developed a calibration method that compensated for the imperfect cylindrical shape of the thermometer stem. Although he knew of Joseph Gay-Lussac's method for constructing a scale based not on equal divisions, but on equal amounts of mercury falling between each division, Bessel rejected it because the scales turned out to be unevenly divided and could not be transferred from one thermometer to another.[13]

The precision with which he analyzed instrument construction was matched by a deepened theoretical understanding of the accidental errors afflicting data. He had analyzed astronomical observations using the method of least squares for some years, and by numerous examples he supported and extended Gauss's theory of error distribution. In addition to analyzing irregularities in data and in instruments, Bessel had examined errors contributed by the observer in the act of measuring, and in 1823 he identified a third kind of error, the so-called personal equation. His use of all three kinds of errors was marked by balance and restraint and by a preference for their quantitative determination over their absolute elimination. Although he believed one could not work completely effectively with deficient instruments, he also believed it was better to compensate for instrumental errors by calculating their effect than to perfect instruments endlessly by mechanical means. For him, "good observations" were those where one could "eliminate every possible error of the instrument from the result."[14]

When Neumann arrived at Königsberg in 1826, Bessel was already engaged in an investigation that required him to extend his exacting techniques of error analysis from astronomy to a special long-standing problem in geodesy and physics: the length of the simple seconds pendulum. Neumann observed firsthand the meticulous care with which Bessel conducted this exemplary experiment, a project initially commissioned by the Berlin Academy of Sciences as part of the reform of weights and measures in Prussia. Although at first reluctant to be distracted from his work in astronomy, he took on the investigation because he thought he "might well have the opportunity to treat the problem in a new way." He very quickly became absorbed in it, even to the point of neglecting his other work and responsibilities. Determining the length of the seconds pendulum at Königsberg occupied Bessel starting in 1823, dictating the

13. F. W. Bessel, "Methode die Thermometer zu berichtigen" [1826], in *AFWB* 3:226–33; idem, "Reduction beobachteter Barometerhöhen auf Pariser Maass und die Dichte des Quecksilbers beim Eispunkte" [1827], in *AFWB* 3:236–37.

14. Stigler, *History of Statistics*, pp. 203, 230n, 240–42; F. W. Bessel, "Persönliche Gleichung bei Durchgangsbeobachtungen" [1823], in *AFWB* 3:300–304; idem, "Methode die Thermometer zu berichtigen," p. 226.

course of some of his minor experiments and finding expression in his teaching at the observatory. In preparation for the experiment, he sent his students, most notably Carl Theodor Anger, to Paris to conduct precise pendulum observations under the guidance of Arago; upon return they briefed Bessel on what they had learned.[15] The technical and administrative details of his own investigation were complicated; there was much to consider and coordinate before he could begin. He had to find a mechanic of first rank; ascertain the geographic conditions; run preliminary trials; determine the dangers of transporting such a delicate instrument; develop the theory of the seconds pendulum with respect to experimental conditions; and finally, despite the academy's commission, obtain funding.

In 1824, Bessel wrote to East Prussian authorities about his plan. He reported that although the astronomer Heinrich Schumacher was willing to test some of Bessel's ideas for the design of the instrument, Schumacher thought the best way for Bessel to realize his plans was to go himself to Hamburg, where he could supervise the construction of the seconds pendulum by the mechanic Johann Georg Repsold. Bessel, who considered Repsold one of the best German instrument makers responsible for the revolution in astronomical observations at the beginning of the nineteenth century, explained that only by having contact with Repsold could he determine "whether or not the entire thing corresponded precisely to my idea."[16] That there should be a correspondence between his "idea"—which was not just the theory behind the seconds pendulum but also the rationale for the experimental protocol he designed—and the apparatus itself was the only way, Bessel believed, to be confident both of his numerical observations and of their potential applications (e.g., measuring gravity, determining the exact shape of the earth, or creating a standard unit of length for the reform of weights and measures). Successful in obtaining funding, Bessel traveled to Hamburg, where he oversaw the construction of his apparatus, which was completed early in 1825. In August it was transported to Königsberg, and Bessel began trials that lasted just over two years. On 5 January 1828 his account of the completed investigation arrived at the Academy of Sciences in Berlin.

Bessel divided his "Investigation of the Length of the Simple Seconds Pendulum" into three sections. In the first, he described his apparatus, compared it with similar ones, and examined the conditions under which he employed it. Technical appendixes constituted the second section. In

15. F. W. Bessel, *Untersuchungen über die Länge des einfachen Secundenpendels*, besonders abgedruckt aus den Abhandlungen der Akademie zu Berlin für 1826 (Berlin: Kgl. Akademie der Wissenschaften, 1828); Bessel to Gauss, n.d. [January–March 1823], GN; on Anger see his obituary in *Programm*, Städtisches Gymnasium zu Danzig, 1858–59.

16. F. W. Bessel to Oberpräsident, 24 November 1824, ZStA-M, Acta betr. die Sternwarte bei der Universität zu Königsberg, Rep. 76Va, Sekt. 11, Tit. X, Nr. 16, Bd. II: 1814–25, fols. 162–63.

the third, he reported his measurements, not only of the seconds pendulum but also of the "environment" of his experiment, such as temperature and barometric pressure. Although he was able to determine the value of gravity at Königsberg with greater precision and so to double the eighteenth-century third-decimal-place limit on precision (his measure of the gravity was precise to 1/60,000, or over half way between the fifth and sixth decimal place, whereas Newton's result had been precise to 1/1,000), the significance of his study lay less in its quantitative result than in its method. Bessel's investigation was a model for the execution of experimental protocol that sought precision through the painstaking determination of features of the "environment" of the experiment (temperature, barometric pressure, and especially air surrounding the swinging pendulum) or of the apparatus itself (irregularities in its construction) that affected the final result and could be expressed in quantitative terms.

Bessel's model protocol can be divided into two parts. In the first, embracing what Bessel called section one, he described in great detail both the mechanical construction of his apparatus and his method of measuring. A drawing of the apparatus (Figure 4), he believed, added to the technical clarity of his discussion. His complex, involved, and thorough measuring methods included an analysis of subsidiary measuring instruments (micrometer, chronometer, thermometer, and barometer) as well as of the seconds pendulum itself. He calculated the effect of surrounding conditions, such as temperature, on the seconds pendulum and on the execution of his measuring operations. He utilized measuring techniques that increased precision and reduced errors. One technique was coincidence observations; another, to calculate the length of the seconds pendulum by measuring the difference in length between two pendulums, thereby eliminating certain constant errors from consideration because they canceled each other out. Computing the constant errors of his experiment was the single most important feature of his investigation; he believed he had accounted for all of them with the exception of the "possible uncertainty in the presumed thermal expansion of the *toise du Pérou* [his standard of measure, which equaled 1.949037 meters]." Finally, he calculated the accidental errors of the averages of his measurements by using the method of least squares, a method still largely unknown in physics. In the end, Bessel ventured "no opinion of the probable error of the [final] result, because although one to be sure can probably estimate the influence of individual causes of error, this way alone seldom gives a correct estimate of the result of combined operations." He believed his observations "were repeated often enough" to render any remaining errors "innocuous."[17]

In what might be considered part two of his protocol, which included section two and the appendixes of his paper, Bessel examined several

17. Bessel, *Untersuchungen über die Länge des einfachen Secundenpendels*, pp. 56, 57.

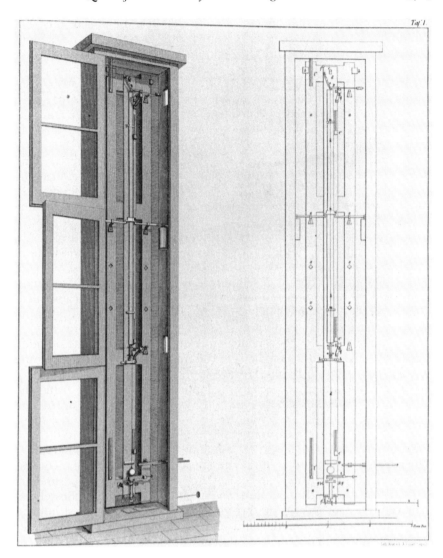

Figure 4. Bessel's seconds pendulum apparatus. From Bessel, *Untersuchungen über die Länge des einfachen Secundenpendels*, Table 1, courtesy of Yale University Photographic Services.

experimental sources of error. In this part, Bessel gave a thorough theoretical analysis of the mechanical principles underlying his investigation by comparing his method with those of Borda and others. His real achievement was in computing the effect of air on the motion of the pendulum. To Newton's hydrostatic correction he added a hydrodynamic

one, demonstrating that air not only retarded the motion of the pendulum but also added to its mass while it moved. Some air, he demonstrated, actually traveled with the pendulum and thus changed its center
of gravity and moment of inertia. In each case, his consideration of the
theory of his apparatus led him to a deeper understanding of the constraints upon his investigation: how his experiment did not satisfy ideal
laws and how his protocol deviated from an ideal execution of technique.
"The idea" of his method, Bessel explained, was to design a protocol
where "the accuracy of the result is limited not by the apparatus but above
all by the diligence that one applies to the observations and in repeating
them."[18] Yet although he argued for a protocol that transcended the
material limitations of the apparatus, it was precisely those limitations,
coupled with several in the experimental "environment," that became
objects of study in themselves.

In terms of its comprehensive consideration of constant errors in a
physical investigation, Bessel's study of the seconds pendulum was without
equal when it appeared. Around the same time, the physicist Wilhelm
Weber was independently developing ways of accounting for constant
errors and achieving precision in physics. But the geomagnetic investigations he undertook with Gauss, which also became a model for precision
measurement, still lay in the future. (Weber did not join Gauss at Göttingen until 1831; they began their investigations later in the decade.)
Bessel later extended his methods to the calculation of the length of the
seconds pendulum in Berlin. He also continued to deepen his technical
mastery of instruments, especially the Frauenhofer heliometer, which he
brought to perfection as a precision measuring instrument in the late
1820s. As one biographer wrote, Bessel was not only driven to achieve
"accurate micrometric measurements"; he also had "a certain predilection
for complicated instruments that stimulated and tested the observer."[19]
The advantages his studies of instruments offered for physical investigations went beyond the model protocol they set for an exact experimental
physics. In his seconds pendulum investigation, for example, Bessel had
achieved a way of measuring directly 0.02 second, the smallest interval
that could then be achieved and one that still had not been further
subdivided almost a decade later, except by indirect means.[20]

From time to time Bessel commented on epistemological issues surrounding his exact experimental procedure, but never with great elaboration. Although he pondered the relation between mathematical truth and
observational data, wondering if there were an instrument that could yield
absolutely certain measurements, he was never convinced that the way to

18. Ibid., p. 2.
19. F. W. Bessel, "Lebensabriss [Fortsetzung vom Herausgeber {Rudolf Engelmann}],"
in *AFWB* 1:xxiv–xxxi; on xxvii; see also p. xxvi.
20. Heinrich Wilhelm Dove, *Ueber Maass und Messen*, 2d ed. (Berlin: Sander, 1835), pp.
56n–57n, 59.

achieve more accurate results was by improving instruments.[21] Instead, in his view, one had to analyze instruments exhaustively, calculating every possible source of error. "The task of the present-day art of observation," he instructed, was "to eliminate the apparatus from the results." Now all too aware of the errors generated by instruments and of the difficulties in identifying and calculating them, Bessel considered "too many instruments" to be "just as detrimental as too few."[22] Especially between 1832 and 1840 in his popular lectures, he pointed to the approximate nature of knowledge based on observations, a knowledge whose degree of certainty was linked to the probabilities contained in the method of least squares, and he distinguished that knowledge from mathematics, where truth was certain and absolute.[23] As he explained to Alexander von Humboldt in 1838, "Results, which are based on observations, may never be found with the certainty that mathematical truth claims by right of law. Therefore I consider it essential that results always appear together with information that provides the criteria for evaluating them."[24] So imperious had Bessel become on the central role of measurement in science that even the mathematician Carl Jacobi, much to the surprise of his brother Moritz, a physicist, became impressed by "the repetition of simple results [and] steadfast observations."[25] Bessel's techniques were also known at Königsberg through his students, who were trained at the observatory in his own methods of instrument and data analysis.

Neumann, who earlier had not chosen any particular way of pursuing quantification in his research, found in Bessel's study of the seconds pendulum exemplars for measuring operations, data analysis, and experimental protocol. His interest in Bessel's exact methods is evident in the scattered notes that remain from his 1827/28 course on the physics of the earth, but those methods were not as yet so dominant as to indicate a decisive preference for them. In his 1829 letter to Weiss, he had been sensitive to the need for accurate measurements but still believed accuracy was attained by using the purest crystal samples possible. A year later, in his first publication since his arrival at Königsberg, "Das Krystallsystem des Albites und der ihm verwandten Gattungen," Neumann de facto admitted that such ideal material conditions could not be realized.

21. F. W. Bessel to C. F. Gauss, 18 July 1816, in *Briefwechsel zwischen Gauss und Bessel*, ed. A. Auwers (Leipzig: W. Engelmann, 1880), p. 242.

22. F. W. Bessel to A. v. Humboldt, 10 April 1844, A. v. Humboldt Nachlaß, SBPK-Hs; Bessel to Gauss, 10 March 1811, GN.

23. F. W. Bessel, "Ueber Wahrscheinlichkeits-Rechnung," in *Populäre Vorlesungen über wissenschaftliche Gegenstände*, ed. H. Schumacher (Hamburg: Perthes-Besser & Mauke, 1848), pp. 387–407, esp. pp. 387–88; idem, "Ueber den gegenwärtigen Standpunkt der Astronomie," in ibid., pp. 1–33, on p. 18; idem, "Ueber die Verbindung der astronomischen Beobachtungen mit der Astronomie," in ibid., pp. 408–57, on p. 431.

24. F. W. Bessel to A. v. Humboldt, 24 January 1838, A. v. Humboldt Nachlaß, SBPK-Hs.

25. Moritz Jacobi to Carl Jacobi, 5 October 1826, in *Briefwechsel zwischen C. G. J. Jacobi und M. H. Jacobi*, ed. W. Ahrens (Leipzig: B. G. Teubner, 1907), p. 3.

Instead he pursued an investigative strategy whose protocol closely resembled Bessel's in his study of the seconds pendulum. Neumann subtitled his report "Part I: The Method and Error of Measurement"; a projected second part never appeared. What he did publish closely followed the format of Bessel's experiment.[26]

Neumann made the issue of measurement paramount by focusing especially on establishing the reliability of his data. He recognized that any judgment of the reliability of his measurements rested in part upon an exhaustive understanding of his procedure. He pointed to the need to look for variations in measurements taken under differing conditions, to consider deviations among crystals, and to calculate errors in crystalline measurements. His goal was not only to find the procedure most effective in reducing errors but also to understand the conditions under which measurements could be combined. As Bessel had done, Neumann first treated the mechanical construction of his instrument, a reflection goniometer, and then turned to an examination of the sources of error in the goniometer and in his protocol, especially his method of measuring. Like Bessel, he gave a complete description of his apparatus, with diagrams. Remarkably, he had used a goniometer many times before, but he now gave it a thorough theoretical treatment. "First," he explained, "the instrumental sources of error that are independent from one another should be sought [, and] the value of these errors and their collective influence on the final result will be given."[27] To ensure a technical execution that minimized errors, he employed, for example, a telescope equipped with cross hairs to make sure that the goniometer axis was horizontal, and he adapted Bessel's method of coincidence observations to his study. Neumann either computed the effect of experimental errors that could not be eliminated or minimized errors to the point where their effect was negligible. He even described a way of reducing the errors stemming from a crystal's surface imperfections.

Criticizing others for taking only a few measurements, he showed the value of taking many readings and of knowing the conditions under which they were taken. The care he now devoted to instrumentation and to the errors introduced during the act of measuring led him to reconsider his earlier belief that the influence of external disturbances could be minimized if measurements were repeated in rapid succession. Now, he "never used the method of repetition. In my view, this method must be thrown out entirely. It is quite useless because the errors of the instrument . . . must be ascertained directly. [The method of repetition] is harmful [in this investigation] because the displacement of the disk, which occurs in every initial rotation of the internal axis, is summed up in

26. FNN 32: Physik der Erde; Neumann, "Das Krystallsystem des Albites und der ihm verwandten Gattungen" [1830], in GW 1:377–418.
27. Neumann, "Krystallsystem des Albites," p. 383.

the result achieved through [the method of] repetition, and this produces a noticeable imperfection." Just as Bessel had done in the second part of his protocol, Neumann followed his analysis of his instrument and of his method of measuring with a discussion of how well the conditions described in certain natural laws were fulfilled in the experiment. But whereas Bessel had been able to utilize well-developed theories for, say, the motion of an object in a resisting medium, Neumann had to evaluate from scratch the "natural errors" occurring in crystals. Although in the end he could not determine the exact nature of these errors, he admitted that their effect upon his measurements was greater than that of other errors produced in the execution of the experiment. But he needed some quantitative rule for dealing with natural variations in crystals. Undoubtedly to facilitate his calculations, he held as "equally probable every correction needed for each measured deviation [in the crystals], that is to say, to give them all equal weight." He was able to combine measurements taken under different conditions by using the method of least squares. He ended the essay by reproducing his measurements of angles. Neumann's discussion of the goniometer, his method of measuring, his procedure, and his error analysis constituted one-half of his essay.[28]

The appropriation of Bessel's protocol and exacting techniques, especially the adaptation of error analysis and the method of least squares, marked a turning point in the evolution of Neumann's investigative style. Most significantly it placed Neumann among a very small group of investigators who were applying the method of least squares to problems outside astronomy and geodesy. Although there were attempts in the late eighteenth century to develop ad hoc procedures for combining measurements, it was not until Legendre published the method of least squares in 1805 that investigators had a uniform and consistent procedure for doing so. Gauss, who gave the method a rigorous derivation in 1809, claimed that he had been using it for some time in astronomical observations and that before 1805 he had discussed the method with acquaintances.[29] His later commentaries on its use in combining measurements helped to distinguish constant and accidental errors. He still continued to believe, however, that accidental errors, like constant errors, could stem in part from material conditions such as the "imperfections of instruments which are unavoidable and follow no rules" and from the disturbances in external conditions afflicting experiments. Still, he believed that making "the uncertainty of the result as small as possible" was "uncontestably one of the most important tasks in the application of mathematics to the natural sciences."[30]

28. Ibid., pp. 390–91, 392, 393, 394.
29. Stigler, *History of Statistics*, pp. 12–15, 55–61, 140–43, 157; *Carl Friedrich Gauss Werke* 4:98–99.
30. *Carl Friedrich Gauss Werke* 4:95, 98.

Few in the natural sciences followed Gauss's suggestion. Among the first to outline how the method of least squares could be used in physical investigations was the gymnasium teacher Magnus Georg Paucker, who taught in the vicinity of Dorpat. In an 1819 essay, he identified the introduction of the method of least squares into physics as "a new period in the history of the natural sciences." "The time is perhaps no longer distant," he wrote, "when every experiment of the physicist, as complicated as it appears, will be calculated according to number, mass, and weight. Only then, when observation supports mathematical construction, will physics be able to lay claim to the name science [*Wissenschaft*]."[31] To demonstrate how the method could be used in physics, he took three relations—between thermal expansion and temperature, between the specific heat of water and temperature, and between the elasticity of steam and its degree of heat—which had been expressed mathematically by Jean-Baptiste Biot but only on the basis of a few observations (that Biot had believed precise but in fact were not), and showed how Biot's mathematical formulas had to be altered when one used many observations analyzed by the method of least squares. Suggestive as Paucker's essay was, it went largely unnoticed because it was published in a secondary school program.

But the Heidelberg physicist Georg Muncke, in his article "Observation" in Gehler's *Physikalisches Wörterbuch*, published in 1825, took note of Paucker's essay and based his explanation of the method of least squares upon it and Gauss's publications. Calling the method "one of the most important and most significant extensions of the use of calculus in astronomical and physical observations," Muncke also warned his readers that it was "not entirely easy," but "copious" and "laborious."[32] Despite its difficulties, by the 1820s a few investigators had used it to analyze data in physics. Laplace applied it to barometric measurements taken to determine the effect of the moon on the tides of the atmosphere, and Bessel used it in his studies of the thermometer.[33] But the physical applications by Laplace and especially Bessel were the result primarily of their continuing study of the theory of the method rather than of an interest in developing rigorous investigative techniques in physics.

The tediousness and sheer computational difficulty of the method may have inhibited many from employing it in their investigations. But a far

31. Magnus Georg Paucker, "Ueber die Anwendung der Methode der kleinsten Quadratsumme auf physikalische Beobachtungen," in *Programm zur Eröffnung des Lehrkursus auf dem Gymnasium illustre zu Mitau für das Jahr 1819* (Mitau: Steffenhangen & Sohn, 1819), p. 3. On Paucker see *Dr. Magnus Georg von Paucker (1787–1855)* [Sonderabdruck aus dem "Inland"] (Dorpat: Laakmann, 1855).

32. Georg Muncke, "Beobachtung," in *Johann Samuel Traugott Gehler's Physikalisches Wörterbuch*, revised by [H. W.] Brandes et al., 11 vols. (Leipzig: Schwickert, 1825–45), 1.2:884–912, on 901, 909.

33. Stigler, *History of Statistics*, pp. 148–57; Bessel, "Methode die Thermometer zu berichtigen."

more likely reason for its limited use in physics around 1830 was the lack of a clear understanding of why it or a similar procedure was needed at all. The limitations of instruments in producing precise data and the statistical nature of data were two different issues. At the very least, a sensitivity to crude probabilistic considerations in measuring observations was necessary to understand the latter. In dispensing with the method of repetition and assigning equal probabilities to natural deviations in samples of the same type of crystal, Neumann indicated that he had at least a part of that sensitivity. What this newly imported technique meant to Neumann for the determination of precision and accuracy in physical investigations and even for the construction of scientific knowledge, however, remained to be demonstrated. Although he had known of the method of least squares as early as 1823, only after his first years at Königsberg, after he became acquainted with Bessel's investigations, did he acknowledge its usefulness in analyzing data. Neumann was probably the first to apply it to mineralogy.[34]

With increased frequency between 1830 and 1834, concerns for instrumentation, measurement, and error analysis entered Neumann's publications, publications that now embraced the physics of crystals and minerals that he had mentioned to Weiss. In some cases, considerations of exact experimental procedure became so central that Neumann devoted entire essays to them. In 1831 he published his first investigation on the specific heats of minerals, a study he had mentioned to Weiss. Cognizant of the "obstacles which every investigation of specific heat has," Neumann examined and compared three different methods for determining specific heats—mixtures, cooling, and calorimetry—for the purpose of identifying and computing the constant errors that afflicted the final result and that made the calculation of specific heat uncertain.[35] In contrast to the year before, when he focused on an instrument, the goniometer, and the errors inherent in it, he now concentrated on the analysis of experimental procedure and the "environment" or conditions of the experiment. He especially looked at the method of mixtures and carefully scrutinized the limitations in practice of the theoretical assumptions behind the method, recognizing the need for corrections for inaccuracies in the temperature of the heated substance at the moment of plunging, for the possibility that the substance and the water into which it was plunged did not come to the same temperature after mixing, and for the loss of heat during the mixing process. He developed correction formulas out of theoretical equations,

34. Albert Wangerin, *Franz Neumann und sein Wirken als Forscher und Lehrer* (Braunschweig: F. Vieweg & Sohn, 1907), p. 18, on the application of least squares to mineralogy; see also pp. 54–55.

35. Neumann, "Untersuchungen über die specifische Wärme der Mineralien" [1831], in *GW* 2:1–36, on p. 3; see also idem, "Theoretische Untersuchung über die zur Bestimmung der specifischen Wärme dienende Methode der Mischung" [unpubl. ms. c. 1831], in *GW* 2:53–64.

such as Newton's for the passage of heat from a heated substance to water and Fourier's for absolute internal conductivity (because no use could be made of the data that accompanied Fourier's equation, Neumann thought), and recognized the accuracy of his results improved with each correction.

Theoretical excursions such as these enabled Neumann to assess precisely under what conditions the method of mixtures was optimally suited. After computing the constants of the equations so corrected, and noting the difficulties in eliminating all constant errors, Neumann then put his modified theoretical equation to an experimental test, making sure that his observations and the new approximations based on them produced a small accidental error by the method of least squares. Finding that his corrections were not yet adequate, he then questioned not the technical aspects of his procedure but the theoretical assumption that the specific heat of water decreased with increasing temperature. In other ways he refined the theory of his protocol and improved the precision of his results. Like Fourier, Neumann calculated the internal conductivity of his container, and he developed a mathematical expression for the absolute internal conductivity of a substance. Conducting enough trials to reduce his error overall, he then compared the method of mixtures to the method of cooling. Finally, on the basis of his more accurate results, he concluded that the molecular heat of a compound substance was equivalent to the sum of the atomic heats of its separate constituents, thereby extending the Dulong-Petit law from simple to compound substances. Although he achieved accuracy primarily by analyzing theoretically both his equations and his protocol, he mentioned in a postscript that his research had been aided by using an apparatus that reduced errors considerably.

Before 1834, aspects of this investigation on specific heats continued to occupy Neumann as he worked to perfect his exact experimental procedure. He examined in greater detail several additional factors, including the specific heat of water, the internal conductivity of a substance, and the advantages of the method of mixtures for determining specific heats. In so doing, he systematically improved the precision of his investigative techniques through the mathematical analysis of both theoretical assumptions and experimental conditions. An anomaly that occurred when equal amounts of warm and cold water were mixed—the final temperature was not the arithmetical mean of the two initial temperatures—motivated him to go beyond the practical matter of examining thermometers and thermometric scales and to examine an empirical issue, the changes in the specific heat of water near its boiling point. By uncovering errors in previous experiments, he was able to show that the specific heat of water was higher, not lower as had been assumed, than its value at colder temperatures. He was not able, however, to give an exact expression for how the specific heat of water increased with temperature and could only remark that "this is from a theoretical point of view a very complicated phenomenon."[36] He examined the internal conductivity of a substance in greater analytic detail by integrating Fourier's equations for the distribution of temperature in a homogeneous sphere with a consideration of the empirical boundary conditions that had to be satisfied. Only then could he refine his experimental procedure and move away from Fourier's theoretical equations, which he regarded as a "structure" but not one whose form had to be strictly preserved.[37] Despite the success at his efforts at quantification—his application of Fourier's formulas was novel—Neumann concentrated on improving the experimental procedures used in thermal investigations, especially the method of mixtures. Nevertheless, one can see in his studies on specific heats that he was developing an investigative style that moved back and forth between theoretical and experimental considerations, using one to perfect the other, and that he relied heavily on exact mathematical and quantitative techniques, especially approximation methods and the calculation of constant and accidental errors, in order to attain precise experimental results.

Particularly in his analysis of the method of mixtures, Neumann demonstrated the kind of exchange between theory and experiment which could take place when errors were rigorously considered. In the method of mixtures, a solid body was heated and then sunk into a fluid, usually water, which was at a lower temperature. When the fluid reached its maximum temperature, one could assume that the solid substance was at that temperature as well. Provided the heat given up by the solid substance equaled the heat taken up by the fluid, and provided there was no loss of heat during the process of mixing, one could then compute the specific heat of the substance. None of these conditions was actually fulfilled, so Neumann corrected them by using Fourier's equations, which described heat flow more precisely than did Newton's. That appeal to theory, Neumann explained in his unpublished notes for this study, was crucial in the experimental investigation of thermal phenomena because

on the one hand one can, on the basis of theory, determine in advance the most suitable construction of the procedure. On the other hand, one can determine by means of the theory the possible limits of certain errors that originate from influences that are removed from direct observation in the procedure. Understandably it is thereby assumed that one already possesses some kind of approximate knowledge about the value of the constants contained in the theory. [Neumann continues,] The most suitable construction of an experiment consists in allowing those elements, which one

36. Neumann, "Bestimmung der specifischen Wärme des Wassers in der Nähe des Siedepunktes gegen Wasser von niedriger Temperatur" [1831], in *GW* 2:37–51, on p. 49.

37. Neumann, "Wie man durch geeignete Beobachtungen den absoluten Werth der innern Wärmeleitungsfähigkeit eines homogenen Körpers zu bestimmen vermag" [unpubl. ms. c. 1831], in *GW* 2:65–78, on p. 71.

wants to study more closely, to have a predominant effect—a complete independence from the remaining elements is not to be achieved. But one will have to strive to make the disturbing elements as small as possible. To do that it will be necessary to express the effect in question in a theoretical way through a definite analytic expression that gives more precise information about the dependency of the effects upon all principal elements coming into consideration.[38]

So by 1834, Neumann had integrated fully into his research style the experimental methods used by Bessel. Like Bessel, he corrected ideal formulas and experimental procedures based upon them, acknowledging that certain assumptions, built into theoretical equations, were not always met. He was thus able to compare more rigorously the relative merits of what might otherwise be considered equivalent experimental procedures. Hence when determining specific heats, he could be confident that the method of mixtures yielded the most precise data and that it was the procedure whose results set the standards of precision for all others. It was for his promotion to *Ordinarius* for which Neumann wrote his 1834 investigation of the determination of specific heats by the method of mixtures. Two of his students served as his principal opponents at its public defense.[39]

Between 1830 and 1834, Neumann also worked on the wave theory of light, a second investigation also related to his study of the physical properties of crystals and minerals. As in his investigation of the specific heats of minerals, where he replaced one theoretical foundation— Newton's formulas for thermal conductivity—with another—Fourier's equations—he began his study of optical phenomena by first improving upon theory in preparation for formulating, theoretically understanding, and executing his experimental protocol. His most substantial result during this period was the creation of a theory of double refraction based on the equations of mechanics, published in 1832.[40] Working from Augustin Jean Fresnel's optical theory, Neumann sought to make it more rigorous by exploring its theoretical foundations and empirical consequences.

Neumann's objective was to extend Fresnel's wave theory of light to crystalline substances where the wave surface was not the same in all directions. Assuming that light vibrations could be treated analogously to elastic vibrations in a solid body, Neumann used Claude Navier's elasticity equations as the foundation of his theory. To satisfy boundary conditions

38. Neumann, "Aus Neumann's hinterlassenen Manuscripten [on specific heats and the method of mixtures, c. 1834]," in *GW* 2:114–17, on p. 114.

39. Neumann, "Commentatio de emendanda formula, per quam calores corporum specifici ex experimentis methodo mixtionis institutis computantur" [1834], in *GW* 2:79–113.

40. Neumann, "Theorie der doppelten Strahlenbrechung, abgeleitet aus den Gleichungen der Mechanik" [1832], in *GW* 2:159–90.

in a crystalline substance, Neumann assumed that light vibrated in the plane of polarization, not perpendicular to it as Fresnel had assumed. His adaptation of elasticity equations to the study of optics united two different physical phenomena through analogy; enabled him to interpret the theoretical meaning of the constants in his equations (they characterized the crystalline substance under investigation); and helped to shape his experimental procedure, because his development of Navier's equations yielded approximation techniques and useful hints for measuring operations. Unlike his thermal investigations, however, Neumann's optical investigations did not yet lend themselves to a high degree of perfection in experimental protocol. In subsequent investigations he did seek to examine the elasticity constants of crystals empirically, as well as to explore the meaning of the optical, thermal, elastic, and main axes of crystals. But these investigations did not yet display the more rigorous techniques of error analysis of his earlier studies; for Neumann concentrated his efforts first on the mathematical expression of optical theory. He revealed in them that he had received mathematical assistance from Bessel.

Although his thermal and optical investigations were not each equally developed in their experimental and theoretical parts, together they demonstrated how the exacting and laborious mathematics of error analysis was becoming integral to the "mathematical physics" Neumann was developing and, conversely, how important the sophisticated mathematical techniques of analytical mechanics were to the theoretical analysis of experimental procedure. Thus by the early 1830s Neumann had chosen the pathway of an exact experimental physics as the route to mathematical physics, a path on which he was guided strongly by Bessel's example even as he was borrowing from various traditions in French mathematical physics, including the phenomenological approach of Fourier and the more "practical" mathematical physics of Poisson and Navier. Throughout, it was Bessel who encouraged Neumann to consider more carefully the experimental arrangements whose results would be used to expand or modify theory. When, for example, Neumann sent him a balance that was not working properly, Bessel calculated the precision of the needle, concluded that the balance was an "insensitive instrument" that "could not be trusted," and recommended it be sent back to the manufacturer to be realigned—mentioning, along the way, that more precise instruments could be found in England. He urged Neumann to take a close look at the scale Repsold had cut on the seconds pendulum apparatus.[41]

Yet even as Neumann's investigations became more sophisticated mathematically, he seems not at first to have established strong cooperative ties to Carl Jacobi. Jacobi, who went to Königsberg after the death of the

41. F. W. Bessel to Neumann, 26 April 1839, FNN 53.IA: Briefe von Collegen.

mathematician E. F. Wrede, arrived there with Neumann and, like him, received an initial salary of two hundred taler as *Privatdozent*. Initially, Jacobi was characteristically critical of Neumann, skeptical of his abilities as a teacher, certain that his personality was disagreeable, but admiring of his talents as a physicist. (They were later to develop a cordial working relationship.) Much of Jacobi's mathematical research up to 1834 had potential applications to astronomy and physics, especially his work on variational calculus and its applications to analytical mechanics, integral calculus, elliptic functions, first-order partial differential equations, and the theory of differential equations in general. Jacobi developed a close friendship with Bessel, frequently sharing a "stimulating daily exchange" with him. Bessel, who often found useful mathematical techniques for his own calculations in Jacobi's research, hoped at first that Jacobi would turn his efforts to applied areas. By 1829, however, when Jacobi was working on elliptic transcendentals, Bessel resigned himself to the belief that Jacobi "did not have the interest or the knack to work with lasting consideration of astronomy or of any other applications."[42] Jacobi's reluctance to work more closely in areas of applied mathematics meant that others, like Neumann and Bessel, had to sift through his results if they were interested in using them in the physical sciences.

Despite the potential advantages for the development of mathematical physics that might have arisen at this time from an exchange between Neumann and Jacobi, Neumann seems between 1829 and 1833 to have been concerned more about the related issues of experimental protocol and precision in measurement. Hence he directed his efforts at improving his material environment, seeking equipment that would facilitate the exacting nature of his investigations, rather than at securing approval for the natural sciences seminar. On 7 May 1829 he requested from the ministry better physical instruments for his investigations. He argued that Hagen's equipment did not allow for the "exact investigations" he wished to make and, moreover, that his experimental investigations required a degree of precision "now demanded by science" that could not be achieved with the instruments available at Königsberg. To purchase instruments, he had been drawing upon his small salary of 300 taler but found that he could extend himself no longer. Since, when he had arrived as *Privatdozent*, he had "had the opinion that the means and materials for scientific investigations would not be lacking at Königsberg," he asked the state for 1,210 taler for instruments, 150 taler per year for their upkeep, and funding for other research supplies. He also requested something "a real experimentalist must have," a location where experiments could be performed and equipment stored. A small portion of the 1,210 taler was

42. Carl Jacobi to Moritz Jacobi, 15 December 1831, in *Briefwechsel zwischen C. G. J. Jacobi und M. H. Jacobi*, p. 11; Leo Koenigsberger, *Carl Gustav Jacob Jacobi* (Leipzig: B. G. Teubner, 1904), p. 28; F. W. Bessel to C. F. Gauss, 2 January 1829, GN.

for instruments related to his planned optical investigations: a polarization apparatus, instruments for double refraction, and prisms. But most was intended for the purchase of measuring instruments suited for "sharp quantitative determinations": a chronometer, water balance, goniometer, spherometer, heliostat, and two microscopes equipped with micrometers.[43]

Neumann received permission to purchase at least some of these instruments, which he did by placing an order with the instrument maker Karl Philipp Heinrich Pistor; and for this initial order, he requested a 300 taler advance from the ministry. About the same time, 1829–30, he also became heir to the instruments previously ordered by Hagen and Dove, including a magnetic declinatorium and inclinatorium. Apparently the unpaid balance on these prior requisitions cut into Neumann's allotment because the ministry had assumed that Neumann's order would not be filled for at least a few years and that the state would spread its payment over that time. By August 1831, enough of the budget for physical apparatus (150 taler annually, plus 8 taler to pay an attendant to clean the room in which the equipment was stored) had been spent to place in doubt the award of additional funds, despite earlier promises and despite orders already placed. Although it is not known how much of his original request was denied, it is certain that Neumann did acquire some equipment and that at least a part of his material needs were met during the early 1830s. But he seems not to have made many additional purchases because by the spring of 1834 he had accumulated a surplus in the budget for physical equipment, which now had over 500 taler in reserve, indicating either that the 150 taler annual budget from 1832, 1833, and 1834 for the physical cabinet had not been spent or that the ministry had continued to deposit funds in his account in anticipation that Neumann would eventually provide the receipts for the equipment he received, which he was required to do in order to be reimbursed.[44]

His quest to create favorable material conditions for his research was somewhat complicated by changes in the faculty positions closest to his own. In 1829 when Dove left Königsberg, Ludwig Moser asked to teach physics there. Trained in medicine at Berlin, Moser came to Königsberg as *Privatdozent* for physics in 1831, two years after Dove left. Neumann undoubtedly greeted the appointment with mixed feelings. Because Moser became responsible primarily for teaching general courses in

43. Neumann to Kultusminister Karl Altenstein, 7 May 1829, ZStA-M, Acta betr. den mathematisch- physikalisch- und chemischen Apparat der Universität zu Königsberg, sowie die Errichtung eines physikalischen- und eines chemischen Laboratoriums und eines magnetischen Häuschens, Rep. 76Va, Sekt. 11, Tit. X, Nr. 15, Bd. I: 1809–76, fols. 63–65.

44. Letters and notes from 1829 through 3 March 1834, esp. Königsberg University Curator Reusch to Neumann, 15 February 1830, 4 April 1830, 20 April 1830, 18 January 1831, 20 August 1831, 14 October 1831, and 3 March 1834; and Neumann to Curator Reusch, 29 April 1831; all in FNN 62: Kassensache.

physics to medical students, his appointment probably freed Neuman to intensify the mathematical content of his own courses in physics, unencumbered by the constraints of appealing to a broad student clientele. Yet different clienteles did not prevent other problems from emerging. Shortly after Moser arrived, he and Neumann clashed over the right to use Königsberg's modest physical cabinet. Both needed instruments and equipment for their lectures. Moser especially depended upon demonstration experiments to embellish his lectures, and he had a stronger case than Neumann for using the equipment, not just because his teaching style was different but because he proved to be a more popular teacher. Jacobi reported that during Moser's first year as *Privatdozent*, he had "30–40 students in experimental physics, which is unheard of here."[45] Even though Neumann's physics lectures became more sophisticated in the mathematical treatment of physical problems, enrollment in them remained stable. Neumann never had more than one-third to one-half the students in Moser's courses.[46]

By the mid-1830s, when attendance at Moser's lectures began to drop, Moser thought the trend could be reversed if confusion did not always result when both he and Neumann wanted to use the same instrument from the physical cabinet. Moser had been given support (the amount was not specified) for demonstration apparatus in 1833, but it was apparently insufficient; for by 1835 he claimed to have only duplicate instruments from the physical cabinet at his disposal (an electrical machine, air pump, and Atwood's machine) plus a small collection of demonstration apparatus formerly used in Herbart's pedagogical seminar. Neumann had made some purchases for the physical cabinet, but evidently not enough to improve it substantially, because both Neumann and Moser agreed that "overall, the collection is little suited both for the current standpoint of [physics] and for the present needs of academic lectures in it." Acting on the requests of both Neumann and Moser, the Königsberg University curator Reusch suggested to the ministry that it either make Moser an *ordentlicher* professor and raise his salary enough to cover needed physical apparatus or give him 150 taler annually, the amount Neumann had. The ministry did not act immediately on either suggestion.[47]

What was at first a bureaucratic matter quickly evolved into one of the means by which Neumann and Moser were able to define their teaching responsibilities more clearly. Both of them knew that much more was at stake in settling their differences than just the peace of mind of knowing

45. Carl Jacobi to Moritz Jacobi, 15 December 1831, in *Briefwechsel zwischen C. G. J. Jacobi und M. H. Jacobi*, p. 11.

46. On the number of students in Neumann's courses, see below, App. 2.

47. Neumann to Curator Reusch, [2] July 1833, 21 May 1835, FNN 61.7: Kampf um das Laboratorium. Neumann to Reusch, 10 October 1835; and Curator Reusch to Kultusminister K. Altenstein, 16 October 1835; both in ZStA-M, Math.- phys.- und chem. Apparat Königsberg, fols. 128, 126–27.

that prisms would be available for an optics lecture. Controlling Königsberg's physical cabinet (or establishing a second cabinet, which is what they had in effect proposed) meant administering its budget and enhancing its holdings to accommodate pedagogical and other needs. In turn, the instruments on hand shaped the composition of lecture courses and hence their appeal and relevance to students. Moser's presence may have given Neumann the freedom to develop a mathematical physics without regard for lower-level students or those students for whom physics was just an elective, but it also restricted Neumann's access to scarce funding for equipment more expensive than the demonstration apparatus Moser needed. Eventually in 1839 the ministry appointed Moser professor of experimental physics (but without a separate equipment budget) at the recommendation of Neumann and Jacobi, a decision that officially acknowledged the differences between his approach to physics and Neumann's. Königsberg's two physics professorships distinguished it from most other German universities. The ministry thus strongly endorsed Neumann's pedagogical program and reinforced the institutionalization of mathematical physics.

What differentiated Neumann's and Moser's physics pedagogy most strongly was their divergent views on the role of quantification in physics instruction. By 1833/34, Neumann had achieved considerable sophistication in several quantitative techniques. His courses on theoretical optics and the theory of heat not only incorporated the methods of Fourier and Fresnel but went considerably beyond them in the deployment of approximation techniques and the more detailed consideration of experimental results. Moser's lecture courses, in contrast, remained broader in scope but less mathematical. His topics were more characteristic of an earlier time when physics, or *Naturlehre*, embraced meteorology, physical geography, and related issues. Even in his research, where he was sympathetic to quantification, Moser did not draw upon the wide variety of mathematical techniques that Neumann did. Nonetheless, Moser was sensitive to the limitations involved in viewing mathematical expressions or the results of numerical calculations as the guarantors of unquestionable truth. Commenting in 1835 on the work of a colleague, he criticized the tendency to base conclusions too heavily on the results of numerical calculations, pointing out that "through the agreement between numerical values, which have been computed with the assistance of an hypothesis, and observed values, only the numerical part of the hypothesis can be confirmed, not its physical part."[48] Concerned also that many people "place the smallest possible weight on numbers and on the conclusions based upon them, considering these conclusions to be trivial shadows, or nests from which truth has flown," he nevertheless acknowledged the need for

48. Ludwig Moser, "Ueber den Magnetismus der Erde," *AP* 34 (1835): 63–84, 271–92, on 68.

a balance between those who believed that everything should be derived from numerical calculations and those who believed that nothing could.[49] He seemed thereby to acknowledge that the natural sciences had a long way to go before mathematical methods and numerical computations were generally accepted and, more important, understood.

It is ironic, therefore, that Moser did little to promote that public understanding. Not only was he acutely aware of the difficulties of teaching mathematics and quantitative techniques to students of the natural sciences, but he also questioned whether teaching quantification was even necessary. So did other directors of the natural sciences seminar when Neumann and Jacobi proposed their mathematico-physical seminar in 1833.

By the time 1,060 taler from Herbart's pedagogical seminar was released in early 1833, proposals for both a natural sciences seminar and a mathematico-physical seminar had taken shape. By then, Neumann had undoubtedly been impressed by Jacobi's success in bringing teaching and research more closely together and in working closely with his students. Jacobi's courses were from the start extremely specialized, and he used them to "give structure" to branches of mathematics, such as elliptic transcendentals, in which he was working.[50] The demanding nature of his courses, Jacobi recognized, necessitated a greater interaction with his students. Even though he had only a few students each semester, Jacobi began in the winter semester of 1826/27 to offer a supplement to his lecture courses, analytic exercises or problem sessions, usually held weekly. His extra teaching prepared students for advanced investigations. By the early 1830s he submitted student investigations to the *Journal für die reine und angewandte Mathematik*.[51] His innovative teaching and successes undoubtedly motivated him to establish a more formal setting for his exercises, a mathematical seminar.

Jacobi was not, however, the first to hold exercise sessions, which were appearing in both mathematics and natural science courses at several Prussian universities. Jacobi's predecessor, E. F. Wrede, had instituted exercises in higher geometry and differential and integral calculus at Königsberg in 1825. A year later, Bessel's former student and teaching assistant Heinrich Ferdinand Scherk established a mathematical society at Halle, partly to train secondary school teachers and partly to encourage students to take up more difficult problems. Greifswald's mathematical society was also intended for future secondary school teachers, and it operated very much like a seminar: with a lower division to review mathematics as taught at a good Prussian gymnasium and an upper division for

49. Ludwig Moser, *Die Gesetze der Lebensdauer* (Berlin: Veit, 1839), pp. xviii–xix.
50. Koenigsberger, *Jacobi*, p. 115.
51. Ibid., pp. 144–45.

more advanced topics, including applications to the natural sciences. The upper division was intended for students devoting themselves exclusively to the study of mathematics. At Berlin, the mathematician Gustav Lejeune Dirichlet held private exercises beginning in 1834, but his were probably not intended for secondary school teachers.

The exercise sessions Prussian mathematicians founded were emerging elsewhere in the German states. Mathematical exercises had been considered at Göttingen in Lower Saxony in the eighteenth century. At Munich in Bavaria they were firmly established in the 1820s, but a mathematical seminar was rejected in 1832, as was a mathematico-physical society in 1834. Hessian officials rejected in the early 1830s a proposal for a mathematical seminar at Giessen. In 1824, Baden officials likewise rejected a proposal from the mathematician Franz Ferdinand Schweins for a mathematical seminar at Heidelberg, even though Schweins argued that the seminar would promote the study of mathematics throughout Baden's educational system; that it would be advantageous for Baden to have the first such institute in Germany; and that in it would be trained teachers, men from commerce and business, civil servants, as well as scientists—all professions and occupations important to the state.[52] These societies, sessions, and seminars indicate that in the first few decades of the nineteenth century, plans for the improvement and expansion of mathematical instruction were not unique to Prussia.

Well before Neumann and Jacobi had proposed their mathematico-physical seminar, the Prussian ministry had toyed with the idea of establishing an institute to train students for the *Mathematiklehrerberuf*, so prominently did mathematics figure on the state teaching examinations of 1810 and after. The best-known attempt originated in the ministry itself, which tried in the early 1820s to establish in Berlin a polytechnic institute modeled upon the École Polytechnique in Paris. As Gert Schub-

---

52. Wilhelm Lorey, *Das Studium der Mathematik an den deutschen Universitäten seit Anfang des 19. Jahrhunderts* (Leipzig: B. G. Teubner, 1916), pp. 112–13, 116–17; L. F. Kämtz and L. A. Sohncke to Kultusministerium, 8 June 1937, ZStA-M, Acta betr. die Errichtung eines Seminars für Mathematik und die gesammten Naturwissenschaften auf der Universität zu Halle, Rep. 76Va, Sekt. 8, Tit. X, Nr. 36, Bd. I: 1837–89, fols. 3–4; W. Langhammer, "Some Aspects of the Development of Mathematics at the University of Halle-Wittenberg in the Early Nineteenth Century," in *Epistemological and Social Problems of the Sciences in the Early Nineteenth Century*, ed. H. N. Jahnke and M. Otte (Dordrecht: Reidel, 1981), pp. 235–54, esp. p. 242; "Die Königliche Universität Greifswald: 9. Die mathematische Gesellschaft," in *Die preußischen Universitäten: Eine Sammlung von Verordnungen*, ed. J. W. F. Koch, 2 vols. (Berlin: Mittler, 1839–40), 2.2:713; Kurt R. Biermann, *Die Mathematik und ihre Dozenten an der Berliner Universität, 1810–1920* (Berlin: Akademie, 1973), p. 31; Hellfried Uebele, *Mathematiker und Physiker aus der ersten Zeit der Münchner Universität* (Munich: Francke, 1972), pp. 120–21; Karl Neuerer, *Das höhere Lehramt in Bayern im 19. Jahrhundert* (Berlin: Duncker & Humblot, 1978), pp. 106–7, 128; "Antrag des Hrn. Professor Dr. Buff auf Errichtung eines mathematisch-physikalischen Seminariums an der Landes-Universität," 20 January 1861, UAGi, Phil. H, Nr. 36; F. F. Schweins to Heidelberg University Curator, 19 December 1823, GLA 235/3228.

ring has shown, the Berlin polytechnic was designed not only to train more competent mathematics teachers but also to provide practical instruction in mathematics and physics for professions and occupations of importance to the state, including public security and welfare, the military, and several in industry and trade.[53] A major feature of the original proposal was teaching by the seminar method. When plans were revived in 1828, instruction in mathematical physics and practical exercises in physics and chemistry were also included. But when the mathematician and ministerial advisor August Leopold Crelle was asked his opinion, the scope of the proposed institution was narrowed considerably: physics was no longer treated as an independent subject; pure mathematics was emphasized; and a pure scientific approach—including in teacher training—replaced a practical one. Schubring's story of the polytechnic shows that much was at stake in establishing the institute; for especially its teacher training function was viewed as essential for maintaining the strength of the mathematics and science curricula in secondary schools. Even though in the early 1830s plans for the polytechnic fell through for financial reasons, the ministry never lost interest in an institute for training mathematics teachers.

Strong support for mathematical exercises was not matched by similar enthusiasm for exercises in the natural sciences. Bonn's natural sciences society and later its seminar, Greifswald's chemical society directed by D. C. Hünefeld, and Halle's physico-chemical and mathematico-physical societies were the most prominent ones that the philosophical faculties of the six Prussian universities had to offer. Not all were founded to further the state's interest in secondary school teaching. J. S. C. Schweigger, who directed the Halle physico-chemical society, modeled his pedagogical methods partly on the tutor system of Oxford and Cambridge. He was extremely successful in promoting student publications (Wilhelm Weber based one of his first published investigations on exercises done in this society), and he operated his society like an institute, with a 520 taler budget and student assistants. Halle's mathematico-physical society probably grew out of Scherk's exercises. Although it too was also designed to train teachers, some students completed exercises in mathematics and physics which the directors hoped would lead to larger and more original investigations. But most students undertook their own trials and experiments only to become acquainted with the physical apparatus they would use in the secondary school classroom.[54] Early exercises and seminars in

53. Gert Schubring, "Mathematics and Teacher Training: Plans for a Polytechnic in Berlin," *HSPS* 12 (1981): 161–94. My discussion of the Berlin polytechnic follows Schubring's.

54. R. Steven Turner, "Justus Liebig vs. Prussian Chemistry: Reflections on Early Institute-Building in Germany," *HSPS* 13 (1982): 129–62, on 154, n. 14. Kämtz and Sohncke to Kultusministerium, 8 June 1837; J. S. C. Schweigger to Dean of the Halle Philosophical Faculty, 21 January 1838, 30 January 1838, 26 February 1838; O. A. Rosen-

the natural sciences, though, did not strongly emphasize quantitative techniques, which were generally confined to mathematical exercises alone.

Despite this local interest in exercises in the natural sciences, the Prussian government remained most interested in supporting mathematical exercises because of the prominent role of mathematics on the teaching examination and in the secondary school curriculum. So by cooperating with Jacobi, Neumann stood a greater chance of obtaining official ministerial approval for an institute dedicated in part to mathematical physics. The seminar they suggested in 1833, and then officially proposed in February 1834, incorporated in a more formal way Jacobi's innovative pedagogical techniques. Although proposed as one seminar, it was really two, with its separate divisions for mathematics and physics. Jacobi's method of instruction—using exercises both to train students and as preparation for independent investigations—could be institutionalized in the seminar, offering Neumann a novel template for physics instruction. Just as Jacobi's exercise sessions assured him of a more competent clientele in mathematics, so would Neumann's seminar exercises in physics. The discipline building that Jacobi had been able to accomplish in his lecture courses could be enhanced, and it was accessible to Neumann in mathematical physics because the seminar was designed to promote strong intellectual, pedagogical, and institutional bonds between mathematics and physics. They made no mention in their preliminary statutes of training teachers (although they intended to and the ministry assumed they would). Instead they seemed to establish more scholarly objectives by setting prerequisites for entry—including knowledge of the calculus and a level of accomplishment in physics—and by making the principal seminar activity the execution of independent investigations intended for scholarly publication. Neumann intended that his students investigate problems in "mathematical physics that either are purely theoretical or require specific observations and measurements on the basis of a mathematical theory."[55]

Despite the obvious need for a seminar like this one which would train teachers, neither local nor ministerial officials approved it immediately. In addition to the mathematico-physical seminar and the natural sciences

---

berger to Halle University Administration, 3 January 1839; all in ZStA-M, Math. und naturw. Seminar Halle, fols. 3–4, 50–51, 53–61, 43–49, 75–76. Wilhelm Weber considered Schweigger's physico-chemical society to be a seminar; see his "Auszug aus den die Theorie des Schalles und Klanges betreffenden Aufsätzen von Felix Savart," in *Wilhelm Weber's Werke*, ed. Kgl. Gesellschaft der Wissenschaften zu Göttingen, 6 vols. (Berlin: Springer, 1892–94), 1:3–28, on p. 3.

55. "Vorläufige Statuten des mathematisch-physikalischen Seminars an der Königsberger Universität," ZStA-M, Acta betr. das mathematisch-physikalische Seminar an der Universität zu Königsberg, Rep. 76Va, Sekt. 11, Tit. X, Nr. 25, Bd. I : 1834–61, fols. 4–5; rpt. with changes in *Die preußischen Universitäten* 2.2:858–59. See the text of the preliminary statutes below, in App. 1.

seminar, one other seminar competed for funding. Wanting to instill "a true love of scholarly activity" among their students, the Königsberg law faculty prepared a strong and lengthy argument for a juridical seminar. Law professors wanted to change study habits and alter students' cavalier attitude toward scholarly training for the practice of law. "It is unlikely to be denied," their argument began, "that not only at our university but also at others the students of jurisprudence in general show an eagerness that is inferior to that shown by students of other faculties." They held the structure of the educational system itself, and with it, the state examination system, responsible for this state of affairs. Pointing out that the state examination was not designed to encourage a close interaction between students and professors (they mistakenly argued that other faculties were accustomed to quite the opposite), the law faculty believed that institutionalizing such an exchange would curtail the deplorable student practice of reading compendiums to prepare for the examination. Furthermore, because it was impossible in lecture courses to clarify the relationship between *Rechtswissenschaft* and the practice of law, and because lecture courses alone offered insufficient training for practice, the law faculty wanted to establish a seminar that would mediate theory and practice, or education and profession.

Although practical exercises in preparation for the legal profession, like apprenticeship exercises, were consistent with the utilitarian tasks of the university to train functionaries for the state, Königsberg law professors explained that there was "another kind of exercise that belongs to the university, without which the theoretical instruction given in lectures would for the most part remain unused: namely, specific exercises in the acquisition and use of *wissenschaftlich* instruction." The distinction they tried to make here was subtle but nonetheless important. They wanted to dissociate instructional exercises from the notion of strict practicality, while extending the relevance of these exercises to learning scholarly issues of practical import.[56] But the university curator was not convinced. Writing to the ministry, he pointed out that the purpose of the juridical seminar was not like others—that is, it appeared to be insufficiently utilitarian—because it was designed "to educate scholarly lawyers and university teachers [of law] and would have only an *indirect* influence on [educating] practicing lawyers, while the remaining [proposed] seminars directly educate clergymen and school teachers."[57] Denying that the practice of law could have a scholarly foundation, as the law faculty had argued, the Königsberg University curator removed the juridical seminar from consideration.

56. Dean of the Königsberg Law Faculty to Kultusministerium, 20 February 1834, ZStA-M, Math.-phys. Seminar Königsberg, fols. 10–12.

57. Königsberg University Curator Reusch to Kultusminister Karl Altenstein, 28 February 1834, ibid., fols. 2–3 (emphasis added). A juridical seminar was not established until 1872 ("Reglement für das juristische Seminar bei der Universität zu Königsberg," *Centralblatt für die gesammte Unterrichtsverwaltung in Preußen* 14 [1872]: 534–36).

The curator still had to decide what was to be done with the proposals for the two seminars in natural sciences and mathematics. Months earlier, in October 1833, the state had asked the directors of both proposed seminars to consider a combined seminar for mathematics and all the natural sciences. By the end of February 1834, after Neumann and Jacobi had submitted their preliminary statutes, the university curator knew that the proposed union would not work. Days later, in early March, the directors explained why. Neumann, von Baer, Dulk, Meyer, and Moser expressed rare unanimity in rejecting the ministry's proposal because the kind of physics represented by Neumann's seminar was a pedagogical novelty.[58]

Two issues dominated their responses. The first was the opportunity Neumann and Jacobi offered for the pursuit and publication of original investigations, which representatives of the natural sciences seminar interpreted as the principal and only activity of the mathematico-physical seminar. In the natural sciences seminar, by contrast, students undertook original research only in the sense that seminar exercises were intended, in Meyer's words, "to *introduce* [students] to their own investigation"; that is, the exercises engaged students actively in learning and mimicked research techniques, but they did not necessarily lead to novel results. As Meyer explained, these exercises "consist in covering practices that are only taught theoretically in academic lectures." So important—and different—did he consider these exercises that he identified them as the "essential feature" of the natural sciences seminar, a "special method that probably would be pursued with difficulty in a mathematical seminar." On these grounds, he considered the union of the two seminars to be "thoroughly impossible." Yet as significant as this distinction between seminar exercises and original research was, Meyer failed to take into account that the mathematico-physical seminar had the same foundation in exercises. He also neglected to consider the possibility that even Neumann's and Jacobi's exercises might remain introductions to investigations rather than mature into original research. Moser also rejected independent investigations as a possible activity for students, pointing out that "the natural sciences seminar does not aspire to such a high goal," while von Baer believed many students only wanted to be trained in seminars to pursue teaching, not to undertake original research. Directors of the natural sciences seminar thereby dissociated the pursuit of original research from teacher training.[59]

Their criticism of original research did not so much identify a novelty of the mathematico-physical seminar as a difference in pedagogical objec-

58. Reusch to Altenstein, 28 February 1834; Karl Ernst von Baer to Kultusministerium, 6 March 1834; and Franz Neumann, Friedrich Philipp Dulk, Ernst Meyer, and Ludwig Moser to Kultusministerium, n.d. [c. 6 March 1834]; all in ZStA-M, Math.-phys. Seminar Königsberg, fols. 2–3, 6–8.

59. Ibid., fols. 6–8.

tives. Original student investigations had been promoted in principal by several other Königsberg seminars. Herbart, for example, had required an annual paper illuminating an aspect of pedagogical theory through the student's teaching experiences; he even considered it desirable to publish these essays. The philological and historical seminars likewise placed a strong emphasis on scholarly objectives, achieved by students working from introductory exercises to more complicated investigations.[60] What von Baer and his colleagues implied was that one need not go so far in instructing science students and that original student investigations were not an automatic part of any seminar's operation. Hence von Baer could argue that the two seminars were "rather different in their tasks"; Dulk, that with their differing purposes, they could not be united; and Moser, that "the two institutes do not differ from one another by degrees, but they are throughout heterogeneous."[61]

For all involved, the second issue—the prominent role of mathematical and quantitative techniques in Neumann's division of the seminar—was far more serious. From the beginning, Jacobi was convinced that little would be accomplished for mathematics if the mathematico-physical and the natural sciences seminars were combined.[62] Conversely, von Baer and Moser believed instruction in the natural sciences would be hindered by the addition of mathematics. Assuming that the seminar for mathematics and physics, like his own, accepted students immediately from the gymnasium, von Baer questioned whether students would come prepared in differential and integral calculus, which most secondary schools did not teach. He not only believed one could learn the natural sciences without knowing mathematics, but he also considered it "unreasonable" to expect those who wanted to study the natural sciences to learn mathematics too. Surprisingly, Moser, the only director of the natural sciences seminar with professed mathematical leanings, went further than von Baer. Because members of the natural sciences seminar were supposed to "cultivate *all* branches of the natural sciences equally, and to a certain degree," learning mathematics was in his view a disturbing element. The "spirit of contemplating nature" (*Geist der Naturbetrachtung*) was, he believed, "entirely foreign" to the "spirit of mathematics." The natural sciences seminar, he told the ministry, "will train capable teachers of the natural sciences, and it

60. K. Kehrbach, "Das pädagogische Seminar J. F. Herbarts in Königsberg," *Zeitschrift für Philosophie und Pädagogik* 1 (1894): 31–47, on 35; "Reglement für das philologische Seminar bei der Universität zu Königsberg: Vom 10. November 1822," in *Die preußischen Universitäten* 2.2:850–55; "Reglement für das historische Seminar bei der Universität zu Königsberg: Vom 13. Dezember 1832," in ibid., 2.2:855–58.

61. Von Baer to Kultusministerium, 6 March 1834; and Neumann, Dulk, Meyer, and Moser to Kultusministerium [c. 6 March 1834]; both in ZStA-M, Math.-phys. Seminar Königsberg, fols. 6–8.

62. Ferdinand Lindemann, *Gedächtnisrede auf Ludwig Philipp von Seidel* (Munich: Kgl. Bayer. Akademie, 1898), p. 50.

would be difficult to achieve this [goal] if it drew into its instruction parts of differential and integral calculus."[63]

More than pedagogical considerations, though, motivated their objections. Although von Baer and Moser were not generally opposed to quantification in the sciences, they did object to incorporating *certain* kinds of mathematics into science teaching. They singled out for criticism what they perceived to be a curricular subject quite different from other natural sciences, namely, French mathematical physics. For von Baer, the two seminars were not alike partly because the physics each handled was different. Moser expressed the difference in more dramatic terms: "The only connection between mathematics and the natural sciences is to be found in physics. But with the exception of the calculation of numerical results, which the natural sciences seminar does not at all exclude, there remains only the so-called French mathematical physics, which displays the heterogeneity of physics and mathematics as their actual coalescence." By juxtaposing calculation and mathematical physics, Moser artificially separated quantitative experimental investigations both from the mathematical laws that might be drawn from them and from the more general pursuit of mathematical expressions of theory. Neumann's response to the ministry's proposal was that although both seminars had a common object, physics, his was concerned with its mathematical aspects, while the natural sciences seminar dealt "with those parts [of physics] that assume the least mathematical knowledge."[64]

The statements by Neumann and Moser are especially revealing of assumptions about the relation between teaching and research. Moser distinguished between what was pedagogically suitable for students—especially teaching candidates—and what was not, and hence he implicitly differentiated techniques taught in the classroom from those used in research. Neumann, in contrast, seemed to make that distinction less sharply. Yet of the two types of investigations in mathematical physics—one "purely theoretical" and the other drawn from an exact experimental physics—that his seminar students would presumably undertake, only the fundamentals of the first, in the form of the French mathematical physics of Fresnel and Fourier, had he formally incorporated into his teaching. Although he mentioned measuring operations in the seminar's statutes, it was not clear whether he meant the rigorous techniques he had adopted from Bessel (much as that was implied) or the simpler forms of measurement hitherto practiced. His colleagues in the natural sciences, who admittedly dealt with a much less specialized student clientele, seemed to think he meant the latter and hence objected only to the

63. Von Baer to Kultusministerium, 6 March 1834; Neumann, Dulk, Meyer, and Moser to Kultusministerium [c. 6 March 1834]; both in ZStA-M, Math.-phys. Seminar Königsberg, fols. 6–8.
64. Ibid.

difficult mathematical techniques associated with French mathematical physics.

More important, their statements illuminated different views of the epistemological function of mathematics in science instruction. Moser's belief that higher mathematical and natural scientific thinking were antagonistic implicitly drew sharp distinctions between mathematical and scientific evidence, reasoning, even truth. Thus he classified French mathematical physics as mathematics. Moser's views, both explicit and implied, found support in the institutional classification of mathematically oriented sciences that identified French mathematical physics as a part of mathematics, especially applied mathematics. As the mathematician H. W. Brandes wrote in his article "Mathematics" in Gehler's *Physikalisches Wörterbuch*, a source consulted by Bessel and Neumann, "Soon the theories of heat, electricity, and magnetism will probably also be considered branches of applied mathematics, because Fourier, Ampère, Poisson, Murphy and others already have made very significant contributions to a mathematical construction of these theories."[65] So on the basis of common practice, Moser thus was justified in viewing Neumann's physics as different from his and in arguing for the propriety and even the necessity of grouping it institutionally with mathematics in the proposed mathematico-physical seminar. The statutes of the mathematico-physical seminar reinforced the classification of mathematical physics as a branch of mathematics in that Jacobi's section included applied topics such as mechanics and physical astronomy.

That classification is not, however, what Neumann intended, because his section of the seminar, where French mathematical physics was discussed, also included instruction in experimentation. It is not that Neumann did not also distinguish mathematical from scientific thinking, but rather that he thought differently about the role of mathematics in science, particularly the mathematical expression of physical relationships. Neumann's innovation was to define and classify mathematical physics as physics, thereby raising questions about what French mathematical physics shared with mathematics, especially, it would seem, with the epistemological certitude that French physicists customarily imputed to mathematically expressed results.

These opposing opinions on quantitative techniques in physics instruction were instrumental in convincing the ministry to establish two seminars for the natural sciences and mathematics at Königsberg. Days after the ministry received the director's responses, on 17 March 1834, it approved the natural sciences seminar. It founded the mathematico-physical seminar three months later, on 8 June 1834. Its decision to inaugurate two science seminars at a provincial university with relatively

65. H. W. Brandes, "Mathematik," in *Gehler's Physikalisches Wörterbuch*, 6.2:1473–85, on 1478; Bessel to Neumann, 10 July 1838, FNN 53.IÄ: Briefe von Collegen.

low enrollments constituted a substantial risk, but one it presumably wanted to take in order to improve science instruction in its secondary schools.

In proposing to extend to science teaching the quantitative and mathematical techniques he had employed in his research, Neumann began a pedagogical experiment. His conviction that teaching and research were intimately related—in particular, that pedagogy could, on a smaller scale, duplicate practice—was not one widely accepted, as the directors of the natural sciences seminar indicated. Only in a few places—Justus Liebig's chemical laboratory in Giessen or Jan Purkyně's physiological "institute" at Breslau, for example—had teaching and research been successfully brought together, and how they interacted changed constantly.[66] Moser had not sought to align his physics so closely with research practices.

Nor had Heinrich Dove, then teaching at various institutions in Berlin. At work on a major study on measurement in the early 1830s, he came to realize what Neumann had through Bessel: that "at the present time the art of observation owes its high perfection just as much to the development of mathematical methods appropriate to it [i.e., the method of least squares], as it does to the technical improvement of the means of observing."[67] But Dove appears not to have strongly promoted quantitative and technical measuring methods in instruction. His primary responsibility was a course in experimental physics designed for a diverse clientele preparing for professions such as medicine. A former student (from the 1850s) described his lectures on optics—generally acknowledged to be one of the more thoroughly mathematical areas of physics—as "not mathematical . . . but more popular" and as illustrated by stunning experiments. His lectures on meteorology were delivered in so "humorous, popular, and such a clear and understandable manner" that they often brought him the gratifying applause of his students.[68] Constrained by institutional conditions to view teaching as intellectually dominated by research rather than as a possible forum for disseminating new ideas or for promulgating new research methodologies, Dove never seriously considered or pursued the symbiotic relationship between teaching and research which Neumann had announced in the mid-1820s and which now seemed possible to implement in the mathematico-physical seminar.

The major advantages the mathematico-physical seminar offered Neu-

---

66. Frederic L. Holmes, "The Complementarity of Teaching and Research in Liebig's Laboratory," *Osiris* 5 (1989): 121–64; William Coleman, "Prussian Pedagogy: Purkyně at Breslau, 1823–1839," in *The Investigative Enterprise: Experimental Physiology in Nineteenth Century Medicine*, ed. W. Coleman and F. L. Holmes (Berkeley and Los Angeles: University of California Press, 1988), pp. 15–64.

67. Dove, *Ueber Maass und Messen*, p. 166.

68. The remembrances of this student, Karl Emil Gruhl, were published by Gert Schubring as "Die Errinnerungen von Karl Emil Gruhl (1833–1917) an sein Studium der Mathematik und der Physik in Berlin (1853–1856)," *Jahrbuch Ueberblicke Mathematik* (1985): 143–73, on 156–57.

mann were that it allowed him to cultivate physics as a mathematically exact science and it sanctioned the creative teaching that the natural sciences seminar had excluded from its purview. But how and to what extent his institute would succeed depended on his ability to make mathematical physics not only accessible but meaningful. He had no pedagogical model on which to draw. His seminar statutes did not as much outline the operation of his institute as depict an ideal state to which he could aspire. He undoubtedly realized his pedagogical innovation was risky: students might not register; it could prove impossible to create appropriate exercises; cooperation with the mathematical division might not work; the fruits of his labor might not be accepted by the scientific community.

The real question, of course, was how many students would be drawn to an institute with such a specialized intellectual focus. Not only did Prussia not have in 1834 the scientific culture for which Germany as a whole later became known, but there were then no real careers for persons trained in physics other than university professorships, academic pursuits, or secondary school teaching. The last was the most probable career choice, and it would seem to offer little opportunity to make use of highly specialized knowledge in physics. The state teaching examination was certainly not very demanding; for even though teachers of the two upper forms, *Prima* and *Secunda* (where physics was taught), had to have "a scholarly grounded knowledge of physics," that meant enough "to be able to lecture on it to the two upper forms." The primary responsibility of anyone who taught physics was the more important and extensive mathematical instruction; thus students attracted to the seminar would be inclined to be more interested in mathematics than physics. There was nothing about the 1831 teaching examination that required or even promoted a deeper study of physics because the natural sciences, grouped together with mathematics, were only tested on the oral part.[69]

Certainly Neumann knew from his own experiences the benefits of upgrading the qualifications of teachers of mathematics and the sciences as well as the family and social obstacles a student might encounter when pursuing a teaching career in physics or mathematics. He had studied at the Friedrich Werder Gymnasium in Berlin under Gottlob Nordmann, who was the first mathematician to take the initial state teaching examination in 1810. Although he did well under Nordmann, "surpassing the prescribed goal in mathematics" according to the comments on his 1817 gymnasium-leaving certificate, he later sought tutoring outside the educational system, part of it with a Berlin mechanic, to acquaint himself with more advanced methods. Even at the university, he oscillated between

---

69. *Das Reglement von 20. April 1831 für die Prüfungen der Candidaten des höheren Schulamts in Preußen mit den späteren Erläuterungen, Abänderungen, und Erweiterungen* (Berlin: Enslin, 1865), pp. 28–29, on p. 29.

learning mathematics formally in the classroom and informally through other contacts.[70] These experiences no doubt impressed upon him the need not just for more qualified teachers but for a more comprehensive program in mathematics instruction. Through the activity of the seminar, what had been for Neumann ad hoc procedures of learning could become more standardized and institutionalized. But having immersed himself in mathematics and the natural sciences and been drawn to creative and original problems in them, Neumann only reluctantly considered taking the state teaching examination in 1825 because he thought it "might no longer be possible to exchange pure scholarly pursuits for those of the school."[71] Even the civil service status accorded secondary school teaching in 1810 did not always make it an appealing career.

Although the state teaching examination did not require or encourage specialization in physics, one aspect of teaching physics in a secondary school did make it attractive to attend a seminar like Neumann's: physics instructors, like upper-level mathematics instructors, were *Oberlehrer*, teachers in the upper forms who had salary, professional position, and social status second only to those who had achieved the highest rank, professor, or become directors—a significant consideration for the professionally minded. When Neumann began his seminar he not only had to work out a manner of instruction suitable for his students and to define a mathematical physics appropriate for the classroom but also to relate instruction to the professional aspirations of his students and to enhance both the appeal and the definition of the *Physiklehrerberuf*.

The relation between teaching and research in the sciences has never been monolithic, as the debate over quantification in the mathematico-physical seminar demonstrates. Interaction occurs in various ways, from the relative absence of research concerns in service courses to the dominance of research over teaching in advanced study. But these two examples represent extremes. Historical circumstances usually fall at various points on the spectrum, points where there can be degrees of fruitful symbiosis and antagonistic exchanges. The debate at Königsberg demonstrates that educational reform at the university level was not always motivated or determined by research imperatives. Walls could easily be erected to prevent the institutionalization of certain techniques or ideas, as they were at Königsberg to prevent advanced mathematics from playing a role in the natural sciences seminar even though the scientists involved by and large supported quantification. A more critical view of the relation between teaching and research suggests, too, that it is not

---

70. LN, p. 85. On Neumann's mathematical education, see LN, pp. 18, 67, 71, 79, 80, 84–85, 89, 90, 94.

71. Neumann to Kultusminister Karl Altenstein, 2 January 1825, rpt. in LN, pp. 225–27, on p. 227.

even possible to view the mathematico-physical seminar as having been shaped by concerns of pure *Wissenschaft*, much as the terms of the debate reinforced that perception. Neumann may have been more open than Moser was to the interaction between teaching and research and to the limitations of classifying French mathematical physics as mathematics, but despite his early introduction of French methods into his lecture courses, he had not yet created a solid physics curriculum. Influential as Bessel had been on the evolution of his investigative strategy, it was not yet clear that Bessel's techniques—especially error analysis, which had proven so effective in uniting education and practice in the observatory— would or even could have the same impact on Neumann's teaching. The seminar offered Neumann the opportunity to find a way to teach physics as a mathematically exact science, but in 1834 he had only an ideal vision, not a well-formed plan.

CHAPTER THREE

Shaping a
Pedagogical Physics

BEFORE THE FIRST MEETING of the Königsberg seminar for mathematics and physics was held in the winter semester of 1834/35, Neumann and Jacobi had a good idea of how they wanted their institute to operate. Notions of an ideal student seem to have guided them in the construction of the seminar's statutes; for one of their central concerns was to delineate the advance preparation a student should have: little more, in fact, than what a student would ideally have learned in the gymnasium. Admission was to be based on an entrance examination, administered by Neumann and Jacobi, that tested familiarity with calculus and Ernst Gottfried Fischer's *Lehrbuch der mechanischen Naturlehre*, one of Prussia's most influential secondary school physics textbooks.[1] Neumann and Jacobi seemed therefore to be expecting primarily younger students, but they also wanted to attract older ones and even encouraged the participation of postgraduates. They also expected students to be highly motivated and willing to attend regularly, do the work for both divisions, and shape their own course of study; for neither Neumann nor Jacobi suggested a curriculum in the form of a sequence of mandatory lecture courses in either mathematics or physics. Students were to register for whatever courses they thought they needed. Generally, students would teach themselves in the seminar through the lectures they were to prepare and present to all members. Neumann and Jacobi did not specify the level of these lectures

1. Ernst Gottfried Fischer, *Lehrbuch der mechanischen Naturlehre*, 3d ed., 2 vols. (Berlin: Nauck, 1826–27). Although the second and subsequent editions of this textbook were intended for university students, the book was used in gymnasiums.

nor whether they were to be based on textbook material or journal articles, but they did view them as a means for students to engage in dialogue and to share critical assessments of their peers' work.

The principal activity of the seminar, however, was to be the preparation of an original investigation. Funds were available from the seminar's 150 taler budget to support original investigations, especially in physics, and to reward exceptional performance with seminar premiums. Students who reworked their investigations for submission to the *Journal für die reine und angewandte Mathematik*, *Astronomische Nachrichten*, and the *Annalen der Physik und Chemie* were to be eligible for an additional ten taler remuneration. Neumann and Jacobi thereby encouraged going beyond the narrow confines of the institute and participating in the larger community of scholarship. In this way the seminar was designed not only to bridge the gap between learning and scholarship but also to mediate between apprenticeship and professional activity.

As conveyed by its statutes, then, the seminar was to be less a teaching than a research institute, a conclusion confirmed by the projected roles of Jacobi and Neumann. Although they were to receive a small stipend from the seminar's budget for their contributions, their purpose was to advise, to guide, and to supervise but not necessarily to instruct, as was customary in lecture courses. Jacobi (Figure 5) planned to have Ludwig Adolf Sohncke, *Privatdozent* in mathematics, direct sessions in which students worked on exercises and delivered mathematical lectures; his own participation was to be limited to assigning homework ("small problems . . . solved by all . . . members") and suggesting themes for larger independent projects. Both types of assignments were to be reviewed by all students in special weekly meetings that presumably would function as Jacobi's exercise sessions had. Neumann (Figure 6) had not yet thought through the operation of his division in such detail and expected it would operate on the basis of student presentations and individual investigations in mathematical or measuring physics. He also did not perceive his institute as a self-contained unit, offering students all they needed to know in physics, because his students were to be allowed to cross register for Moser's division on experimental physics and Moser's students, to attend his division without registering for Jacobi's.[2]

In principle their expectations were reasonable. In Prussia's officially recommended secondary school curriculum, parts of the calculus were supposed to be taught in *Prima*, the upper form of the gymnasium, and Fischer's textbook was popular, especially in metropolitan Prussian gymnasiums. That the seminar requirements could indeed be satisfied also seemed to be assured by the recent transition in all gymnasiums from

2. "Vorläufige Statuten des mathematisch-physikalischen Seminars an der Königsberger Universität," ZStA-M, Acta betr. das mathematisch-physikalische Seminar an der Universität zu Königsberg, Rep. 76Va, Sekt. 11, Tit. X, Nr. 25, Bd. I: 1834–61, fols. 4–5. Sohncke's session is discussed in ibid., fol. 18. (The statutes are translated below, in App. 1.)

Figure 5. Carl Gustav Jacob Jacobi. From Leo Koenigsberger, *Carl Gustav Jacob Jacobi* (Leipzig: B. G. Teubner, 1904), courtesy of Library of Congress Photoduplication Services.

subject classes (in which students accelerated in certain subjects and abandoned others altogether) to year classes (in which every student took every subject in the curriculum for each form). Even their plans to engage students in original research were reasonable given that by 1834 both Neumann and Jacobi had supervised investigations by advanced students. Shortly before the seminar began, Neumann had permitted at least one advanced student, Carl Eduard Senff, who had not matriculated at Königsberg but who had come there in 1833/34 expressly to study

Figure 6. Franz Ernst Neumann. From *GW*, courtesy of Library of Congress Photoduplication Services.

physics, to conduct polarization experiments in a simple laboratory in
Neumann's home. Senff's objective was to determine the relation between
the plane of polarization and the optical axis of a crystal, and for that
purpose he drew upon the "theoretical observations" in Neumann's optical work as well as Neumann's measurements of polarization planes at
various temperatures. In line with the investigative protocol Neumann

had adopted by then, Senff repeated his measurements before ending his investigation, which gained for him an *außerordentlicher* position at Dorpat University.[3]

In practice, however, their expectations turned out to be optimistic and naive. Students interested in mathematics and physics—a minority at best—were expected to gravitate naturally toward the seminar and to have a well-defined reason for attending. *Innerer Beruf* might not have brought them there, but they were expected to know what to make of their training, even though the statutes were conspicuously silent on the professional objectives of the seminar exercises. Although the required exercise of lecturing related directly to teacher training, and the practice of debate and discussion shaped professional scholars, neither profession was mentioned in the statutes. Overall, the research-oriented exercises promoted qualifications for university teaching and independent scholarship, but few students could realistically plan for such careers.

When the seminar convened for the first time in November 1834, Neumann and Jacobi discovered just how unrealistic their plans were. Younger students were not at first attracted; six older students participated in both divisions: Ludwig Otto (Anton) Hesse, 23 years old, in his third year at the university; Julius Czwalina, 24, in his fifth year; Theodor Schönemann, 22, in his third year; Gustav Kade, 22, in his second year; Louis Eduard Carl Pahlen, 24, in his sixth year; and Carl Albert Busolt, 24, also in his sixth year. A seventh student, J. H. C. E. Schumann, was allowed to register for the physical division only, even though he was not a member of the natural sciences seminar, which had not yet opened. These students were career-oriented; Czwalina, for example, was preparing for the teaching examination.

Jacobi's mathematical exercises dealt with conic sections and spherical trigonometry but had little to do with his lecture course for that semester on the theory of partial differential equations. While not all the students' work was original, a good part was advanced. Jacobi reported that ten of the eighteen exercises or problems had dealt with "little known" theorems. Frank in evaluating performance, Jacobi held back student, Busolt, because his work was not up to par and reported that Pahlen's was also deficient. Although Sohncke's exercise session was designed to help students like Pahlen and Busolt, neither participated in it. That only three students—Hesse, Schönemann, and Kade—took part in Sohncke's session suggests that the activities of the regular divisions were time consuming enough. In Sohncke's sessions students discussed foundational, but difficult, topics: Moebius's treatment of compound fractions, Dirichlet's treatment of the convergence of series, Taylor series, and Lagrange's analytical mechanics. Originally required to deliver their presentations in Latin, participants in this session soon found that obscure points had to be

3. Carl Senff to Neumann, 19/31 January 1838, FNN 53.IIA: Briefe von Schülern.

discussed in German. Enthused by what they had gained from Sohncke's session and perhaps in greater need of guidance than Jacobi had at first anticipated, these three students agreed that in the following semester they would meet with Sohncke for four hours per week rather than the one presently allotted.

In contrast, Neumann reported that only one student, Hesse, completed an independent investigation, which was on the optical properties of thin films. Hesse was also the only student who chose a topic related to Neumann's lecture course on the theory of light from that semester, an indication that his success was the result of coordinating what he had learned in the course with what transpired in the seminar. To other students not yet ready to undertake independent investigations, Neumann assigned "chapters in physics," which they used as the basis of their seminar presentations. He thereby began to work more closely with his students than originally planned. In fact, so time consuming had the seminar become—he reported that he devoted his "entire effort" to it (student presentations alone took two hours per week)—that he canceled his participation as director of the mineralogical division of the natural sciences seminar, the first meeting of which was yet to be held, so he could direct his own seminar exercises twice a week.

Hence, from the start, Neumann was drawn more deeply into teaching than he had expected. The pedagogical task before him was not easily defined because the "chapters in physics" that he assigned were drawn neither from textbooks nor even from the kind of French mathematical physics he had already incorporated into his lecture courses. Instead the students' seminar presentations were based on exemplary and usually recent physical investigations featuring measuring techniques. Far from being ordinary experimental investigations, these had produced substantially more precise measurements through the improved analysis of errors or improved instruments. Bessel's investigation of the simple seconds pendulum inspired Czwalina's discussion of the motion of the simple pendulum in a vacuum and in air. Kade talked about Henry Cavendish's experiment on the attraction of lead balls and showed how the density of the earth could be computed from Cavendish's measurements. Busolt examined Pierre Simon Laplace's use of the barometer to measure height. Schönemann worked on hygrometry, especially as it had been done by Joseph Gay-Lussac, and he examined in particular the theory of the hygrometric scale, while Pahlen demonstrated how to calibrate and correct the scale on a thermometer, for which he probably consulted Bessel's work.

Although it appeared that Neumann based his physics pedagogy in the seminar on the standards set by Bessel's investigations, the students accomplished much less than their topics implied because all exercises were conducted at first as oral presentations. Each topic may have entailed knowing how to use measuring instruments in refined ways, but it is not at all clear that students actually deployed instruments or even whether they used them for demonstration purposes. Neumann and Jacobi reported using the seminar's funds to support physical investigations, but there is no evidence that all students had in fact gained practical experience with instruments. Further circumscribing the didactic merits of these oral presentations was their failure to involve students in quantitative determinations. None of the topics involved the simple determination of a physical constant; almost all involved the analysis of instruments and the production of data that could be used as the basis for functional relations. But only Schumann, who tried to ascertain the relation between the length of the connecting wire of a galvanic circuit and the decline in efficiency of the circuit, combined measurement with higher forms of mathematization. (Notably, Neumann pointed out that Schumann had chosen his own topic.) Moreover there was nothing about the physical exercises that required special coordination between the mathematical and physical divisions of the seminar as presumably would have been the case had Neumann taught primarily the French mathematical physics exemplified by Fourier and Fresnel.

To conclude that Neumann's division of the seminar engaged students in original research would be to misrepresent what was actually involved in these exercises. Neumann had not, in fact, planned to construct them. To the ministry, he expressed surprise about his students: he "regretted [their] lack of physical knowledge." Once he saw that original investigations were not immediately possible, he had to consider the pedagogical alternatives available. Not limitless, they were constrained by what he thought his students should understand about physics: not its concepts, but its technical procedures. Yet even though the set of experiments he had chosen for his students to review was distinguished by its precise measuring techniques, his students did not mindlessly report on or reproduce them. Instead, Neumann seems to have prompted his students to extract from them their essentials, thus prompting them to shape pedagogical exemplars or model experiments that encapsulated the fundamentals of experimental protocol and measuring techniques. These were not original investigations, as Neumann indicated by singling out only Hesse's work as fulfilling his expectations. At this time it was also unclear if these exercises were preparatory to undertaking more substantial original investigations or merely intended to acquaint students with basic investigative procedures. Whereas Jacobi had always used his exercises to train students for original research, Neumann's appeared at first to have been ends in themselves.

Neumann must have viewed these oral presentations as temporary; for he told the ministry that he hoped that "the progress of the seminar [would result] in an earnest study of physics." So much in need of instruction did he consider his students that he did not even find it worthwhile to give entrance examinations and so abandoned the practice outlined in the statutes. He must have been disappointed. Perhaps he was also frustrated at being drawn more deeply into teaching. After all, the students he had admitted into his division had been at the university for some years and presumably had already taken courses with him. Their lack of knowledge, then, reflected the inadequacies of his own teaching.

The disparity between vision and reality forced Neumann to cast about for reasons why his students were ill-prepared and did not always take the study of physics seriously. He found the principal reason not in scientific pedagogy but in the treatment physics received on the state's teaching examination. To enhance the curricular importance of physics, Neumann boldly told the ministry he "could not conceal from [the ministry] that it could contribute to [a more earnest study of physics] if it added to the teaching examination the requirement of a substantial written work in physics," as was already the practice in mathematics.[4] But whereas he blamed the state's teaching examination for his students' lack of interest, the ministry apparently suspected otherwise. In May 1835 it asked the Königsberg University curator about Neumann's lectures and, in response, learned that attendance was low and that although Neumann's lectures were "distinguished by great thoroughness and mathematical elements," Moser's "were more appropriate for the needs of students"; for they were "stimulated by great liveliness."[5] Rather than shaping a pedagogical physics suited to his students, Neumann expected them to adapt to him.

When the seminar opened for its second session in the summer semester of 1835, eight students registered for Jacobi's division. Six chose to participate in the physical division, to which Busolt also returned without taking part in Jacobi's exercises. Two of Neumann's students were new; one of them was Bessel's son, Karl Wilhelm Bessel. Jacobi offered lecture courses on variational calculus and second-order surfaces but designed his seminar exercises to be more elementary than his lectures, covering in the exercises algebraic problems and their geometric interpretation.

Neumann also changed the operation of his seminar. The oral presentations remained, but the time devoted to them was shortened to about an hour a week and their content changed radically. In contrast to the first semester, when the oral presentations were drawn from a variety of investigations, now they were closely aligned with Neumann's lecture course from the previous semester, on the theory of light, and they overlapped only slightly with his current lecture course, on crystallography. The earlier emphasis on learning measuring techniques through oral presentations now proved untenable. Students instead discussed

4. "Bericht über das mathematisch-physikalische Seminar zu Königsberg i. Pr. im Winter-semester 1834/35," n.d., ZStA-M, Math.-phys. Seminar Königsberg, fols. 15–18.
5. Königsberg University Curator Reusch to Kultusminister Karl Altenstein, 5 May 1835, ibid., fol. 14.

more general topics from optics: the structure of the human eye and how vision could be corrected; the derivation of formulas for reflection and refraction; paths of light rays and the shapes of shadows; and so on. Only Schumann, who had been in the seminar the previous semester, worked on a measuring exercise: weighing and the corrections that had to be made for temperature and air pressure. Although these exercises were less sophisticated, they had the advantage of being more coherent because they were linked to one another and to Neumann's lecture courses.

Neumann also added to the exercises a written assignment of the kind he had hoped would be included on the state's teaching examination. He made the completion of this assignment dependent upon an understanding of the material in his lecture courses by requiring that it "be related to earlier studies." Emerging from his efforts to draw students into the study of physics was the conviction that topics best followed one another in order of complexity. Schumann, for example, took up a set of closely related optical issues seriatim: he examined the path that light follows in a prism, including in double refraction; he determined indices of refraction, including the index of refraction of a liquid found by using a solid prism; discussed the achromatic prism; and then examined the thermal dilation of crystals. Schumann thus knitted together the material from two of Neumann's lecture courses, on crystallography and optics. He also returned to the procedures in weighing that he had discussed in his earlier oral presentation. Again, there is no direct evidence that all students routinely performed the procedures discussed in their essays. These "exercises" were mainly expository, aimed at the mental and mathematical review of physical processes and, to a lesser extent, experimental procedures. In the seminar's second semester, then, Neumann chose not to pursue the exact experimental physics that had been a central theme in the first semester but instead turned to simpler topics that permitted him to make instruction more systematic and more closely related to his lecture courses.[6]

Yet even as he turned in the winter semester of 1835/36 to simpler topics and exercises in order to accommodate his students, Neumann did not give up his vision of what his institute could ultimately be. In his report on the third semester, he again singled out Hesse, this time for his "experimental determinations" of the path of polarized light. Moreover, during that semester, Neumann had seen in Hesse's exemplary performance the opportunity to use an advanced student as an assistant in his own investigations. While actively working on several problems in optical theory in 1834 and 1835, especially on the optical properties of crystals, Neumann had had Hesse carry out several measurements on polarized

6. ["Bericht über die physikalische Abtheilung, 1835/36"], n.d., ibid., fols. 27–29; "Bericht über die mathematische Abtheilung . . . von Ostern 1835 bis Ostern 1836," 10 June 1836, ibid., fol. 30.

light.[7] These observations may in fact have been the ones to which Neumann now referred in his report, although he did not say. Even though Hesse's assistance must have buoyed Neumann's spirits about the potentialities of his institute, he could not have forgotten that other students were not yet ready to undertake such tasks.

From the beginning of his career, Jacobi had been more enthusiastic about bringing students to the point where they could undertake original research than about accommodating instruction to the needs of students. Writing to his brother, the physicist Moritz Jacobi, on 28 December 1832, he boasted, "I have reason to be pleased with my academic performance. I have recently submitted, with one of my own papers, three papers by my students." Proud that he had already supervised three doctoral students (among them two who became his assistants, Sohncke and Friedrich Julius Richelot), Jacobi continued, sounding much like Gauss: "This compensates to a certain extent for the fatigue of university lecturing, over which well-founded complaints are raised as long as the world exists and there are professors."[8] Once the seminar began, he strove, sometimes with too much effort and expectation, to increase the difficulty of mathematical instruction, believing this was the way to promote original research among students. That approach proved to have a deleterious impact upon the operation of his seminar.

Jacobi's contribution to the seminar was substantially less in the winter semester of 1835/36, when he decided to lecture on the difficult topic of elliptic transcendentals. Not only did he thereby bring his research into the classroom and require that students spend ten hours a week listening to him; he also found that he did not have enough time to hold seminar exercises and canceled most of them (although he reported that in the few he did hold, he had covered the principles of statics and the rotation of solid bodies). Despite the fact that he had once made his seminar exercises more elementary than his lectures, he continued to use his lecture courses as forums for consolidating recent advances in mathematics, teaching mathematics in ways that did not appear in contemporary textbooks, and thus promoted pure mathematics in Prussia. His pedagogical tactics, however, loosened the ties that were supposed to exist between the mathematical and physical divisions of the seminar. Several of his lecture courses were directly relevant to solving problems in physics (including variational calculus, partial differential equations, series, and analytical mechanics). Some of his research would also have been suitable for the physical division, such as his study of elliptic integrals, in which he achieved a complete mathematical solution of several physical problems,

7. "Ueber die optischen Eigenschaften der hemiprismatischen oder zwei- und eingliedrigen Krystalle" [1835], in *GW* 2:341–55, on 354.

8. Carl Jacobi to Moritz Jacobi, 28 December 1832, in *Briefwechsel zwischen C. G. J. Jacobi und M. H. Jacobi*, ed. W. Ahrens (Leipzig: B. G. Teubner, 1907), p. 8.

including the vibration of a pendulum and the force-free rotation of a rigid body.[9] But in the seminar, physical applications of mathematics were of secondary concern to him.

As Jacobi began to move away from the seminar to concentrate his pedagogical efforts on creating more specialized lecture courses, Neumann continued to base seminar exercises upon material already covered in his lecture courses from the previous semester. With five students (one of them new) registered for his seminar in the winter semester of 1835/36, Neumann offered a lecture course on Fourier's analytical theory of heat for the second time. Only one of the students' oral presentations derived from this course (it was on the thermal expansion of crystals); the rest covered problems from optics. With a firm understanding of the principles taught in the lecture course on the theory of light, students could return to the examination of simple instruments. Kade, for instance, showed how microscopes and telescopes could be constructed out of two lenses and how to use a microscope to determine the coefficient of refraction of a liquid. He determined experimentally the coefficients of refraction of several crystals. In contrast to the previous semester, when measuring operations had all but disappeared, the phrase "experimental determination" now occurred often in Neumann's report, indicating that students actually performed practical exercises with instruments. They worked more closely with experimental data. Using the background from the lecture course on the theory of heat, they also tried to determine a formula for daily changes in temperature from a series of thermometric observations. They sought to express other thermal properties mathematically as well and to work out the analytic problems they encountered in bringing theoretical calculations and experimental data into agreement. Topics in mechanics also appeared, including the principle of least action and aspects of elasticity theory. Schumann, who had worked on problems in elasticity, earned Neumann's praise for his study of small vibrations.[10]

After only three semesters, the seminar appeared to be relatively successful. Although attendance had dropped somewhat during the winter semester of 1835/36, to five students in each of the two divisions, Neumann and Jacobi had no reason to be overly concerned about contributions of the seminar to the study of mathematics and physics. The viability of Neumann's division was due, however, less to the execution of his program as outlined in the statutes than to the changes he had made to accommodate its participants. He had modified his original intentions to

9. L. Kronecker, "Verzeichniss der Vorlesungen, welche Jacobi an den Universitäten zu Berlin und Königsberg gehalten hat," in *C. G. J. Jacobi's Gesammelte Werke*, 7 vols., plus supple. (Berlin: Reimer, 1881–91), 7:409–11.

10. ["Bericht über die physikalische Abtheilung, 1835/36"], n.d., ZStA-M, Math.-phys. Seminar Königsberg, fols. 27–29; see also A. L. Crelle to Kultusministerium, 8 September 1836, ibid., fols. 23–24.

introduce students to instrument construction and measuring operations in order to teach students more systematically and to link the elements of his teaching—his lecture courses and seminar—more closely. Gradually he had learned to arrange material in a form suitable for his students. Through experience he had also learned that the best path to student research was through his lecture courses: Hesse's success showed him that. During the first semester, the students' seeming lack of preparation to study physics and lack of effort had been a sign not of their inability to undertake original research but of Neumann's unreasonable expections and his failure to assume responsibility for guiding them to the point where they could undertake original research. Still, his curricular program lacked a tight organization because each semester it was re-created in response to current circumstances. His pedagogical physics was just beginning to take shape. The principal source of his modifications was his deepening understanding of what his students could and could not do. Only to the extent that the seminar exercises were beginning by 1836 to employ what Neumann considered to be central techniques and methods of physics—measuring methods and the analysis and construction of instruments—was research the foundation for his teaching. His students began to cultivate an exact experimental physics similar to that exemplified by Bessel, but there was no indication of just how skillful they were supposed to become in the execution and analysis of measuring operations.

Just when Neumann had come to the point where he could adapt his teaching style to his students, circumstances changed dramatically. By February 1836 the ministry had taken note of what it perceived to be the more lively student participation in Jacobi's division in comparison with Neumann's. In time, though, it hoped Neumann would be able "to awaken a lively participation in physics," but it admitted there was some truth to Neumann's observation that in order to accomplish that, the state's teaching examination would have to be altered.[11] Also, in the early spring of 1836, funding for the seminar was withdrawn, a surprising decision given that most other Prussian universities had substantially more support for physical instruments or for science instruction in general: 500 taler for mathematico-physical apparatus at Berlin; 400 for the physical cabinet and 400 for the natural sciences seminar at Bonn; 348 for physical apparatus at Breslau; 520 for the physico-chemical laboratory at Halle; but only 60 taler for physical instruments at Greifswald and 158 for the same at Königsberg.[12] The university curator tried to regain the 150 taler-per-year budget by proposing that the 350 taler budget for the

11. Kultusministerium to Königsberg University Curator Reusch, 10 February 1836, ibid., fol. 19.

12. For the budgets of these university institutes see *Das Unterrichtswesen des preußischen Staats in seiner geschichtlichen Entwicklung*, ed. L. M. P. von Rönne, 2 vols. (Berlin: Veit, 1854), 2:430, 438, 444, 449–50, 456, 460.

natural sciences seminar, which had opened one year late in 1835, be transferred to Neumann and Jacobi because the directors of the natural sciences seminar had made no use of their funds, but his request was refused.[13]

Neumann appears to have been ambivalent about whether his recent pedagogical modifications would make a difference. He thought that still more had to be done. With his report for 1835/36 he sent a separate letter to the ministry, dated 17 June 1836. The contrast in the quality of students between the first and second years of the seminar led him to consider how a student came to study physics in the first place. He noted that the only students who had thus far distinguished themselves in the seminar, Kade and Schumann (in his haste, it seems, he neglected to mention Hesse), were those who had "inclination and talent for physical studies." Discouraged, he found that he could not

> conceal that little hope is on hand that it will come to be otherwise if the ministry by its resolution persists in not placing the requirements for physics on the teaching examination on the same level as those for other disciplines, especially mathematics. As long as the state examination board does not include an examiner in physics and does not demand written work for physics as it does in other subjects, the study of physics will more or less continue to be left to inclination and talent.[14]

For the second time in as many years, Neumann was complaining to the ministry about the teaching examination. That he expected the state to enhance the importance of physics to students so soon after the establishment of his seminar indicates the urgency he felt about problems in the classroom. Once again, his initial reaction to a pedagogical crisis was to blame factors extrinsic to instruction itself: first students and now the state teaching examination. Through the teaching examination, Neumann hoped motivation to study physics might be strengthened in two ways. The first was to identify a career option more clearly by defining the *Physiklehrerberuf*; students might then find reason to engage seriously in learning physics.

The second motivation, he seemed to suggest, was the sheer pressure of the examination. In recommending that a physicist serve on the examining board, he was in essence proposing himself, because Königsberg was a regional center for administering the examination. Once on the examining board, Neumann could become a gatekeeper controlling students' entry into the profession of science teaching. Students would be compelled to register for his seminar for purely bureaucratic and official reasons. One wonders, then, if his suggestions to the ministry were in-

---

13. Curator Reusch to Kultusministerium, 14 March 1836, ZStA-M, Math.-phys. Seminar Königsberg, fol. 20.
14. Neumann to Kultusministerium, 17 June 1836, ibid., fol. 26.

tended to enhance the study of physics or to increase his student enrollment, or both. Even though his hope of using the teaching examination to
establish a dedicated student clientele was a gamble, his perception that in
the absence of a strong relation between the seminar and the requirements of the teaching examination, both the enrollment in and the vitality
of his institute would decline was valid. But in his thinking on the shortcomings of the seminar, he failed to acknowledge that the changes in his
teaching were as yet only subtle and ad hoc ways of obviating problems.
He persisted in his conviction that the seminar should function at a
certain intellectual level, even though few students could meet his expectations. Despite what Neumann perceived as inadequacies in the promotion of physics instruction, the ministry, by late 1836 and early 1837, was
pleased with the visibility and encouragement Neumann and Jacobi gave
to the study of physics and mathematics.[15]

Yet problems continued to plague the seminar in the summer semester
of 1836, when student enrollment dwindled to three in each division:
Hesse, Johann Georg Rosenhain, and G. S. H. von Behr in the mathematical division; Hesse, Karl Otto Meyer, and Robert Hermann Heinrich
Hagen in the physical division. The state of the seminar had "suddenly
changed," Neumann and Jacobi noted in their joint report written in
Jacobi's hand. Because funding had been withdrawn, they had "reached
the conclusion that, in order not to allow the current restricted activity of
the seminar to become the norm, this seminar, like others at the university, requires support from the ministry as we had originally proposed at
the time the seminar was founded. This support is necessary partly to
stimulate greater diligence through small remunerations, and partly to
engage the students in experimentation and observation in outstanding
and successful ways." Jacobi and Neumann thereupon canceled the seminar for the winter semester of 1836/37 and until such time as funding
should be reinstated. The seminar remained open only provisionally
during the summer semester of 1836. They hoped that by placing pressure on the ministry to support an institute it had recently praised, they
would be able to reopen in the summer of 1837. Their tactic proved
fruitless. The seminar did not officially meet again until five semesters
later, in the summer of 1839, still without the funding that enhanced
instruction and stimulated student participation. In the interim Neumann held informal exercise sessions in physics.[16]

The loss of funding was only a small part of their reason for canceling
the seminar. They were also concerned about the declining enrollment
and, more important, the diminished quality of enrollees. "The outstanding members of the seminar left our university," Jacobi explained, "partly

15. Kultusminister Altenstein to Curator Reusch, 26 September 1836, ibid., fol. 31;
Johannes Schulze, "Promemoria," 15 February 1837, ibid., fols. 36–37.

16. ["Bericht über das mathematisch-physikalische Seminar, 1836/37"], 24 August
1837, ibid., fols. 45–46, on fol. 45.

to take up positions as teachers, partly to devote themselves to the initial work required for the teaching examination." Neumann and Jacobi had found that because students joined the seminar voluntarily, they could not in fact be required to complete assignments in it. They discovered in practice that students were stimulated to work seriously only in the presence of very good students who inspired by example.

By the summer semester of 1836, there was a sharp difference between the two divisions. In the mathematical division Jacobi offered exercises from Gauss and Lagrange in mapping curves and surfaces and other related problems. Two student exercises from his division were accomplished enough to receive the seminar's prizes: Hesse's on conic sections and second-order surfaces and Schumann's on elasticity. In contrast, Neumann, left with unexceptional students, had to adjust his teaching to fit their needs. With even more poorly prepared students than he had earlier had, there was a stronger pedagogical imperative for him to create simpler tasks for them. Neumann found that he now had to assign themes for both the oral and written assignments. For the oral presentations he created a hierarchy of exercises—he called it a "more integrated plan"—executed by students in common and designed to teach measuring techniques. Students were introduced to two measuring instruments, the thermometer and the barometer; they learned how to make absolute determinations of measurement; and finally, they ran through measuring exercises on specific gravity and thermal expansion. In a similar vein for the written assignments, Neumann worked up "small problems" related to the oral presentations but apparently so elementary he did not even mention them as he had in previous reports.[17]

This change in clientele placed Neumann in more of a quandary than the lack of funding did. Since the inauguration of his institute, matters had only become worse. Each year he had been forced to modify some aspect of the ideal vision embodied in the statutes. In the first semester, he had had to abandon the entrance examination and shape measuring exercises from exemplary investigations. In the second semester, he had had to introduce written assignments and tie his lecture courses and the seminar activities more closely together. He had also begun to guide students through exercises in order of complexity. In the third semester, the bond between the seminar and his lecture course had become even tighter, and for the first time there is evidence that all students did indeed perform measuring observations. But each student still performed a different set of exercises from his classmates. Now, in the fourth semester of the seminar, Neumann was forced again to rethink the function of seminar exercises. Given the quality of his students, he could have canceled the seminar altogether.

17. Ibid., fol. 45; see also Crelle to Kultusministerium, 9 November 1837, ibid., fols. 41–42, on fol. 41.

Instead, Neumann brought the exercises of the seminar to their most elementary level yet, thereby significantly shifting the original intent and purpose as expressed in the statutes. Now his seminar exercises more noticeably took on the character of a structured learning program, of *Ausbildung*; for he had shifted his pedagogical emphasis from the monitoring of individual performance on customized problems to the construction of common tasks executed by all students. Furthermore, he chose to develop his pedagogical program not so much by teaching a sequence of interrelated ideas as by teaching a hierarchy of techniques or skills essential, he thought, for doing physics. With his introduction of a written assignment back in the second semester, Neumann had begun to construct problem sets not then customarily found even in textbooks. Unlike the kind of student research he initially envisioned, which would expand the frontiers of physics, these reformulated seminar exercises cultivated soil that would have lain fallow. Their significance was therefore twofold. Their elementary nature clearly "lowered" the intellectual level of the seminar. At they same time, however, they were in principle accessible to a larger and more varied student audience, and hence they held out the possibility of recruiting more students to the study and, more important, practice of physics.

With the creation of exercises executed in common, what had previously been ad hoc changes in Neumann's teaching plan now became purposeful adjustments. In one sense, common exercises—problem sets as we would call them today—were like the state teaching examination in that both were agents that worked on students to keep them in the field. They were not like *innerer Beruf*, which originated in a student's predisposition or talents. The creation of these exercises signaled a fundamental pedagogical revolution. Rather than allowing students to participate as individuals, Neumann created the conditions for them to participate first as a group. Rather than remaining a passive director of what he hoped would be an institute for student research, Neumann assumed responsibility for constructing a syllabus of exercises for guiding methodically through the study of physics those for whom it did not come naturally. Lacking gifted or sufficiently well-prepared students, he made a more concerted effort to appeal to average students, to a less exceptional clientele. Jacobi, in contrast, took as a matter of course the minor adjustments he made in his teaching program; appears to have made these adjustments less often; and more frequently pursued topics of greater sophistication with more success than did Neumann.

The modifications Jacobi and especially Neumann made in their original teaching plans were not as yet a matter of concern to the ministry. It took note of the difference in levels of student enthusiasm in the two divisions but was generally admiring of the "scholarly value" of the students' work, which it interpreted as a sign of the "influential effectiveness of the seminar and the insightful guidance of its directors."[18] But the ministry's own internal evaluator of the seminar, the mathematician August Leopold Crelle, did not always view the seminar so positively. Having worked with the ministry since 1828 as an advisor on mathematical instruction and teaching, Crelle, whose identity remained unknown to both Neumann and Jacobi, was first asked to evaluate the seminar in August 1836. He submitted his first two reports on 8 September 1836 and 9 November 1837. Impressed by the quality of the assignments he reviewed, he was nevertheless guarded in his praise of the seminar in 1836; for in his opinion it was not possible to judge from the student assignments "whether and in how far students have comprehended their *Wissenschaft*, and to what extent they have mastered the whole." Even in his report for 1837, which reviewed the basic exercises in physics that Neumann found necessary to create in 1836, Crelle circumscribed his praise with caution: "To be sure, it is not possible to judge unconditionally from these assignments how far the students have overall penetrated their *Wissenschaft*." That alternating concern for breadth and depth showed that Crelle, who was looking at how well the seminar trained teachers, valued mastery of material over learning of techniques.[19]

Alone, Crelle's evaluations did not mean much. But their ambivalence was undoubtedly magnified in the context of renewed discussion of reform of university teaching in the late 1830s. By 1836 some members of the ministry were not only convinced that lecture courses had to be supplemented by classes that promoted dialogue between professors and students in order to encourage independent activity but also that such dialogue would strengthen students morally and stem any further political action on their part. Professors spoke of renewing the pedagogical practices of Pestalozzi and of reinstituting Socratic learning as ways of sustaining a discursive element in learning and of achieving independent activity among students. But, like Crelle, other professors proposing educational reform generally started from the premise that students had to master a subject before participating in discussion. Hence, seminars were viewed not only as a way for students to achieve the conceptual understanding and mental self-control necessary for "self-activity" but also as a way for educational authorities to control students (in more ways than one) by subjecting them to rigorous forms of learning. The empirical disciplines, especially the natural sciences, were persistently problematic in these plans because it was not clear what role dialogue could play in instruction. Moreover, much as discussion was valued, it was not clear

18. Ministerial evaluations of 10 February 1836, 26 September 1836, 15 February 1837, 14 December 1837, ibid., fols. 19, 31, 36, 47.
19. Kultusminister Altenstein to Crelle, 16 August 1836, ibid., fol. 22; Crelle to Kultusminister Altenstein, 8 September 1836, 9 November 1837, ibid., fols. 23, 42.

how disorder, and possibly political conflict, could be avoided if all enrolled students participated in it; such problems were much less likely to
arise in seminars, with their limited enrollment.[20]

Other measures to curb student political activity were then in force and
of consequence for the mathematico-physical seminar at Königsberg.
Although the ministry approved seminars through the 1830s, not until
1834 did it permit other kinds of student meetings outside lecture courses,
and then only with the assurance that the gatherings would be scholarly.
The July 1830 revolution in France did not noticeably affect politics at the
university, but in the 1830s there emerged student corporations—loosely
tied to lecture courses—which privately promoted the tricolors of German unification and strengthened the "Young Germany" movement.
Throughout the decade, a ministerial edict of 28 April 1823, which denied
seminar membership to students who had taken part in secret societies,
was still in effect; it was not rescinded until 1842. The Königsberg that had
been a seat of reform around 1810 was becoming intolerant of change in
the late 1830s. Suspicions arose about liberal leanings on the campus in
late 1837, when the medical faculty awarded honorary degrees to two
members of the liberal "Göttingen Seven," including the physicist Wilhelm Weber, recently exiled from Hanover. The crown prince and rector
of the university, Friedrich Wilhelm IV, refused to recognize the degrees,
prompting Neumann and Jacobi (but not Bessel) to assure him that
nothing political had been intended in awarding them.

Finally, student migrations—the privilege of moving freely from university to university—were restricted between 1833 and 1838. Neumann
and Jacobi blamed that restriction in part for the diminished enrollment
in the unofficial exercises they held after the meetings of the mathematico-physical seminar were suspended, and they even wrote to the
ministry about the problem in June 1838, when they pointed out that
most of their students had thus far come from East Prussia or adjoining
provinces. If the ministry were serious in its intent to promote the study of
mathematics and physics at Königsberg, they argued, then it would have
to offer financial support to students from more distant provinces. Although they mentioned that the modifications they had made in their
teaching plans were costly, they were less concerned with being reimbursed than with creating a stimulating learning environment, which in
their view required a diverse and enthusiastic student clientele.[21]

20. Erich Feldmann, *Der preußische Neuhumanismus: Studien zur Geschichte der Erziehung
und Erziehungswissenschaft im 19. Jahrhundert*, vol. 1 (Bonn: F. Cohen, 1930), pp. 149–61.
Feldmann's study is based on Prussian educational ministry archives.

21. Hans Prutz, *Die Königliche Albertus-Universität zu Königsberg i. Pr. im neunzehnten
Jahrhundert* (Königsberg: Hartung, 1894), pp. 98, 121, 126–29, 131–33, 137–38; Ferdinand
Falkson, *Die liberale Bewegung in Königsberg (1840–1848)* (Breslau: Schottlaender, 1888), pp.
12–13; Fritz Gause, *Die Geschichte der Stadt Königsberg in Preußen*, 2d ed., 3 vols. (Cologne:
Böhlau, 1972), 2:493; Dr. Stettiner, *Aus der Geschichte der Albertina (1544–1894)* (Königs-
berg: Hartung, 1894), p. 65; Neumann and Jacobi to Kultusministerium, 17 June 1838,
ZStA-M, Math.-phys. Seminar Königsberg, fol. 51.

Their student enrollment also suffered by comparison with the natural sciences seminar at Königsberg. When that seminar finally opened in the winter semester of 1835/36, more students—twelve—registered than had ever registered for either division of the mathematico-physical seminar. Representatives of the natural sciences, however, considered even this number insufficient for their institute. As at Bonn, not all students registered for all divisions. Moreover, several students had "inadequate theoretical [i.e., lecture-based] preparation," so coordinating the various divisions proved impossible. The directors consequently devoted their efforts to making each division work separately, even though here, too, they encountered difficulties because they did not have the apparatus needed to conduct exercises. In the physical division Moser taught four students, one each from medicine, philosophy, law, and mathematics (the last being G. S. H. von Behr). Not having time to go over many parts of physics, Moser covered primarily gravity and the laws of motion. In their written assignments students discussed free fall, the pendulum, and At-wood's machine; he found their oral presentations to be less than desirable. Moser's division appears to have been the strongest because the seminar's prize question for its first year was in physics: to perform galvanic experiments on a frog and to assess the results in light of the current understanding of galvanism.[22]

How differently the natural sciences seminar operated can be determined partly from the published version of its first-year report, which differed substantially in tone and emphasis from the handwritten version. Written by Meyer, the report defended the existence of the natural sciences seminar on the grounds that the cultivation of the natural sciences went hand and hand with the "great and sudden movements" then taking place in all aspects of life. Meyer argued that while it was true that the university had done much to promote the study of the natural sciences in the construction of institutes,

> only one thing almost remained unconsidered, the need to give young investigators in the natural sciences exercises in their own investigations. This was all the more strange because in most disciplines, academic courses ended with practical exercises. Here [in the natural sciences] where there was perhaps the greatest need, exercises were sometimes neglected entirely [or] sometimes out of necessity bound to theoretical instruction in lectures. Individual professors sought in their own circles to fulfill this need and at most universities chemical and zootomical exercises and zoological, botanical, and

22. "Gehorsamster Bericht über das naturwissenschaftliche Seminar zu Königsberg von Eröffnung desselben zur Michaelis 1835 bis Ostern 1836," n.d., ZStA-M, Acta betr. eines Seminars für die naturwissenschaftlichen Studien auf der Universität zu Königsberg, Rep. 76Va, Sekt. 11, Tit. X, Abt. X, Nr. 21 [1827–76], fols. 26–27; Ernst Meyer, *Erster Bericht über das naturwissenschaftliche Seminar bei der Universität zu Königsberg: Nebst einem Vorwort über den Zweck und Plan der Anstalt, einem Auszug aus dem vorläufigen Reglement derselben, und einer Preisaufgabe für die Theilnehmer* (Königsberg: Hartung, 1836), p. 19.

mineralogical field trips were arranged for students. But [these exercises] still lacked systematization, a thorough plan, and the methodical sequencing that was so necessary.

On the surface, Meyer expressed the instructional philosophy for which Neumann had been groping over these years. But as Meyer continued to elaborate upon these exercises, we find that he was not so much arguing that a student should learn disciplinary knowledge in structured ways but that exercises should lead them to "regulated self-activity." "Seminar students," he continued, "should not only become acquainted with instruments, they should also use them; [they should] not only handle topics in the natural sciences, but also learn to investigate. An exercise of this type is far from being mechanical. It involves hands, eyes, and soul together. Only before them does nature lift the veil of her secrets, which are drawn out with the spirited hand, the feeling eye, and the contemplative mind." In an overwritten and poetic style, Meyer conveyed an image of the activity of the seminar that was an ode to *Bildung* and to the neohumanist philosophy of the Prussian educational system. Because this report was a public document, one could argue that Meyer was posturing, but there is much to indicate that he meant what he said. He acknowledged the freedom that students (but not those training to be teachers) had in choosing their courses, investigations, and methods. In a way that seemed to assume the existence of something akin to *innerer Beruf*, he argued that students learned in a Socratic fashion. He also denied that his seminar existed primarily to train investigators in the natural sciences and insisted that the success of an institute, his own included, should be measured by its student enrollment.[23] His report, addressed to the ministry, thereby implicitly challenged the philosophy and operation of the mathematico-physical seminar.

Examination of the educational practices of the natural sciences seminar over the next three years demonstrates in other ways how different the two Königsberg institutes were. During the first four years, Moser, for example, like the other directors, catered to individual student desires, allowing students to choose their own issues for presentations and investigations. Likewise, the topics he chose for instruction formed no particular pattern and covered a variety of subjects including electricity, magnetism, galvanism, optics, statics, and thermomagnetism. In line with Meyer's thinking, Moser also disavowed "practical activity" as the seminar's principal purpose, emphasizing instead the vaguer "self-activity," which stood at the core of *Bildung*.[24] Like Neumann, Moser had problems

23. Meyer, *Erster Bericht über das naturwissenschaftliche Seminar*, pp. 7, 9.

24. ["Bericht über das naturwissenschaftliche Seminar, 1836/37"], 19 June 1837, ZStA-M, Naturw. Seminar Königsberg, fols. 46–49, on fol. 49; ["Bericht über das naturwissenschaftliche Seminar, 1837/38"], 14 July 1836, fols. 84–90, on fol. 90; Ernst Meyer, *Zweiter Bericht über das naturwissenschaftliche Seminar bei der Universität zu Königsberg: Nebst einer*

with inadequate student preparation, but he mentioned no compensating pedagogical adjustments. He persisted in allowing students to pursue separate problems rather than tasks in common, and he seems not to have been sensitive to the difference between teaching ideas and teaching the skills of investigation. Another weakness of his division—indeed of all divisions of the natural sciences seminar—was the absence of mathematical instruction and coordination with the state's teaching examination, in which mathematics figured prominently. Students of mathematics who attended the natural sciences seminar were most often thinking of becoming *Oberlehrer*, so they participated only in the physical division, which was "closer to their main endeavor" but still a hardship because they worried, "not without reason, about an extension of their years of study [in order to learn mathematics] and about the demands placed upon them by the mathematical part of the examination."[25]

In contrast to the seminar for mathematics and physics, the natural sciences seminar, formed to train teachers, had yet to show the fruits of its labor. Funds were never fully expended, students rarely participated in all sections, and a disparity existed between the requirements of the teaching examination and what was learned in the seminar. On this last point, the mathematico-physical seminar was only slightly better off. But whereas the mathematico-physical seminar included candidates for teaching among its student clientele, the natural sciences seminar catered to a more diverse group: teachers who had already taken the state examination and were returning to learn one or more of the natural sciences; theology, law, pharmacy, and medical students and others who were uneducated in one or more branches of the natural sciences; a few students who wanted to devote themselves to an academic life; and finally, two students who wished to become gymnasium teachers and intended to avail themselves of instruction in all divisions of the seminar. Both of them failed to complete their course of study. One gave up his seminar work; the other was reported to have a doubtful future because he had more courage than talent for teaching. Despite their mixed student clientele, the directors maintained that the natural sciences seminar best served the purpose of cultivating the study of the natural sciences, especially in the gymnasium.[26]

Students essentially agreed with them. The natural sciences seminar

vergleichenden Erklärung eines bisher noch ungedruckten Pflanzen-Glossars (Königsberg: Hartung, 1837), p. 24; Heinrich Rathke, *Dritter Bericht über das naturwissenschaftliche Seminar bei der Universität zu Königsberg nebst einer Abhandlung über den Bau und der Entwicklung des Venensystems der Wirbelthiere* (Königsberg: Hartung, 1838), p. 21; Ludwig Moser, *Vierter Bericht über das naturwissenschaftliche Seminar bei der Universität zu Königsberg: Nebst einer Abhandlung über die Entwicklung des Schädels der Wirbelthiere* (Königsberg: Hartung, 1839), p. 33.

25. ["Bericht über das naturwissenschaftliche Seminar, 1836/37"], 19 June 1837, ZStA-M, Naturw. Seminar Königsberg, fols. 46–49, on fol. 46.

26. Ludwig Moser, Ernst Meyer et al. to Kultusministerium, 15 January 1839, ZStA-M, Math.-phys. Seminar Königsberg, fols. 52–53.

continued to enroll more students than the mathematico-physical seminar: 14 in the summer semester of 1836; 9 in the winter of 1836/37 (8 in
the physical division alone); 13 in the summer of 1837; 9 in the winter of
1837/38; 8 in the summer of 1838; and 15 in the winter semester of
1838/39. Moser's section generally drew considerably fewer students than
the seminar as a whole. With the instructional apparatus they needed, the
directors believed they could have attracted even more students. Nevertheless, when in 1837/38 Moser mentioned that students worked with
galvanometers, barometers, and other simple instruments and when in
1838/39 students were allowed to use a Plössl microscope for their own
anatomico-morphological investigations, no substantial increase in enrollment occurred.[27]

The ministry praised the natural sciences seminar for its effectiveness in
1836 and gave explicit support to the average level of instruction offered,
arguing that students need not learn everything, only what they would
teach in the secondary school. For that reason, too, the ministry maintained that the directors need not place such emphasis on instruments,
when much could be accomplished with "few and simple means."[28] Moreover, the ministry appears to have been impressed by the seminar's enrollment. When it came time in late 1838 to reconsider reopening the mathematico-physical seminar, the ministry seems to have wanted the best of
both worlds: the higher enrollments of the natural sciences seminar and
the intellectual challenge of the seminar for mathematics and physics. It
was also interested in saving funds. So it proposed to the directors of both
seminars that once again they reconsider uniting their two institutes, and
once again the answer was a resounding no. Neumann and Jacobi simply
rejected its proposal, while the directors of the natural sciences seminar
reiterated their objection to students engaging in original observations
and independent work.[29]

By 1839, Neumann's perspective on seminar teaching had changed
considerably. When the seminar had last been held, the original concept
had been altered beyond recognition. Although little is known about how
he interacted with his students in the unofficial exercise sessions from
1836 to 1839, we do know that during that time he set up a geothermal
measuring station and on 18 August 1837 began to record thermometric
and barometric observations. According to Bessel, Neumann used a thermometer at a depth of 24 feet to obtain what Bessel called "unexpected
practical knowledge" about geotemperatures. Part of the motivation for
undertaking these readings came from East Prussian officials who needed

27. ZStA-M, Naturw. Seminar Königsberg, fols. 46–49, 90, 109, 157.
28. Curator Reusch to Kultusminister Altenstein, 14 March 1836, ibid., fol. 24; Altenstein to Reusch, 23 August 1836, ibid., fol. 31.
29. Moser, Meyer et al. to Kultusministerium, 15 January 1839, ZStA-M, Math.-phys.
Seminar Königsberg, fols. 52–53; Jacobi and Neumann to Kultusministerium, 6 March
1839, ibid., fol. 54.

them for monitoring the province's agricultural performance during a period of unexpected population growth. The physics and mathematics teacher J. A. Müttrich, who had studied with Bessel and who now worked at the Altstädtisches Gymnasium in Königsberg, had kept such records for the province before Neumann; he and Neumann were close friends. But the nature of Neumann's readings indicates that he was also involved in a much more complicated project, one stemming from Fourier's suggestive remarks on geotemperatures and focusing on the theoretical analysis of thermometers and his data; for he carefully recorded data from both dry and wet thermometers. He took temperatures two or three times a day for about a year; thereafter, his students took readings as part of their work in the seminar, although it is not clear how continuously they kept records.[30]

Thus after he canceled his seminar, Neumann created an experimental laboratory of sorts, a geothermal station, in which students could study a simple instrument, take precision measurements, analyze data, and compare theory and observation, much as Bessel's students did in the observatory. In that sense, Neumann's geothermal station of the late 1830s was similar to the geomagnetic station Gauss and Weber had already created at Göttingen, its scientific and educational functions also modeled on the observatory. Both stations provided an opportunity to explore the empirical foundations of French mathematical physics more deeply—Fourier in the case of Neumann and theories of electricity and magnetism in the case of Weber. Each station viewed the problem of measurement differently, however: at Göttingen, largely in terms of precise measurement through improved instruments; at Königsberg, in terms of data analysis. Also, whereas Gauss and Weber published extensively, Neumann did not and so did not exploit to its fullest the scientific and educational advantages of the station. Although the geothermal project played an important role in physics instruction at Königsberg, its disciplinary significance was less than its complement at Göttingen.[31]

During the years the seminar was suspended, Neumann created problems that his students could execute in common, a daunting and time-consuming enterprise. In judging what students could handle, he undoubtedly considered what he taught them in lectures and what they had learned elsewhere; for in the seminar he had encountered varying backgrounds in physics. Most of his students had attended the gymnasium

30. F. W. Bessel to Paul Erman, 18 August 1837, Darms. Samml. Sig. J 1844; FNN 61.1: Meteorologische Beobachtungen.

31. The observations of the Göttingen geomagnetic station are published in *Resultate aus den Beobachtungen des magnetischen Vereins im Jahre 1836–41*, 6 vols., ed. C. F. Gauss and W. Weber (Göttingen: Dieterich, 1837–38; Leipzig: Weidmann, 1839–43). On geomagnetic studies at Göttingen see Christa Jungnickel and Russell McCormmach, *Intellectual Mastery of Nature: Theoretical Physics from Ohm to Einstein*, 2 vols. (Chicago: University of Chicago Press, 1986), 1:63–77.

continuously; all were graduates of the last form, *Prima*. Among the sciences taught in the gymnasium, physics had the strongest curricular identity, but its proportional share of classroom time, like that of other sciences, was small: two hours per week in *Prima* and one or two hours in *Secunda*. Pedagogical trends in teaching physics were generally in Neumann's favor because physical cabinets were expanding, active participation through discussion and homework problems was more in evidence, and measuring exercises performed by students were noticeable by the late 1830s. None of these trends dominated physics instruction as yet, but they could be found in the several Königsberg secondary schools most of Neumann's students had attended.[32]

One of Neumann's best students, Ludwig Otto Hesse, had attended the Altstädtisches Gymnasium in Königsberg between 1830 and 1832. In his last two years in *Prima*, Hesse took classes in mathematical geography, acoustics, and mechanics—all as a part of his physics requirement while his fellow students in *Secunda* covered experimental physics and chemistry.[33] The Friedrichs Collegium generally had the most progressive science classes in Königsberg. They were broadly defined—students could take classes in zoology and anthropology in *Prima*, if they wished—and by teaching students the fundamentals of observation in the botany and natural history classes of the lower forms, they prepared students for more intensive involvement in science learning later on.[34] When Bessel's son was there, collegium students in *Secunda* and in *Prima* worked from Fischer's *Lehrbuch*, which was more common at Prussia's various *Realschulen* than at gymnasiums.[35] But at the Kneiphöfisches Gymnasium, where the Königsberg physicist Ludwig Moser held an auxiliary position, a spectrum of topics from experimental physics—from demonstrations with electrical machines to atmospheric physics and climatology—was offered to both upper forms. Moser made no attempt to construct a systematic syllabus, even though his predecessor had covered major topics in statics and dynamics in a graduated fashion.[36] The experiences of Hesse and Bessel's son suggest that there was a connection between the quality of secondary school physics instruction and the decision to study it at the university level. Hence it made sense for Neumann to argue for better-trained *Physiklehrer* so that university physics instruction would benefit.

The differences that characterized physics instruction at Königsberg

32. Kathryn M. Olesko, "Physics Instruction in Prussian Secondary Schools before 1859," *Osiris* 5 (1989): 94–120. On Prussian secondary school curricula see Max Nath, *Lehrpläne und Prüfungs-Ordnungen im höheren Schulwesen Preußens seit Einführung des Abiturienten-Examens* (Berlin: W. Pormetter, 1900).

33. *Programme*, Altstädtisches Gymnasium zu Königsberg, 1830–32.

34. *Programme*, Friedrichs Collegium zu Königsberg, 1831–37.

35. As at the Petri-Schule in Danzig, where an exceptional program in physics instruction included laboratory exercises; *Programme*, Petri-Schule, Danzig, 1826–32.

36. *Programm*, Kneiphöfisches Gymnasium zu Königsberg, 1832/33.

secondary schools were repeated throughout Prussia; sometimes physics courses were overloaded with a wide variety of seemingly unrelated topics; other times an astute instructor offered his students a glimpse into controversies that plagued the field or gave them an opportunity to try their hand at the operation of basic measuring instruments.[37] Local variations were to be expected, but the wide variety of practices was also an indication that physics was unsettled pedagogically. Teaching a course in physics involved more than choosing topics: it also involved style and even ideology. That the study of physics, like the study of mathematics, could contribute to *Bildung* by training the mind was a common justification for secondary school instruction in physics, but it did not necessarily mean that physics was taught from a mathematical perspective. Often it meant quite the opposite; for a neohumanist physics was most frequently a philosophical physics that began with the metaphysical characteristics of matter as a foundation for understanding physical properties such as extension, cohesion, and porosity. Some textbooks then moved on to the theory of imponderables, which were considered the foundation for understanding the experimental areas of physics: light, heat, electricity, and magnetism. Textbooks that incorporated this approach were the major competitors to Fischer's, which shunned metaphysical considerations and hypothetical constructions, as did Neumann in his teaching. Residues of this philosophical approach to physics remained until midcentury, tending to detract from the exact nature of physics as Neumann and others saw it.[38]

Neumann maintained close contact with physics instructors in Königsberg. Yet there is no indication that he enlisted their help in preparing students to study in his seminar. In the end, he did not try to solve the problem of student preparation by suggesting changes in the state-regulated gymnasium curriculum or in the *Abitur*, the state examination taken by all students in *Prima* who expected to go on to university study and which did not include a written examination in physics but did require students to know the "principal laws of nature" without using calculus.[39] Instead, Neumann gradually accepted the task as his own. The changes he made in the seminar in the mid-1830s enabled him to attract to the study of physics not only those who had "inclination and talent" but also those who entered the seminar inadequately prepared.

The continued reform of his physics pedagogy weighed heavily on

37. See, e.g., the descriptions of the physics courses offered at the Royal Gymnasium in Marienwerder and the St. Johannes Schule in Danzig: *Programm*, Königliches Gymnasium zu Marienwerder, 1831/32; *Programm*, St. Johannes Schule zu Danzig, 1837/38.

38. See, e.g., Rudolf Arndt, *Der Anschauungsunterricht in der Naturlehre* (Leipzig: L. Voss, 1869), and Helmut Müller, *Die geschichtliche Entwicklung des naturwissenschaftlichen Weltbildes in den Schulfächern des preußischen Gymnasiums* (Würzburg: Mayr, 1934).

39. Quoted in Franz Pahl, *Geschichte des naturwissenschaftlichen und mathematischen Unterrichts* (Leipzig: Quelle & Meyer, 1913), p. 282. Students who did not pass the *Abitur* could still enter the university but could not take the state's teaching examination.

Neumann's mind even during the years when the seminar was not officially held. Changes in his lecture courses proved even more crucial to the survival of his institute than the alterations he had already introduced. His older lecture courses remained unchanged. They included courses on mineralogy, crystallography, the theory of heat, optical theory, and topics in mathematical physics. He viewed his courses on mineralogy and crystallography as essential components of his physics curriculum. For him, these were subjects "associated with the investigation of nature, not its history" and hence they did not focus on "individualities," "special natures," and "particulars" as did courses on natural history.[40] These courses were especially useful in reinforcing his phenomenological approach to physical phenomena. Following the broad outlines of his own investigations, in class he did not introduce unnecessary hypotheses, especially about the constitution of matter. So in order to discuss concrete examples of certain physical processes, Neumann drew upon their manifestations in minerals and crystals. The mathematical methods of Fourier and Fresnel, with Neumann's modifications, proved useful in discussing the thermal and optical properties of crystals and minerals, especially the complex phenomena of diffraction, polarization, and the appearance of colors in crystals. As early as 1835 he ended his course on optical theory by applying known analytic equations for light to reflection and refraction in crystalline media, thereby extending the scope of the mathematical theory of light to phenomena it had not previously treated. Always the enrollments in his courses were low: two to eight students in his two-hour private lecture courses and two to ten students in his four-hour public lecture courses.[41]

Only belatedly did Jacobi's teaching even offer the possibility of complementing Neumann's. During this hiatus in the seminar, Jacobi first offered a course on number theory, which did not relate at all to either Neumann's unofficial exercise sessions or his lecture courses. A lecture course on variational calculus—relevant to the treatment of certain physical problems—was canceled in the summer semester of 1837. Jacobi offered it again, one semester later, along with a course on analytical mechanics.[42] Then, in 1838/39, the curriculum in mathematics and physics offered by Neumann and Jacobi changed dramatically. During the summer semester of 1838, Jacobi turned to the foundations of analytic geometry and taught it as a fundamental course. In the winter semester of 1838/39 he taught the theory of surfaces, but he also offered a course on the application of differential calculus to series. Techniques

40. FNN 40: Mineralogie: F.N.'s Ausarbeitungen von Einleitungen zu seinen Vorlesungen.
41. FNN 21: Theorie des Lichtes. Wahrscheinl. Vorlesung von F.N. vom W[intersemester] 1834/35; FNN 24: Theorie des Lichtes nach Fresnel.
42. On Neumann's lecture courses, see below, App. 2; on Jacobi's, see Kronecker, "Verzeichniss der Vorlesungen."

learned in this lecture course could have been deployed with profit by students studying the theory of heat, elasticity theory, and other issues with Neumann.

During the winter semester of 1838/39, Neumann also offered new courses, the first time he had done so in five years. Publicly he offered a course on capillary theory, which drew eight students. Privately he offered a course he called "theoretical physics," which covered mechanics in a particular way and drew ten students.[43] In several respects these two courses differed substantially from others he had offered. His previous physics courses (with the exception of the ones on "general physics" and "experimental physics") were drawn from his own research interests; the new ones had no direct bearing on the subject matter of his research. Why he chose to teach them at that time deserves further scrutiny.

All along, Neumann had struggled, slowly making pedagogical adjustments so his institute could operate. Most of the time he concentrated on teaching the techniques of physics, but the content of his seminar exercises generally had little to do with his lecture courses except when students made connections between them, and then the connection was conceptual, not technical. What seminar students could draw from Neumann's lecture courses was actually very limited in scope because until 1838/39 he had offered only two courses in physics proper: the theory of light and the theory of heat. Despite his problems in teaching the seminar, Neumann showed little interest in expanding the *content* of his instruction. It was Moser who had taught magnetism, electricity, galvanism, and other related topics; Jacobi, who had taught parts of analytical mechanics and topics concerning the mathematical treatment of motion. But now, after several troublesome semesters in the seminar, Neumann decided to expand his curriculum.

His decision to teach mechanics as an introduction to the theoretical physics probably did not derive from his own educational experiences; for if it had, Neumann presumably would not have waited so long to try it out. The fact that it took him so much time to create his pedagogical physics, both in his seminar and in his lecture series, indicates that he did not borrow it from a canonical format for physics instruction, if one even existed at the time. Since 1834 he had been drawn more deeply into teaching, dealing in the best way he could with inadequate student preparation. All the modifications he made in the seminar up to 1838/39 were reactions in response to student needs, formulated piecemeal. But his modifications gradually fell into a pattern: in the seminar he had operated on the principle that mastery of subject matter was not essential for

---

43. See below, App. 2. Neumann's course on theoretical physics was announced in *Index Lectionum in Academia Albertina 1838* (Königsberg: Hartung, 1838), p. 9. In the same lecture catalog, the mathematico-physical seminar is listed with all other seminars even though Neumann and Jacobi were then holding only informal exercises.

performing the simple practical exercises he assigned. So without abandoning his initial plan to teach measuring operations, he probably came to realize that the best foundation for learning physics was to become acquainted with the techniques of physics and its instruments, most of which were then constructed according to mechanical principles. He probably reasoned further that if a student understood more clearly the principles by which instruments operated, they would be able to work more effectively, and creatively, with them. Neumann's pedagogical modification was thus not merely to offer mechanics but to create a course that was appropriate for learning the investigative techniques of a measuring physics, or what he came to call "practical physics."

His teaching program was thus shaped not only from above, through the pedagogical transformation of the techniques and results of research, but also from below, by considering what a student needed to study physics in the first place. Neumann's creation of this course and its accompanying exercises illustrates a particular moment in the evolution of science curricula when there was a pedagogical imperative to meet the needs of students. His course on mechanics as an introduction to theoretical physics was his most substantial and innovative pedagogical modification to date; for it not only created the foundation for a physics curriculum but also shaped a particular approach to the study and practice of theoretical physics. It also constituted a major turning point in the history of his seminar because, once the physical division of the seminar reopened permanently one semester later, in the summer semester of 1839, its subsequent stability was due to Neumann's decision to offer mechanics as an introductory course for beginning students.

The history of the physical division of the seminar revolves around this "theoretical physics" course and its related exercises. Together they created a solid physics curriculum by providing the material for an introductory course; by bringing mechanics squarely within the boundaries of a pedagogical physics; and by making physics instruction a matter of *Ausbildung*, or systematic training. They prepared students for the critical measuring and mathematical techniques that Neumann had begun to emphasize in other seminar exercises, thus bridging education and practice. Finally, they offered the possibility of attracting to the seminar not only students drawn to study physics by personal vocation or calling but also those who were inadequately prepared but trainable.

But for the time being, the introduction of this course was just one more pedagogical modification like all others. Neumann could not have foretold the pivotal role his course on theoretical physics would play in sustaining the operation of his seminar. Moreover, whatever plans he may have had to refine his pedagogical physics still further were put aside on 29 December 1838, when his wife of almost eight years died suddenly, leaving him with five small children, ages five months to six years. He had described his life up to then as family centered, simple, and involved with only a small circle of friends. Now the peaceful tranquillity of his home, and of his career, was shattered. Five days after the death of his wife, he wrote to Christian Weiss that he now knew "that it is not knowledge and scholarship which binds us, but love and devotion."[44]

44. Neumann to Christian Weiss, 3 January 1839, rpt. in LN, p. 337.

CHAPTER FOUR

Mechanics and the Besselian Experiment

In THE YEARS BETWEEN the founding of the seminar in 1834 and the appearance of his course on mechanics in 1838/39, Neumann focused his research efforts more decisively on optics, in particular on the optical properties of crystals. In 1835 he wrote to Johann Christian Poggendorff about his work on the optical properties of hemiprismatic crystals. At first Neumann was interested in determining the optical axes of these crystals and in measuring the physical and optical changes, especially in dispersion phenomena, that occurred when the crystals were heated.[1] Later, however, he turned his attention to deriving analytic expressions for optical intensity and to measuring optical intensity experimentally. As he explained in a major study presented to the Academy of Sciences in Berlin in December 1835, the high "degree of perfection" that had been attained in explaining, theoretically from dynamical principles, the direction of reflected and refracted light for isotropic media was not matched by equal success in explaining optical intensity, an area in which accurate experiments were difficult to achieve. Furthermore, he pointed out, the equations for the direction and the intensity of light in crystalline media were still undeveloped. He was concerned in particular about an empirical condition: the passage of light from one crystalline medium to another and the boundary conditions that had to be satisfied at the interface.[2] After studying Augustin Jean Fresnel's wave theory of light closely, and finding again that Fresnel's laws could not be extended to crystalline

1. "Ueber die optischen Eigenschaften der hemiprismatischen oder zwei- und eingliedrigen Krystalle" [1835], in GW 2:341–55.
2. "Ueber den Einfluss der Krystallflächen bei der Reflexion des Lichtes und über die Intensität des gewöhnlichen und ungewöhnlichen Strahles" [1835; publ. 1837], in GW 2:359–550.

substances without modification, Neumann believed he could bring his expertise in crystallography to bear upon these problems, thereby not only extending the mathematization of optics but also creating an optical theory based on dynamical principles which would have far greater scope than earlier theories that had ignored crystalline substances.

In Neumann's view the gaps in optical theory were considerable. Still wanting were laws concerning polarization angles and the planes of polarization in crystals, the properties of crystalline reflecting surfaces, and the behavior of polarized light produced by crystalline surfaces. To develop a theory that embraced these phenomena (Fresnel's laws, Neumann concluded, could not be applied in their present form to crystalline substances), he made his departures from Fresnel's assumptions more explicit, and in particular argued that it was the elasticity of the optical ether, not its density as Fresnel had assumed, that changed. Altering Fresnel's assumptions was once again necessary, Neumann argued, to achieve continuity in the mathematical equations describing the passage of light from one medium to another. Although he wanted to construct a more comprehensive optical *theory*, he did not ignore experimental considerations in his investigation. But he seems to have had some misgivings about his own trials (his seminar student Ludwig Otto Hesse had assisted him) because in comparing theoretically calculated to experimentally observed values, he relied not on his own data but rather on data on polarization angles taken by the Berlin physicist August Seebeck, who had not yet published his own findings. Neumann took Seebeck's data at face value, neither correcting nor reducing it. He marveled at the agreement between theory and experiment that Seebeck had achieved, considering it a sign not only of the "correctness of the theory [i.e., Neumann's own formulas]" but also of the "great skill of the observer."[3]

Over the next two years Neumann developed experimental methods for determining "the intensity of the reflected, refracted, ordinary, and extraordinary rays with great precision." He described the principle of his photometric method as one in which "the rays, whose intensity we wish to determine, act upon one another" in such a way that either (1) the azimuth of the plane of polarization can be measured, from which can be determined the relative intensity of the two rays, or (2) the two rays can interfere until both disappeared, and then the ratio of their intensities can be deduced since "the proportion in which one ray is diminished comparatively to another is known."[4] In his continuing study of photometry he relied on Seebeck's data and found again that "the numerical results, which I have derived out of my formula, agree so completely with See-

---

3. Ibid., pp. 363–64, 394, 403.

4. "Photometrisches Verfahren, die Intensität der ordentlichen und ausserordentlichen Strahlen, sowie die des reflectirten Lichtes zu bestimmen" [1837], in *GW* 2:575–85, on 577, 578.

beck's observations on complete polarization" that it was impossible to "doubt the reliability of the basic principles" he had developed. He reported in 1837 that he had conducted experimental trials of his own in order to measure the intensity of light reflected and refracted from a crystalline surface, including the intensities of the ordinary and extraordinary rays. But whereas in his earlier investigation he seemed satisfied that a reasonably good agreement (which he did not quantify) between theory and experiment confirmed his theory, now he proceeded more cautiously, questioning in what sense a theory was "actually proven through observations" and admitting that a theory was confirmed only within the limited range within which his data were reliable.[5]

The major alteration in Neumann's investigative techniques between 1830 and 1834 had been his adaption of Bessel's style in experiment, especially its consideration of constant and accidental errors. In his optical investigations an even more complex interaction between the mathematical expressions of theory and the numerical data of experiment was emerging by the late 1830s and the early 1840s. Although Neumann's work in optics remained dominated by theoretical considerations, it now began to incorporate more frequently the concerns for exact experiment, including the analysis of instruments and protocol, which he displayed in his work on specific heats. An 1841 study of the effect of thermally induced deformations of uncrystalline media upon double refraction and dispersion included a theory of the bimetallic thermometer. For Neumann these deformations carried both theoretical and practical interest because when Siméon-Denis Poisson had considered the problem, he assumed that the thermal expansion was proportional to the temperature, which Neumann found to be inaccurate. Neumann therefore examined the errors stemming from the nonuniform distribution of temperature. The agreement between theory and experiment he achieved was, he claimed, "as good as can be expected from the inaccurately known coefficients of internal and external conductivity." Precise results were thus dependent upon "the verification of equations on which the motion of heat depends, especially in poorly conducting media," without which measurements of optical angles could not be properly understood or interpreted.[6]

In his optical studies Neumann also brought his analysis of errors to bear more closely on theoretical matters. Earlier he had deployed the method of least squares as Bessel had: in an experimental context to determine the range of error in data and to combine observations. Now he began to use least squares for more complex tasks. He calculated

5. "Beobachtungen über den Einfluss der Krystallflächen auf das reflectirte Licht, und über die Intensität des ordentlichen und ausserordentlichen Strahls" [1837], in *GW* 2:591–618, on 594, 599, 615, 617.

6. "Die Gesetze der Doppelbrechung des Lichts in comprimirten oder ungleichförmig erwärmten unkrystallinischen Körpern" [1841], in *GW* 3:1–256; on 17, 19. Cf. ibid.

accidental errors, "the limits of probable uncertainty in data," to determine the limits within which theory itself was reliable.[7] His heightened concern for errors in data made him aware of just how difficult it was to confirm theories with certainty. Attaining accuracy in observational data was not sufficient, he believed, for ascribing *complete* certainty to the formulas derived from numerical results. In his 1832 investigation of the elliptical polarization of light on metal surfaces, for instance, he remarked that although there seemed to be enough reason to consider Brewster's formula for retardation produced by reflection to be "more than an interpolation formula," only more exact observations could confirm that it indeed was. He eventually offered a correction for Brewster's formula in 1837.[8]

"Interpolation formulas" were thus for Neumann merely empirical formulas that provided raw expressions of mathematical relations among numerical results. In his view, an "interpolation formula" was clearly not a law, nor was it necessarily a transitional stage in the construction of one, although it could be. Instead we find Neumann here, as in other places in his teaching and research, preferring to construct mathematical laws on a theoretical foundation. He did not always succeed, however, and so continued to work cautiously with interpolation formulas. But there were problems. As he improved his skills in detecting and computing both accidental and constant errors, the gap between the mathematical expression of a law and the numerical data of an experiment widened. The ability to determine error more rigorously, especially by application of the method of least squares, meant that it became more difficult to construct or confirm a theory because of the imperative to qualify laws either by correction factors determined through error analysis or by specifying exactly the range within which empirical data were precise and reliable. It is significant that once Neumann began to relate theory and experiment in more sophisticated ways, he mentioned less often his numerical data and the extent to which a theoretically derived result also rested on patterns in empirical data.

At the end of 1838, after his wife's death, Neumann set aside his research agenda in optics and his continuing work on specific heats. Even though by the fall of 1839 he had found someone to help him care for his five children, he was still unsettled. "Concerning myself," he explained to Christian Weiss, "I am not yet in the state of mind to have a scientific thought or to hold and follow one through. I have taken up and worked through many earlier investigations, but it is going slowly and disjointedly."[9]

7. "Ueber die Hornblende und den Augit" [unpubl. ms. c. 1840], in *GW* 1:374–76, on 374.
8. "Theorie der elliptischen Polarisation des Lichtes, welche durch Reflexion von Metallflächen erzeugt wird" [pt. 1, 1832; pt. 2, 1837], in *GW* 2:199–250, on 209, 228–29.
9. Neumann to Christian Weiss, n.d. [fall 1839], rpt. in LN, p. 343.

Neumann may have felt unsettled, but he could only have been encouraged by the reopening of the seminar in the summer semester of 1839. Jacobi, who was then on leave, appointed his former student Friedrich Julius Richelot interim director of the mathematical division. Richelot, who had studied with Bessel and Jacobi from 1825 to 1831, received his doctorate and became *Privatdozent* in 1831, *außerordentlicher* professor the next year. Devoted to teaching, he worked easily with students, preparing them systematically for learning more difficult mathematics. Foreseeing a more permanent role for Richelot in the seminar, Neumann and Jacobi included a small stipend for him in their requests for renewed financial support. In May 1839, Königsberg's university curator Reusch pressed the ministry again for funding for the seminar, reporting that it had drawn foreign students to Königsberg and had raised the level of mathematical instruction. Funds for the seminar had already been reappropriated in the budget for 1840–42, but Reusch requested that the 350 taler budget—200 taler over the original appropriation—be made available sooner because Neumann and Jacobi had held exercises the past few years without any funding at all. In November, Neumann and Jacobi reported no further need for Sohncke's position and the special exercise session and requested just 250 taler for 1839: 50 for Richelot and 200 for instruments and other expenditures. Also in November, Reusch repeated his request for 350 taler, arguing that because the directors of other seminars were reimbursed for their efforts, so should Neumann, Jacobi, and Richelot be, at 50 taler each; the remaining 200 taler was for instruments. Finally, on 27 January 1840, the ministry, on Reusch's recommendation, awarded the seminar 350 taler retroactively but changed the budget slightly, giving Neumann 50 taler, Jacobi 30, Richelot 20, and appropriating 250 taler for instruments, student premiums, and other small items.[10]

In December 1839, five years after Neumann first expressed his concern about the state's teaching examination and his conviction that it could be used to encourage a more serious study of physics, the ministry revised the examination, enhancing the role of the natural sciences on it by adding a natural scientist to each of the state's examining boards and by more clearly specifying the natural science requirements. A candidate who wished to teach physics in the upper forms had to demonstrate not only a broad acquaintance with physics and "a close acquaintance with the construction and use of the most simple and common physical instruments" but also "a more comprehensive and penetrating knowledge of all parts of physics, including modern discoveries and tools, and of the most

10. Königsberg University Curator Reusch to Kultusminister Karl Altenstein, 17 May 1839, 13 November 1839, 20 November 1839, ZStA-M, Acta betr. das mathematisch-physikalische Seminar an der Universität zu Königsberg, Rep. 76Va, Sekt. 11, Tit. X, Nr. 25, Bd. I: 1834–61, fols. 57, 62, 63–64; Neumann and Jacobi to Curator Reusch, 14 November 1839, ibid., fol. 65.

important chemical theories," as well as the ability to express mathematically those parts of physics that could be. The close relation between mathematics and physics on the examination, however, proved problematic because candidates who wished to teach the natural sciences, including physics, were not always competent in mathematics, and those who wished to teach mathematics and the mathematical part of physics were not always well versed in all areas of the natural sciences. To accommodate these two kinds of teaching candidates, the ministry offered restricted or conditional certification, enabling them to teach either mathematics and the mathematical parts of physics in all forms, or the natural sciences in all forms and mathematics in the lower forms.[11] Although the conditional certification meant that the examination did not promote a mathematical approach to physics exclusively, physics did remain the most specialized natural science on the examination and the only one strongly associated with techniques of quantification, two points that enhanced the importance of Neumann's seminar. What the examination did not yet incorporate was a written requirement in the natural sciences, including physics, something Neumann had earlier believed was essential for promoting the study of physics.

Both the reinstatement of funding and the changes in the examination occurred after the seminar had reopened in the summer of 1839. Richelot had directed the mathematical division well enough so that when Jacobi returned in the winter semester, he found that students had progressed to the point where they could undertake problems from integral calculus. Two of the three students registered for the mathematical division, Carl Borchardt and Ferdinand Joachimsthal, did distinguished work (Borchardt's was submitted for publication); the third, Philipp Wilhelm Brix, just managed to get by.[12]

Although Neumann neglected to mention his new course on theoretical physics in his seminar report, the subsequent stability of the seminar's operation was due in large part to that course's role in his curriculum. It provided what exemplars drawn from his own ongoing research could not: a simpler and more systematic presentation of the practical methods of physics. Although he did draw upon both his investigations and those of others in constructing seminar exercises, it was not the integration of research and teaching that stabilized the curriculum. From 1839 on, it was the exercises drawn from his course on theoretical physics that were

11. *Das Reglement vom 20. April 1831 für die Prüfungen der Candidaten des höheren Schulamts in Preußen mit den späteren Erläuterungen, Abänderungen, und Erweiterungen* (Berlin: Enslin, 1865), pp. 30–31. On the debates on specialization in this examination see Erich Feldmann, *Der preußische Neuhumanismus: Studien zur Geschichte der Erziehung und Erziehungswissenschaft im 19. Jahrhundert*, vol. 1 (Bonn: F. Cohen, 1930), pp. 35–39; *Das Unterrichtswesen des preußischen Staats in seiner geschichtlichen Entwicklung*, ed. L. M. P. von Rönne, 2 vols. (Berlin: Veit, 1854), 2:21, 37n–38n.

12. ["Bericht über das mathematisch-physikalische Seminar"], 30 November 1840, ZStA-M, Math.-phys. Seminar Königsberg, fol. 75.

crucial to the seminar's operation, providing the foundation for all its advanced work.

Seven students, a respectable number, enrolled for the physical division in the summer semester of 1839. Along with Brix and Borchardt, Wilhelm Ebel, A. Haveland, F. C. T. Brandis, and two previous members of either the physical or mathematical divisions, K. O. Meyer and G. S. H. von Behr, participated. It seems reasonable to assume from matriculation records that all but Borchardt had heard the lecture course on theoretical physics from the previous year. Four of the seven had been at the university one year or less, while Ebel, von Behr, and Meyer had been there since 1837, 1835, and 1834 respectively. The experience of these last three made a difference in their seminar performance. Neumann focused primarily on teaching his students the principles of a measuring physics. He integrated into the exercises, apparently for the first time, Gauss's methods for measuring geomagnetism. Together with Neumann, the students worked through Gauss's 1832 essay on the intensity of terrestrial magnetism, first acquainting themselves with Gauss's mathematical methods and then interpreting them physically. Students then repeated Gauss's method of measuring the intensity of the earth's magnetism, discussed the theory of geomagnetic instruments, and learned how to calibrate certain instruments. Geomagnetic topics and exercises did not, however, assume the central pedagogical role they had at Göttingen. The "independent" investigations of the older students concerned measuring techniques and the analysis of instruments, but only one drew upon the new seminar assignments drawn from Gauss's investigations. Ebel wrote on the theory of the balance; Meyer, on the achromatic lens; and von Behr, on the theory of bifilar suspension in a magnetometer (introduced by Gauss in 1837), the theory of magnetic inclination, and measurements of horizontal magnetic intensity taken at Königsberg. Tersely written, Neumann's report did not describe in detail what the other students had done. He canceled the seminar for the winter semester of 1839/40, when he taught another course in mechanics—this one on elasticity theory—which drew upon mechanical principles as a foundation for understanding optical phenomena mathematically (especially the optical ether) and which was related to a study of double refraction he was then completing.[13]

In the years immediately before and after Neumann first introduced his course on mechanics as an introduction to theoretical physics, mechanical considerations assumed a more prominent role in his own investigations and facilitated the theoretical treatment of several issues in optics. But the types of exercises he assigned after 1838/39 indicate that he defined the pedagogical function of mechanics differently. In subsequent seminar reports he described in greater detail the tasklike orienta-

13. Ibid.

tion of the exercises and what he intended students to accomplish in completing them. In the summer semester of 1840, when he taught his course on theoretical physics for the second time, students in the seminar learned the theory and operation of Etienne Louis Malus's goniometer. Neumann taught them more than merely how to measure angles in crystals by treating the goniometer

> as an example of how an instrument must be discussed when one wants to achieve the kind of results for which one can give the limits of their reliability. The various sources of error must be identified, the boundaries within which these sources of error could possibly operate must be determined, and the influence they exert on the final result must be calculated. In this way members [of the seminar] conducted measurements [on several crystals] and these were computed according to the method of least squares.

Even in the following semester, when the rooms for the mineralogical collection, which had been used for the practical exercises of the seminar, were no longer available, Neumann insisted on teaching his students how to analyze errors. He supplemented "theoretical exercises, which were primarily mathematical in nature," with small exercises on the thermometer: calibrating its scale, determining its fixed points, and calculating corrections for readings.[14]

Although it might appear that Neumann was merely using his course on mechanics as preparation for practical exercises with instruments, more was involved. A deeper and more systematic understanding of mechanics certainly would facilitate the determination of constant errors afflicting the apparatus and "environment" of an experiment, but the calculation of constant errors had been handled in the seminar since 1835 without the benefit of students having taken a course on mechanics. What was different after the inauguration of this course is that by 1840 Neumann was also requiring his students to compute accidental errors by the method of least squares. This was a significant turning point in the operation of the seminar because it shifted the purpose of the calculation of constant errors: instead of a step in determining the material limitations of an instrument or experiment, it became a step in the preparation for a deeper understanding of the reliability of data, and thus of theory. In the process, the analysis of instruments and their constant errors became, to the extent possible, much more theoretical, almost to the point where the instrument itself—as a material object—disappeared from view as the student tried to understand experimental apparatus and protocol solely in analytic terms. When A. L. Crelle evaluated the seminar for this year, he reacted to this shift by criticizing Wilhelm Ebel's theoretical study of

14. "Bericht über das Königsberger mathematisch-physikalische Seminar im Jahre 1840," 8 January 1842, ibid., fols. 89–90, on fols. 89, 90.

the balance, remarking that it needed "a more accurate description or drawing because [Ebel's] essay gives no idea of [the balance's] construction."[15] The method of least squares was thus a way to encourage students to consider broad epistemological issues rather than concentrate merely on either technical problems (as would have been the case had Neumann emphasized only the analysis of constant errors) or narrow conceptual issues (as would have been the case had he viewed mechanics solely as a means of teaching students the conceptual foundation of physics).

Although the practical exercises of the seminar were basic—designed to teach the fundamentals of investigation in an exact experimental physics—they were also intended to prepare students for more complicated projects. As students moved through stages of greater difficulty in the deployment of those techniques, Neumann found that the physical division had to be organized in introductory and advanced sections, thereby modifing the operation of his institute in yet another important way. Whereas in the past he had sometimes differentiated between the performance of the "older" and the "younger" students, now for the first time he separated them formally for instructional purposes. The stepwise learning that was emerging had not only an intellectual expression insofar as Neumann's courses and exercises began with mechanics but also an organizational one in his division's elementary and advanced sections. Neumann directed the work of introductory students but allowed advanced members to dispense with assigned exercises if they were engaged in ones of their own design. In that case, Neumann explained, "the student only has to report orally or in writing about his work, its progress, or about the difficulties he has encountered." Even though it was not always easy to find a suitable location for students to perform seminar exercises (he and Jacobi noted again "the lack of a standing location" where the physical exercises of the seminar could be held), some students were not only able to pursue advanced investigations but to do so at a publishable level. Yet their investigations stemmed less from Neumann's own than from the seminar exercises he assigned. Of the five students permitted to pursue problems on their own, three produced what Neumann viewed as exemplary results. He considered K. O. Meyer's work on achromatic lenses to be publishable, but it never appeared in print. Von Behr, who by then was in his sixth year at the university and had taken almost all the courses Neumann had offered, including the lecture course on mechanics as an introduction to theoretical physics, was by 1840/41 deeply immersed in the geomagnetic observations he had begun a year earlier. Not only did he continue to analyze the principle of bifilar suspension in Gauss's magnetometer theoretically, but he also organized interested seminar students into a group that regularly took observations on

15. A. L. Crelle to Kultusministerium, 31 January 1841, ibid., fols. 71–74, on fol. 73.

magnetic inclination and declination in Königsberg, observations Neumann considered precise enough to replace older data.[16]

The third independent investigator was Brix, who despite his unexceptional performance in the mathematical division, turned his seminar exercise in the physical division into a doctoral dissertation, the first that Neumann supervised, and was promoted in 1841. In his investigation we begin to see just how Bessel's model for experiment shaped Neumann's seminar exercises and the values they cultivated. Brix's interest in physics had predated his seminar years. Born in Berlin, he had attended the Friedrich Werder Gymnasium and Cöllnisches Realgymnasium, schools particularly strong in the sciences. While in *Secunda*, he had had private lessons in mathematics and physics. After attending Berlin University for three semesters, he moved to Königsberg.[17] In spite of the strength of the physics courses he had taken elsewhere, Brix's approach to physics was shaped at Königsberg. Neumann saw in Brix's investigation an opportunity to demonstrate to the ministry what could be done in the seminar. He urged Brix to publish his dissertation, an investigation into the latent heat of steam, and told the ministry to expect an offprint in the following year. Appearing in Poggendorff's *Annalen der Physik und Chemie* in 1842 under the title "On the Latent Heat of Steam of Various Liquids at Their Boiling Points," Brix's investigation was the first student publication to issue from the seminar.[18]

The study of latent heat—the amount of heat a substance absorbed or released while undergoing a change of state—was rife with experimental and numerical discrepencies. Brix knew, therefore, that before he could examine the relationship between latent heat and other variables such as the density of a liquid, he had "to measure as accurately as possible the latent heat of steam at the boiling point of as many homogeneous liquids as I can create."[19] But what unfolded in the course of his fifty-page essay was not a set of decisive single values for the latent heats of various substances but a procedure for measuring the latent heat by taking into account the disturbances arising in the course of the experiment. Brix isolated three main sources of experimental error: ϕ, the thermal change in a liquid as its steam was conveyed through a curved tube connecting two containers, one holding the liquid and the other holding water; α, the heat retained by the apparatus; and γ, the heat lost to the surrounding air. A critical review of the extant literature convinced him that the variation

---

16. "Bericht . . . im Jahre 1840," ibid., fols. 89–90; see also Crelle to Kultusministerium, 12 March 1842, ibid., fols. 84–88, esp. fol. 86.

17. E. Lampe, "Philipp Wilhelm Brix," *Verhandlungen der Deutschen Physikalischen Gesellschaft* 1 (1899): 125–35.

18. P. W. Brix, "Ueber die latente Wärme der Dämpfe verschiedener Flüssigkeiten bei deren Siedetemperatur," *AP* 55 (1842): 341–90.

19. Ibid., p. 341.

in values for latent heats resulted largely from the neglect of ϕ and γ, and he tried to achieve a correction formula for each one.

Brix had chosen a topic ripe for error analysis. Several types of errors plagued his experiment, and he tried to account for all of them. Some of the constant errors of his experiment could be attributed to his instruments; to the environment of the experiment; and to variations in the execution of the protocol. For instance, he explained in a footnote that he employed arbitrary units of length and weight, measured time in minutes, and used Réaumur's scale for temperature.[20] Not all dimensions were of equal importance in the experiment. Brix's most important instrument was the thermometer, so he examined its construction carefully. "The Réaumur degree," he explained,

> possessed almost the length of ¼ inch on the scale and was directly divided into six parts. I ascertained the internal variations of the tube according to a method of calibration regularly used by Professor Neumann, which coincided essentially with Bessel's method. Taking this into account, I accurately determined the value of the individual dimensions of the scale. So I believe that an error of 0.016°R in the observation [of the final temperature minus the initial temperature] will not be exceeded.[21]

Brix's examination of the "internal variation of the stem" and of thermometer reading in general drew upon Neumann's detailed lectures and accompanying seminar exercises on capillary theory.

Despite his analysis of experimental errors, Brix did not completely succeed in obtaining values for latent heats. His extensive analysis of the ways in which uncertainty was built into the act of measuring took over the course of his investigation. He explained that "certain errors and uncertainties that depend on the nature and dimensions of the apparatus . . . could not be circumvented." One could only reduce these errors and uncertainties through the careful construction of the apparatus and experimental protocol to compensate for, but not entirely eliminate, them. Brix seemed to argue for a sensitivity to the delicate balancing act one had to achieve between mathematical correction and experimental construction. To hyperdevelop one or the other might prove counterproductive because after a point, the limited returns would not warrant the extra effort. Such was the case with ϕ because, according to Brix, "there would be just as little success in restricting ϕ to such small values, in order to be able to neglect it in comparison to the other uncertainties of the experiment, as there would be in calculating directly the necessary correction." Although he knew that ϕ was a function of the duration of the experiment and he tried to develop a formula for it, Brix eventually abandoned his efforts to determine ϕ quantitatively and opted instead "to

20. Ibid., p. 361n.
21. Ibid., p. 378.

eliminate the error through a combination of many experiments with the same apparatus under different conditions."[22]

The problem, though, was that even after Brix considered the constant errors afflicting the experiment, he still had to calculate the reliability of his results. He acknowledged that measurements yielded only "probable values," and he computed "the most probable value" of the error by using the method of least squares. But he seemed to want to reduce probabilities and uncertainties as much as possible—a goal that in fact had been tacitly cultivated in seminar exercises—by generating more data, which he hoped would increase his accuracy. He justifiably believed that "the smaller [the error] is in an individual experiment, the closer the result of the calculation is to the truth," but he never believed his errors were small enough.[23] At times he came dangerously close to admitting that one could not know anything with certainty, as when he explained that even in his corrected formula for latent heat, errors could still enter. Brix ended his investigation without attaining decisive values for latent heats; at best his investigation isolated several sources of error and set a course for future investigations.

Brix's *Annalen* paper was indicative of more than how seminar exercises prepared students for advanced work. It also suggests that Neumann's seminar exercises cultivated a particular epistemological approach to the construction of scientific knowledge. Error and data analysis became, in Brix's investigation, the tools of the skeptic. Brix was not seeking something entirely new in his investigation (a "discovery"), nor was he trying to synthesize what was known about latent heat in order to create a comprehensive theory. Instead, he was trying to articulate—and to express in analytic terms—what was *not* known and, more important, why. We might say that he was not trying to bring light to darkness but to show the darkness in what was considered to be light. Appropriately, he did not end his paper by discussing how his investigation settled problems in the study of latent heat but rather by mentioning cases where his approach—his detailed examination of errors—met the limits of its applicability. He thereby suggested that higher order corrections would be necessary for more difficult studies of latent heat, such as when one dealt with a non-homogeneous liquid or when two substances, such as alcohol and water, were mixed.[24] His investigation thus revealed a characteristic limiting feature of Neumann's exercises when they evolved into investigations: that the pursuit of higher and higher degrees of numerical precision through error and data analysis could prove endless.

That Brix and von Behr had both been at the university several years before embarking on their investigations indicates that much prior preparation, both in Neumann's courses and in his seminar, was necessary

22. Ibid., pp. 354, 355.
23. Ibid., p. 374.
24. Ibid., pp. 385–90.

before a student could exercise independent critical judgment. That Neumann's and Jacobi's methods of teaching were in fact laying the foundations for *Ausbildung* or training in mathematics and physics, rather than *Bildung* or inner cultivation, can be seen in the patterns of their teaching practices over the next four semesters, from the summer semester of 1841 through the winter semester of 1842/43. Jacobi, who earlier had pursued advanced topics with his students, was now finding it difficult to sustain the high intellectual level of his division. When Neumann was supervising the likes of von Behr and Brix in 1840/41, Jacobi had students whose preparation was wanting and to whom only elementary exercises could be assigned. So in the first semester he created problems of "pedagogical interest" on descriptive geometry, and in the second, on the polynomial theorem because "the scientific standpoint of the students did not allow, as in earlier years, penetration into deeper subjects."[25] Over the next two years he had to coordinate his lecture courses and seminar exercises more closely, using discussions in the seminar as a way to "fill in the gaps" left by material covered in his lectures and to clarify points that he went over quickly.[26]

Confined to more elementary topics now than in the late 1830s, Jacobi's seminar exercises were not accomplishing all that he had hoped; for during 1840s he took aside his better students, including Ludwig Seidel and Eduard Heine, for private lessons. Seidel's decision to join the seminar was "dependent upon what kind of problems [Jacobi] gave," he wrote to his parents in 1843, "because last semester, in order to accommodate the majority of Königsberg students, Jacobi assigned such easy problems that they carried no interest" for him. Jacobi's enthusiasm for the seminar had in fact waned by late 1843, by which time he had taken nine semester-long sabbaticals, mostly for health reasons. Suffering from as yet undiagnosed diabetes and discontent with the provincialism of Königsberg, Jacobi, after lengthy negotiations with the ministry, left Königsberg in October 1844 for Berlin, where he took up a position at the Academy of Sciences with a salary of 1,000 taler above his 1,667 taler Königsberg salary and with no obligations to teach, although from time to time he offered lecture courses. For some time Jacobi had tried to leave Königsberg—a year after the seminar was founded he had asked the ministry if he could be transferred to Bonn not only to establish a school of mathematics but also to be closer to the Parisian mathematical community—and for just as long, Bessel had worked to keep him at Königsberg. But when Jacobi finally decided to leave in 1844, Bessel realized that no institutional features could keep him; for Jacobi "did not need an observatory or any

25. "Bericht . . . im Jahre 1840," ZStA-M, Math.-phys. Seminar Königsberg, fol. 89.
26. ["Bericht über das mathematisch-physikalische Seminar, 1841–43"], n.d., ibid., fol. 105.

other equipment that was bound to one location—the location is, therefore, irrelevant to the problem."[27]

Despite the fact that his mathematical division had turned toward elementary matters in the early 1840s, earlier it had met Jacobi's expectations and contributed, along with his lecture courses, to upgrading considerably mathematical instruction in Prussia and bringing international recognition to Königsberg, achievements he had thought would help him bargain for his transfer to Berlin in 1841. Although that effort had failed, his earlier success eventually made his departure easier because, as he noted to the crown, two of his former students, Richelot and Hesse, "teach my field in fundamental and solid lecture courses and make the replacement of my position at Königsberg unnecessary."[28] So when he left Königsberg, the transfer of leadership to his former student Richelot was smooth and natural and created no disruption in the operation of the seminar. In fact, Richelot's promotion to full professor in 1844 and appointment as codirector of the seminar in 1845 strengthened the mathematical division of the seminar as a teaching institute and ensured the harmonious interaction of the two divisions.

Equally sensitive to the needs of beginning students, Richelot and Neumann created special sections for them in both divisions in the second half of the decade. Richelot divided topics up according to their level of difficulty. Older students most often covered difficult mathematical problems in dynamics; the written work of the seminar in 1848/49 included applications of elliptic transcendentals to mechanical problems, a discussion of Euler's treatment of partial differential equations, and techniques of double integration. The introductory section seemed to concern Richelot more, however. As foundational topics, he offered analytic geometry, trigonometry, the theory of algebraic equations, and introductory calculus. So that at least a part of his task of training newcomers might be alleviated, he wished that mathematics teachers in secondary schools would teach more geometry, especially in a way that would reflect some of the recent advances in the field.[29] Pedagogically, Richelot preferred concrete subjects over abstract ones; intellectually, he underscored the geo-

---

27. Ferdinand Lindemann, *Gedächtnisrede auf Ludwig Philipp von Seidel* (Munich: Kgl. Bayer. Akademie, 1898), pp. 10, 44; Leo Koenigsberger, *Carl Gustav Jacob Jacobi* (Leipzig: B. G. Teubner, 1904), pp. 324–25, 173–75; F. W. Bessel to C. Jacobi, 8 July 1844, Bessel Nachlaß, Briefband 15, ZAA. Before 1844, Bessel spent a considerable amount of time trying to keep Jacobi at Königsberg; see Bessel to H. Schumacher, 13 October 1827, Bessel Nachlaß, Briefband 7, ZAA; Bessel to Jacobi, 21 July 1834, ibid., Briefband 15; Bessel to C. F. Gauss, 11 September 1841, GN. After his departure from Königsberg, Jacobi told the mathematician Gustav Lejeune Dirichlet about his unhappiness there (Jacobi to Dirichlet, 4 January 1845, Dirichlet Nachlaß, SBPK-Hs.)

28. Koenigsberger, *Jacobi*, p. 278.

29. ["Bericht über die mathematische Abtheilung, 1848/49"], 17 June 1849, ZStA-M, Math.-phys. Seminar Königsberg, fol. 165.

metric foundation of more difficult mathematical topics. For him, differences in the abilities of students led to an understanding of how mathematical topics should be arranged hierarchically in a curriculum.

Although Neumann was not yet as explicit in his understanding of the ordering of topics in a physics curriculum as Richelot was in mathematics, his teaching before the summer semester of 1843 had begun to show a pattern. Roughly every three semesters (the winter semester of 1838/39, the summer semester of 1840, the winter semester of 1841/42, and the summer semester of 1843), he offered his course on mechanics as an introduction to theoretical physics, thereby establishing a cycle. Each time his course on theoretical physics was offered, it coincided with the arrival of a new cohort of students, confirming the foundational role of mechanics in his curriculum. But as his curriculum tightened and the role of practical exercises in his seminar became more secure, Neumann had felt more strongly than ever the imperfect fit between the operation of the seminar and the material facilities at his disposal. At least eight students attended his seminar between 1841 and 1843, and although he had sent their seminar exercises to the ministry for its review, he had not been entirely pleased with what they had done. Their exercises were "in part experimental, in part theoretical, investigations about physical apparatus or experimental methods." But, Neumann explained, "these latter subjects must unfortunately still function as improvisations, because the space assigned for the seminar does not permit the pursuit of experimental concerns on a regular basis." So, to compensate for the lack of space at the university, Neumann told the ministry, he allowed students who could be trusted to take instruments and other apparatus back to their own living quarters, where they could execute their experimental trials. "This inconvenience, which hinders the effectiveness of the seminar," he remarked, "can only be removed by the construction of a physical laboratory."[30]

Whenever Neumann had commented on seminar exercises, including those that had led to original investigations, he had always emphasized how well students deployed techniques. To be sure he also told the ministry that these exercises had "scientific value," but that was only "part" of their significance.[31] His evaluation of his students' work is important because in the absence of an explicit description of what kind of physical laboratory Neumann wanted, it suggests that he was not asking the ministry for a physical laboratory equipped for original research, either by him or by his students, but rather one where more basic exercises could be conducted. Only through his teaching in the seminar had Neumann come to that position; for in his initial requests for a

---

30. ["Bericht . . . 1841–43"], ibid., fols. 105–6; see also Crelle to Kultusministerium, 13 September 1843, ibid., fols. 97–101, esp. fol. 98.
    31. ["Bericht . . . 1841–43"], ibid., fols. 105–6.

laboratory, made in 1834, he had in mind something like Bessel's observatory—a place where he could live and conduct research because "in science one works not as in business, according to specific hours"—and he had specifically mentioned that "the university is not just an educational institution for young people; it is also a pure scholarly institution that participates in the development and progress of science."[32]

But later requests reflected Neumann's deepening involvement in the seminar and especially in the pedagogical tasks that had accompanied it. So in his 1840 response to Justus Liebig's polemical essay on the state of chemistry in Prussia (Neumann was among the few who supported the laboratory-based science instruction Liebig advocated),[33] he viewed the function of a physical laboratory primarily in terms of his practical exercises, as a place where training could be had in, among other things, "judging the limits within which [scientific results] are true" and "evaluating the protocols used in quantitative determinations."[34] On other occasions in 1840 and 1841 he emphasized the educational function of a laboratory, including its importance in training secondary school teachers and hence for enhancing school science instruction and for creating a "nursery for mathematical physics."[35] It was the absence of a laboratory, he had claimed, that prompted him to turn students away to study physics elsewhere.[36] Even though the crown had acknowledged that Königsberg needed an institute for chemistry and physics and admitted that it would not be difficult to find a location that would allow for vibration-free measurements, nothing came of Neumann's request. He had been compelled to continue relying on his own resourcefulness in creating and conducting the practical exercises of the seminar. "My own scientific activity," he claimed around 1841, "must be restricted to what a physicist can do for science in an attic."[37]

Through 1843, Neumann's students had not pursued original research as much as might have been expected when the seminar was founded. In June 1843, A. Emil Schinz had become Neumann's second doctoral student, with a dissertation on the dependence of the height of fluids in capillary tubes upon temperature—a topic compatible with Neumann's seminar exercises in that it involved a modification of a known law through the analysis of actual experimental conditions and errors—but little is known about Schinz's investigation because he did not publish it. Of the four students thus far who had completed investigations Neumann reported

32. LN, p. 347.

33. R. Steven Turner, "Justus Liebig vs. Prussian Chemistry: Reflections on Early Institute-Building in Germany," *HSPS* 13 (1982): 129–62, on 135–36.

34. LN, pp. 348–49. I thank R. Steven Turner for checking the original document.

35. LN, p. 352.

36. Neumann to Kultusminister Johann Albrecht Friedrich Eichhorn, 13 July 1843, ZStA-M, Math.-phys. Seminar Königsberg, fol. 104.

37. LN, pp. 348, 353.

to be publishable—von Behr, Meyer, Schinz, and Brix—only Brix's made it into print. In the early 1840s Neumann had struggled with the discordance between his intentions for the seminar and its performance to date. He had begun to blame material conditions for the quality of his students' work. He had alleged around 1841 that because his budget for instruments was small—150 taler from the budget for the physical laboratory, a small portion of the 350 taler seminar funds, and occasional extraordinary purchases made with funds not spent from the 100 taler mineralogical collection budget—"the choice of subjects [in seminar exercises] was often restricted."[38]

Yet financial records for 1834 to 1843 indicate that Neumann made consistent purchases for investigations in thermometry (including geothermal work), specific heats, optics, and to a lesser extent in electricity and that the available funds were not always expended as swiftly as they could have been because from time to time a surplus accumulated, which was later spent.[39] In short, by 1843, more by default than by design, Neumann was not promoting original investigations as strongly as he perhaps could have. Even though the instruments he purchased for the seminar were related to his own investigations, they were very simple instruments—thermometers, barometers, and the like—used in practical exercises. In 1843 he just did not have a strong enough argument for a better equipped laboratory where the original investigations of a select few could be pursued. The laboratory he wanted was primarily a *teaching* laboratory, one where his seminar students could conduct practical exercises on a more frequent and regular basis. Although such a laboratory did not preclude the possibility that some students, at least, might undertake original investigations, as Neumann still hoped they would, it was not to be a facility designed with only those few students in mind.

By the early 1840s, Neumann's pedagogical definition of theoretical physics had taken shape primarily within the context of his course on mechanics and the seminar exercises associated with it. Intended to initiate beginning students into the practice of physics, it also exerted a powerful influence on his students' investigative styles. Although very little is known about the course as it was first taught in the late 1830s or even as it was offered in the 1840s, the course as taught after then can be reconstructed from a combination of lecture notes later published by Neumann's students and unpublished lecture and seminar notes his students took.

His student Carl Pape published a version of the course based on notes he took in the winter semester of 1858/59.[40] This version was first taught,

38. LN, p. 353.
39. FNN 62: Kassensache.
40. F. E. Neumann, *Einleitung in die theoretische Physik*, ed. C. Pape (Leipzig: B. G. Teubner, 1883).

at the earliest, in the mid- to late 1840s, after Neumann inserted a section on the conservation of living force, which became the new foundation for his exposition of hydrodynamic and aerodynamic principles. Other new discoveries and investigations published in the late 1840s and especially in the 1850s did not change the organization of his course but were merely integrated into a structure that had been in place for some time. Most of the new material Neumann later introduced concerned refinements in theory based on more precise measurements and more comprehensive analyses of errors. Neumann's course was not one on analytical mechanics in which the equations of motion were developed through refined mathematical techniques such as Lagrange's or Hamilton's equations; nor was it one that consolidated and assembled previous accomplishments in mechanics. It is not even clear that he intended to portray mechanics as the conceptual foundation of his theoretical physics.

Analytical mechanics, to be sure, was never far from consideration in Neumann's exposition of theoretical physics. "We begin with the mechanical part of physics," he explained, "because it builds the foundation for all remaining branches of physics and contains the principles that are used in it. Knowledge of analytical mechanics will not be assumed. The basic ideas of analytical mechanics are clarified and derived in a series of considerations. They are not developed systematically, but only when there is a need for them. In this way, a secure foundation for the deeper penetration of analytical mechanics, the most important resource [*Hilfsmittel*] in physics, is achieved."[41] De-emphasizing some of the complicated mathematical formulations in mechanics meant that his mechanics was not necessarily organized like a mathematical treatise, although it did convey important mathematical methods in physics.

The course's table of contents listed several topics, most of them applied: gravity, hydrostatics, aerostatics, the conservation of living force, hydrodynamics, and aerodynamics. Gravity, the most important topic, occupied one-third of the course. From the lectures on gravity, students learned fundamental equations of motion. Neumann discussed the properties of gravity, which he defined simply as "a force that acts constantly and continuously." But he qualified his definition, noting that gravity only appears to act continuously; for if small enough time intervals could be measured, he explained, gravity probably could be shown to operate by small impulses.[42] In his treatment of gravity, Neumann mentioned the cause of motion, force. "We have the inner necessity," he stated, "to explain every change by a cause."[43] Yet in spite of the deterministic nature of mechanics as then largely understood, he did not discuss causality or force in a sustained way. Instead he used the mathematical equations of

41. Ibid., p. 1.
42. Ibid., p. 2.
43. Ibid., p. 5.

mechanics as vehicles for introducing students to the properties of mechanical instruments and apparatus. After discussing gravity, Neumann turned to Atwood's machine. All subsequent topics followed the same pattern: he exposed a few basic theoretical ideas and then illustrated them through application, especially through a well-chosen instrument or apparatus. His discussions of parabolic motion, the parallelogram of forces, and motion in a curve were followed by an analysis of the inclined plane. Very quickly, then, the conceptual side of his theoretical physics and especially of mechanics became submerged, hidden by detailed expositions of practical problems. Neumann nonetheless had an important and sophisticated goal in mind; for his discussion of the inclined plane was the foundation for the most important practical example in the course, the pendulum. It is here that we begin to see how Neumann created a pedagogical physics that used the principles of exact experiment, as they appeared in Bessel's seconds pendulum investigation, to explore the limitations of mathematical expressions of physical theory critically.

The strategic role of the pendulum in Neumann's course is obvious. The pendulum was the largest topic, occupying two-thirds of his discussion of gravity and one-quarter of his entire set of lectures on theoretical physics. It functioned as a paradigm in his mathematical physics. As Neumann explained,

> Next to the balance, the pendulum is the most important physical instrument. It offers the means to measure forces, to express intervals of length and time, and to study these. It serves as well to measure gravity and to study its properties; to prove that its effect is proportional to mass, that g is independent of the quality of the material, and that it has different values at different places on the earth. Moreover, no other instrument offers the required accuracy that the pendulum does. Not only gravity, but also magnetic, electric, torsional, and frictional forces are measured with the pendulum. One can say that the observation of the pendulum's period is the only method by which to measure forces directly. The remaining instruments either give only a relative measure or are based on the pendulum.[44]

As he unfolded the theory of the pendulum's properties, especially its period, Neumann introduced tools useful in the assessment of theoretical expressions. He discussed techniques of quantification, especially series expansions, approximation techniques, and solutions to simultaneous equations, as well as techniques of exact experiment, including error analysis, the method of least squares, measurement, instrument construction, and experimental protocol. Always he emphasized the mutual interaction of theoretical formulations and experimental results, and the dependency of both upon the material conditions of an experiment and the quality of the data it yielded. Although he used the pendulum to

44. Ibid., pp. 32–33.

develop important physical ideas such as the equilibrium motion of a system of masses, the center of gravity, and the moment of inertia, it was the usefulness of the pendulum as a measuring instrument that defined its role in his pedagogical physics. Bessel's experiemntal methods, as exemplified in the seconds pendulum investigation, thus became the fulcrum of Neumann's course on mechanics as an introduction to theoretical physics and the foundation of his conception of the practice of theoretical physics.

In teaching his students the techniques of exact experiment, Neumann advised them to construct an experimental trial so as to utilize the techniques and equations of pendulum motion because, he believed, the reliability of the pendulum in measuring gravity produced the greatest accuracy and precision in measurement. "Of all variables," Neumann reminded his students, "g is known with the greatest precision. Therefore, other variables are more accurately determined if they are measured in terms of g."[45] He believed that one could achieve greater precision by manipulating the mathematical formulas of the phenomenon under investigation so as to obtain a gravity-dependent expression. If that were not possible—a common enough occurrence—then there were two other ways to achieve the precision of the pendulum: to derive a time-dependent expression (for time was second only to gravity as the variable most precisely measured by the pendulum), or to employ an instrument the operation of which could be viewed as analogous to pendulous motion, such as the vibrations of a magnetic bar in a multiplicator.

The most important function of the pendulum in Neumann's mechanics, though, was as an exemplar from which to learn error analysis. In his discussion of the pendulum's period, he explained Huygens's cycloidal pendulum. But because the measurement of many physical quantities, especially time, depended upon knowing the period of the pendulum as precisely as possible, he introduced the method of least squares; for only through that method could "the most probable value of [the period]" be known. The pendulum served just as appropriately for teaching students how to compute constant errors, eliminating or diminishing them by adjusting experimental protocol or by using techniques developed for analyzing pendulumlike motion. For example, through the pendulum, he introduced students to technique of bifilar suspension and then used the principles of the physical pendulum to analyze the geomagnetic-induced vibrations of a magnetic bar. For the latter, Neumann explained that one could determine the period of oscillation of the magnetic bar by noting the number of vibrations that took place in a specified time T; the period was then equal to T divided by the number of vibrations. He told his students that the period obtained was only approximately correct, for the first and last vibration of the bar did not precisely coincide with the

45. Ibid., p. 60.

first and last strike of the chronometer. He then showed how the approximate value of the period of oscillation could be refined by taking into account the errors of the chronometer, and then he demonstrated again how to apply the method of least squares. But, as Neumann pointed out, in the end one could only obtain "from several observations the most probable value of an unknown," the most probable value being one that satisfied the condition that "the sum of the squares of the observational errors had to be a minimum." The resulting period was therefore the most precise in that it was the most probable. To simplify their data analysis, he suggested his students ignore observations with very large errors, provided a sufficient number of observations still remained to justify using the method of least squares.[46]

An especially important technique associated with the pendulum was coincident observations. Determining the period of oscillation of any pendulum, for example, depended on comparing it with an astronomical clock. Ideally, the length of the investigated pendulum could be adjusted to match the length of the clock's pendulum so that both had an equal number of vibrations in equal periods. Neumann explained how to achieve precision by determining the number of vibrations, n, such that when one pendulum passed through $n + 1$ vibrations, the other would pass through n vibrations. Both pendulums would coincide at two points in time, t_i and t_f, where $t_f - t_i$ was the time in which one pendulum experienced $n + 1$ vibrations and the other n vibrations. As n increased, so the precision of the measurement of the period of the investigated pendulum increased—or so it would appear. Actually, as Neumann pointed out, the increase in precision always depend on the precise observation of those times at which the positions of both pendulums coincided, no matter how large n was.[47] Because that observation was itself "very difficult" (the two pendulums were not close to one another), Neumann introduced Bessel's method of coincident observations.

Neumann explained to his students that the two pendulums Bessel used were an astronomical clock and a pendulum consisting of a thin silver wire and a platinum ball. To detect the coincidence accurately, both pendulums were equipped at their lower ends with very thin wires upon which the observer could focus attention. Observing through a telescope from some distance away, Bessel placed a lens between them so that an image of the platinum pendulum was thrown into the plane of vibration of the clock pendulum, thereby reducing the observational error resulting from parallax. Neumann took pains to show his students in addition how the precision of Bessel's measurements was increased not only by such material refinements but also through proper data reduction and the analysis of errors. Of central importance were Bessel's reduction of

46. Ibid., pp. 38, 70.
47. Ibid., pp. 74–75.

pendulum vibrations to a vacuum and his hydrodynamic correction for the motion of a pendulum in air, which was a vast improvement over Newton's hydrostatic correction. These topics in particular had obvious theoretical interest, but Neumann drove his point home by demonstrating how the accuracy of observations increased through them. So, in addition to Bessel's seconds pendulum investigation, he also mentioned Bessel's 1830 determination of gravity based on pendulum observations that had an accuracy of 1/60,000.[48] These and other correction methods were object lessons on how the proper design of an experiment could eliminate or reduce errors. So vital to his pedagogical physics were Bessel's techniques and correction factors that one could view the overall structure of Neumann's course as dictated by the divide that Bessel had created in the seconds pendulum investigation: gravity, hydrostatics, and aerostatics were in a sense "Newtonian" topics, the basis of first-order corrections; while hydrodynamics and aerodynamics were "Besselian" ones, used for second-order corrections. (The conservation of living force made the division between the two conceptually, but not experimentally or technically, sharper.)

Having isolated the principal sources of error in measuring pendulum vibrations and having shown how to deal with them, Neumann then discussed what use could be made of the seconds pendulum, such as a precise measurement of *g* and a refined proof that all bodies are equally affected by gravity. Other topics followed: the variation of *g* on the earth's surface; centrifugal force, Kepler's laws, and the attraction of masses in general; and the density of the earth and torsional forces. The final pages of the section on the pendulum were devoted to a discussion of a related instrument, the balance, which was, "next to the pendulum, the most important physical instrument." His analysis of the balance was particularly useful for students, who were required to construct their own in the seminar. Neumann not only pointed out the sources of error in a scale and how one could assess the sensitivity of its readings, but he also gave many hints on how to improve its construction. He ended this section with a description of two balances: his own and Wilhelm Weber's. His own, he claimed, was characterized by "great accuracy"; Weber's, by "great reliability."[49]

Neumann's discussion of other topics followed a similar pattern. Just as he had introduced gravity to explain the measuring advantages of the pendulum, he introduced aerostatics to review the design and construction of the barometer, its sources of error, and various methods for correcting them. His lectures on hydrostatics dealt with capillary theory and the analysis of the thermometer. Even in his discussion of hydrodynamic principles, instrument construction was never far from view. In

48. Ibid., pp. 74–78, 83–84.
49. Ibid., pp. 78–109, 110–16, on pp. 110, 116.

addition to instruments he also discussed errors that afflicted the measurement of fundamental constants, such as the density of air, because of their "influence on countless physical observations."[50] Years after leaving the seminar, Neumann's former student Johann Pernet recalled that "in his introduction to theoretical physics" Neumann lectured "not just on the mechanical parts of physics but also on the most important physical measuring instruments belonging to it, to treat in detail the pendulum, the balance, and the barometer, and to confirm precisely the corrections that had to be considered for accurate observations."[51]

The published version of Neumann's lecture course on theoretical physics included almost three hundred pages of text. Paul Peter's lecture notes from the winter semester of 1870/71 indicate that in class Neumann covered almost all this material. Nevertheless the course had evolved; for Peters's notes reveal that in 1870 Neumann referred even less to causality and that his course became even more focused on the analysis of errors for the purpose of observational and theoretical refinements. "We will seek as a measure of motion not its cause, force," Neumann told his students, "but its effect, velocity, . . . and so reduce the mathematical (quantitative) definition of the idea of force to that of velocity." (This turning away from a discussion of causes was, as we shall see, consistent with other aspects of the pedagogical physics he shaped over the next few decades, which was phenomenological in orientation; concentrated to an even greater extent upon the techniques of practical physics; and acknowledged the approximate nature of theoretical knowledge in physics.) He continued to introduce mechanical principles in order to analyze instruments, but he was more explicit now about defining instruments used for "scientific purposes." "What matters in a physical instrument," he pointed out, "is not so much that it be error free, but that one can make the following assumptions about it: (1) that the instrument is reliable; (2) that the errors of the instrument be small in comparison to the variable to be measured; (3) that the error be constant; (4) that [the error] can be determined or be eliminated through some combination of many observations."[52] Even as German scientists and technicians constructed precision instruments that yielded data of higher and higher degrees of accuracy, Neumann still considered error analysis important enough to physical investigation to continue to teach it through discussions of the limitations of simple instruments. Year after year he repeated his analysis of certain instruments—the pendulum, the balance, the method of bifilar

50. Ibid., p. 158.
51. Johann Pernet, "Ueber die bei Messungen von absoluten Drücken und Temperaturen durch Capillarkräfte bedingten Correctionen und über die Arago'sche Methode zur Bestimmung der Luftspannung im Vacuum der Barometer," Zeitschrift für Instrumentenkunde 6 (1886): 337–83, on 377–78.
52. FNN 14: Theoretische Physik (F. N.'s Vorlesungen vom Wintersemester, 1870/71), pp. 3, 97.

suspension, the multiplicator, the barometer, the thermometer—until he had codified the technical procedures for evaluating their results. The importance of error analysis in his curriculum meant that he did not merely teach classical experiments useful only for the purpose for which they were designed. He taught standardized techniques transferable from one experimental setting to another, which were particularly useful in understanding data and the limits of certain theoretical expressions.

When we turn from Peters's lecture notes to his seminar notes from 1870/71, we see even more clearly the prominent role of instruments and error analysis in Neumann's scientific pedagogy (Figure 7).[53] In the seminar, students derived with greater thoroughness the equations for the influence of gravity and roll in Atwood's machine; free fall and vertical motion in a resisting medium; the cycloidal and ballistic pendulums; bifilar suspension; the determination of small intervals of time with the pendulum; and the application of hydrostatic equations, especially in capillary tubes. Neumann gave his students very simple differential equations to integrate in order to determine the particular form of the equation, including the relevant constants, for each application under discussion. The ultimate objective was to achieve an analytic expression suited for determining its constants through measurements and suited for comparing theoretical values with observational ones. Students thereby became facile in the use of mathematical methods in physics, although in this course they did not have to know much more than ordinary and partial differential equations, series solutions and expansions, logarithmic functions, and fairly simple integrations. More important than the ability to work mathematically was the knowledge of what measuring observations were not only possible but also necessary in order to assess the limits to the mathematical expression of theory. Hence what students really had to learn was how to determine what terms in their equations could be eliminated without significant effect upon their final result; and to do that, they had to know how to analyze errors.

Especially through Neumann's course on mechanics, then, we can see how, in his practical exercises, students were not so much learning physical theory as learning how to do physics, execute experiments, and quantify results. They were not merely learning precision measurement; they were learning how to make their numerical results tractable. They were not necessarily conducting original research but learning what to be conscious of when they did do research. They were also learning that any conclusions about the physical world that they wished to draw from numerical results had to take into account the conditions under which those results had been achieved.

This particular function of mechanics had not at first been present in

53. FNN 15: Physikalisches Seminar, Wintersemester 1870/71, Sommersemester 1871, zur theoret. Physik.

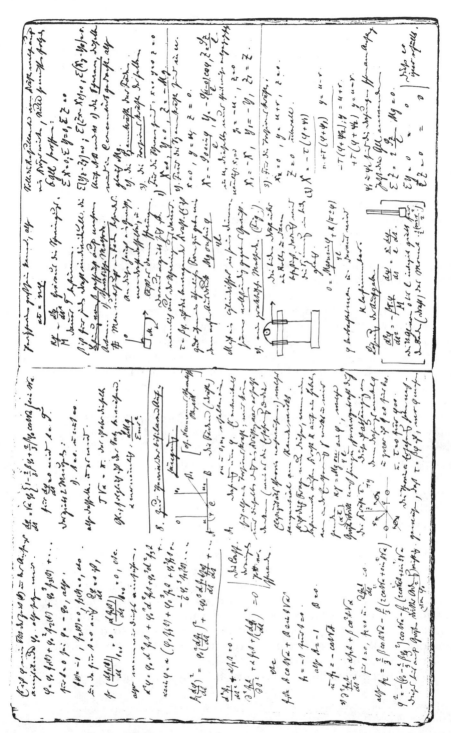

Figure 7. Page from Paul Peters's seminar notes on theoretical physics, 1870/71: two pages primarily on the theory of bifilar suspension. From FNN 15, with permission to reproduce granted by UBG-Hs.

Neumann's pedagogical physics, even though problems in mechanics were. In the maiden session of the seminar, during the winter semester of 1834/35, Julius Czwalina presented an oral report on the motion of a pendulum in a vacuum and in a resisting medium. One year later, Schumann also took up the study of the pendulum. Although Bessel's investigation of the seconds pendulum probably inspired these exercises, Neumann had not as yet attributed any special significance to these problems, nor did he indicate the need for a special course on mechanics. Only when the problem of student preparation became pressing in subsequent semesters did Neumann formulate more clearly his ideas on the role of mechanics in physics instruction. Still, it does not appear that he understood immediately the role his course on theoretical physics played. Although in 1838/39 he had a potential foundation for his curriculum, it was only after years of experience—and trial and error—that he acknowledged his course on mechanics as that foundation. Neumann taught theoretical physics as a private course, thereby affirming its function as a specialized introduction to physics. Although he taught the course regularly during the 1840s (seven times in that decade), he did not comment on the exercises derived from it in his seminar reports of that time. Nor did he as yet explicitly link the course to the needs of beginning students, even though it had obviously been created for them.

By the 1850s and 1860s, the practical exercises in his seminar expanded to embrace all the topics later found in the published version of his lecture course on theoretical physics. He offered the course seven times in the 1850s, five times in the 1860s. In seminar reports from the 1850s to the 1870s, Neumann finally explained the functions of his practical seminar exercises on mechanics. At first he only linked his exercises to the lecture course in a general way. When new students entered the seminar in the winter semester of 1850/51, he offered his course on theoretical physics, and in the seminar he expected that "the mechanical and physical principles developed in the lecture course" would be "clarified and made familiar to the students through an independent application in the seminar [exercises]." "I believe that through these problems," he continued, "I made the [seminar] members responsive to and well prepared for a deeper study of mechanics."[54] In the following year he explained that exercises in mechanics "first of all are meant to bring to vivid clarity the ideas of mass, of force, of the influence of the distribution of masses upon motion, and so on." Then, however, he acknowledged that learning mechanics was not an end in itself but a means to a higher goal. In his view, learning mechanical ideas was secondary to achieving "certainty" in the

54. "Bericht über die Leitung der physikalischen Abtheilung des mathematisch-physikalischen Seminars während des Zeitraums von Ostern 1850 bis Ostern 1851," n.d., ZStA-M, Math.-phys. Seminar Königsberg, fols. 179–80.

"theoretical judgment" that was supposed to be attained in the course of completing the seminar's practical exercises.[55] So when he taught the theory of the pendulum in 1852/53, he explained to the ministry that although the observations undertaken by the seminar members were supposed to demonstrate factors influencing the period and amplitude of the pendulum, they were also intended to inspire students to extend their theoretical understanding of the instrument; to apply its theory; to learn "general principles of the art of observation"; "and to demonstrate the most favorable use of the data obtained."[56] In the draft version of the same seminar report, Neumann mentioned that through pendulum observations, students learned how to judge the limits within which their data were reliable.[57]

As a foundation for the study and practice of theoretical physics, Neumann's course on mechanics marked the beginning of systematic training in physics at Königsberg. Neumann did not compensate for the poor preparation of his new students by cramming into their heads what was then considered to be doctrine in physics; he wanted instead to engage them in independent thinking and investigations. As long as students were able to choose their own topics and to work on problems haphazardly, there was no assurance they would systematically acquire the tools, and the values, for undertaking independent investigations; from time to time, though, a student would appear who by dint of his own talents performed well. It was only through his course on mechanics that Neumann could guide his less gifted students in the practice of physics. Practical exercises in error and data analysis helped inculcate the standards of judgment Neumann thought most useful in doing theoretical physics. Success gave students the confidence not only that they understood mechanics but also that they had acquired skills and values transferable to other areas of physics. His course on mechanics thus disciplined students.

Neumann's course on mechanics as an introduction to theoretical physics exhibited both agreement with and deviation from existing patterns in both mechanics instruction and prevailing institutional definitions of "mechanics" and "theoretical physics." Pedagogical traditions in mechanics partially explain why Neumann called his course an introduction to *theoretical* physics, even though it provided the foundation for learning the methods of an *exact experimental* physics as well. During the eighteenth century, mechanics appeared from time to time in university lecture catalogs of the German-speaking areas of Europe as a course on

55. "Bericht über die Leitung der physikalischen Abtheilung . . . während des Zeitraums von Ostern 1851 bis Ostern 1852," 31 July 1852, ibid., fols. 187–88.
56. "Bericht über die Leitung der physikalischen Abtheilung . . . während des Zeitraums von Ostern 1852 bis zum Ostern 1853," 8 July 1853, ibid., fols. 196–97.
57. Draft report, physical division, 1852/53, FNN 48: Seminar Angelegenheiten.

theoretical physics. Mechanics was then also called "dogmatic physics" in contrast to what was regarded as the exploratory nature of experimental physics. Covering the mechanical and dynamical work of Descartes or Newton or their popularizers, or even just the properties of matter (especially minerals and crystals), such courses were usually taught during the winter semesters, when the cold chill of lecture halls foreclosed demonstration experiments. Yet even during the eighteenth century, the opposition between "theoretical" and "experimental" was also considered artificial, constructed for pedagogical purposes but not necessarily found in practice. Although someone like Johann Fischer argued in his *Physikalisches Wörterbuch* in 1800 that mathematics—principally the mathematics of measurement—should be used to make lecture courses in experimental physics rigorous, he (like others) had a very narrow conception of what those mathematical techniques constituted. He maintained that for physics to become a *Wissenschaft*, it had to be mathematized and quantified, but he seemed not to have been certain about how to classify those parts of physics that had been given mathematical form. He included mechanics in general *Naturlehre*; light, heat, electricity, and magnetism in special *Naturlehre*. But in his view, mechanics was also applied mathematics, along with optics and astronomy.[58]

By the early nineteenth century, especially after the development that the French gave to mechanics and other areas of physics, analytical mechanics had become so mathematized that it appeared in German university catalogs as a part of mathematics. We have already seen that in his 1836 article "Mathematics" in Gehler's *Physikalisches Wörterbuch*, the mathematician H. W. Brandes asserted that as a consequence of how Joseph Fourier, Siméon-Denis Poisson, and others had mathematized physics, heat and electricity would soon be considered a part of applied mathematics as well. In his view, the extension of mathematical methods into physics was to be watched cautiously because it often entailed the introduction of hypotheses. Two years later, in the same *Wörterbuch*, we see signs of a breakdown in the distinction between the two types of quantification (exact experimental and mathematical). The physicist Georg Muncke wrote in his article "Physics" that although "one differentiates *theoretical* and *experimental* physics as opposed to one another and uses the expression *mathematical physics* without implying an opposition to the experimental," theoretical and experimental physics were not necessarily different types of physics. Instead, according to Muncke, when applied to specific examples, they merely identified the dominant manner of treatment. In the end, Muncke thought, "there is no *theoretical* physics in the strict sense of the word." Referring to Bessel's seconds pendulum investigation and other examples of an exact experimental physics, he argued

58. Johann Karl Fischer, *Physikalisches Wörterbuch*, 10 vols. (Göttingen: Dieterich, 1798–1827), 2:302–3 ("Experimentalphysik"); 3:887–903 ("Physik").

that "[in] the execution of observations and experiments, measurements as well as the determination of size and relation are entirely indispensable. In these cases, therefore, one can dispense with mathematics as little as in the formation of general laws."[59]

On this point, Neumann would have agreed with Muncke. For like Muncke, he defined "theoretical" broadly, especially in his course on theoretical physics. But he also adhered to an older pedagogical definition of theoretical physics as mechanics. Despite his partial retention of old traditions and his apparent alignment with current definitions, he was also doing something new and significant in his course. At a time when analytical mechanics was more strongly identified institutionally—that is, pedagogically—with mathematics, Neumann brought mechanics more firmly within the curricular domain of physics by making it a foundational course and by stressing its advantages for practical instruction in an exact experimental physics. At a time when equations in mechanics and French mathematical physics were believed by some (especially the French) to possess the truth of mathematics, Neumann was teaching his students that they did not. His intent in teaching mechanics was to train students to think critically about physical theory, especially about the uncertainties contained in ideal mathematical expressions of physical phenomena, uncertainties that could be determined through data and error analysis, especially through the computation of the most probable error by the method of least squares. Mechanics supplied the technical and analytic means, not the concepts, for that critique. Neumann continued to use mechanics in this way through the 1870s, when others either viewed the pedagogical function of mechanics otherwise—as the conceptual foundation of physics—or constructed mechanics courses from an entirely different perspective.[60]

When Neumann first offered his course in 1838/39, and for some time thereafter, neither was physics so thoroughly mathematized nor were its separate branches so strongly based on mechanical principles that his choice of mechanics as an introductory course was an automatic one. Indeed, the textbook he recommended his students know before entering the seminar, Ernst Gottfried Fischer's *Lehrbuch der mechanischen Naturlehre*, was at first not well received in the German-speaking areas of Europe, in part because it rejected the speculative tradition in *Naturphilosophie* and argued so strongly that the more closely physics was bound to mathematics, "the more one penetrates knowledge of nature." Fischer was intent on defining physics more sharply and on eliminating the "uncritical mixing of certain with hypothetical views" found in older

59. H. W. Brandes, "Mathematik," in *Johann Samuel Traugott Gehler's Physikalisches Wörterbuch*, revised by H. W. Brandes et al., 11 vols. (Leipzig: Schwickert, 1825–45), 6.2:1473–85, on 1478, 1480; G. W. Muncke, "Physik," in ibid., 7.1:493–573, on 505, 513.

60. E.g., as did Hermann von Helmholtz, who viewed it from a more philosophical perspective; see Hermann von Helmholtz, *Einleitung zu den Vorlesungen über theoretische Physik*, ed. A. König and C. Runge (Leipzig: B. G. Teubner, 1922).

textbooks. Physics proper, Fischer maintained, was "mechanical *Naturlehre*," which was "in its essential parts almost entirely mathematical." (In Fischer's classification, mechanical and chemical *Naturlehre* dealt with inorganic forces, which together with organic forces—then known "incompletely"—comprised theoretical *Naturlehre* or theoretical physics. Applied *Naturlehre* included such subjects as physical geography.)[61]

Yet despite his strong support of mathematical physics, Fischer was not arguing for a *purely* mathematical physics, which in any case was impossible to teach to gymnasium students. Moreover, the pure mathematical theory of motion was, in his view, still a part of applied mathematics, and physics was fundamentally oriented toward empirical investigations: "The essential business of physics is to find out through observations and trials the foundations on which the mathematician can further build."[62] Hence laced throughout his book were references to recent experiments and data, which included the instruments used. Fischer's and Neumann's approaches to teaching physics overlapped in their emphasis on the principal measuring instruments of physics. Through mechanical principles Fischer explored the theoretical foundations of instruments such as the balance and the pendulum. In other ways, he provided a rudimentary introduction to measuring operations. Students learned the theory of scales on instruments through his discussion of how thermometers and barometers were calibrated through both testing and calculation. He showed students how to evaluate two versions of the same instrument— the calorimeters of Rumford and Lavoisier, for instance—and how to compare the measurements taken with them. Although Fischer stressed the need to determine physical variables with the highest possible degree of accuracy, he also demonstrated in simple ways how "accurate" measurements were compromised by the constant errors built into instruments and by corrections for those errors. Fischer, who intended that his textbook be used primarily for upper level gymnasium forms, nevertheless dealt with all these issues only in a cursory fashion. Although Neumann had known the book for some time, he did not seem to draw upon it immediately or directly in constructing his curriculum.

The approach to mechanics Neumann eventually took was not the only one available to him. A few of his own investigations, especially one from 1841 on double refraction, involved reductionist descriptions of physical processes at the microscopic level in terms of mechanical principles. In them he also relied more on hypotheses than his course on mechanics might suggest he would, especially on hypotheses concerning the alleged "optical ether" and its interaction with matter.[63] But he chose not to teach in mechanics in this way. Nor did he elect to teach the analytical mechanics

61. Ernst Gottfried Fischer, *Lehrbuch der mechanischen Naturlehre*, 3d ed., 2 vols. (Berlin: Nauck, 1826–27), 1:2, vi–vii, ix, 2.
62. Ibid., 1:3.
63. "Die Gesetze der Doppelbrechung des Lichts."

of mathematicians. Until Neumann offered his course on theoretical physics in 1838/39, his students could have learned analytical mechanics with Jacobi, who began his course with the mathematical equations of motion, using Lagrangian formulations. In 1834 and 1835, Jacobi had introduced into his lectures Hamilton's formulation of dynamics, but when he lectured on mechanics in the winter semester of 1837/38, he concentrated on techniques of integration (a novel feature of his approach to mechanics, as most were of his own creation) and in general approached mechanics as the French, especially Lagrange, had done: through virtual velocities, D'Alembert's principle, and Lagrangian formulations in general. Both he and the four students who took his course on dynamics, introduced in 1842/43, considered it difficult, largely because of Jacobi's extension of Hamiltonian formulations. Seidel, who had taken Jacobi's course on dynamics the first time it was offered, remembered it as being "pure, theoretical astronomy" where the main interest was "in its difficulty and in the interesting mathematical considerations to which it led."[64] The role of Jacobi's course in understanding the physical side of analytical mechanics was small; for he treated physical issues abstractly and dealt not at all with instruments and other experimental considerations. Had Jacobi's course on analytical mechanics been the required foundation for the study of physics in the seminar, the seminar exercises in the physical division would have been radically different, both conceptually and technically, and theoretical physics at Königsberg would have evolved in quite another way.

In 1826, when Neumann began to teach at Königsberg, and even eight years later, when he opened the seminar, he had not thought of offering an introductory course or a course on mechanics, nor had he considered a curriculum composed of courses arranged hierarchically. Only after dealing repeatedly with inadequate student preparation in the seminar, where he interacted more closely with his students, did he create a special introductory course that, he soon realized, served his students and his pedagogical objectives well. But although necessity can account for the timing of his introductory course, didactic issues only partially explain its content (mechanics, as the most thoroughly mathematized science to date, helped to teach a mathematized physics) and its strong emphasis on error analysis.

Effective science pedagogy at advanced levels also depends on an awareness of the ideas and techniques of scientific practitioners and on the ability to reshape them into a form suited for the classroom. There was, however, no simple relation between teaching and research, between

---

64. Ludwig Seidel to his parents, 2 December 1842, quoted in Lindemann, *Seidel*, pp. 35–36; see also *C. G. J. Jacobi's Vorlesungen über Dynamik* [1842–43], ed. A. Clebsch (Berlin: Reimer, 1884); Koenigsberger, *Jacobi*, pp. 37–38, 173–75, 240–42, 295–96.

students' needs and disciplinary demands, in Neumann's course on mechanics. Unlike his courses on light and heat, in which he extended to new cases the mathematical methods developed by others, his course on mechanics was not designed for creating new scientific ideas, even though it incorporated advanced research techniques. The mathematical and measuring methods Neumann had begun to employ by the late 1830s were found in it, as were techniques from several other scientific investigators, especially Bessel. That Neumann's pedagogical style paralleled that of Bessel's investigations of instruments suggests that Bessel continued to exert a powerful influence on his perception of scientific practice, a point that is amply confirmed by the central role of the pendulum in his course. Neumann referred to Bessel frequently in his seminar exercises on mechanics, recommending Bessel's seconds pendulum investigation "as the best model to study when combining a theoretical investigation with observation."[65] He constructed his entire course on theoretical physics around the principle that the pendulum, as studied by Bessel, was a paradigm for physics instruction. It was not primarily a paradigm for solving problems in the study of motion (as it had been for Galileo and others), or even in the analysis of forces. It was a paradigm for the theoretical analysis of instruments, of errors, and of experimental protocol because, as Neumann pointed out, it was the most accurate measuring instrument then available. Dominating the quantification of physics, as Neumann undertook it in his teaching, was the notion of "*astronomische Genauigkeit*," the accuracy of astronomy.[66] Bessel's analysis of instruments thus conveyed the skills but, more important, also the values, including the epistemological ones, that guided scientific investigations.

Error analysis, especially the method of least squares, was a matter of intense concern at Königsberg in the late 1830s. By then Bessel had extended the techniques he had used in his seconds pendulum investigation to other instruments with equally satisfying results. In 1835 he also published a theoretical analysis of how to use a barometer to measure height, thereby building on one of the well-established problem areas in which exact methods of quantification had first entered physical inves-

---

65. Neumann, *Einleitung in die theoretische Physik*, p. 78. Seconds pendulum investigations had been important in several countries since around the mideighteenth century, primarily in the context of weights and measures reform. But their impact varied. For instance, David Miller ("The Revival of the Physical Sciences in Britain, 1815–1840," *Osiris* 2 [1986]: 107–37, on 126) has discussed the role of seconds pendulum investigations in bringing mathematical, error, and data analyses and a cooperative spirit into British experimental physics after 1815. Yet although these new methods transformed British physics and led to a "fuller understanding of the pendulum and its uses in determining the figure of the earth," British interest in the pendulum died out by the mid-1830s. Judging from the evidence presented by Miller, seconds pendulum investigations and their methods seem not to have had a deep and lasting impact on exact experiment in Britain.

66. F. E. Neumann, *Vorlesungen über theoretische Optik*, ed. E. Dorn (Leipzig: B. G. Teubner, 1885), p. 57.

tigations in the eighteenth century. After beginning observations with his heliometer in 1837 with the intention of measuring parallax, he was convinced by the spring of 1838 that there was a parallax of about 1/3 minute and that 61 Cygni was a double star—results apparent only after an extensive analysis of errors. Other types of measuring techniques occupied him at the end of the decade, too. He was heavily involved in triangulation, in a study of the perturbations in the path of Uranus, and in the reform of Prussia's weights and measures.[67] Bessel also became more deeply drawn into what the method of least squares meant when applied to observational data and especially into the meaning of the most probable error. In 1837/38 he examined the probabilities involved in the errors found in telescopic observations, a study that interested Jacobi so much that by 1840 both he and Bessel published methods for calculating least squares.[68] By the late 1830s and early 1840s the interest at Königsberg in accidental errors and the method of least squares was thought to distinguish the exact sciences there. It was Jacobi's belief that at Göttingen, the other most likely location for the cultivation of these techniques, that Gauss no longer considered the method of least squares so important and no longer lectured on it, as Bessel continued to do at Königsberg. Although Jacobi was wrong about Gauss (who did in fact continue to lecture on the method of least squares), it was his *perception* that interest in the method was far more intense at Königsberg that is significant.[69]

In his own way, Neumann also contributed to the heightened concern for the method of least squares. Sometime between 1834 and 1838 he reviewed lecture notes taken in Bessel's course on geodesy, especially Bessel's discussion of how the method of least squares was used in observations in general.[70] We have already seen that once he had used the method of least squares to understand irregularities in observational data, he next began to use it as a tool for refining and criticizing theory. Then in 1838 he developed new analytic methods for computing the constants of

67. F. W. Bessel, "Ueber Höhenbestimmungen durch das Barometer," *AN* 12 (1835): cols. 241–54; idem, "Bemerkungen über barometrisches Höhenmessen" [1838], in *AFWB* 3:249–65; idem, "Lebensabriss [Fortsetzung vom Herausgeber {Rudolf Engelmann}]," in *AFWB* 1:xxiv–xxxi, on xxvii. On the introduction of rigorous quantification into barometric experiments see Theodore S. Feldman's exemplary study, "Applied Mathematics and the Quantification of Experimental Physics: The Example of Barometric Hypsometry," *HSPS* 15 (1985): 127–97.

68. F. W. Bessel, "Untersuchungen über die Wahrscheinlichkeit der Beobachtungsfehler" [1838], in *AFWB* 2:372–91; idem, "Ein Hülfsmittel zur Erleichterung der Anwendung der Methode der kleinsten Quadrate" [1840], in *AFWB* 2:398–400; idem, "Neue Formeln von Jacobi, für einen Fall der Anwendung der Methode der kleinsten Quadrate" [1840], in *AFWB* 2:401–2.

69. Jacobi to his cousin Schwinck, 31 October 1837, quoted in Koenigsberger, *Jacobi*, p. 237. Gauss's courses on least squares are listed in *Index Scholarum publice et privatim in Academie Georgia Augusta*, 1837–44.

70. FNN 57: Vorlesung über Geodäsie, wahrscheinl. von Bessel, pp. 144–56 (on least squares).

equations involving longitude and latitude (such as for geotemperatures or geomagnetism) which were as accurate but not as tedious or time consuming as the method of least squares. Impressed, Bessel urged Neumann to apply his theoretical results "as soon as possible."[71] Finally, by 1838/39, Neumann introduced the method of least squares into physics pedagogy via his course on mechanics, and by 1840 it surfaced in his seminar exercises.

By adapting error analysis to his pedagogical physics, Neumann could discipline students just as Bessel had done in the observatory. Through the analysis of constant errors, students could correct the imperfections in instruments and apparatus, compensate for them, and put them into the proper shape for conducting an investigation. Mechanics was an essential foundation for these tasks because most instruments were constructed according to mechanical principles and because the disturbing conditions in the environment of the experiment could most often be handled through mechanical equations. To this end, the topics in Neumann's course on mechanics could be viewed as conditions that could be isolated and quantified in an experimental investigation. In 1862/63, when Neumann discussed the pendulum (this time the ballistic pendulum), he explained that its study "provided the opportunity to investigate how to extend many theoretical principles, such as the moment of inertia and the main axes of rotation," which students needed to know in order to understand the principle of bifilar suspension and the disturbances that could affect it.[72]

As important as Bessel's example was to Neumann's pedagogical physics, though, Neumann's adaption of Bessel's exact experimental techniques was neither uncritical nor purely imitative. Neumann did not, for instance, consider it necessary to introduce personal equations into physical investigations. Introducing the method of least squares and error analysis in general into physics pedagogy would have been considered a first-rank achievement even if Neumann had confined their use to experimental concerns. But Neumann, unlike Bessel, also viewed them as strategic tools in the analysis of theory. To understand why, we must turn from the short-term causes of his pedagogical innovation (student needs and his adaptation of Bessel's methods) to one long-term trend shaping part of the problem structure of physics. Neumann's adaptation of Bessel's methods and values would have been slavish and even meaningless were it not for compelling disciplinary needs.

The mathematization of physics that had been accomplished by the late

---

71. "Ueber eine neue Eigenschaft der Laplace'schen $Y^{(n)}$ und ihre Anwendung zur analytischen Darstellung derjenigen Phänomene, welche Functionen der geographischen Länge und Breite sind" [1838], in *GW* 3:425–37; F. W. Bessel to Neumann, 10 July 1838, FNN 53.IA: Briefe von Collegen; see also Bessel to Alexander von Humboldt, 29 July 1838, A. v. Humboldt Nachlaß, SBPK-Hs.

72. Draft report, physical division, 1862/63, FNN 48: Seminar Angelegenheiten.

1830s was mostly the product of French efforts. Much of Neumann's own work and his physics courses centered on discussions of French theories, including René-Just Haüy's crystallography, Joseph Fourier's theory of heat, several works by Siméon-Denis Poisson, Augustin Jean Fresnel on optics, and others. Although by the late 1830s Neumann was drawing from the works of Bessel, Cavendish, and Gauss, it was Fourier, Fresnel, and Poisson who figured most prominently in his pedagogical physics, with references to Arago, Biot, Gay-Lussac, Laplace, and others. The works of each scientist certainly offered Neumann the kind of material he needed for constructing exemplars of a mathematical physics. But from the perspective of his seminar exercises, what was important about these French theories was, first, their often exaggerated claim to certainty and, second, the relative absence of the kind of solid empirical verification achieved through a careful examination of experimental conditions using rigorous techniques of error analysis. That is not to say that French investigators did not confirm their theories or calculate and reduce their errors—although several did not—but that their experimental style did not measure up to the standards or deploy the techniques that Neumann had learned from Bessel.

Although much remains to be told in the story of the relation between experimental trials and mathematical formulations in French mathematical physics, enough is known to suggest that Neumann was not alone in judging French efforts at confirmation as inadequate. Bessel himself criticized Laplace's treatment of observations, claiming that he represented "only the average observation accurately."[73] Despite Poisson's slightly greater attention to experiment in comparison to Fourier, Poisson took few measurements that he could compare with theory and did not necessarily develop theory that lent itself to measurements. Although Fourier attempted to quantify errors, he gave more importance to experimental trials at the beginning of his investigation than at the end. His style was more one of deducing formulas from general principles.[74] Haüy's work generally lacked data, estimates of accuracy, and an accounting of errors.[75] As Jed Z. Buchwald has so admirably demonstrated, French experiments in optical theory crossed a distinct divide in the 1810s. Because of the shift from geometric reasoning to more sophisticated forms of mathematization, the cultivation of higher standards of experimental accuracy at the École Polytechnique, and improvements in instrumentation, Malus, Biot, but especially Fresnel reported their numerical

73. Bessel to Gauss, 23 October 1805, GN.

74. Kenneth L. Caneva, "Conceptual and Generational Change in German Physics: The Case of Electricity, 1800–1846" (Ph.D. diss., Princeton University, 1974), pp. 345–46, 349; see also James R. Hofmann, "Ampère, Electrodynamics, and Experimental Evidence," *Osiris* 3 (1987): 45–76; and Caneva, "Conceptual and Generational Change," pp. 355–63.

75. Jed Z. Buchwald, *The Rise of the Wave Theory of Light: Optical Theory and Experiment in the Early Nineteenth Century* (Chicago: University of Chicago Press, 1989), pp. 12, 18, 19.

data in tabular form, included estimates of errors, and in general wrote more sophisticated experimental reports than their predecessors had.[76]

Yet even though the French improved upon the accuracy of their results in optics, they had clearly not achieved the precise level of experimental protocol that Bessel had. They eliminated errors by artificial means: by artful manipulations of the experiment, by combining trials, or by averaging sets of data. Their accounting of errors was not always rigorously quantitative, and they did not always go back to theory afterward and use quantitative determinations of error, including the method of least squares, to modify theory as Neumann was beginning to do in both his teaching and his research. In contrast to Neumann, who was guided by the example of Bessel's seconds pendulum investigation, French investigators, Neumann's student Woldemar Voigt recognized, adjusted their protocol and instruments to reduce error and increased precision primarily by taking averages of ever more measurements.[77] As accurate as French quantitative results may have been, French theories neither satisfied the rigorous accounting of errors nor were sufficiently qualified by error and data analysis, especially by the application of the method of least squares, to satisfy the standards Neumann and Bessel advocated.

By the time French mathematical physics came to be considered in German academic circles, skepticism about the use of mathematics in physics was still strong enough to provoke critical probes of the empirical and analytic foundations of French theories and the limits of their validity. In Neumann's pedagogical physics, examples drawn from French theories provided the opportunity to apply the techniques of an exact experimental physics to accomplish that critical assessment. So although Neumann's courses and seminar exercises were enriched by their integration of the literature on physical theory, they did not convey the sense that the proper object of investigation in theoretical physics was the synthesis of theory. Instead his courses and especially his seminar exercises suggested that the investigative enterprise in mathematical physics consisted in a more analytic task, the comparison of theory and experiment. Neumann introduced examples from French mathematical physics and other theories for this purpose in the winter semester of 1838/39, when, in addition

76. Ibid., pp. 12, 41, 84, 89, 99, 118, 123, 167. Most French investigations did not exceed a fourth-decimal-place precision, as compared with the fifth- or sixth-place precision of Bessel.

77. Woldemar Voigt, "Gedächtnissrede auf Franz Neumann" [1895], in *GW* 1:1–19, on 11–12; cf. Buchwald, *Wave Theory of Light*, pp. 30, 36, 84. Why the method of least squares was not as popular among French experimentalists as it was in some German circles awaits closer scrutiny. Part of the answer might be found in French debates over the method's applicability; see Ivo Schneider, "Laplace and Thereafter: The Status of Probability Calculus in the Nineteenth Century," in *The Probabilistic Revolution*, vol. 1, *Ideas in History*, ed. L. Krüger, L. J. Daston, and M. Heidelberger (Cambridge, Mass.: M.I.T. Press, 1987), pp. 192–214, esp. pp. 193, 201–2, 209.

to his new course on mechanics, he also offered a new course on capillarity, followed a year later by a new course on elasticity theory.

In his course on elasticity, for instance, he taught Augustin-Louis Cauchy's formula for dispersion in uncrystalline media, noting that "Cauchy
also compared his formula with numerical data." But Cauchy "found that
the first three terms of his infinite series completely satisfied reproducing
the measurements accurately. The series converges quickly enough to
prove itself as practically useful. In spite of that, however, one may not
conclude from this favorable success *that the foundation of the theory corresponds just as certainly to reality because the same [result] . . . could probably have
been derived by many other suitably chosen interpolation formulas.*"[78] For this
reason, Neumann argued, one could doubt the reliability of the foundation of Cauchy's explanation, which rested in part on the assumption that
"distant ether particles" caused dispersion and which predicted that dispersion could then occur in empty space. So Neumann integrated into the
course his own interpretation of dispersion, worked out in full between
1841 and 1843. He assumed that optical ether and matter interacted with
one another, an assumption that agreed with the numerical data as well or
better than Cauchy's but did not lead to impossible predictions.[79]

That more than one mathematical expression or theoretical censtruc-
tion could account for the same data was problematic for Neumann. We
can view his deployment of error analysis, as well as the dialectical exchange between theory and experiment which he was developing, as ways
of limiting the choices among theories that were proposed to explain data.
One had to have a way of deciding among alternative theoretical expressions because not all were equally likely. For Neumann it was not just that
a mathematical formulation of theory had to agree with data. More important, it must be possible through experiment to determine the range
within which it agreed and to measure the theory's parameters, including
its constants. That critical comparison between theoretical results and experimental data, made possible especially by the method of least squares,
constituted the core of Neumann's advanced seminar exercises, which
were intended in part to result in refinements of theory. In the winter
semester of 1860/61, for instance, when seminar students worked on
capillary theory, Neumann assigned a problem whose mathematical solution required a close examination of empirical conditions: "the general
conditions that are filled at the boundary of two liquids." This exercise was
supposed to facilitate a comparison with theory and the determination of
capillary constants. Students also examined the motion produced by capillary forces, the force needed to stop such motion, and the equilibrium
figure of two fluids mixed together and set in rotating motion. The real
focus of their attention, however, was on refined procedures in experi-

78. F. E. Neumann, *Vorlesungen über die Theorie der Elasticität der festen Körper und des
Lichtäthers*, ed. O. E. Meyer (Leipzig: B. G. Teubner, 1885), p. 289 (emphasis added).
   79. Ibid., pp. 290, 296.

mentation that aided the assessment of theory, such as the influence of capillarity on the determination of specific gravity and methods for determining the depression of mercury in a barometer. At the same time, they learned the advantages of moving from Laplace's capillary theory to Gauss's and finally to Neumann's, which could better handle, but not entirely solve, certain empirical conditions, such as the boundary conditions that had to be filled at the intersection of the surfaces of three fluids.[80]

Neumann attributed the cool reception that Georg Simon Ohm's theory of the galvanic circuit met in Germany not to its mathematical formulation per se (as historians since then have proposed) but rather to the fact that physicists were "still unaccustomed to viewing the correctness of theoretical consequences as proven through accurate measuring observations."[81] Since error analysis was among the techniques Ohm had used for confirming his mathematical formulation, Neumann's comment on Ohm's theory (like his many discussions of French works) suggests that he viewed error analysis as useful for establishing the veracity of theory and the best among several competing theoretical expressions. It is this application of error analysis that marks Neumann's creative advance over Bessel. Whereas Bessel applied error analysis largely to instruments and used his results to modify them and his experimental protocol, Neumann applied it to theory as well, which was then to be amended according to the analytic expression of error. Just as for Bessel an instrument manifested an incomplete reality that required correction factors, so for Neumann a theory did not quite represent reality and required correction factors too. The importance of error analysis in Neumann's investigative and pedagogical style also explains why he advocated going back to theory after making measurements; for theory had either to be so modified or be considered an interpolation formula or an abstraction. The French either omitted this last step or minimized its importance.

From time to time, Neumann advocated solving problems "from a theoretical point of view" by developing equations "more rigorously."[82] Yet he never wanted pure mathematical deduction or aesthetic criteria alone to dictate the shape of theory. He also did not view measurement simply as the precise determination of physical constants. The course on mechanics cultivated more complex conceptions of both mathematization and measurement as well as interactions between theory and experiment. In his course he advocated paying constant attention to the conditions under which measurement was performed, especially to those conditions

80. Draft report, physical division, 1860/61, FNN 48: Seminar Angelegenheiten; A. Wangerin, *Franz Neumann und sein Wirken als Forscher und Lehrer* (Braunschweig: F. Vieweg & Sohn, 1907), pp. 161–62.

81. F. E. Neumann, *Vorlesungen über elektrische Ströme*, ed. K. Von der Mühll (Leipzig: B. G. Teubner, 1884), p. 1.

82. Neumann, *Einleitung in die theoretische Physik*, p. 215.

that did not fulfill ideal assumptions or laws. Mathematical laws of physics, in turn, were to be developed or modified with a consideration of the instruments that produced measurements; for instruments helped to determine, by the degree of accuracy they allowed, what terms could be neglected. The comparison of theory and experiment was thus an exercise in deploying approximation techniques and in determining significant terms.[83] Insofar as Neumann's course on mechanics had an epistemological role in his pedagogical physics, it was not one of supplying the conceptual foundation of physics but, rather, of providing the quantitative experimental tools that could be used to shape physical theory and judge the limits within which it could be confirmed. Hence although in some respects the course resembled Friedrich Kohlrausch's later "practical" physics, Neumann, unlike Kohlrausch, always went back to theory after analyzing the results of an exact experiment.[84]

Neumann's course on mechanics also suggested that exact experiment could help in the refinement of theoretical formulations in more than technical and analytic ways. For instance, one problem taken up in seminar exercises as well as in later investigations by former seminar students (because of its usefulness in studying the kinetic theory of gases) was the motion of fluids in tubes, especially narrow tubes.[85] According to Jean Léon Marie Poiseuille, whose theoretical and experimental investigations provided the foundation for all subsequent studies on the topic, the velocity, u, of a fluid in a tube was directly proportional to its pressure, h, and the square of the tube's radius, R, but inversely proportional to the tube's length, l: $u = mhR^2/l$, where m was the coefficient of friction (or viscosity coefficient) of the fluid. Poiseuille, however, had only used narrow tubes in his experiment (0.03mm to 0.6mm) and had not taken into account the effect of temperature. When Neumann and his students used Poiseuille's law, then, they made several modifications, adjusting it to actual experimental conditions and reworking its theoretical form. This was a step common to almost every example Neumann presented in his course; what was unusual was the opportunity Poiseuille's law offered for examining a theoretical matter.

Neumann pointed out that because m was an internal friction (or viscosity) that fluid particles experienced in passing by one another, there was a "great interest" in pursuing further the motion of fluid in a tube "because there is in this m the measure of a new kind of force, the force with which the fluid particles affect one another in motion."[86] To examine further the nature of this force, Neumann did not assume point atoms acted on by short range forces but rather constructed, as was characteris-

83. See in particular Neumann's discussion of the determination of specific gravity, ibid., pp. 146–51.

84. Friedrich Kohlrausch, *Leitfaden der praktischen Physik* (Leipzig: B. G. Teubner, 1870).

85. Neumann, *Einleitung in die theoretische Physik*, pp. 246–64.

86. Ibid., p. 247.

tic of his approach to physics, a more phenomenological model. He thought of the fluid in the tube as composed of cylindrical pencils having a finite cross section, each pencil moving with a different velocity parallel to the main axis of the tube. With this model he was able to develop a modified law that preserved the general features of Poiseuille's law (the velocity remained proportional to the pressure and square of the radius and inversely proportional to the tube's length) but that took more precise account of the behavior of the fluid in the tube, including the errors stemming from the measurement of the velocity of each pencil and the effect of friction. Although Neumann did not carry out the implications of his model any further, in particular to show how the fluid particles interacted with one another at a more refined level, he was one step closer to so doing.

Neumann certainly wanted his students to apply mathematics to physical phenomena, but they had to learn when it was appropriate to do so. Error analysis gave him a way to teach quantification in stages so that students learned the limits of mathematization. The first step involved the application of mathematical equations, for which mechanics provided the model. Neumann often mentioned in his seminar reports that mechanical principles were "certain," "clear," and belonged to a "stronger domain" in physics. As he remarked in his 1857/58 report, instruction in mechanics was intended "to give an introduction to how physical phenomena are transferred into the domain of more rigorous principles and the calculus."[87] A more important step followed: determining the limits to those mathematical expressions. In 1853/54, Neumann showed students how measurements were tainted by "various assumptions" about the alleged accuracy of instruments, and then he showed his students (presumably after calculating constant errors) how to apply the method of least squares to the data they had taken. Teaching quantification in this way was especially important for those students for whom "mechanical principles . . . were still very vague and every ability to subject a physical phenomenon to a mathematical treatment was lacking."[88] Error analysis was thus a check on the compulsion to apply mathematics in a speculative fashion.

Careful experimentation in the seminar was essentially an exercise in attaining a small enough value for the most probable error, computed by the method of least squares. If the most probable error were too large, then that was an indication that the experimental skill of the investigator was somehow deficient. Error analysis was thus a powerful mechanism for normalizing judgment among students. Since every student did not per-

87. "Bericht über die Leitung der physikalischen Abtheilung . . . während des Zeitraums von Ostern 1857 bis Ostern 1858," n.d., ZStA-M, Math.-phys. Seminar Königsberg, fols. 243–44, on fol. 244.
88. "Bericht über die Leitung der physikalischen Abtheilung . . . während des Zeitraums von Ostern 1853 bis Ostern 1854," n.d., ibid., fols. 204–5.

form equally well, error analysis provided a means for "measuring" the differences between students on the basis of just how well each had attained precision and accuracy in experiment; how effectively students carried out their error analyses constituted a test of what they had learned and how well they had learned it. Error analysis, especially the computation of the most probable error, thus contributed to the internalization of both the values and the work habits of scientific practice. It suggested the sequence of tasks in an investigation and thus shaped the labor of science. Brix, for instance, had not only adopted Bessel's and Neumann's methods for calibrating thermometers but had also essentially followed the protocol of Bessel's seconds pendulum investigation in his thorough analysis of errors. For Brix and other students, the size of accidental errors indicated if they had gone far enough in determining constant errors and in refining experimental procedure. The method of least squares enabled students to quantify the last remaining residue of uncertainty after constant errors were calculated, and so to value the attainment of higher degrees of certainty.

Enhancing the importance of error analysis in Neumann's physics pedagogy were considerations that transcended the intrinsic and strategic contributions of mere furthering of precision in physical investigations. Insofar as error analysis criticized the foundations of mathematically expressed physical theories and assessed degrees of certainty in them, it strengthened the status of mathematical physics as a rigorously exact science rather than a branch of mathematics, including applied mathematics. Insofar as Neumann trained prospective secondary school teachers in error analysis, he promoted a more scholarly and professional orientation among them by supplying them with the tools for assessing knowledge in physics. Finally, the appearance of error analysis in his seminar exercises coincided with the reform of weights and measures in Prussia. Through Bessel's efforts, the Prussian foot received a scientific foundation in the seconds pendulum investigation and was established as the state's unit of length on 10 March 1839, just months after Neumann had initiated his course on mechanics in which the seconds pendulum investigation figured so prominently. Thus, the professional values cultivated in his seminar exercises, especially scientific exactitude, resonated with larger social and economic interests.[89]

That Neumann's course on mechanics played a central role in his curriculum meant that an *introductory* course, not his advanced ones,

89. F. W. Bessel, *Darstellung der Untersuchungen und Maassregeln, welche, in den Jahren 1835 bis 1838, durch die Einheit des Preußischen Längenmaasses veranlasst worden sind* (Berlin: Akademie der Wissenschaften, 1839); idem, "Ueber das preußische Längenmaass und die zu seiner Verbreitung durch Copien ergriffenen Maassregeln" [1840], in *AFWB* 3:269–75. Faced with the prospect that the Prussian foot might be replaced by the meter in 1862, Neumann clung to the standard of length grounded in Bessel's investigation; "Neumanns Bericht an die Preußische Regierung betr. die Einführung eines einheitlichen Maß- und Gewicht-Systems" [1862], rpt. in LN, pp. 448–52, on p. 452.

shaped the character of his institute. It also meant that the Prussian university reforms that brought teaching and research closer together did not necessarily mean that research dominated teaching, dictating the content of courses. The widespread curricular changes in Prussia before 1850 embraced the construction not only of advanced courses but of introductory ones as well, a sign that there was a pedagogical imperative to create courses for particular, and sometimes unexpected, classroom situations. Many professors, like Neumann, accepted the burden of creating introductory courses as a way to achieve higher goals. But in so doing, they came to define their fields pegagogically in ways that might not have occurred had classroom conditions been ignored: as a hierarchy of courses through which students moved systematically. Students had to have taken Neumann's course in theoretical physics before proceeding any further in the study of physics. Science pedagogy thus helped to form the institutional structure of disciplinary knowledge.

There is always a difference between a pedagogical physics and a practiced physics. It seems nonetheless relevant to reflect upon the conceptual significance of Neumann's use of mechanics as an introduction to theoretical physics and as a pedagogical foundation for the study of physics in general. The choice of mechanics as a foundational course might in retrospect appear to be a natural one, especially when considered from the perspective of the history of ideas, where it is often assumed that mechanics supplied a conceptual framework for interpreting the physical world and for organizing and unifying knowledge about it. The reasons for that assumption are not hard to find. When viewed from the vantage point of the spasms plaguing physics at the end of the century, there appears to have been before then an intellectual continuity in the function and role of mechanics within physics since the time of Newton: that mechanics was the foundation of a dominant, deterministic worldview in physics. But Neumann's choice of mechanics as an introduction to theoretical physics was not dictated by intellectual or philosophical convictions. More practical and immediate reasons shaped his decision: the needs of his students and the necessity to assess critically the limitations of French and other mathematical theories in physics. Neumann's course, by making precision measurement and the assessment of theory a matter of the probabilities of results, and by taking a phenomenological approach that de-emphasized causes, was not a simple, unqualified example of the deterministic, causal mechanics so often thought of as the foundation of physics for most of the nineteenth century.

In choosing the pedagogical strategy of emphasizing practical exercises in measuring techniques, especially error analysis, as a means of achieving these goals, Neumann had been guided by what I shall call the Besselian experiment. That experimental style included a quantification of error that went beyond the estimation or numerical determination of constant errors to the theory of the physical source of the error and the determination of accidental errors by the method of least squares. In the Besselian experiment, analytic means had priority over material ones in the pursuit of accuracy and precision; better instruments were not necessarily the best means to better results. For Bessel, the best experimental protocol transcended the material conditions of the experiment by using the observer's mathematical and technical skills to their fullest. Yet despite the prominent role of analytic and mathematical expressions in the Besselian experiment, the epistemological certainty of experimental data, as well as results based on it, was nevertheless considered less than the certainty of a mathematical theorem.

As Brix's investigation demonstrated, however, the adaptation of the Besselian experiment to physics pedagogy, especially to a pedagogy concerned with the construction of physical theory, was not without modification and even difficulties. Brix tried to determine correction formulas by considering theory, as when he applied capillary theory to his thermometer rather than relying merely on calibrating and testing it. But in the end, the problem he had chosen proved unamenable to this experimental style; for ϕ could not be readily expressed in analytic form. So Brix resorted to what the French had done and eliminated errors (or so he thought) by a combination of trials. In other ways, too, he seemed to go beyond what the Besselian experiment required. Rather than admit the limited certainty achieved by experimental means, Brix seemed driven to achieve more.

Neumann's repeated invectives against interpolation formulas as well as Brix's desire for greater accuracy indicate that in adapting the Besselian experiment to the pedagogy and practice of physics, Neumann departed from Bessel in more than one way. We have already seen that whereas Bessel used the techniques of error analysis only to enhance the precision of his results, Neumann used them also to modify theory or to set limits on its confirmation, both in his own research and, after 1840 or so, in seminar exercises. Our knowledge of physical reality, he seemed thereby to say, was not what was expressed in ideal analytic laws but what actually happened under the constraints of observation, be it in the laboratory, the observatory, or even the geothermal station.

A far more significant but subtler departure, however, concerned the relation between scientific and mathematical truth. The distinction Bessel drew between mathematical truth and truth based on experimental data was in one sense upheld by Neumann's skeptical attitudes both toward the analytic expression of theory uncorrected for experimental errors and toward interpolation formulas based on empirical data. In Neumann's view, analytic expressions as well as interpolation formulas had to be refined before they could be considered adequate expressions of theory. Neumann's departure from Bessel began in that process of refinement. In contrast to Bessel, who admitted that there were limits to the truth that could be achieved with experimental data (even when corrected for errors), Neumann's students—to the extent that Neumann's seminar reports and Brix's investigation portray the techniques used in seminar exercises and suggest the values cultivated by them—seemed driven to achieve higher and higher degrees of certainty. So although Brix's investigation upheld the difference between mathematical and scientific truth, it also blurred the distinction between them because he felt compelled to reduce and quantify error further before he could end the investigation and draw a conclusion. That is not to say that Neumann's students actually expected to achieve the certainty of mathematical truth but, rather, that mathematical-like accuracy remained an ideal useful for constructing the next step in an investigation and for arguing that still more had to be done in the analysis of error.

Neumann believed his course on mechanics as an introduction to theoretical physics had created a "practical" physics wherein the application of mathematics to physical phenomena could not take place without the checks and balances of an exact experimental physics. How "practical" that physics actually turned out to be, however, depended on the ability of his students to execute its methods so as to achieve its goals of refining measurements, expressing theory mathematically, and designing experimental protocol rationally and objectively. The difficulties Brix encountered in his investigation suggest that not all were able to measure up to the task.

CHAPTER FIVE

Successes and Realizations

W HEREAS BY THE early 1840s Neumann's course on theoretical physics provided the foundation for introductory exercises in the seminar, no comparable source of independent student projects was yet in place. Even though his seminar students took geothermal, geomagnetic, and optical measurements by the end of the 1830s, their observations were merely a part of seminar exercises. Presumably he continued to use advanced students as research assistants, as he had done with Hesse. But it seems his students generally viewed neither their participation in his research nor their original measurements in seminar exercises as potential sources of more involved investigations. They tended to view exercises principally as a way of acquiring technical skills in physics, not as stages in creative research. Brix's work was the exception. Thus it was effective teaching, not guiding student research, that absorbed Neumann for more than five years after the inauguration of the seminar.

By the late 1830s Neumann was working on the theory of induced electric currents in addition to his optical investigations. His interest in electricity found expression first in his teaching. Previously, only one student, Schumann, had undertaken simple galvanic exercises in the seminar in 1834/35. After Neumann introduced Gauss's geomagnetic investigations in 1839, exercises in electricity and magnetism appeared more frequently. Although Neumann never drew upon Gauss as much as Bessel, Neumann considered Gauss's techniques for measuring geomagnetic intensity both exemplary and suitable for attaining an accuracy comparable to what could be achieved by using free fall or a pendulum for measuring gravity.[1] By the winter semester of 1843/44, Neumann

---

1. F. E. Neumann, *Vorlesungen über die Theorie des Magnetismus*, ed. C. Neumann (Leipzig: B. G. Teubner, 1881), p. 19.

offered a new course on magnetism, followed by a new course on electromagnetism and electrodynamics in the summer semester of 1844. This period of pedagogical creativity (which turned out to be his last) preceded the publication, in the fall of 1845, of the first part of his seminal investigation of induced electric currents.

We do not know what these courses were like at first, but it is probable that Neumann used them to work out ideas he was still developing and that his research and teaching moved in tandem. In all likelihood the first version of his course on magnetism grew out of his earlier introduction into the seminar of Gauss's work and focused primarily on magnetic measuring instruments, their errors, and attaining "the accuracy of an astronomical observation."[2] Later, Gauss's work and the analysis of instruments figured prominently in the course, which Neumann is reported to have taught in a variety of ways. The version of his lectures that was eventually published notably began with the work of Siméon-Denis Poisson and also incorporated the work of André-Marie Ampère; neither French investigator had been particularly meticulous about the quantitative confirmation of his theory—Ampère, almost not at all. Neumann preferred neither the hypotheses of Poisson—that there were two magnetic fluids—nor of Ampère—that magnetism was the effect of small electric solenoid currents—because neither could be proven directly. After his elaboration of potential theory in his induced electric current studies in 1845 and 1847, he used the potential, not the French work he had earlier found useful, to develop the concepts of magnetic poles, moments, and axes.[3]

Although Neumann's course on magnetism proved useful in providing students with the background they needed for conducting geomagnetic observations, it was his course on electromagnetism and electrodynamics that was more thoroughly developed both theoretically and experimentally. Again we do not know how the course was initially offered except that the work of Georg Ohm, Michael Faraday, Alessandro Volta, Emil Lenz, and especially Ampère formed its conceptual core; later Neumann added the discussions of the quantitative experimental work of Gustav Fechner, Rudolph Kohlrausch, and Wilhelm Weber. Weber's work was particularly important. Although Neumann believed no explanation could yet be given of Weber's velocity-dependent "fundamental law," he considered it advantageous in the construction of theory. He showed how Ampère's law, which had not been rigorously confirmed, could be derived from Weber's, which had been extensively compared with experiment. The strong empirical foundation of Weber's law led Neumann to believe that "the building constructed on the basis of [Weber's] law can suffer no

---

2. Ibid., p. 53.
3. Ibid., pp. 29ff.; F. E. Neumann, *Vorlesungen über die Theorie des Potentials und der Kugelfunctionen*, ed. C. Neumann (Leipzig: B. G. Teubner, 1887); cf. FNN 25–28: Potentialtheorie [lecture notes from 1852/53, 1856/57, and 1872/73].

change." Weber's law, however, did have a crucial drawback: it did not lend itself easily to the computations Neumann required of his students in the comparison of theory and experiment.[4]

By the early 1850s, Neumann's course represented a synthesis of all existing theoretical discussions of the electric current, including by then his own explanation using potential theory, with conscientious references to the experimental confirmation of the laws of the electric current. As in his course on magnetism, the principal focus was on the relative advantages of measuring instruments and techniques. Students learned how to use, calibrate, and obtain data by means of Fechner's method for determining current strength, Gauss's and Weber's geomagnetic instruments, Becquerel's multiplicator, Neumann's differential multiplicator, rheostats, and other instruments and procedures. The lesson here was clear: scientific facts and theories do not exist independently of the specific instruments and experimental protocols that produced them. Students reviewed original data if Neumann had it on hand. About Volta's experiments on the pile, for instance, Neumann pointed out that Volta, although he had sought proof of his formulas for the "tension" of the pile through direct electroscopic observations, had had to resort to an indirect proof because "the value of the observed data was so small that his observations could scarcely be considered decisive."[5] Students compared Thomas Seebeck's data on thermoelectricity to Becquerel's; evaluated several methods for measuring current strength; and in other ways learned how to achieve accurate results.

Like his course on mechanics, Neumann's course on electricity projected a particular view of the investigative enterprise in physics. The mathematical expression of theory—whether of a physical phenomenon or of an instrument—was valid only insofar as the difference between calculated and observed values was smaller than the probable errors of observation. "If this condition is not found to be satisfied," Neumann warned his students, "then the formula does not possess the accuracy that corresponds to the sharpness of the measurements and consequently is not useful."[6] Practicing physics thus meant determining degrees of precision in part by describing how instruments operated in well-defined experiments and in particular by subjecting data to the method of least squares. One could broach the question of how the natural world operated in the abstract, Neumann seemed to say to his students, only after an extensive analysis of how physical phenomena were manifest in the manufactured, artificial world of the laboratory, which did not conform to ideal mathematical laws but was subject to constraining conditions often not taken into account in theory.

4. F. E. Neumann, *Vorlesungen über elektrische Ströme*, ed. K. Von der Mühll (Leipzig: B. G. Teubner, 1884), esp. pp. 2, 308; cf. FNN 29.I, 29.II, and 30: Elektrische Ströme [lecture notes from 1858, 1858/59, 1870, and 1872/73].

5. Neumann, *Vorlesungen über elektrische Ströme*, p. 6.

6. Ibid., p. 28.

Because these new courses so nicely applied the principles of investigation that students had learned in the introductory course on mechanics, they offered students the opportunity to go beyond the conceptually unexciting but technically challenging exercises in mechanics to new areas of investigation in electrodynamics and electromagnetism, where there was a greater likelihood of producing an original investigation. Opportunities for original research also expanded in the mid-1840s because Neumann himself was so heavily involved in a stimulating theoretical investigation on the induced electric current. His attempt to construct a general mathematical theory of induced currents promised to be a major contribution to his own efforts to mathematize physics; to the mathematization of Faraday's electromagnetic induction; and to the generalization of Gauss's potential function. Its advantage for his pedagogical physics, however, was found less in its achievement than in its shortcomings. The difficulties Neumann encountered in his mathematical formulation of induced currents inspired measuring exercises and projects for his students in the seminar.

Neumann's study of induced currents was his most theoretically oriented investigation to date.[7] Read before the Academy of Sciences in Berlin on 27 October 1845, it drew upon the investigations of Lenz, Ohm, Faraday, and Ampère to create a mathematical theory of the "simplest" case, linear induction in closed currents. Combining Lenz's law with the assumptions that the inducing cause entered with a velocity that was small in comparison with the propogation velocity of electricity (so as to apply Ohm's law) and that the intensity of the induced current was proportional to the velocity with which the conductor moved, Neumann constructed a general induction law, $E \cdot Ds = -\epsilon vCDs$, where Ds was an element of the induced wire having a unit of current; $E \cdot Ds$, the electromotive force induced in Ds; v, the velocity of Ds; C, the component of the force acting on Ds; and ϵ, a constant. To find the strength of the induced current, he divided the right hand side of his general induction equation by the resistance of the conductor and then integrated over the circuit to find the induced differential current, $D = -\epsilon\epsilon' \int vCDs$, where ϵ' is the inverse of the resistance. Since D could be observed directly only if it were constant, Neumann integrated D with respect to time to find the induced integral current, $J = -\epsilon\epsilon' \int dt \int vCDs$, where t is the time.

The mathematical equations for D and J suggested analogous ways of viewing the physical process of induction: the electromotive force of the differential current could be viewed as the negative virtual moment of the force exerted on the conductor; the electromotive force of the integral current could be viewed as the loss of living force in the conductor; and finally, the electromotive force of the integral current could be viewed as the difference in pressure of two equilibrium surfaces in a liquid of

7. Neumann, "Die mathematischen Gesetze der inducirten elektrischen Ströme" [1845], in *GW* 3:257–344.

density ϵ. These analogies helped to demonstrate that the integral current was independent of the length and type of the path of the inducing cause, being defined only by the end points of the path.

The most important consequence of Neumann's derivations thus far, though, was that he reconceptualized the problem of induction in terms of a potential function, $V = -\frac{1}{2} \iint ii' \cos (ds, ds') (1/r) \, ds \, ds'$, where i, i' are the intensities of the two currents; ds, ds', the respective path elements of these currents; r, the distance between the two currents; and $\cos (ds, ds')$, the cosine of the angle between the two path elements. Not only did the potential enable him to extend his analysis from electrodynamic to electromagnetic induction and to unite in a single formulation all known induction appearances, but he was also able to make the potential an independent theoretical concept, albeit one whose meaning, especially in relation to force, was still unclear. Consistent with his phenomenological approach that avoided hypotheses, he did not suggest either a mechanical analogy or even a physical interpretation for the potential, which remained an analytic construction. He considered the reason for induction to be "not in motion itself," but instead he identified "the change in potential" as "the *cause* and the *standard of measure [Maass]* of the induced current," pointing out that "it does not matter how this change in value is brought about."[8]

Neumann's discussion of induced currents, fairly harmonious with modern views, was discordant to his contemporaries. According to Jacobi, Johann Poggendorff, who presented Neumann's paper to the academy, found the essay "to be rather like Chinese. He did not even understand the first line." Jacobi suggested that because "in this subject the technical expressions do not appear to be so known or established as fact," Neumann should have included more detailed explanations or at least have sent one of his students, such as Gustav Kirchhoff, as his emissary to the academy. The task of editing the paper fell on Jacobi, who found it necessary to clarify Neumann's variables, especially his peculiar way of representing differentials and partial differentials. Knowing that Neumann's essay was quite different from other physical investigations published by the academy, Jacobi suggested to Poggendorff that it ought to be published in the mathematical rather than the physical class. It was probably as a result of Jacobi's suggestions that Neumann wrote a summary of his investigation, which was published separately and later added as a preamble to the academy version.[9]

Although the exact experimental physics that Neumann had been

8. Ibid., pp. 310, 309.

9. C. G. J. Jacobi to Neumann, 5 December 1845, 6 January 1846, and 28 January 1846, FNN 53.IA: Briefe von Collegen; Neumann to Jacobi, 5 February 1846, partially rpt. in Leo Koenigsberger, *Carl Gustav Jacob Jacobi* (Leipzig: B. G. Teubner, 1904), p. 362. Neumann's summary was first published as "Allgemeine Gesetze der inducirten elektrischen Ströme," *AP* 67 (1846): 31–44.

promoting in the seminar was not evident in this essay, experimental and even measuring considerations were nonetheless important in it. Occasionally he warned that the conditions expressed in the mathematical formulas might not be realizable experimentally. A condition might have an "analytic meaning" and a "calculational advantage," he thought, but not be reproducible experimentally. This was a troublesome prospect because according to the precepts of physical practice that he was busy teaching his students, one could not be comfortably confident of a theory, no matter how logical or internally consistent, without a rigorous, quantitative experimental assessment of the limits of its reliability. Conditions that brought the assumptions of his investigation to their limits or that did not satisfy them were especially problematic. How far, for example, his potential formulas applied to certain cases, "such as where the galvanic current suddenly changes or is broken," he pointed out, "still needed experimental proof." He did not report any data in this investigation, but he did try to demonstrate through alternative analytic formulations that certain conditions were indeed experimentally possible. He also tried to show how his theory related to other investigations that did involve measuring instruments.[10] So although he did not report any numerical results, the imperative to measure partly guided his analytic derivations. He claimed that he formulated his induction law in part "to determine numerically" the intensity of the induced current. Indeed he interpreted the differential and integral currents in terms of measurement: the differential current could be observed only when constant, while the effect of an integral current could be measured by a magnetic needle.[11]

The variable ϵ (the so-called induction constant) in his general induction law would appear to have been particularly suited for a numerical determination, but Neumann was not especially clear about what it meant. He claimed that ϵ could be "thought of as a constant" that had been shown by Faraday and Lenz to be "independent of the condition of the conductor; its numerical value therefore depends only on the units of length, time, and current strength." He also had thought that ϵ was in some sense a function of time, since it effectively disappeared when its argument reached a certain value. But he also referred to ϵ as a "current," and in his fluid flow analogy, it would have to be interpreted as the density of a liquid. When Jacobi asked him about ϵ, Neumann was vague, claiming it to be a "concept [*Begriff*]" that was the essence of all induction phenomena. At best he could say that its meaning was "still wrapped in deep mystery" and tied to the "mysterious connection" that existed between bodies.[12] His less than satisfying explanation of the constant ϵ

10. Neumann, "Die mathematischen Gesetze der inducirten elektrischen Ströme," pp. 294, 317, 331–40.
11. Ibid., pp. 271, 276.
12. Ibid., pp. 260, 272, 273, 261; see also Neumann to Jacobi, 5 February 1846, in Koenigsberger, *Jacobi*, pp. 361–62.

reveals the limitations in Neumann's ability to use philosophical principles creatively, to incorporate hypothetical assumptions, or to elaborate
upon the physical meaning of analytic analogies in the construction of
physical theory. These limitations were manifest in his lecture courses as
well and so formed a part of the image of the practice of physics that he
projected to his students.

Despite the novelty of Neumann's new courses on electricity and magnetism and his recent investigation on induction, not all the activity of the
physical seminar was oriented toward problems in electrodynamics and
electromagnetism during this period of pedagogical and intellectual creativity. He submitted a combined report for both divisions of the seminar
on 11 June 1845, covering the period from the summer semester of 1843
to the winter semester of 1844/45. The report was brief because Neumann's official responsibilities, especially as the university prorector preparing for the university's tricentennial celebration, absorbed his time. In
Jacobi's absence, Richelot had conducted the mathematical exercises once
again. When Jacobi left for Berlin in 1844, Richelot took his place as
director of the mathematical division.[13]

The quality of his students determined what Neumann offered in the
seminar. At least six students attended the physical division, and their
abilities were sufficiently different to require two sections, elementary and
advanced. Students in the elementary section completed exercises on
problems from mechanics, including the balance, the pendulum, Atwood's machine, and the steam engine. Neumann pointed out to the
ministry that the members of the advanced section—Gustav Kirchhoff,
Siegfried Aronhold, Carl Friedrich Boehm, and Hermann Westphalen—
all demonstrated the level in training in mathematical physics that seminar students could attain. About the younger members—C. G. Krause
and S. Brandeis—Neumann could only say that they exemplified the
kinds of "weak abilities" that one had to deal with in the seminar.[14]

Neumann was in fact making progress with his advanced students; all
those from 1843 to 1845 were awarded one of the seminar's premiums. In
1845, Theodor Ebel became Neumann's third doctoral student. Nothing
is known about Ebel's dissertation on borax, but he did study at the
university for six years and had taken most of Neumann's courses. Hermann Westphalen, one of Bessel's doctoral students, Bessel described as a
"young talented astronomer" who had worked on data taken on the path
of Halley's comet during 1835 and 1836 in order "to calculate the most
probable path of the comet" and to investigate if the difference between

13. ["Bericht über die physikalische Abtheilung, 1843–45"], 11 June 1845, ZStA-M,
Acta betr. das mathematisch-physikalische Seminar an der Universität zu Königsberg, Rep.
76Va, Sekt. 11, Tit. X, Nr. 25, Bd. I: 1834–61, fols. 133–34; Königsberg University Curator
Reusch to Kultusminister J. A. F. Eichhorn, 19 June 1845, ibid., fols. 131–32.

14. ["Bericht über die physikalische Abtheilung, 1843–45"], fol. 133; see also A. L.
Crelle to Kultusministerium, 20 August 1845, ibid., fols. 126–30, on fol. 126.

the observed and calculated values were "small enough to be considered an observational error" or large enough to indicate the existence of a force not considered. Westphalen appears to have taken Neumann's seminar only to supplement his astronomy courses, but he set an example for other students in how error analysis was done. Quoting Bessel, Westphalen emphasized that "one must be very careful in determining the most probable error of a calculation based on observations. One must not begin the determination before knowing the causes of the error and how to express their influence theoretically."[15]

But it was Gustav Kirchhoff about whom Neumann was most pleased. Drawn to the study of physics through Neumann's teaching, Kirchhoff was one of the two students (the other was Jakob Amsler) to flourish in the midst of Neumann's pedagogical creativity. Around the time Kirchhoff took the *Abitur* from the Kneiphofisches Gymnasium in Königsberg in 1842, he began to doubt his "calling to mathematics," the subject which until then had given him "so much joy." Inspired by his brother Otto, he took experimental chemistry at the university, but his "enthusiasm" for the subject "cooled" rather quickly. He then turned to physics. Within a short time, Kirchhoff considered Neumann to be his "principal teacher" to whom he listened "with great pleasure and zeal."[16] Kirchhoff's dedication to physics was achieved in spite, not because, of the way Neumann trained his students.

"A great part of my vacillation concerning which science I should apply myself to," Kirchhoff wrote, had "come to an end" upon beginning to study physics with Neumann, "even though with [physics] comes tedious observations and even more tedious calculations." Yet to his surprise he found himself suited to the task of taking observations, on one occasion taking geomagnetic readings "from 10 in the evening until 2 [in the morning] in the main university building behind a telescope, and [I] observed, at a temperature of 1°, [the oscillations of a magnet]. Still, with a cigar and several of the small Madiera cakes, which mother had carefully put in my pockets, the four hours passed as quick as an arrow before I even knew it." While at Königsberg between 1842 and 1847 when he received his doctorate under Neumann, Kirchhoff took astronomy with Bessel; higher mechanics with Jacobi; analysis with Richelot; integral calculus, differential calculus, geometry, analytic geometry, the theory of differential equations, variational calculus, elliptic functions, and Abelian functions with Ludwig Otto Hesse (who had become *Privatdozent* in 1840 and *außerordentlicher* professor of mathematics in 1845); and theoretical physics, theory of light, elasticity theory, magnetism, electricity, galva-

15. Hermann Westphalen, "Ueber die Bahn des Halley'schen Cometen in den Jahren 1835 und 1836," *AN* 24 (1846): cols. 333–47, 365–79, on 333, 335, 369n–70n; 25 (1847): cols. 165–92.

16. Gustaf Kirchhoff to Otto Kirchhoff, n.d., quoted in Emil Warburg, "Zur Erinnerung an Gustav Kirchhoff," *Die Naturwissenschaften* 13 (1925): 205–12, on 205.

nism, capillarity, and hydrostatics with Neumann. He also participated in
the mathematico-physical seminar between 1843 and 1847, and although
he probably attended the seminar continuously over eight semesters, the
records are not accurate enough to confirm that he did.[17]

By the summer of 1845, Kirchhoff had completed a major electrodynamic investigation drawn from what Neumann had taught. In his 1845
report, Neumann described Kirchhoff's seminar investigation to the ministry as being the product of a "thoroughly trained talent" but did not
send the ministry a copy, choosing instead to send it immediately to
Poggendorff's *Annalen* for publication.[18] When Kirchhoff's article appeared in 1845, it was the first published seminar investigation in which
the author identified himself as a "member of the physical seminar at
Königsberg" (Figure 8). His study was a theoretical and experimental
examination of the distribution of electrical currents on a plane, in particular on a circular disk. As he knew from Ohm's law, electricity entering
and leaving a plane from fixed points would distribute itself according to
certain characteristic patterns; what he sought was an experimental confirmation of the "curves of equal tension [*Spannung*]" predicted by theory,
which, for a conducting disk with one entry and one exit point for the
current, should be circles. Using a thin copper disk, Kirchhoff confirmed
their existence by measuring along points of the disk with a multiplicator.
Although he found some deviations from theory, he considered them "so
small that they probably could be explained satisfyingly by the irregular
conductivity of the copper disk and by the observational errors." In
conveying his results, the "curves of equal tension," through a graph
(Figure 9), he became the first seminar student to represent experimental
results visually in this way.[19]

On the basis of his results, Kirchoff regarded his theory as "adequately"
proven, although he communicated only a few deviations (without precisely indicating their cause or source), and he did not exhaustively determine the quantitative value of his observational errors as Brix had done,
by deriving correction formulas for his measuring instruments and for
the "environment" of his experiment. Instead, in order to obviate experimental errors, Kirchhoff varied his experimental procedure, but not as
earlier French investigators had done; for Kirchhoff sought to express
experimental conditions explicitly in quantitative terms. In the course of
trying to determine just how irregular the conductivity of the disk was, he
constructed circuits of various types so as to be able to observe small

17. Warburg, "Kirchhoff," p. 205; Kirchhoff's vitae, n.d. [c. 1849], Darms. Samml. Sig.
Flc 1859(2).
18. ["Bericht über die physikalische Abtheilung, 1843–45"], fol. 133.
19. Gustav Kirchhoff, "Ueber den Durchgang eines elektrischen Stromes durch eine
Ebene, insbesondere durch eine kreisförmige" [1845], in *Gesammelte Abhandlungen* (Leipzig:
Barth, 1882), pp. 1–17, on p. 12 (herafter cited as *Ges. Abh.*); idem, "Nachtrag zu dem
vorigen Aufsatze" [1846], in ibid., pp. 17–22.

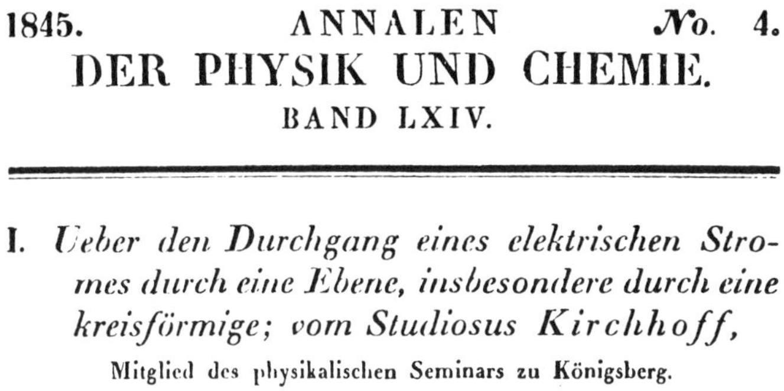

1845.   ANNALEN   *№.* 4.
DER PHYSIK UND CHEMIE.
BAND LXIV.

I. *Ueber den Durchgang eines elektrischen Stro-*
mes durch eine Ebene, insbesondere durch eine
kreisförmige; vom Studiosus Kirchhoff,
Mitglied des physikalischen Seminars zu Königsberg.

Figure 8. Portion of title page from Kirchhoff's first investigation. From Kirchhoff, "Ueber den Durchgang eines elektrischen Stromes durch eine Ebene, insbesondere durch eine
kreisförmige," p. 497.

changes in resistance. It is this part of his essay that proved to be the
enduring contribution of his investigation; for in order to determine his
experimental errors exactly, he formulated what later became known as
Kirchhoff's laws: that the algebraic sum of currents at a junction is null,
and that the sum of the electromotive forces in a branched closed circuit is
zero. In 1846 he published a "postscript" to his essay in which he compared theory and experiment once again—he seemed to think more
accurately—but this time he more fully developed the theory behind the
magnetic deflections by taking into account the magnetic moment of the
needle and the horizontal component of the earth's magnetism. Constant

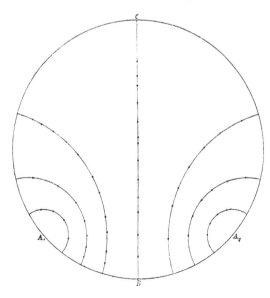

Figure 9. Kirchhoff's
drawing of the curves of
equal tension. From
Kirchhoff, "Ueber den
Durchgang eines elektrischen
Stromes durch eine Ebene,
insbesondere durch eine
kreisförmige," Table V,
Figure 3.

errors were still present, but once again Kirchhoff did not say how rigorously he had determined them or how large they were. He reported the
calculated and observed values of the deflections in tabular form; the
experimental values deviated from the calculated ones by 0.78 percent to
1.1 percent.[20]

For his achievement, Kirchhoff was awarded 30 taler from the prize
funds of the seminar and 15 taler to support his investigation (one other
student received 45 taler; another, 30; and three others, 15 taler each).[21]
Yet despite his accomplishments, Kirchhoff did not demonstrate in his
seminar investigation the same kind of attention to practical matters that
other students had. Brix, for example, would have derived correction
formulas for the conductivity of the copper disk; developed a theory of
the instruments used; and explained in other ways the limitations of the
experiment. In the published version of his investigation, Kirchhoff mentioned neither these nor the routine operations in error analysis he had
learned in the seminar. His analysis of experimental data was simple and
direct: he calculated averages, reported his data in tabular form, and let
the difference between theoretical and experimental values stand as his
measure of accuracy. As his derivation of the laws still linked to his name
demonstrates, in trying to understand his experiment more closely,
Kirchhoff varied experimental conditions by formulating mathematical
descriptions of them, not by varying immediately the material conditions
of the experiment (although that followed). In general he was more
interested in his mathematical formulas than he was in his measurements.
Despite these differences between him and Brix—differences that could
in a certain sense be regarded as falling short of the techniques considered elementary in seminar exercises but advanced in terms of where
those techniques were supposed to lead—Kirchhoff, unlike Brix, reached
a conclusion.

In scientific circles, Kirchhoff's achievement brought fame to him and,
by association, to the seminar. Representatives of the exact sciences in
Berlin found much to praise in Kirchhoff's fledgling investigation. Poggendorff immediately recognized its novelty, its nonhypothetical foundation, and its importance for the study of measuring instruments in electricity. Six months after he received Kirchhoff's essay, he presented a
"glorifying lecture" on it to the Academy of Sciences. Jacobi in particular
hoped that Kirchhoff would come to Berlin, but he thought Kirchhoff
would not profit from his visit "if he does not want to learn chemistry." Yet
Jacobi knew that in ministerial circles, enthusiasm for Kirchhoff's type of
investigation was not widespread, and so he urged Neumann to press
government authorities—the *Kultusminister* Eichhorn and if necessary
Alexander von Humboldt himself—for funds to support Kirchhoff's

20. Ibid., pp. 1, 14–15, 17–22; cf. Neumann, *Vorlesungen über elektrische Ströme*, p. 255.
21. Reusch to Eichhorn, 19 June 1845, fol. 131.

Berlin sojourn. Although Kirchhoff eventually received support to go to Berlin, the ministry did not discuss his investigation in their response to Neumann's seminar reports, much to Neumann's disappointment. (In his evaluation of the seminar from 1843 to 1845, Crelle merely lumped Kirchhoff's work together with that by other advanced students on geomagnetism, the rainbow, and the polarization of light.) Along with his seminar report for 1845/46, Neumann sent the ministry an offprint of Kirchhoff's first *Annalen* paper, but as he reported to Jacobi, "the result was that the minister said nothing at all about the work of the seminar but was only pleased with its good progress."[22]

The ministry's apparently unsatisfying response to Kirchhoff's work is on the surface curious. Kirchhoff more than anyone else represented the epitome of what the seminar was designed for. But there are underlying reasons for the ministry's response, some of them having to do with Neumann's own evolving understanding of his pedagogical strategy. For although Neumann praised Kirchhoff's continuing investigations in June 1846, he also knew they were not representative of the bulk of the student activity in the seminar, and the difference was considerable. Rather than concentrate on Kirchhoff's achievements in his report to the ministry, Neumann decided to justify his own pedagogical strategy. He reconfirmed his commitment to elementary instruction, to a graduated training program, and to what he had learned from experience in the early years of the seminar. "The written work" of the students, he explained, was closely related to his lectures and had two purposes: first, "that what had been known, assumed, or otherwise gone over in the lectures was produced by the [seminar] members themselves"; and second, that "the principles given in the lectures were worked out in exercises."[23] Most students were thus involved in much less than Kirchhoff, whose continuing accomplishments showed that Neumann was trying to deal with the few students who could move quickly ahead by themselves even though he concentrated more on those who had to be helped each step of the way.

The higher priority he placed on introducing students to the study of physics did not entirely, however, prevent Neumann from promoting original student research more strongly in the mid-1840s, especially now that he was deeply involved in his own investigations of the induced electric current. The prize question he posed for the philosophical faculty in 1846—the experimental determination of the constant ϵ that had remained enigmatic in his theory of induced currents—was a classic

---

22. J. C. Poggendorff to Neumann, 20 June 1845, FNN 53.IA: Briefe von Collegen; Jacobi to Neumann, 5 December 1845, 6 January 1846, ibid.; Neumann to Jacobi, 13 January 1846, partially rpt. in Koenigsberger, *Jacobi*, pp. 360–61; "Bericht des mathematisch-physikalischen Seminars von Ostern 1845 bis Ostern 1846," 4 June 1846, ZStA-M, Math.-phys. Seminar Königsberg, fols. 151–52, on fol. 151; see also Crelle to Kultusministerium, 20 August 1845, ibid., fols. 126–130, on fol. 127.
23. "Bericht . . . von Ostern 1845 bis Ostern 1846," fol. 151.

example of what historians have generally considered to be the modus operandi of German university research institutes: professors directing research by meting out to advanced students unsolved problems from their own investigations, the solutions for which necessitated deploying the research techniques cultivated in class.

Kirchhoff's response to this prize question became his second major investigation as well as his doctoral dissertation. But in contrast to his first investigation, where he had not been explicit about the conditions of his experiment, his work was now more in keeping with what was emphasized in seminar exercises. Although he had to design his circuits so that he could calculate their potentials, the mathematical aspects of the problem concerned him less than the execution of the experiment—how he would measure the inducing and the induced currents. He used a magnet to which a mirror was connected by a wire, the two together hung from a silk thread. About twelve feet away from them stood a telescope directed at the mirror. A multiplicator was connected to the magnet, and a cabinet similar to the one Bessel had used in his seconds pendulum investigation protected the magnet and the mirror from the disturbances caused by air currents. Before beginning his measurements, Kirchhoff calculated the errors afflicting the experiment, or "at least," he claimed, "established a limit over which they could not go." He chose solenoids as his two current carrying wires, because he at first thought that their potentials were easy to compute. By the end of his calculations, he found they were not. The strength of this first of a series of three different answers Kirchhoff wrote on the prize question was in its experimental parts.[24]

Kirchhoff, who "had no confidence" in the quality of his work, considered it "a great surprise" that he had won the prize question competition and even thought that Neumann hinted beforehand that he had won the prize only to relieve Kirchhoff's chronic hypochondria. "I have never before worked so strenuously in my life," he wrote to his brother in January 1847. "I believed that when I had finished it I would collapse. But not so! I never felt happier or more cheerful." The award bolstered his ego and his spirits, and he felt confident that he could go on. He wanted to submit his investigation as his doctoral dissertation, but he knew that some parts, especially his derivation of the potentials and their expression in a form suitable for performing measuring observations, would first have to be refined. He thought that he "could finish [the calculations] in several weeks," but he became sick. When he returned to his work by late January 1847, he proceeded to spend five months on "entirely uninteresting numerical calculations" until he "finally had to admit" he "was not on the right path." Kirchhoff then fretted that he had not reflected as carefully as he should have on the course of his investigation so that such detours could have been avoided.[25]

24. FNN 49.I: Kirchhoff's Preisaufgabe; quote on p. 24.
25. Gustav Kirchhoff to Otto Kirchhoff, 16 January 1847, 9 June 1847, quoted in Warburg, "Kirchhoff," p. 206.

Between 1847 and 1849, Kirchhoff reworked his prize essay twice, once for his dissertation and later for publication. Unlike his prize essay, which began with experimental matters, his dissertation opened with more general theoretical considerations. He thought the "most natural way" to compute the constant ϵ was to give the conductors a form for which the potential could be computed. The most tractable form, he determined, was a solenoid, where the windings of the wire were assumed to be close enough to be treated as circles at infinitely small distances from one another. Computing the potential once again proved difficult; unable to calculate one elliptic integral, he had to use mechanical quadrature. After spending eighty pages calculating the potential, he turned to his experiment, arranging his circuits and multiplicator so that his own laws of circuits could apply. He computed the horizontal component of the earth's magnetism and adjusted his protocol to compensate for other disturbing factors. Kirchhoff then took simple averages of his results and expressed the value of ϵ in terms of the units used in his experiment: "The constant that is sought, K [as he called ϵ in this investigation], equals 1, if one uses as a unit of velocity the velocity of 1,000 *Fuss* per second and as a unit of resistance the resistance of a copper wire of 1 square *Linie* cross section and 0.4337 *Zoll* length."[26]

Among Kirchhoff's accomplishments in this study was his calculation of the induction constant in electromagnetic terms. But, as in his earlier investigations, accurately measuring the resistance of his wires was extremely difficult. So despite the beauty of his calculations of potentials, he knew that his methodology was still flawed in its experimental parts. At this point, Kirchhoff, the exceptional student eager now to cross the threshold into the world of scientific practice, denied himself the immediate satisfaction that might be obtained by rushing into print, choosing instead to try to fulfill the standards of the exact experimental physics in which he had been trained. He had wanted to publish his dissertation immediately, but, as he wrote to Neumann in October 1848, "it had not escaped me that many criticisms could be directed at it, in particular that I had given no consideration to how the resistance of the wire depends on temperature." Moreover, after Kirchhoff had visited Leipzig in August 1848, and as a result of conversations he had with Wilhelm Weber while there, he gave up his plans to publish because, he reported to Neumann, Weber was planning to bring out a work "in which he shows that there exists an absolute measure for electromotive force and resistance." Not only did Kirchhoff believe that "Weber's work therefore treated the same subject as mine" but also that Weber had conducted his investigation "in a much simpler way than I because he used the current induced by geomagnetism." The "main reason" that Weber's work "appears to make mine superficial," he concluded, was that Weber had a much better way

26. FNN 49.II: Kirchhoff's Dissertation; quote on pp. 2, 109. Kirchhoff recopied his dissertation for Neumann's files; see Kirchhoff to Neumann, 13 October 1848, FNN 53.IIA: Briefe von Schüler.

of expressing and empirically determining the resistance of the wire. Kirchhoff had only been able to "give the resistance of a copper wire of a definite length and thickness, by which no accurate determination of the various conductivities of the copper could be made."[27]

Although Kirchhoff was unable to solve as accurately as he might have liked the problem of how to measure the resistance of the wire, he nevertheless published his determination of ϵ in 1849 on the strength of its improved experimental protocol and of the theory of the multiplicator he developed. In contrast to his dissertation, which had been laden with equations relating to his theoretical derivation of the potential, the published version had only one paragraph that discussed the potential (he did not even reproduce his formulas). He reported that the sixfold integral for the potential could be reduced to a twofold integral with the help of elliptic transcendentals, but the remaining expression had to be calculated using mechanical quadrature, "and this required rather boring calculations."[28] The bulk of his essay concentrated on the experimental issues that had constituted his prize essay of 1846. He corrected for errors in his multiplicator readings by computing the horizontal component of the earth's magnetism; the horizontal magnetic moment of the multiplicator needle; and the horizonal moment of rotation cause by a unit current on a multiplicator needle, when the needle is in the magnetic meridian. All these corrections he had made from theoretical considerations. In practice, he admitted, other adjustments should be made as well, including computing the effect of the current produced in the multiplicator. But he could not complete all of them. The problem of determining the resistance accurately for all parts of his circuit remained unsolved, and so he had to manipulate his equations in order to minimize the error. Kirchhoff designed his experiment as in his prize question, using a telescope directed at a suspended magnet and mirror.

Kirchhoff still expressed his result in the same cumbersome way he had in his dissertation; that is, he described what experimental conditions would lead to a unit value for ϵ. He changed those conditions slightly, though, by reducing the number of significant figures in the length of the copper wire to three, reporting a length of 0.434 *Zoll*. He did not say if the change had been the result of calculating errors more accurately, of executing the experiment more carefully, or of realizing the limits within which his calculations were accurate (he did not mention using the method of least squares). Nor did Kirchhoff state directly the certainty in his calculation of ϵ but only mentioned that because "the conductivity of the copper varied between certain values," the numerical value of ϵ had only a "limited accuracy."[29]

27. Gustav Kirchhoff to Neumann, 13 October 1848, ibid.
28. Kirchhoff, "Bestimmung der Constanten, von welcher die Intensität inducirter elektrischer Ströme abhängt" [1849], in *Ges. Abh.*, pp. 118–31, on p. 131.
29. Ibid., p. 118.

In his seminar report for 1846/47, Neumann had mentioned that Kirchhoff, who had been the only student who had returned to the upper section of the physical division that year (the remaining students were in it for the first time), was engaged in a "great experimental investigation, which I will not neglect to send to the ministry when it is completed."[30] Neumann could have been referring to either the continuation of Kirchhoff's efforts to determine the value of ϵ or to Kirchhoff's further development of the circuit equations (Kirchhoff's laws) that he had introduced in 1845. In 1847, Kirchhoff had published a more thorough treatment of the circuit equations for systems of linear conductors, and in 1848 he had extended his analysis to systems that consisted in part of nonlinear conductors.[31] By 1849, then, all his published essays either had originated in seminar investigations or were problems that had arisen in the course of solving them.

It is worth reflecting on Kirchhoff's education and academic performance in the seminar because he is so often cited as an examplar of the first generation of professionally trained physicists in Germany. He was in fact "well trained," but it cannot be forgotten that his investigative strategy did not conform completely to what he had learned. He dealt with practical matters in a way similar to what had been taught in seminar exercises, as the investigations of other students demonstrated, but he never allowed practical matters alone to dominate the course of his investigation. He took into account certain constant errors, but his analysis was not always exhaustive, and small remaining differences between theory and experiment were not grave issues for him. Although the shortcomings of his investigation were at first troublesome enough to inhibit Kirchhoff from publishing his results, he eventually did publish them, even though the problem of how to calculate the resistance of his wires more accurately remained.

What we see in Kirchhoff is an urge to move quickly beyond experimental conditions and deal in more sustained ways with the mathematical construction of theory without being encumbered by a consideration of empirical results. Not only did the analysis of constant errors not become an obsession with him, but he also did not indicate that he relied to any great extent on the measure of uncertainty in data provided by the method of least squares. Sometimes he took into account his less than absolutely rigorous treatment of errors when he compared theory and

30. "Bericht über das math. physikalische Seminar der Königl. Universität zu Königsberg von Ostern 1846 bis Ostern 1847: Physikalische Abtheilung," n.d. [c. 1 July 1847], ZStA-M, Math.-phys. Seminar Königsberg, fol. 157.

31. Gustav Kirchhoff, "Ueber die Auflösung der Gleichungen, auf welche man bei der Untersuchung der linearen Vertheilung galvanischer Ströme geführt wird" [1847], in *Ges. Abh.*, pp. 22–33; idem, "Ueber die Anwendbarkeit der Formeln für die Intensitäten der galvanischen Ströme in einem Systeme linearer Leiter auf Systeme, die zum Theil aus nicht linearen Leitern bestehen" [1848], in ibid., pp. 33–49.

experiment. In 1845 he said that his theory was "adequately" proven (he did yet not use the term *rigorous* when talking about the confirmation of theory); at other times he did not. The graph of his "curves of equal tension" (Figure 9), allegedly drawn from measured points, depicted perfect circles, just as theory had predicted, without any indication of the range, determined by the method of least squares, within which those points were accurate. It thus seems that other standards of theory construction guided Kirchhoff, telling him when enough was enough in the perfection of instruments, the refinement of experiment, and the analysis of data, including the determination of error.

That partial rejection of what he had learned—one may also call it risk taking—proved crucial in his shaping as a creative professional physicist. As the evolution of his investigations on the determination of ϵ indicates, Kirchhoff, who did not interpret theory exclusively in terms of the analysis of instruments or experimental protocols, felt himself pulled in two somewhat contrary directions: toward experiment but also toward "pure" theory. In the end, he adhered only within limits to the technical values and standards promoted in the seminar and so was able to be more purely "theoretical" than Brix had been. Although Kirchhoff subsequently conducted refined experiments, it was his theoretical orientation that became stronger. He took up the study of nonlinear conductors in 1848, for example, not only to generalize his own equations for circuits but also to generalize Ohm's law. At the end of his 1848 investigation of nonlinear conductors, he also tried to interpret his equations in terms of Joule's law for the production of heat in a galvanic current.[32]

At least part of Kirchhoff's concern for theory may be attributed to Neumann's success in shaping seminar exercises according to his research interests in induced currents in the 1840s. Between 1845 and 1847, Neumann generalized his theory of induced electric currents to cases where either the form of the conducting circuits was altered or the intensity of the electric current in one of the conducting elements changed. After the publication of Wilhelm Weber's *Electrodynamische Maassbestimmung* in 1846, Neumann also attempted to bring bring Weber's theory—which was based on Ampère's law and had assumed the existence of positive and negative electricities—and his own into agreement.[33] At least one other seminar student, Jakob Amsler, who had come to Königsberg University from Switzerland in 1844, drew upon Neumann's continuing theoretical investigations in shaping his own. Notably, he, like Kirchhoff, emphasized theory rather than experiment. Amsler, who also studied with Richelot in the mathematical seminar, was first mentioned in Neumann's seminar reports in 1847.

32. Kirchhoff, "Ueber die Anwendbarkeit der Formeln," pp. 33, 43.
33. Neumann, "Ueber ein allgemeines Princip der mathematischen Theorie inducirter elektrischer Ströme" [1847], in *GW* 3:345–409; Wilhelm Weber, *Elektrodynamische Maassbestimmungen*, in *Wilhelm Weber's Werke*, ed. Kgl. Gesellschaft der Wissenschaften zu Göttingen, 6 vols. (Berlin: Springer, 1892–94), 3:25–214.

During the winter semester of 1847/48, Amsler began an experiment (we only know it concerned electromagnetism) in the seminar which he had to end when revolutionary political events in 1848 forced him to return home. He hoped to find the materials and equipment he needed to continue his investigation in Zurich, but in the meantime he "laid out and sketched the apparatus in such detail" that he thought he would be able to finish the experiment quickly so that he could publish his investigation. He even suggested that Neumann send it along as an "appendix" to Neumann's annual seminar report, even though Amsler had been out of the seminar long enough to be disqualified from receiving a seminar premium for it.[34] Matters did not work out as well as he had hoped, so he put his experiment aside and turned to a theoretical derivation of the potential of a free magnetic fluid at a point on the surface or inside of an iron object, a project he had also started in the seminar. His derivation was guided by the analogies he saw between the mathematical expression of the potential in heat, electricity, and gravitational attraction; he believed that "up to a point, these disciplines can be handled together." His demonstration of the limitations of Poisson's derivation of the potential, for which an exact solution was impossible except for a sphere, impressed upon him the need for a balanced interplay between theory and experiment. Amsler pointed out that in the treatment of practical questions, "a more general solution to the induction problem than analysis alone can provide" was needed. He indicated that Neumann, who was dissatisfied with practical methods of determining magnetic inclination, developed his own method from a "completely rigorous theoretical foundation." Neumann's formula for the magnetic potential enabled Amsler to determine certain coefficients "with the help of several observations" of the dimensions and structure of the iron body. Even though Neumann's formula had certain restrictions—the body had to be homogeneous and bounded by a surface of rotation—it nevertheless enabled Amsler to calculate potentials more general than Poisson's formula would have and to move from ideal cases to real ones in spite of the experimental errors that were not easily eliminated. More analytically oriented than Brix had been in his study of latent heat, Amsler did not discuss ways in which those errors could be reduced. Although doubtful at first that his derivation would allow for practical applications, less than a year later he was able to use his results for a pressing navigational problem: how to compensate for the effect of a ship's iron on a compass needle.[35]

Despite the successes of Kirchhoff and Amsler, Neumann did not direct

34. Jakob Amsler to Neumann, 29 February 1848, FNN 53.IIA: Briefe von Schülern.
35. Jakob Amsler, "Zur Theorie der Vertheilung des Magnetismus im weichen Eisen," *Neue Denkschriften der allg. Schweizerischen Gesellschaft für die gesammten Naturwissenschaften* 10 (1849): 3–26; idem, "Methode den Einfluss zu condensiren, welchen die Eisenmassen eines Schiffes in Folge der Vertheilung der magnetischen Flüssigkeiten durch den Erdmagnetismus auf die Compassnadel ausüben," *Verhandlungen der Schweizerischen Naturforschenden Gesellschaft* 33 (1848): 146–53. (Amsler completed his theoretical essay first.)

more of the seminar's activity into problems of a pure theoretical nature, much less into aspects of his own ongoing investigations. In 1846/47 he still had to divide his seminar into two sections, one for new members and one for returning students. There was a "great difference" in the work of the two sections, he told the ministry. Whereas the work of the upper section was characterized more or less by "independent" exercises, carried out with "outstanding success," the work of the lower section concentrated on learning how to apply mathematics to physics. Perhaps to distinguish the work of his lower section from the broad-based survey courses in experimental physics taught at Prussian universities, Neumann emphasized that for students who were "lacking talent," "clarity and skill in a limited area [of physics] is the best preparation" for advanced study.[36]

What is remarkable, in fact, is that, in the face of the auspicious success of two of his students in original research, Neumann instead directed his pedagogical efforts more than ever at basic matters. Between 1843, when Kirchhoff first entered the seminar, and 1849, when Amsler published his investigation, Neumann's views on the function of university instruction, especially in his seminar, evolved to the point where, although he did not completely abandon his hopes for original student research, he clearly had assigned a higher priority to more foundational and systematic means of instruction. In public and before his success with Kirchhoff, Neumann supported the notion that university instruction was designed in part for the promotion of pure *Wissenschaft*.[37] But at a time when he had begun to move his students systematically through certain courses in physics and had stressed the necessity of mastering certain fundamentals before moving on to independent projects, it seems in retrospect contradictory for him not only to have stressed "intellectual freedom" but also to have condemned the kind of systematic instruction available at a trade school and to have chastised students for taking courses only to prepare for the state's civil service examinations.

Part of his reason for conveying an image of pure *Wissenschaft* to the public had to do with the fact that in 1843/44, Neumann was using preparations for the university's tricentennial celebration—for many reasons a politically sensitive event—as a forum for acquiring a physical laboratory. That celebration was a nodal point in a string of "liberal" events from 1840 to 1848 in Königsberg that shaped the context within which the Prussian ministry viewed the material needs of the natural sciences. A visit from the crown in 1840 became an opportunity for several members of the university community, including Neumann, to speak about the needs of the university. It was also an occasion for local liberals, including the provincial administrator of East Prussia, Theodor

36. "Bericht . . . von Ostern 1846 bis Ostern 1847: Physikalische Abtheilung," fol. 157.
37. Neumann's speeches from 1843 and 1844 are reprinted in LN, pp. 354–55, 356–63. A draft of the 1843 speech is found in FNN 61.4: Ansprachen.

von Schön, to promote liberal causes, especially reform, sufferage, and political representation. Within a short time, Königsberg became sufficiently identified with liberal causes and social disruption to provoke measures to curb intellectual unrest and potentially dangerous social gatherings, including the annual student ball at the university, which was not held from 1842 until the tricentennial celebration in 1844.[38]

Although the official purpose of the university's tricentennial celebration was to honor *Wissenschaft*, the large crowds that descended on Königsberg from 28 August to 1 September 1844 (it was the largest unrestricted gathering the crown had sanctioned since 1840) transformed it into a *Volksfest*. But one historian's description of the tricentennial celebration as a "demonstration of liberalism against reaction" exaggerates the extent to which the festivities were overtly critical of the state's social and political policies.[39] The actual political and social impact was largely symbolic. Restrictions on students were only temporarily lifted. Within the limits of civil order, the festivities suspended the boundaries between social classes: the lower classes mixed with educated classes; eleven hundred students (over half from outside Königsberg) mingled with the professoriate; and the theme of unity—of classes, of religions, of the professions—was so apparent that Karl Burdach, directly evoking a liberal cause, described the celebration as "a picture of the unity of Germany."[40] Kirchhoff, who attended the festivities with other students from Neumann's seminar, viewed the celebration as a world turned on its head: "The otherwise peaceful and proper life in Königsberg was at this time turned completely upside down. The entire day, hordes of philistines and students moved through the streets from one pub to the next, singing and celebrating without being stopped."[41]

The official theme of the celebration was "knowledge is power," and no branch of knowledge played a more prominent role in the celebration and preparations for it than the natural sciences. It was not only that Neumann and Burdach had been chosen prorectors but also that by 1844 the university was perhaps best known for the achievements of its exact scientists—a point emphasized by the physicist Wilhelm Weber, who sent

38. On the restrictions imposed at Königsberg see Hans Prutz, *Die Königliche Albertus-Universität zu Königsberg i. Pr. im neunzehnten Jahrhundert* (Königsberg: Hartung, 1894), p. 141.

39. Fritz Gause, *Königsberg in Preußen: Die Geschichte einer europäischen Stadt* (Munich: Gräfe & Unzer, 1968), p. 169.

40. On the tricentennial celebration see Karl Friedrich Burdach, *Amtliche Nachrichten über die Feier des dritten Secularfestes der Albertus-Universität zu Königsberg* (Königsberg: Gräfe & Unzer, 1844), esp. pp. 2 (quote), 4–5; Alexander Jung, *Die grosse National-Feier des dritten Universitäts-Jubiläums zu Königsberg* (Königsberg: Tag & Rach, 1844), esp. p. 78; Prutz, *Albertus-Universität zu Königsberg*, pp. 144–46; Ferdinand Falkson, *Die liberale Bewegung in Königsberg (1840–1848)* (Breslau: Schottlaender, 1888), pp. 90–105; Dr. Stettiner, *Aus der Geschichte der Albertina (1544–1894)* (Königsberg: Hartung, 1894), p. 74.

41. Quoted in Warburg, "Kirchhoff," p. 206.

greetings from his university in Leipzig. Among the "gifts" to the university was an essay on the solar illumination of the clouds by a *Physiklehrer* at the Altstädtisches Gymnasium in Königsberg, J. H. C. E. Schumann, one of Neumann's first seminar students. The dean of the medical faculty, on the second day of the celebration, honored Karl Ernst von Baer as the individual who had secured a natural scientific direction for the medical faculty. "The natural scientific method has prevailed over medicine," the dean told those assembled before him. In recognition of the strategic importance of the natural sciences to its interests, the medical faculty awarded all nine of its honorary degrees to chemists and physicists, including Neumann and Dulk at Königsberg and Heinrich Rose, Gustav Magnus, Johann Poggendorff, and Heinrich Dove at Berlin.[42]

From the point of view of the faculty in medicine and the natural sciences, especially Neumann, the tricentennial celebration provided a perfect opportunity to try to gain additional improvements for instruction. The ministry funded the celebration at the cost of 12,000 taler (the size of a small institute), and the state was committed to laying the cornerstone of a new university building at the celebration. The ministry seems to have expected additional requests for the improvement of instruction. In December 1843, while prorector, Neumann submitted the university's formal requests for the celebration, listed in rank order, with the new university building first. Next, ahead of all other requests, including that for much-needed clinics for the medical school, Neumann asked for a physical and chemical laboratory for "fruitful instruction according to the present standpoint of these sciences." But in contrast to his public speeches, where he emphasized the pursuit of pure *Wissenschaft*, in his formal requests for a laboratory he justified its need by citing the practical applications that these sciences had produced and by invoking the pedagogical strategy he had adopted in the seminar. The laboratory was required, he argued, so that the rational and technical aspects of physics and chemistry could be cultivated. "Overall," he emphasized, "the termination of uncertainty and of practical incompetency in the exact sciences can be achieved by the construction of the scientifico-practical institute discussed herein."[43]

Neumann spent the following months detailing the needs and purposes of this institute, which he envisioned as for the students in the seminar. "Without a laboratory young men can be trained neither for teaching physics, as service to the state requires it, much less for the service of scholarship," he argued.

> Instruction in physics, as it is held in lectures, must be completed and built upon in the seminar. But the seminar can only achieve its purpose if it gives

42. Burdach, *Amtliche Nachrichten*, pp. 129, 91, 144, 145, 201.
43. Neumann to Kultusminister Eichhorn, 6 December 1843, GSPK, Acta betr. die Feier des dritten Säcularfestes der Universität zu Königsberg, Rep. 76/262, Abt. I, Nr. 12, Bd. I: Okt. 1843–Okt. 1844 (unpaginated).

young men the opportunity to assimilate, through independent and experimental work, the essence and spirit of methods in physics. In this way alone not only will they be in the position to solve a physical problem and to become familiar with the art and techniques of observation but they will also be able to judge what science holds for certain.

In his view, laboratory instruction was accomplished best if the students were provided with a place where they could work at any hour of the day or night, over long stretches of time, leaving it only to attend regularly scheduled lectures and seminars. It was also, in his view, a place where students could *expect* to receive the "encouragement and assistance" of the professor "at every moment." He explained that he had gone to great lengths in creating practical exercises in order not to allow instruction in the seminar to become "illusory." Aware that it would take years for a laboratory to be built, Neumann proposed an interim solution: that for the cost of about 500 taler, a house suitable for student laboratory work and for Neumann's living quarters be rented. He even offered to pay 200 of the 500 taler cost as his portion of the rent in order not to appear "egotistical," even though he was then paying only 150 taler rent for his current house. He also indicated that he would need an additional 100 taler for an assistant, who would also live at the laboratory, and for heating. In closing, Neumann appealed to the ministry's desire to have well-educated physics teachers, for whom he considered laboratory training essential.[44]

Around the time of the tricentennial celebration, Neumann must have received at least some positive reaction to his request, because only days after the celebration, on 10 September 1844, he wrote to the ministry about its acknowledgment that "the hindrances that cripple my academic activity will soon disappear and I will be able to apply my strength, as much as God has given me, to science and to the training of young men eager to learn." He argued that a laboratory "was the only way in which physicists and teachers of physics can be trained" and that "if there is any subject in which just listening to lectures leads to superficial and imaginary knowledge, it is the physical sciences." He thereby suggested that his pedagogical strategy went beyond the narrow confines of his own seminar to embrace physics instruction everywhere. Nothing firm came from the ministry, however, even though Jacobi, by then in Berlin, had spoken to the ministry and to the crown about Neumann's requests.[45]

From the awarding of an honorary degree to Wilhelm Weber in 1837 to

44. Neumann to Königsberg University Curator Reusch, 7 February 1844, 9 February 1844, ZStA-M, Math.-phys. Seminar Königsberg, fols. 117–18, 115–16.
45. Neumann to Kultusministerium, 10 September 1844, ZStA-M, Acta betr. den mathematisch- physikalisch- und chemischen Apparat der Universität zu Königsberg, sowie die Errichtung eines physikalischen- und eines magnetischen Häuschens, Rep. 76Va, Sekt. 11, Tit. X, Nr. 15, Bd. I: 1809–76, fols. 133–34; Jacobi to Neumann, c. 20 January 1845, FNN 53.IA: Briefe von Collegen.

the high profile of the natural sciences in the university's tricentennial celebration to the liberal events of 1848, the natural sciences at Königsberg were associated with liberal causes. It is, however, difficult to determine whether that association affected their level of financial support, especially when the depressed economic conditions of the 1840s could in themselves account for the ministry's fiscal conservativism. Still, it is worth considering the complexion of political views represented by several natural scientists at Königsberg. Both Neumann and Moser remained steadfast in their political convictions over the years. Moser, openly liberal, was favorably viewed by East Prussia's chief provincial administrator, Theodor von Schön. Neumann, however, was royalist, deeply Prussian, and supportive of most state policies; his role later in the 1848 revolution was that of peacemaker, trying one night to quell the unrest of workers.

By the mid-1840s, however, Bessel had modified his earlier liberal views. In 1843 he described political change as analogous to change in the natural world, where he believed there were no sudden alterations. He defined himself as conservative by this definition and did "not doubt that we in Prussia are in general conservative." Bessel's conservative turn was the result of depressed economic conditions and a reevaluation of the ability of commercial classes to change the world through existing institutions that worked slowly. Königsberg's position as an international commercial center had slipped markedly in consequence of the blockades earlier in the century; by the 1840s, with a decline in the demand for German goods, shipping and trade had noticeably suffered. Although East Prussia did not experience the same pressures of population growth that western and southern German states did between 1815 and 1845 and which affected the economic equilibrium of those regions, Bessel was nonetheless troubled by overpopulation because in his view it threatened to perpetuate narrow-minded interests. Formerly believing in the beneficial practical effects of Pestalozzian-inspired educational reform, he now even considered formal educational reform inefficient. He attributed the current ineffectiveness of the Königsberg Physico-economic Society to the decline of commerce and resulting low regard in which the public viewed trade; but he still felt that the goals of the society had been "given up too early." Instead he now urged commercial classes to disseminate public information through broadsides, flyers, and pamphlets—as was done in England—so that the intelligence of the population could be improved, overpopulation circumscribed, and the disastrous consequences of overpopulation stemmed.[46]

Bessel's new political views troubled Schön, who labeled the change the "Besselian sickness." By 1844 other political changes affected the standing of the natural sciences at Königsberg. Schön stepped down as Oberpräsi-

---

46. W. Ahrens, *C. G. J. Jacobi als Politiker* (Leipzig: B. G. Teubner, 1907), pp. 7, 21n–22n; F. W. Bessel, "Schreiben an die Redaction (der Königsberger Allgemeinen Zeitung)" [1843], in *AFWB* 3:486–89, on 488; idem, "Uebervölkerung," in *AFWB* 3:483–86.

dent of East Prussia in 1842, although he retained influence in East Prussian governmental circles. Neither Bessel nor Schön seem to have aided Neumann's efforts to gain a physical laboratory around the time of the tricentennial celebration. The East Prussian recommendation to the ministry on university improvements in honor of the celebration supported Neumann's request, but it was not enthusiastic, and it assigned priority to medical clinics over new science laboratories. It thus contrasted with Schön's earlier strong support for the natural sciences.[47]

After the heady events of the tricentennial celebration had long passed, Neumann renewed his requests for a laboratory but not always in the most politically expedient way. Although he continued to achieve outstanding success in the seminar with Kirchhoff and Amsler, he failed to utilize their accomplishments as ploys for a laboratory. Instead he stressed the importance of the foundational exercises and his desire to provide *all* seminar students with the opportunity to conduct practical exercises— complicated ones, if possible. But as long as he did not have a laboratory, "the few experimental exercises" would continue to concern, he pointed out, "the calibration of instruments," such as the thermometer, and "participation in magnetic observations."[48]

Moser had also submitted a request during the tricentennial celebration, a modest one for 200 taler for physical instruments, claiming it was "an anomaly to be a physicist without instruments." Within two weeks of his request, the ministry granted the funds as a special "jubilee premium," but Moser was able periodically to renew his appropriation directly from university funds, which he did through at least the mid-1850s. One might perhaps conclude on the basis of the ministry's support for Moser that Neumann was ignored, mathematical physics unappreciated, and what really mattered was Moser's political views and his relationship to Schön. But the situation was more complicated. Before the 200 taler award, Moser had very limited funds at his disposal—only a small, unspecified portion of the 350 taler budget of the natural sciences seminar. Neumann, in contrast, controlled over one-third of the 350 taler for the mathematico-physical seminar, 158 taler for the physical cabinet, and 100 taler for the mineralogy collection (which was sometimes used for purchases in physics). In addition, by 1844/45 the operation of the mathematico-physical seminar had become more complex: an attendant, paid 10 taler per year, had been added to care for the seminar's growing collection of instruments; a modest book collection had grown into a reference library; and students occasionally worked as unpaid assistants. Moreover, financial records from 1841 to 1855 indicate that Neumann purchased or

47. Quoted in Ahrens, *Jacobi als Politiker*, p. 6; Oberpräsident [of East Prussia] to Kultusminister Eichhorn, 11 March 1844, GSPK, Die Feier des dritten Säcularfestes (unpaginated).

48. "Bericht des mathematisch-physikalischen Seminars von Ostern 1845 bis Ostern 1846," 4 June 1846, ZStA-M, Math.-phys. Seminar Königsberg, fols. 151–52, on fol. 151.

repaired instruments as needed but did not use all funds available to him. From time to time a surplus grew, was partially spent, and then grew again.[49] Neumann thus not only had more than Moser; he did not always use effectively what was at his disposal.

Neumann's repeated requests for a laboratory were for a permanent *location* where exercises could be conducted, not for instruments per se; he had enough instruments for his students to complete practical exercises as a part of their seminar learning. Over the years, the practical exercises moved across the university: from the rooms for the mineralogical collection to the main university building and sometimes to Neumann's house. But no location was permanent enough to allow the apparatus to stay up all the time. When Neumann was able in 1847 to set up a laboratory in a house purchased with his second wife's inheritance, he did so primarily with his practical exercises in mind, not with the prospect of all his students taking part in original research. Later, in the 1850s, when the ministry was in a position to establish a small chemistry laboratory with a dozen spaces for students, Neumann was already operating his home laboratory with about the same number of spaces and hence needed no additional support. To be sure, he continued to request an official location, explaining that the task of the physical division was not just to develop theoretical knowledge in physics but also to apply it to investigations and observations. But he never asked for compensation for his home laboratory, a point the ministry noted when it agreed to reimburse some students for costs incurred in conducting their practical exercises. Even with the home laboratory, Neumann allowed surpluses to accrue in the various accounts he handled.[50] It was probably his inefficiency and, ironically, his well-intended actions that helped to seal the fate of a physical laboratory at Königsberg.

By the second half of the decade both Neumann's seminar and his plans for a laboratory indicate that his pedagogical strategy was not directed exclusively at the cultivation of elite and talented students. Richelot shared a similar strategy. Like Neumann, he adjusted to a student clientele that was at times less than desirable. In 1845/46 the mathematical division also had two sections owing to the varying ability of the students, and the lower section was assigned "very easy problems" because the students in it "had only very weak school knowledge." In 1846/47, Richelot began to admit auditors, and they participated in the lower section. He assigned "very easy problems in analytic geometry" to the lower section in

49. Ludwig Moser to Staatsminister, 17 October 1843, ibid., fols. 109–10; Kultusminister Eichhorn to Königsberg University Curator Reusch, 31 October 1843, ibid., fols. 111–12. Several subsequent letters also deal with Moser's funding; see, e.g., ibid., fols. 120–24, 168–70, 175–76. On funding for the mathematico-physical seminar, see also FNN 62: Kassensache.

50. LN, p. 373; FNN 61.7: Kampf um das Laboratorium; FNN 62: Kassensache; Königsberg University Curator Eichmann to Kultusministerium, 26 September 1854, ZStA-M, Math.-phys. Seminar Königsberg, fols. 202–3.

1847/48. In the following year he explained that he assigned these problems because, he claimed, Prussian schools rarely taught geometry. By the end of the decade, then, Richelot's students did not all participate uniformly in the seminar: auditors attended irregularly and did no written work at all; introductory students sometimes barely went beyond the material in his lecture courses; and advanced students either worked on more difficult problems or on independent projects. Richelot met with his students for several hours on the appointed day of the seminar meeting, but often he did not meet with all students at the same time and instead divided his time with them throughout the week. Richelot, like Neumann, taught in a labor-intensive way. Although he sent only the work of advanced students to the ministry, it did not represent what went on the seminar as a whole. Furthermore, by the end of the 1840s, he was allowing students to enter his seminar during their first year of university study, thereby committing himself to operating two sections permanently.[51]

Problems within the seminar—in particular the quality of the students—contributed to the shift in pedagogical emphasis. But there were other reasons why an elitist orientation would have been inappropriate in the 1840s. Between 1842 and 1844 the new *Kultusminister* Eichhorn had tried to shape an educational policy designed to strengthen university instruction by providing the means for achieving a certain level of accomplishment in all students: exercise sessions linked to all courses at all universities. Designed as review and question-and-answer classes, these exercise sessions were intended to help students master the material in lecture courses. In July 1843 when Eichhorn asked for opinions on the matter, the response was not uniformly positive, although there was general agreement that a greater exchange between teachers and students had to be promoted. But many professors thought Eichhorn's proposal made schools and universities look too much alike. The ministry wanted exercise sessions in which *all* students could participate; the faculties not only questioned the capabilities of all students to do so but also feared that elite and talented students would be forgotten because instruction would be reduced to its lowest common denominator. The ministry also wanted exercise sessions in all disciplines; the faculties questioned the suitability of one institutional form for all subjects. Furthermore, that Eichhorn's proposed exercise sessions were to be required, they viewed as a violation of *Lehrfreiheit* and *Lernfreiheit*.[52]

51. "Bericht . . . von Ostern 1845 bis Ostern 1846," fol. 152; "Bericht über das mathem. physikalische Seminar der K. Universität zu Königsberg für d. Zeit Ostern 1846 bis Ostern 1847: Mathematische Abtheilung," 1 July 1847, ZStA-M, Math.-phys. Seminar Königsberg, fol. 156; "Bericht über die mathematische Abtheilung des physico. mathem. Seminars zu Königsberg für die Zeit Ost. 47–Ost. 48," 10 June 1848, ibid., fol. 164; "Bericht über die mathematische Abtheilung des phys. mathematischen Seminars in Königsberg für Ostern 1848 bis Ostern 1849," 17 June 1849, ibid., fol. 165–66, on fol. 165.

52. Erich Feldmann, *Der preußische Neuhumanismus: Studien zur Geschichte der Erziehung und Erziehungswissenschaft im 19. Jahrhundert*, vol. 1 (Bonn: F. Cohen, 1930), pp. 163–93.

Natural scientists were not of one mind on Eichhorn's proposal. Karl Burdach at Königsberg endorsed exercise sessions in the natural sciences and in support of his opinion noted the growth of exercise sessions and practica as listed in the Königsberg University catalog. Significantly, he viewed practica in the natural sciences as equivalent to seminars: as settings in which students worked on common problems and had the opportunity to engage in dialogues with professors. He believed, however, that *Privatdozenten* should not be allowed to hold exercise sessions, and he confined younger faculty to directing repetitions and examinations where the material was merely recited rather than applied. Burdach's democratic view of science instruction was not shared by his colleague in anatomy and physiology at Berlin, Johannes Müller. Müller believed that special forms of learning—exercise sessions, practica, and seminars—were useful only for a select few. The proposed exercise sessions seemed to him like the drill sessions of special schools and military-medical institutes. The young men who take part in these sessions, he argued, "lose the courage to think scientifically. They are tired and bored when they are finished. The subjects, which attracted them to the university, become as unpleasant as recitations and examinations themselves." In Müller's view, if all students were admitted to practica or exercises, "the mechanization of teaching" would result, destroying the "scientific spirit."[53]

When Eichhorn finally distributed his ministerial circular on exercise sessions on 17 April 1844, he ignored the older, more conservative views of Müller and instead, within the limits of the freedoms defining the university, strongly promoted a more democratic approach to instruction that gave all students the opportunity to learn actively. Eichhorn noted that although there were some subjects in which it was "necessary" to have a "spontaneous relation" between teachers and students, "the remaining disciplines have for the most part only lecture courses. In this way only the talented and scientifically inspired among the students achieve and preserve a free, scholarly spontaneity. The great majority [of students], who only listen to lectures and write down what is heard, sink all too easily into passivity."[54] Earlier faculty opinions having convinced him that *Lehr-* and *Lernfreiheit* had to be preserved, Eichhorn left the specific form of these exercise sessions up to professors, with the stipulation that they had to be closely connected to lecture courses. Although Eichhorn's circular has often been viewed as an encroachment of the state into university affairs, it could just as well be viewed as a step toward the democratization of university instruction because it was intended to make special forms of learning available to all students.[55] Undeniably his circular made it some-

53. Quoted from Prussian ministerial archives in ibid., pp. 187–88, on p. 188.

54. *Das Unterrichtswesen des preußischen Staats in seiner geschichtlichen Entwicklung*, ed. L. M. P. von Rönne, 2 vols. (Berlin: Veit, 1854), 2:515–19, on 516.

55. Feldmann, *Der preußische Neuhumanismus*, pp. 190, 197, 200–201.

what disadvantageous at a local level to conduct an institute for elite students alone.

The ministry's stance on university instruction was once again transmitted to Neumann and Jacobi indirectly through the mathematician A. L. Crelle. Asked to assess the "scientific value" of the advanced exercises completed in the seminar, Crelle found them accomplished and in some cases original. Although he never reviewed the physical exercises in more than a cursory fashion (he claimed only a "general acquaintance" with physics), he nonetheless took note of the "rather difficult calculations," including the method of least squares, which Neumann's students carried out in conjunction with their theoretical review of instruments and observations. Of special concern to Crelle was how Neumann's, Jacobi's, and Richelot's seminar exercises served teacher training (for which Eichhorn's proposed exercise sessions seemed particularly appropriate). Even though Crelle had acknowledged as early as 1837 "that Königsberg is now, in consequence of the effectiveness and cooperation of the distinguished mathematician and physicist who are there, a first rate school for mathematics and physics,"[56] he was nonetheless skeptical of some of the seminar's exercises because they emphasized depth of penetration over the breadth of coverage he thought a teacher should have.

What Crelle preferred to see in the seminar was more elementary subjects presented methodically and systematically, as Jacobi and especially Richelot did in the early 1840s. Crelle explained that because the seminar was supposed to train teachers,

> it is especially necessary that they know their science as best as possible *throughout*, within its boundaries and at its marginal points. There is little good in giving pupils *inspiring* things, matters at the forefront of scientific research . . . while their knowledge is changing and perhaps still has gaps in it. This kind of methodological introduction [to science] can, to be sure, lead to individual brilliant results, but it does not always guarantee that the prospective teacher will be *overall* skilled in his field. Moreover, it could possibly mean that his knowledge is lacking in parts.

Teaching only difficult subjects, Crelle warned, would only result in reaching the gifted. Because other students would not be able "to find their way themselves," he maintained that a comprehensive coverage of subject matter was more appropriate in science instruction. As a result of his review, the ministry favorably viewed the introduction of exercises with "pedagogical interest" into the seminar, considering them useful for teachers. "If the directors of the institute continually put into operation these rules of instruction proven useful to them," it remarked, "then

56. Crelle to Kultusministerium, 12 March 1842, ZStA-M, Math.-phys. Seminar Königsberg, fols. 84–88, on fol. 86; A. L. Crelle to Kultusministerium, 9 November 1837, ibid., fols. 41–42, on fol. 42.

certainly many future teachers will be educated in the seminar."[57] Systematic instruction in fundamentals was, for the ministry, a model for scientific didactics.

Even though the introductory sections of the seminar strengthened in the early 1840s, Crelle continued to find problems in the way students were taught, especially in what he viewed as the lack of clarity and specialized nature of the exercises. The mathematical exercises, he pointed out, cultivated mathematical authors, not teachers.[58] When he reviewed the seminar's reports from 1843 to 1845, he glossed over the scholarly accomplishments of the students, including Kirchhoff's investigation, and again argued that the point of seminar exercises, no matter how difficult they became, was that they should be designed so as to demonstrate insightful matters to *others*. In this way, he believed, seminar students proved "that they possess the gift and the ability to communicate their views to those who do not know them, as is found in the vocation of teachers." "If this view is correct," he continued, "then it appears to me always to desire clarity in the work of students" because "future teachers instruct in school not by writing but by speaking."[59]

The Bonn natural sciences seminar, with its rigid but comprehensive curriculum, conformed to Crelle's conception of seminar instruction as it related to teacher training. So did the Königsberg natural sciences seminar, whose exercises served as pedagogical exemplars. In the natural sciences seminar's fourth year, 1838/39, for example, students in Moser's physical division performed practical exercises, but "principally those experiments were considered which could be set up with less complicated apparatus and [which] therefore might most often prove useful for the participants in their later *Berufe*."[60]

Neumann and Richelot consistently viewed the relation between seminar instruction and teacher training differently. Returning to fundamentals was for them a way to prepare students for studying more difficult subjects in mathematics and physics. In response to Crelle's criticisms, Neumann defended the deeper, more complicated problems that his students sometimes found themselves working on. "The difficulty of a subject is subjective," he told the ministry, "and if a seminar member is more concerned with the clarification of finer points, then one can conclude that for him the main points are not difficult." Any student working on deeper problems, Neumann implied, had mastered the main points in the lectures. In a draft response that he did not send to the ministry, he

57. Crelle to Kultusministerium, 12 March 1842, ibid., fol. 87; Kultusministerium to Königsberg University Curator Reusch, 1 April 1842, ibid., fols. 91–92, on fol. 92.
58. Crelle to Kultusministerium, 13 September 1843, ibid., fols. 97–101, on fol. 100.
59. Crelle to Kultusministerium, 20 August 1845, ibid., fols. 126–30, on fol. 129.
60. Ludwig Moser, *Vierter Bericht über das naturwissenschaftliche Seminar bei der Universität zu Königsberg: Nebst einer Abhandlung über die Entwicklung des Schädels der Wirbelthiere* (Königsberg: Hartung, 1839), p. 34.

also indicated that the purpose of his seminar exercises was "not the representation of what is learned" but rather, "independent activity."[61] Neither Neumann nor Richelot differentiated between training teachers and training scholars. Their pedagogical strategy was designed in part to bridge the gap between the school and the university in two ways: by compensating for what students had not learned in the gymnasiums and by training better teachers of mathematics and physics, who would, they hoped, send to the university students better prepared to study those subjects.

Despite his defense of specialized exercises, by the end of the decade Neumann viewed the seminar and its laboratory in terms of the basic and systematic instruction both offered students. Hence, to the extent that he used the introductory exercises to "fill in the gaps" of his lectures, Neumann's view of his seminar's operation was consonant with Eichhorn's 1844 circular. On 13 June 1849, Neumann submitted his last seminar report for the decade, one that covered two years. It was not a detailed report; he had been ill, and political events had interrupted seminar meetings in 1848. Yet the circumstances that had unsettled the seminar did not diminish his sense of purpose. He described the purposes as "threefold":

> First, the seminar should make students, especially those who are closer to gymnasium instruction than to university study, aware of the gaps in their knowledge. Then the seminar should direct those who have progressed somewhat toward undertaking the independent mathematical treatment of physical problems. In this way the young men will have to assess, to apply, and so to take actual possession of what they have learned. At the same time, they will also receive what is in my view the most fitting preparatory training for determining physical phenomena through measuring observations. Performing such observations is the final purpose and the proper goal to which the members of the seminar are led.[62]

The operation of the seminar between 1847 and 1849 demonstrated once again how students moved toward those measuring operations. In the summer semester of 1847, the lower section, with seven members, had not yet taken Neumann's course on theoretical physics. He assigned them problems from the lecture course in optics they were attending so that they could learn through simple examples how to use mathematics in physics. Amsler and Jakob Heinrich Karl Durège conducted measuring observations in the upper section; Amsler on various methods for determining geomagnetic inclination, Durège on the experimental determina-

61. ["Bericht über die physikalische Abtheilung, 1843–45"], fol. 133; Draft report, physical division, 1843–45, partially quoted in LN, p. 371.

62. ["Bericht über die physikalische Abtheilung, 1847–49"], 13 June 1849, ZStA-M, Math.-phys. Seminar Königsberg, fols. 162–63, on fol. 162.

tion of current intensities. Because both Durège and Amsler left the seminar before the winter semester began, Neumann was left with beginning students and so offered his course on theoretical physics and, in the seminar, taught students how to take geomagnetic measurements. He had intended to continue these measurements in the summer semester of 1848, but neither he nor his students considered the revolutionary events of 1848 "conducive to scholarship and instruction," so the seminar was suspended for most of the semester. (The one student who attended the seminar that semester determined torsion coefficients.) When the seminar resumed in the winter semester of 1848/49, Neumann returned to fundamentals, "filling in the gaps" left by the lecture course on optics from the year before, and then had the students complete exercises on interference, polarization, and reflection.[63]

Kirchhoff's investigations molded the reputation of the seminar from the 1840s onward, and they epitomized the potential of seminars to cultivate genius. Yet Kirchoff was an exceptional product, one who stood out because he was able to go so far beyond seminar exercises to achieve a balance between theory and experiment and to contribute creatively to physical theory. To be sure, his investigations appeared at a very special time in the evolution of both the seminar (after Neumann had stabilized its curriculum and exercises) and Neumann's researches (while Neumann was actively working on induced currents). It is just because Kirchhoff was outstanding that he is unrepresentative; for Neumann chose not to create his exercises and curriculum for his talented and brilliant students, whom he left to their own resources. To the extent that Kirchhoff's investigations defined the operation of the seminar, he disguised the bulk of its actual functioning, which was to begin with fundamental exercises in mechanics and move on systematically in the investigative techniques of physics, especially error and data analysis. Neumann did not soon forget his own efforts in creating the seminar's course of study; for precisely when he had achieved his most outstanding success in Kirchhoff's string of exemplary investigations, Neumann framed the seminar's operation instead in terms of those activities that had given it stability. In the context of Prussian educational reform of the 1840s, it would have been tactless to do otherwise. His course on mechanics—the only "required" course in his curriculum—and its related seminar exercises gave the seminar a stability that transcended the waxing and waning of student preparation and performance.

Important as Neumann's pedagogical strategy was in stabilizing the seminar, it was, nevertheless, ineffective in stimulating further change. Already in the 1840s there were several missed opportunities as he neglected to use his resources to optimal advantage. By decade's end, his own research had all but ceased. He did not use Kirchhoff's and Amsler's

63. Ibid., fols. 162–63; Prutz, *Albertus-Universität zu Königsberg*, pp. 226–29, 231–32, 239.

achievements to increase funding; he did not use the funding at his disposal in the most efficient way. As the complexion of the natural sciences changed at Königsberg, Neumann kept his seminar insular by not making contact with new scientists on the faculty, even when their research interests were close to those of his assigned seminar exercises. Among the outstanding investigations he ignored were those of Hermann Helmholtz, undertaken at Königsberg in the late 1840s and early 1850s. Appointed *außerordentlicher* professor of physiology in the medical faculty in 1849, Helmholtz worked on the measurement of small intervals of time (for which he found Bessel's work on experimental errors useful), grappled with functional relations expressed in graphical form, tried to understand the constant errors afflicting his instruments, and even found Neumann's work in induced currents and Neumann's personal assistance helpful in his own investigations. Helmholtz praised Neumann's ability as a teacher, but he also found him shy at a more personal level.[64] Yet Neumann did not integrate—or, it seems, even mention—Helmholtz's investigations into the seminar.

Why Neumann never took advantage of all these important and strategic opportunities to improve his institute both intellectually and materially remains puzzling. So little effort would have been required to cultivate both the basic instruction the ministry seemed to want and the advanced research the seminar was originally designed to cultivate. It is difficult to imagine that Neumann's administrative responsibilities, even in the midst of preparations for the 1844 celebration, or the disturbances of the 1848 revolution could have detracted from his desire to promote student research more strongly. Yet the fact remains that even as he began to design and construct his residential laboratory, Neumann failed to make optimal use of the human and material resources readily available to him at Königsberg.

Concentrating on curricular matters, many of them basic, had other consequences as well. In his chemical laboratory at Giessen, Justus Liebig had to deal with some of the same pedagogical problems that Neumann did. But after his laboratory had operated for almost fifteen years, by 1840 or so, he guided teams of students in advanced chemical analyses of fatty acids, an area he himself had not yet investigated, and had gained control of leading chemical journals.[65] Closer to home in Prussia, Gustav Magnus in his private laboratory in Berlin had already supervised a dozen advanced investigations in physics and chemistry, all published between 1842 and 1848.[66] In research productivity, Neumann's seminar fell far

64. Leo Koenigsberger, *Hermann von Helmholtz*, 2 vols. (Braunschweig: F. Vieweg & Sohn, 1902), 1:110, 113–14, 116, 127, 138, 145–46, 158, 160, 162, 179, 190.

65. Frederic L. Holmes, "The Complementarity of Teaching and Research in Liebig's Laboratory," *Osiris* 5 (1989): 121–64.

66. Investigations completed in Magnus's Berlin laboratory are listed in A. W. Hofmann, "Zur Erinnerung an Gustav Magnus," *Berichte der Deutschen Chemischen Gesellschaft* 3 (1870): 993–1101, on 1099–1101. Additional investigations not mentioned by Hofmann are cited

behind these two institutes. Although the absence of comparable material resources at Königsberg accounts in part for the lower productivity of the students there, it was also the case that Neumann seems to have lacked the organizational and entrepreneurial skills needed to direct advanced work consistently.

Neumann's realization in 1849 that the function of his seminar was more basic and fundamental than he had originally intended marks the beginning of the mature stage of the seminar's operation. The second half of the decade was in other respects a turning point. Jacobi's departure proved to have minimal impact on the physical division of the seminar. Neumann's partnership with Richelot worked well, primarily because both had developed similar pedagogical strategies over the course of the decade. But the person whose investigations and ideas on science education had inspired Neumann to create measuring exercises in physics did not live to see all the ways in which the seminar became defined in terms of them. In 1846, Bessel died, leaving behind a legacy that touched every aspect of instruction in the exact sciences at Königsberg.

---

in A. Guttstadt, *Die Anstalten der Stadt Berlin für die öffentliche Gesundheitspflege und für den naturwissenschaftlichen Unterricht: Festschrift dargeboten den Mitgliedern der 59. Versammlung Deutscher Naturforscher und Aerzte von den städtischen Behörden* (Berlin: Stuhr, 1886), p. 140.

CHAPTER SIX

Error Analysis
and Measurement

THE DESIGN OF THE residential laboratory Neumann assembled in 1847 was in all likelihood close to that of the physical laboratory he had proposed to the state a year earlier. In his original plan he had assigned first priority to conditions suitable for conducting precision measurements. He had recommended that instruments be fastened to walls and tables to protect them from disturbances and that there be two workrooms for instrument construction and repair, an auditorium, and a storage room. There were in addition to be three more specialized rooms. One, long and narrow and running from south to north, would house two important instruments from astronomy and geodesy: a heliostat and a theodolite. The second room was to be for galvanic apparatus, to be used and stored separately because of its size. The laboratory, in the third room, was to be bathed with light, dry, and with access to a water supply. Unlike the rooms where refined measurements and observations could be made, the laboratory was to be a true workshop, with an oven, chimney, lathe, glass-blowing apparatus, and other conveniences to enable students to construct their own instruments or to modify what was on hand. Finally, Neumann had recommended that there be a garden where geomagnetic measurements could be taken. This proposed laboratory—actually a small physical institute—bore many of the material features of Bessel's observatory and was even intended to serve teaching and research much as the observatory had done.[1]

For some time, Neumann's house must have been bustling with the activity of craftsmen and students as they assembled this workshop.

---

1. Neumann, "Andeutungen über die Erfordernisse eines physikalischen Laboratoriums" [1846], rpt. in LN, p. 445; see also LN, pp. 372–73.

Rooms were added to his house, and eight were ultimately outfitted for his research, for seminar exercises, and for independent student investigations. One large room, with a high ceiling and three large windows, housed a telescope, balance, air pump, and other bulky apparatus, especially for electrical investigations. Some of the smaller rooms could be used for carpentry, glass-blowing, and woodworking. Books, instruments, and minerals were eventually found everywhere. Neumann had counted on the ministry supporting a small geomagnetic station in his garden, but promises for that went unfulfilled, too. In the end he used his grounds primarily for continuing the geothermal measurements he had begun with his students in the late 1830s.[2]

By 1849, students were at work in this residential laboratory, executing seminar exercises designed to measure and demonstrate induction phenomena. Neumann required that his students return to theory after taking their measurements in order to show the ways in which theory was or was not fulfilled under experimental conditions. In the seminar for the summer semester of 1849 (which met for fewer weeks than had been customary), Neumann reviewed induction phenomena in detail, thereby laying the foundation for a more specialized course on the theory of the electric current in the winter semester of 1849/50. But fundamentals still had to be handled. Now that a regular location for conducting practical exercises was in place, the course on theoretical physics became more strategic as an introduction to the curriculum. That year, students used what they had learned in mechanics as a foundation for designing protocols for measuring thermal constants. Neumann continued to emphasize the determination and calculation of error in experimental trials. In the practical exercises, he pointed out,

> methods of observation were handled in such a way that the limiting conditions, which have a disturbing influence on the results issuing from these methods, are brought into prominence. The members of the seminar are directed to pursue these disturbing influences through mathematical considerations; to determine their quantitative expression; and to derive the means by which these influences can be eliminated. Associated with [this exercise] are pure theoretical investigations, which are developed in such a way that their results can be compared with experience.

He explained two pedagogical objectives in assigning such exercises: to bring what was learned in his lecture course to such "active independent application" that the students could be said to "possess" the material and to prepare his students for the higher mathematical treatment he would

2. On Neumann's residential laboratory see LN, pp. 373–75; A. Amsler and F. Rudio, "Jakob Amsler-Laffon," *Vierteljahrsschrift der Naturforschenden Gesellschaft in Zürich* 57 (1912): 1–17, on 2.

later cover in class.[3] Although Neumann still considered student research to be one sign of pedagogical success, he did not mention any independent investigations or student publications.

No matter how simple or complex the measuring exercises were, seminar students still had to analyze their instruments theoretically, determine the errors afflicting their experimental trials, and analyze and reduce their data to determine its precision and accuracy. As Neumann emphasized during the summer semester of 1850, closing "the gap between theoretical insight and practical realization" was the purpose of his seminar assignments, even when his section had only older students about to leave the university, who had already learned in the introductory exercises how to deal with higher theoretical considerations. Confident that his older students had acquired the necessary foundations, he assigned them practical exercises in magnetism and galvanism. Aware that he again had to report that his students had not produced any publishable work, Neumann paused to explain the pedagogical usefulness of his advanced exercises: "If the numerical results obtained were less significant, and less than I otherwise have attained, then the path to them was not, I believe, without meaning for the students. They were in this way forced to clarify immediately the topics that they had until then put aside as unimportant; to go back to the material they had learned from the lecture course; and to learn these topics thoroughly through the use they were supposed to make of them."[4] He thereby seemed to indicate that numbers were not only the raw products of a practical exercise but also indicators of failure as well as achievement. The attainment of accurate and precise results qualified by a measure of error meant that a student had not only worked through his exercises competently but also mastered manual, mathematical, and intellectual skills. A student whose results were not accurate presumably knew immediately of his failure (by the large size of his data's probable error as measured by the method of least squares) and could take the steps necessary to improve his performance. In this way, Neumann's measuring exercises not only trained students in the techniques of physics but also incorporated mechanisms for helping students to teach themselves.

Neumann reiterated his conviction that original research was not the immediate or even the primary aim of the seminar exercises. In the winter semester of 1850/51, when only new students registered for the seminar,

3. "Bericht über die physikalische Abtheilung des mathematisch-physikalischen Seminars während des Zeitraums von Ostern 1849 bis Ostern 1850," n.d., ZStA-M, Acta betr. das mathematisch-physikalische Seminar an der Universität zu Königsberg, Rep. 76Va, Sekt. 11, Tit. V, Nr. 25, Bd. I: 1834–61, fol. 172.

4. "Bericht über die Leitung der physikalischen Abtheilung des mathematisch-physikalischen Seminars während des Zeitraums von Ostern 1850 bis Ostern 1851," ibid., fols. 179–80, on fol. 179.

he offered his course on theoretical physics. "There is," he remarked, "a definite connection between these lectures and the concerns of the seminar." Through oral discussion and written problem sets, the mechanical and physical ideas exposed in the lectures were brought to "greater clarity." For introductory students, the seminar exercises served to sharpen general concepts such as force and mass; general laws such as the equilibrium of forces; and general equations such as those for motion produced by gravity.[5]

The physical seminar continued to be an extension of Neumann's lecture courses during 1851/52, when he taught topics in mathematical physics, the theory of light, the theory of the electric current, and mineralogy. Because attendance in these lecture courses was low (4, 3, 5, and 7 students respectively), the seminar undoubtedly had few students. Those who attended must have required basic introductory exercises, because in the summer semester students worked through simple problems in reflection and refraction; during the winter semester, on how electrical instruments could be used to understand geomagnetism. Students applied Gauss's theory of focal points and planes to a system of lenses and developed an experimental protocol for determining indices of refraction with a goniometer. In the winter semester they learned how to measure geomagnetic forces and how to use absolute units in measuring an electric current. After these measurements they moved on to theoretical considerations, developing expressions for the motion caused by electric forces and rederiving Fourier's equations for the propagation of heat in discrete and continuous media. They returned to experiment again, determining the constants of their equations through experimental trials. All seminar exercises were executed in common this year; no independent investigations were undertaken.[6]

The low attendance in Neumann's lecture courses and presumably in his seminar resulted in part from a reduction in the number of students enrolling in the Königsberg philosophical faculty from the early to the mid-1850s, when only slightly more than fifty students matriculated per year.[7] That demographic shift had a far greater impact upon the natural sciences seminar at Königsberg, which had not had a stable enrollment for some time and had suffered from lower enrollments more than did Neumann's seminar since the mid-1840s. In 1843/44 no students registered for the physical, chemical, or zoological divisions, so these sections were canceled for that year. The botany division remained barely viable, mainly because botany proved to have marginal relevance to secondary

5. Ibid., fol. 179.

6. "Bericht über die Leitung der physikalischen Abtheilung des mathematisch-physikalischen Seminars der Königl. Universität zu Königsberg während des Zeitraums von Ostern 1851 bis Ostern 1852," ibid., fols. 187–88.

7. Königsberg University, *Amtliches Verzeichniss des Personals und der Studirenden*, 1849–52.

school teaching. One student talented in botany decided to switch to mathematics and physics, where "there were greater prospects for a future position" in teaching. Even Ernst Meyer, one of the directors of the natural sciences seminar, admitted that it was not his seminar, but the mathematico-physical seminar, that "so often" gratified students. Still, of all divisions of the natural sciences seminar, the botany division remained strongest in the late 1840s, followed by the zoological division. Part of the appeal of these two sections derived from the exercises in microscopy that both divisions had offered since at least 1844 and that were useful not only for future teachers of natural history but also for medical students.[8]

By the early 1850s, problems in the natural sciences seminar had become acute. On 17 May 1852 the university curator Eichmann wrote to the ministry, detailing the sorry history of the natural sciences seminar and requesting that its 350 taler budget be reappropriated for a surgical and clinical institute. Although the seminar, he wrote, had had a "lively" beginning, a high level of activity had not been sustained. Its annual budget had frequently gone unspent. In the previous five years, three sections had had no students, while the botany division had had only four, and in the previous semester even that had dropped to two. Eichmann admitted that the natural sciences seminar was not fulfilling its obligations. Moreover, upon talking to the directors, he had found that they had "given up hope that the seminar would be able to fulfill its purpose satisfyingly in the future" and that to continue with the seminar would be to them a "mortifying business." Since the institute "had so little fulfilled its purpose, nothing else remained to be done than to cancel it." A month later the ministry approved Eichmann's recommendation.[9] With the demise of the natural sciences seminar, the only institute left for training secondary school teachers of the natural sciences at Königsberg was Neumann's seminar.

Moser's courses and exercises had always differed substantially from Neumann's. Moser's section of the natural sciences seminar had consistently offered students a wider variety of topics more suited than Neumann's for training teachers, and some students—we do not know how many—would take Moser's seminar before Neumann's for that reason. In 1839/40, when Neumann's students had been learning about bifilar suspension and had begun geomagnetic observations, Moser's students had been completing projects on gases, the specific density of solid and liquid substances, galvanism, optics, the chemical effects of light, and the oscillation of a magnetic needle. Student experiments in the physical division of the natural sciences seminar in the following year had included problems

8. ZStA-M, Acta betr. eines Seminars für die naturwissenschaftlichen Studien auf der Universität zu Königsberg, Rep. 76Va, Sekt. 11, Tit. X, Abt. X, Nr. 21, [1827–76], fols. 162–63, 165–66, 171, 174–75, 178, 180, 198–99, on fols. 179, 180.

9. Königsberg University Curator Eichmann to Kultusministerium, 17 May 1852, ibid., fols. 215–17, on fol. 216.

from acoustics, thermomagnetism, the chemical effects of light, and optics. Not all Moser's exercises were elementary, however. In 1843 he had posed a prize question to members of his seminar "on the various methods for measuring the strength of electricity." Carl Friedrich Boehm, who later was to enroll in Neumann's seminar, won the competition.[10]

But the variety in Moser's physics curriculum had disappeared by the early 1850s, when the natural sciences seminar was canceled. Between the winter semester of 1849 and the summer semester of 1852, he had regularly offered courses in experimental physics and the physics of sense perception. Occasionally he had taught galvanism and magnetism, and from time to time also courses in meteorology; climatology; and mathematics, on the probabilities involved in life expectancies and life insurance.[11] Despite his interest in higher mathematics, Moser did not teach physics with the mathematical rigor Neumann did. Not even his acquaintance with probability calculus facilitated his teaching of an exact experimental physics.

After the cancellation of the natural sciences seminar, Neumann ought to have had a slightly larger clientele in physics, but he did not simplify his curriculum to accommodate those students who might otherwise have taken Moser's courses and seminar. Instead, by the winter semester of 1852/53, Neumann had strengthened the purely analytic and mathematical orientation of his teaching by offering a course on potential theory and its mathematical methods. He had been teaching parts of potential theory since the late 1830s, when he had introduced Gauss's geomagnetic investigations into the seminar; but now he systematized his material, incorporating the results of his electrodynamic investigations and broadening its analytic foundation and expanding its applications. The published version of this course, because it was edited by Neumann's son and former student Carl Neumann, who later concentrated in part on the mathematical methods of potential theory, is more mathematically oriented than Neumann had perhaps intended.

Complex mathematical methods linked together sections of Neumann's course. He first introduced the general features of potential theory, following especially the work of Laplace and Poisson and making sure that students were facile in coordinate transformations. He then went more deeply into the mathematical techniques involved in spherical harmonics before he applied potential theory to several problems, such as the equilibrium figure of an incompressible fluid, the figure of the earth, the variation of gravity on the earth's surface, geomagnetism, geotemperatures, and finally, the problem of the tides. The remainder of the published version dealt mostly with problems in electricity and magnetism:

10. Königsberg University Curator Reusch to Kultusminister J. A. F. Eichhorn, 22 March 1843, ibid., fol. 171.
11. Königsberg University, *Vorlesungsverzeichniss*, 1849–52.

Gauss's theory of geomagnetism and the question of the source of geomagnetic forces; Poisson's theory of the distribution of electricity and surface potentials; and other special problems.

Dominant as these mathematical methods were in his course, however, Neumann did not intend them to become ends in themselves. Potential theory in principle offered several advantages in the construction of physical theory. Through it the phenomenological approach to physics that Neumann had formerly cultivated could be strengthened. Using potential theory meant that students had one more tool to avoid excessive hypothesizing and did not have to consider more deeply the nature of force or, more significantly, the notion of causality in physics. They could instead concentrate on what, for all practical purposes, amounted to static descriptions of physical phenomena. Moreover, potential theory also provided the means for showing them "how the separate disciplines of physics stand in close relation to one another."[12] In the second half of the course, for example, Neumann drew parallels between the distribution of temperature and the distribution of electricity by developing both sets of formulas from Green's theorem.

The mathematical methods of potential theory offered an important advantage for Neumann's instructional program: they offered alternative, and in some cases easier, ways of dealing with observations. Functions that described patterns of physical phenomena on the earth's surface—temperature and magnetic intensity, for instance—were generally inexact because they were based largely on interpolation from observations taken at only a few points. In a chapter on Gauss's 1838 theory of geomagnetism, Neumann mentioned that the differences between theoretical and experimental values for the earth's magnetic intensity were in some cases marked enough to justify a detailed reassessment of both theory and experiment. He furthermore believed that the available observations were in many instances so imperfect that adequate computations could not be made with them, and certainly not by using Gauss's formulas. So after discussing Gauss's methods, Neumann introduced his students to his own method, developed in 1838, for determining spherical harmonics by using observational data. He believed his own method for determining the equation's constants from observations was "noteworthy" because "one could attain the same accuracy" as with the method of least squares, but "with less trouble."[13]

Although Neumann's cycle of lecture courses had achieved considerable mathematical sophistication with the addition of his course on potential theory, his seminar continued to emphasize the same kinds of exercises it had for over a decade. The six students who registered for his

12. F. E. Neumann, *Vorlesungen über die Theorie des Potentials und der Kugelfunctionen*, ed. C. Neumann (Leipzig: B. G. Teubner, 1887), p. 247.

13. Ibid., pp. 128–30, 152, on p. 152.

seminar in 1852/53 were divided into two sections. He assigned measuring exercises in electricity and optics to older students, from whom he required written reports only "if the experimental result and its evaluation especially suggested that it be done." Once again he felt the need to justify his pedagogical techniques because they had not led to original investigations: "It is in the nature of such exercises that the objective, demonstrable result is often very small, while the subjective gain can be significant." New seminar members learned the fundamentals of theoretical physics by completing "simple problems," which "principally had the purpose of bringing to crystal clearness the ideas of mass, force, [how] distribution of mass [affects] motion, etc." Requiring that students achieve certainty in their theoretical judgment by constructing formulas so that the constants in them could be measured, he assigned the younger seminar students practical exercises on small vibrations, which required detailed calculations in order to determine the constants. In a draft for his report where he went over the matter several times, Neumann seemed especially concerned that the ministry know that he was teaching his students how to evaluate the "reliability of the final result" of their measurements. The university curator was well pleased with his manner of teaching and singled out the practical exercises for praise.[14]

By offering the same seminar exercises year after year, Neumann was not only acknowledging his commitment to a pedagogical strategy that emphasized systematic learning and practical exercises in measurement; he was also saying something about his student clientele. Despite the exceptional few like Kirchhoff and Amsler who excelled in their seminar exercises, most students were less capable, a point that is confirmed by Richelot's experiences between 1849 and 1853. Sharing several of the same students, Neumann and Richelot no doubt shared several of the same problems in teaching them. Attendance records for the mathematical division, which are most accurate between 1852 and 1860, indicate that cross-registration between the two divisions was highest in 1852/53, when all six of Neumann's students attended Richelot's division. But altogether, between 1834 and 1860, only about one-third of Neumann's seminar students attended Richelot's division. Most of the cross-registrants began in the mathematical division and only later registered for the physical one.

Between 1849 and 1853, didactics occupied Richelot as he tried to promote intellectual productivity among his students. When he lectured on what he considered a "mostly new *Wissenschaft*"—recent formulations

14. "Bericht über die Leitung der physikalischen Abtheilung des math.-physikalischen Seminars der Königl. Universität Königsberg während des Zeitraums von Ostern 1852 bis Ostern 1853," 8 July 1853, ZStA-M, Math.-phys. Seminar Königsberg, fols. 196–97, on fol. 196; draft report, physical division, 1852/53, FNN 48: Seminar Angelegenheiten; Kultusministerium to Königsberg University Curator Eichmann, 5 September 1853, ZStA-M, Math.-phys. Seminar Königsberg, fol. 198.

in dynamics—during the summer semester of 1849, students worked through related seminar exercises, but little came of them. Although Richelot had expected "fruitful independent work," he considered these exercises "defective and less successful than otherwise." Hoping to correct matters, he adjusted the level of his exercises in the following semester, assigning "very easy problems" in algebra.[15] Because the ability among his students was so varied in 1850/51, he divided his division into two sections and was able to supervise independent work by his older students. The work of one, Heinrich Eduard Schröter, was so satisfying to Richelot that he considered it a "testimony to the success of my efforts in the seminar." Schröter's performance, however, was not the norm. "The remaining participants in this class of the seminar," Richelot explained, "were so close to leaving the university that they believed they were not able to concentrate all their energy on the seminar." In the lower section, which contained a diverse group of beginning students, Richelot met with uneven success. Two students, Rudolph Alfred Clebsch and G. R. E. Kreyssig, devoted most of their time to the physical division of the seminar; two students left the seminar to begin "practical careers"; and other students left to enter military service. When the seminar opened again in the summer semester of 1851, new students were in the majority, so Richelot lectured on geometry. He guided the few remaining older students in conducting original investigations. Along with Schröter, Wilhelm Dumas produced what Richelot considered the first truly independent work of his section of the seminar, thereby reaching "the goal that we recognize as the highest for the effectiveness of our teaching."[16]

Richelot's success was not accidental. He had deliberately guided his students through problems in various areas, mostly higher calculus, analysis, and analytic geometry. He had also reviewed mathematical techniques in physics, as in the winter semester of 1851/52, when he taught students in special afternoon sessions how to apply elliptic transcendentals "to the motion of a pendulum with finite vibrations, and to a planet in a resisting medium."[17] The evolution of his division from one session per week to several special afternoon sessions on special topics had been in response to student needs. These organizational changes brought pedagogical ones that displaced some of the responsibility for teaching from Richelot onto the students themselves. In 1852/53, for example, when the mathematical division again operated in two sections, Richelot called the

15. "Bericht über die mathematische Abtheilung des Seminars für die Zeit von Ostern 1849 bis Ostern 1850," ZStA-M, Math.-phys. Seminar Königsberg, fol. 173.
16. "Bericht über die mathematische Abtheilung des hiesigen mathem. physikalischen Seminars in der Zeit von Ostern 1850 bis Ostern 1851," 5 September 1851, ibid., fols. 181–82, on fol. 181; "Bericht über die mathematische Abtheilung des physik. mathematischen Seminars auf der Universität zu Königsberg in d. Zeit Ostern 1851 bis Ostern 1852," ibid., fols. 189–90, on fol. 189.
17. "Bericht über die mathematische Abtheilung, 1851/52," ibid., fol. 190.

lower one the "elementary section" for the first time and introduced weekly homework assignments that "followed as closely as possible the methods of solution used in the seminar." In the upper division, where Clebsch's work proved exceptional, Richelot assigned problems in the integration of linear differential equations and in approximation methods for integrating dynamical equations. He was disappointed, though, with the poor gymnasium preparation exhibited by students in the elementary section, which made it necessary to assign them easy problems from algebra and analytic geometry. One student, who had "had few problems of this type in school," was assigned even easier "school problems." Although Richelot so modified his teaching, he still believed that the aim of the seminar was not to provide a comprehensive coverage of mathematics but to bring students to "a lively self-activity."[18]

That may have been rhetoric designed to appeal to the ministry and to the by now well entrenched ideological notion of *Bildung* in educational affairs, because in the following year Richelot showed once again how imperative it was to adjust and arrange mathematical topics according to students' needs and abilities. When in the summer semester of 1853 he assigned problems in higher analysis, especially the treatment of definite and indefinite integrals and their application to certain problems (including potential theory), he designed them "as they do not usually appear in textbooks." He did not get very far. Three students demonstrated such an "unclear and uncertain" knowledge that he encouraged them to offer lectures to one another on selected topics so that they could catch up with the other students. Matters turned out no better in the winter semester of 1853/54. He reinstituted elementary and advanced sections to accommodate the students' "very different knowledge" that "they had acquired at various schools," so that students would all be equally well prepared for the special subjects he wanted to teach. Although his lecture course continued to provide the infrastructure of his seminar's activities, Richelot emphasized that he still had to hold extra recitation sections where he engaged students in a dialogue to find out "whether and how my lectures were understood." In constructing his curriculum he also began to take into account the different *Berufe* the students intended to take up after leaving the seminar. Two students were destined for architecture and engineering; two wished to devote themselves to physics and the natural sciences "after they achieved the required mathematical preparation"; the remainder were dedicated to the study of pure mathematics. Only to these last did Richelot assign topics in algebra.[19]

Neumann responded similarly to the needs of his students; for he

---

18. "Bericht über das mathematische Seminar in Königsberg: Ostern 52 bis Ostern 53," 28 June 1853, ibid., fols. 194–95, on fol. 194.

19. "Bericht über das mathematisch-physikalische Seminar im Jahr Ostern 53 bis Ostern 54," 15 September 1854, ibid., fols. 206–7, on fol. 207.

likewise divided them into elementary and advanced sections in the summer semester of 1853, when five registered for his seminar. So that older, more experienced students might "achieve fluency in the application of knowledge already on hand" and become acquainted with "finer methods of observation," he first worked through the theory of various measuring methods and then showed how to apply them to practical examples such as the measurement of electric currents, geomagnetism, and small intervals of time. Related issues, such as calculating the torsion of bifilar wires, were often taken up in conjunction with these measuring exercises. Students determined the constants of electric circuits and independently calibrated differential multiplicators; learned how to measure the resistance of an electrolytic solution separately from the resistance stemming from the effects of polarization; and measured the electrical resistance of wires as well as changes in resistance produced by thermal alterations.

While older students worked on these advanced exercises, new students, whose knowledge of physics was vague and whose ability to use mathematics in physics was weak, worked with simple instruments, learned how to reduce measurements, and gained experience in expressing physical ideas mathematically by using as a model the examples Neumann had worked out in his course on theoretical physics, such as Atwood's machine, the percussion pendulum, motion in a resisting medium, and the motion of a pendulum affected by disturbing forces. Neumann encouraged them to consider "the various conditions, which in the instrument are not completely fulfilled," and to assess "their influence on the final calculated result," thereby teaching them how to compute the constant errors of an experiment. As a final task, Neumann instructed that all observations "had to be calculated according to the method of least squares."[20]

With their seminar reports for 1853/54, Neumann and Richelot included a special note to the ministry. "Before we begin the reports," they began, "allow us to send the following general communication." In this, the fifteenth year of activity after the seminar opened for the second time in 1839, Neumann and Richelot finally believed it had reached maturity. For by 1854, not only had they had some limited success in fostering students' original work, and not only had they assembled a curriculum adjustable to changes in student clientele, but the success of their pedagogy was now manifest in the subsequent academic and professional achievements of their students. Richelot cited one student whose dissertation was on elliptic functions and pointed out that another was now teaching in his probationary year in a Berlin gymnasium. Neumann was

20. "Bericht über die Leitung der physikalischen Abtheilung des mathematisch-physikalischen Seminars der Königl. Universität Königsberg während des Zeitraums von Ostern 1853 bis Ostern 1854," n.d., ibid., fols. 204–5, on fol. 205.

equally pleased with Clebsch's doctoral dissertation on the motion of an ellipsoid in a resisting medium. Clebsch, who had graduated from Königsberg's Altstädtisches Gymnasium, where he had studied with J. H. C. E. Schumann, one of Neumann's earliest seminar students, showed a special talent for developing the mathematical side of Neumann's seminar problems. Although Clebsch's investigation was entirely in the domain of mathematical physics—he included no practical experiments in his derivations, noting that "to compare the result achieved with experience was not very feasible"—he did end his discussion with an examination of the motion of a pendulum in a liquid medium, thus continuing the themes Neumann had introduced in his lectures on theoretical physics. It was their "joy," Neumann and Richelot remarked, that some of their students "have ended their university study with exceptional doctoral and teaching examinations."[21] The effort they had invested had paid off.

Both men had already supervised dissertations and seen students through the *Lehrerprüfung*, but in this year, those landmarks seemed especially significant; for their "student" clientele changed once again. "In addition to the regular students in the weekly seminar meetings," they continued, "earlier seminar members maintained contact with us in a scientific exchange [*wissenschaftlicher Verkehr*] in such a way that in our conversations, the problems of their own independent investigations developed." They indicated that both Clebsch and Carl Neumann, Neumann's son, had inaugurated these exchanges and, under Richelot, completed mathematical essays. It was this exchange, Neumann and Richelot believed, that marked the high point of the seminar's activity and made the seminar worth their effort.[22] Such scholarly exchanges between the directors and older or former seminar members were one indication that the structured training Neumann and Richelot were giving their students was relevant to their subsequent pursuit of original research.

Encouraged by their success, Neumann and Richelot outlined their pedagogical tactics more clearly. Richelot, although he believed in the summer semester of 1854 that his seminar had been run "by the same principles," had actually taken several more steps to integrate his program of instruction. "Working through our more systematic lectures is a very essential *Bildungsmittel*" for the students, he pointed out, because only through the lectures could "the weekly meeting of the seminar members with the director, which was held for several hours, offer not only a suitable opportunity to assist the understanding of the difficult parts of every lecture but also the opportunity to awaken, to exercise, and

21. "Bericht über das mathematisch-physikalische Seminar im Jahr Ostern 53 bis Ostern 54," 15 September 1854, ibid., fol. 206; A. Clebsch, "Ueber die Bewegung eines Ellipsoids in einer tropfbaren Flüssigkeit," *JRAM* 52 (1856): 103–31, on 103; 53 (1857): 287–91.
22. "Bericht über das mathematisch-physikalische Seminar im Jahr Ostern 53 bis Ostern 54," fol. 206.

to place on a suitable path the particular strengths of the students." As Neumann had done, Richelot now also acknowledged the importance of the symbiotic relationship between his lecture courses and the seminar. Without the seminar, he, like Neumann, thought that "only a few exceptional, talented individuals with similar eagerness [would] maintain a lasting and deep interest in the lectures." So in order to draw more students to the study of mathematics, Richelot taught the principles of geometry, assigning problems "from the easier to the more difficult" in the elementary section of the seminar. For example, he discussed Pascal's theorem but did so "without using analytic geometry," because most of the students "possessed only gymnasium knowledge." Two students registered in both the mathematical and the physical divisions proved especially "talented": Louis Saalschütz "for pure geometry" and Wilhelm Ferdinand Fuhrmann "for independent research."[23]

In the winter semester of 1854/55, when Richelot taught Cauchy's algebraic analysis and Euler's finite analysis, he introduced still another pedagogical novelty. Homework assignments, in the form of various kinds of problems, had been common for some time in the physical, and more recently in the mathematical, division of the seminar; now he expanded the function of homework by assigning entire chapters on mathematical topics for students to learn on their own at home. He thereby supplemented the independent application of what had been learned in the lectures (an exercise that led, in the best of circumstances, to original research), with the independent learning of subject matter itself. How well his students understood the assigned material was then evaluated through discussion in the seminar. Although Richelot began his seminar report for 1854/55 with a statement that encapsuled the pedagogical program he had developed—the seminar, he claimed, was partly for the higher *Ausbildung* of individuals who had been at the university for some time—his concluding remarks underscored the sense in which the seminar helped students achieve both scholarly and professional goals. Praising the work of Carl Neumann, now a teaching candidate, Richelot considered "him worthy of being strongly recommended for a secondary school and for the office of teaching" because "in the *Lehrerprüfung*" Neumann applied "ultra-elliptic functions to mechanics for the first time and obtained unconditional certification for teaching."[24]

While Franz Neumann did not, like Richelot, assign material to be read at home in 1854/55, he did continue to strengthen a graduated learning program. In the summer semester of 1854 he assigned, to two older members of the seminar, Anton Müttrich and E. Bardey, exercises in

---

23. "Bericht über die mathematische Abtheilung des math. physikalischen Seminars der Königl. Universität in Königsberg für das Jahr Ost. 1854 bis Ost. 1855," 1 October 1855, ibid., fols. 213–16, on fols. 213, 214.

24. Ibid., fol. 216.

measuring induced electric currents and geomagnetic forces and selected optical phenomena such as the index of refraction of a prism, angle measurements in a crystal, and double refraction. Neumann had apparently not known in advance that new members would outnumber more experienced ones; for he offered topics in mathematical physics and the theory of light as his lecture courses. He noted in his report that although not all the new students had heard his lectures on theoretical physics, he nonetheless had pursued the objectives of that course by using his lecture course on the theory of light, which they *had* heard, as a source of exercises on instrumentation and measurement. For example, he had covered the theory of optical instruments such as the telescope and the microscope by reviewing the principles of refraction; methods for determining the index of refraction; the theory of simple lenses, including practical methods used for deriving focal points and planes; and finally, systems of lenses. The few new students who previously had heard the lectures on theoretical physics occupied themselves by reducing observations from practical exercises concerning small vibrations and the moment of inertia.

In the winter semester of 1854/55, when Neumann offered his lecture course on theoretical physics, several students engaged in exercises designed "to bring the seminar activity into closer connection" with those lectures. Students studied free fall in a resisting medium, several issues concerning pendulums, the influence of the earth's rotation on free fall (which then led to a general theory of the earth's rotation), and finally they worked out the mathematics of Weber's theory for determining electrical resistance and then applied it in measuring exercises. Neumann asked the ministry to examine the student exercises accompanying his report in order to appreciate the "difficulties" involved in achieving "an easy and certain application of the main theorems of mechanics." Other students, who had heard his lectures on optics from the previous semester, deepened their understanding of optical theory by studying the theory of color rings of two-axis crystals, by developing the mathematical formulas representing those rings, and by using the formulas as a guideline for measuring observations. Occupied in both semesters with exercises that just barely went beyond an elementary level, Neumann once again did not succeed in bringing any of his ten seminar students to the point at which they could do original research.[25]

But then in 1855/56, as in 1853/54, Neumann's and Richelot's sensitivity to the intellectual needs of their students, as well as the time they had devoted to fundamental training exercises, produced satisfying results. Richelot was especially pleased that the "method of assigning home-

25. "Bericht über die physikalische Abtheilung des math.-physikalischen Seminars der Königl. Universität Königsberg für das Jahr von Ostern 1854 bis Ostern 1855," n.d., ibid., fols. 211–12.

work, earlier used, was employed in this year with great success." With all the necessary groundwork done at home, he reserved his seminar for demonstrating difficult calculations or for discussing points not well understood. In the first semester, even beginning students learned Cauchy's algebraic analysis in this way. Other students, who had heard Richelot's lectures on analytical mechanics, worked their way through the "most important chapters" dealing with variational calculus in the analytical mechanics of Lagrange. In the second semester, Richelot built upon that foundation by concentrating on systems of linear equations and their applications to dynamical problems. Two students, Müttrich and Jean Charles Rudolph Radau, were sufficiently motivated to read on their own several papers by Lagrange, Gauss, Hermite, and Jacobi. Having mastered those papers, Radau undertook the calculation of a planet's orbit through approximation methods involving elliptic functions and determinants. Overall, Richelot seemed exceptionally pleased with the performance of his students. He registered only one negative remark, about a "physics" student, Paul du Bois-Reymond, who had participated in his mathematical seminar "without, however, submitting any work of significance."[26]

Engaged in his own independent investigation on the surface tension of liquids, du Bois-Reymond did not attend Neumann's physical seminar but worked closely with Neumann under a private arrangement in Neumann's residential laboratory. Before his matriculation at Königsberg, du Bois-Reymond had studied under the physiologist Carl Ludwig at Zurich, where he learned a medical physiology steeped in mathematics and physics. Shortly after arriving at Königsberg in 1854, du Bois-Reymond reconsidered his interest in physiology when he began to study with Neumann. In August, his older brother, the physiologist Emil du Bois-Reymond, wrote to Carl Ludwig, his good friend and close associate in the reform of physiology, that Paul "is working along aimlessly, not knowing if he wants to remain a physicist or a physician, taking courses without attending them." Emil especially worried that Paul did not have the "practical enterprising manner" necessary to embark upon a career.[27]

By December, Paul seemed to have made a decision. According to his older brother, Paul was "occupying himself solely with mathematical physics." Emil told Carl Ludwig that he hoped Paul's "dealings with Neumann and Helmholtz will help him form a proper idea of his abilities."[28] In mentioning in the same sentence both Neumann and Hermann Helmholtz (who had been teaching physiology at Königsberg since

26. "Bericht über das mathematische Seminar in Königsberg in der Zeit von Ostern 1855 bis Ostern 1856," 28 June 1856, ibid., fol. 220–21.

27. Emil du Bois-Reymond to Carl Ludwig, 14 August 1854, in *Two Great Scientists of the Nineteenth Century: Correspondence of Emil du Bois-Reymond and Carl Ludwig*, ed. Paul F. Cranefield (Baltimore: Johns Hopkins University Press, 1982), pp. 84–85, on p. 84.

28. Emil du Bois-Reymond to Ludwig, 27 December 1854, in ibid., pp. 86–87, on p. 87.

1849), Emil underscored the dilemma Paul faced: apparently he could not decide between the two scientific disciplines that, in the eyes of Helmholtz, Ludwig, and Emil, were similar in their emphasis upon measurement. Helmholtz, however, viewed Paul's vacillation more seriously. A few days later Emil received a letter from Helmholtz reporting Paul's frequent visits to the Helmholtz household. Then Helmholtz broke the news that Paul

> is now entirely buried in Neumann's mathematical physics and appears to want to give up physiology. I don't think well of what he is doing for his later career, because as a physiologist he will always be more likely to find employment at a university than as a physicist. And hopefully he will not have the ambition to become a teacher of mathematics and physics at a gymnasium. By the way, I am extremely happy about how well he has matured. He has, in fact, unusually rich talents. What he still lacks from school, Neumann will give to him.[29]

Helmholtz could not have been more ambivalent regarding the study of mathematical physics: he praised Neumann's ability to train students in this new field, but he also knew that *academic* careers in physics were scarce. Paul seems to have felt similarly because although between 1854 and 1856 he attended Neumann's lecture courses on the theory of light, theoretical physics, and selected topics in mathematical physics, he did not completely commit himself to mathematical physics by registering for the physical division of the seminar. Not until his second year at Königsberg did he register for the seminar at all, and then only for Richelot's mathematical division.

Nevertheless, du Bois-Reymond worked closely with Neumann and intended to complete his doctoral dissertation under him until family problems intervened and his father asked him to return to Berlin. Paul tried to work out a compromise, suggesting to his father that he continue his experiments on surface tension in Berlin and then return to Königsberg to be promoted; but his father, who considered Paul's plan unnecessarily disruptive and expensive, asked Neumann if his son could complete his investigation quickly and be promoted by May 1857.[30] That proved impossible. Du Bois-Reymond returned to Berlin, where he finished the experimental investigations he had begun at Königsberg. His study of surface tension forms an interesting contrast to other student investigations that issued from Neumann's seminar.

29. Hermann Helmholtz to Emil du Bois-Reymond, 23 December 1854, Hermann von Helmholtz Nachlaß, SBPK-Hs; rpt. in *Dokumente einer Freundschaft: Briefwechsel zwischen Hermann von Helmholtz und Emil du Bois-Reymond, 1846–1894*, ed. C. Kirsten et al. (Berlin: Akademie, 1986), pp. 151–52, on p. 152.

30. Felix du Bois-Reymond to Neumann, 18 March 1857, FNN 53.IIA: Briefe von Schülern.

Du Bois-Reymond explained that when a drop of one liquid is placed on the surface of another, either the drop will lie stationary on the surface (in which case, its shape can be explained by capillary theory) or it will spread with a certain velocity. In his investigation, he examined the latter case, which was more difficult to control under experimental conditions and to express quantitatively. To calculate the change in the radius of the drop over time, he measured the radii of Newton's rings as they appeared in the drop. From a technical perspective, his accomplishment was considerable, if only because measuring the rings was difficult as the drop slowly expanded.

Du Bois-Reymond's investigation proved difficult from an analytic perspective as well. He had to find the relation between two variables: the mass of the liquid discharged, m, and its radius, r. Finding these variables "very approximately proportional" to one another, he presumed that "the forces that cause the mechanical effect of spreading" originated either on the surface of the fluid upon which the drop had discharged or in the drop itself. But the linear relation he had found between the two variables "troubled" him, prompting him to examine his experimental conditions, especially the evaporation of the drop. After performing another set of observations more "comprehensive" than his earlier ones, he discovered that "the line, which represents r as a function of m, is very weakly curved," but he considered his observations "inadequate" to describe it accurately "through interpolation." "I do not make the claim," he stated, "to have explained this mysterious phenomenon." Du Bois-Reymond concluded instead that "through the communicated investigation, the phenomenon of spreading is sufficiently *defined*."[31]

Like the seminar students before him, du Bois-Reymond found his techniques of investigation permitted him to define an experimental protocol and to examine a phenomenon but not to reach a conclusion. Also like his fellow students he refused to propose hypotheses and believed that investigative strategies in physics should be guided by "experiment and accurate observation." But what he did not use—probably as a result of not having attended the seminar and thus not participating in Neumann's introductory exercises—were all the exacting techniques of error analysis that other seminar students had deployed in assessing the accuracy of experimental results in comparison with theoretical ones. Despite his considerable mathematical talents, he occasionally suppressed mathematical expressions in this investigation, indicating them only in footnotes. Du Bois-Reymond's treatment of errors was similar to how errors were handled in other investigations completed in Gustav Mag-

31. Paul du Bois-Reymond, "Experimentaluntersuchung über die Erscheinungen, welche die Ausbreitung von Flüssigkeiten auf Flüssigkeiten hervorruft," *AP* 104 (1858): 193–234, on 222–23, 223, 224, 232 (emphasis added).

nus's private Berlin laboratory, where students used the techniques of an exact experimental physics but stopped short of considering theoretical issues.[32]

So although du Bois-Reymond was sensitive to the need for precision, he never indicated that he knew how to calculate the reliability of his measurements. Had he taken Neumann's seminar, the sharpness of his measurements and his sensitivity to error analysis probably would have been enhanced considerably. In the end, though, neither shortcoming was of concern because in the months that followed, he did not pursue technical problems in physics but devoted himself to a study of mathematics proper. By the time he defended his dissertation in Berlin in March 1859, another of Neumann's students then in Berlin, Georg Quincke, wrote to Neumann that du Bois-Reymond's dissertation "is supposed to be purely mathematical."[33] Du Bois-Reymond's excursion in physics suggests that there was a difference between working with Neumann informally in his laboratory and taking the seminar, where rigorous techniques in error and data analysis were learned more systematically.

When eight students registered for the physical division of the seminar in 1855/56, Neumann took advantage of their previous exposure to theoretical physics by showing them how the principles and objectives of measuring operations in mechanics could be carried over to electrodynamics. In covering "the effect of an induced electric current on a moving magnet," he wanted to develop "the theory of pendulum motion for completely unusual conditions and to carry out this development so it could be used to calculate [electromagnetic] observations. For this case, the ideal magnetic state was assumed." Seminar students had to determine the accuracy of their experimental data by taking into account the actual magnetic state on hand in their experiment so that they could "free the observational results from hypothetical assumptions." In characteristic fashion, Neumann guided the students back and forth between theory and experiment, for students next had to "develop the ideas of magnetic axes, the magnetic moment, and so on, and out of that to derive methods for the experimental determination of these. Thereby the foundation of the theory of the potential was clarified and was applied by the members of the seminar to the development of the magnetic potential, to the derivation of the conditions under which magnetic forces could be measured, and so on."[34] In these and similar exercises, students moved from theoretical physics to experimental physics, from practical considerations back to theoretical ones, and from the abstract to the concrete. Neumann

32. Ibid., pp. 233, 205, 204–5. On the student investigations stemming from Magnus's Berlin laboratory, see below, Chap. 7.

33. Georg Quincke to Neumann, 27 March 1859, FNN 53.IIA: Briefe von Schülern.

34. "Bericht über die Leitung der physikalischen Sektion des math. physikalischen Seminars während des Zeitraums von Ostern 1855 bis Ostern 1856," 27 June 1856, ZStA-M, Math.-phys. Seminar Königsberg, fols. 222–23, on fol. 222.

could have, at this point, developed theories of the physical constitution of electricity and magnetism. But with that, hypotheses would have entered into consideration, and they cannot be measured. The mathematical theory of the potential, elegant and simple in its expression, supplied a conceptual foundation that could be expressed in both number and measure. In the seminar, physics was *operational*: necessarily devoid of hypotheses, it dealt principally with parameters that could be measured with instruments.

The culmination of this exercise in the mathematization and quantification of electromagnetism in the summer semester was the application of the derived principles to "the execution of magnetic observations. Geomagnetic forces were measured, thereby applying various methods whose merits had to be discussed. The handling of observed values led to many kinds of small theoretical investigations." Among these "smaller" considerations were the influence of mirror images upon the accurate reading of scales and the effect of torsion in bifilar suspension. Then "the seminar members were primarily required to consider every condition that could have an influence on the results and to account for these in the calculations"—in other words, to conclude their seminar exercises with an analysis of errors.[35]

For younger students, theoretical and practical exercises on the electric current followed in the winter semester of 1855/56 when "the various methods for determining the constants of the current were discussed and worked out; the method of absolute current measurement was developed and applied; the theory of measuring instruments developed; the methods for determining the value of multiplicator readings, in particular the differential multiplicator, were developed and applied." That was not all. At the end of the semester, these students turned to the theory of heat and developed "pure theoretical considerations" concerning the propagation of heat in discrete and continuous solid media.[36] In contrast, the advanced section in 1855/56 focused on problems more closely tied to Neumann's lecture courses on mineralogy and the theory of light. Müttrich, who by the winter semester of 1855/56 had been in the physical division of the seminar for eight consecutive semesters (and was the only student in that section),[37] performed crystallographic and optical investigations that Neumann claimed were of *wissenschaftlich* interest. Nev-

---

35. Ibid.
36. Ibid., fol. 223.
37. Müttrich's Nachlaß, destroyed in World War II, is reported to have included lecture notes from Neumann's courses on optical theory (1852), elasticity theory (1854), electric current (1853), potential theory (1854), theoretical physics (1852), hydromechanics (1852/53), and crystallography (n.d.). He also took courses on differential and integral calculus, variational calculus, analytic geometry, analytical mechanics, number theory, differential equations, and elliptic functions. (Stadtbibliothek Frankfurt am Main, *Katalog der mathematischen Abteilung* [Frankfurt a.M.: Englert & Schlosser, 1909], pp. 35–36.) I thank Gert Schubring for this reference.

ertheless, Müttrich's seminar exercises did not lead immediately to the publication of original results.

By the mid-1850s, Neumann's seminar had been in operation for more than five years since a student investigation had last appeared in print. (Jakob Amsler's investigation appeared almost two years after he had left the seminar, in 1849.) Although Neumann had diligently accommodated his students during the early 1850s by shaping exercises in measurement and error analysis to their levels of ability, and although he and Richelot had experienced some success with returning students, their accomplishments in the seminar turned out to be more social, one might say, than intellectual. Students returned to discuss their work; they did not return especially to conduct research as one might have expected in an "institute" of this type, especially now that Neumann had a laboratory where students could regularly do research. Moreover, the conditions that had stimulated the investigations of both Kirchhoff and Amsler—Neumann's ongoing electrodynamic and electromagnetic investigations—were no longer present. In fact, Neumann himself had not published anything since the late 1840s. Without the stimulation and creative thinking that had accompanied his research, the exercises in measurement and error analysis that he had designed to *prepare* students for undertaking research by introducing them to fundamental techniques threatened now to become ends in themselves when students were not inspired by example to apply those techniques to unsolved problems in physics.

One sign that technical performance was becoming more important than original theoretical thinking was the transformation of exercises into didactic exemplars. Although Neumann continued to believe that the attainment of accuracy and precision marked a student's level of achievement, he nevertheless acknowledged that even without such results, seminar exercises could yield "subjective gain." By not even requiring written reports from his advanced students, he further threatened to make his exercises ends in themselves rather than avenues of intellectual development. The original research objectives of the seminar had receded even more by the early 1850s, when his seminar exercises involved either the measurement of fundamental constants, where precise results were important but not necessarily required, or the comparison between theoretical calculations and experimental results, where accuracy was sought.

Neumann's exercises had promoted the quantitative, analytic determination of how technical imperfections in instruments and shortcomings in experimental protocol affected measurements. This goal was in full accord with his initial adaption of the Besselian experiment to physics pedagogy. That did not mean that the material improvement of instruments and apparatus had been ignored but only that it had been secondary to the quantitative determination of errors. But now in the mid-1850s, with the more advanced goals of research less evident in the seminar, what had been minor or subsidiary tasks in seminar exercises assumed

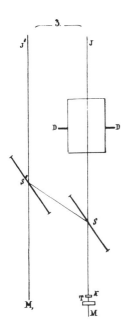

Figure 10. Schematic drawing of Wild's polarimeter. From Wild, "Ueber ein neues Photometer und Polarimeter nebst einigen damit angestellten Beobachtungen," Table II, Figure 3.

greater importance and were perceived as potential foundations for "original" publishable investigations. Whereas Kirchhoff's and Amsler's published investigations of the 1840s had been in part theoretically oriented, the next published student investigation from the seminar, by Heinrich Wild, concerned the material improvement of an instrument. Wild, who had never registered for the mathematical division of the seminar, had completed four semesters in the physical division by 1856 and had also taken Neumann's lecture courses on theoretical physics and the theory of light.

In his investigation Wild took one instrument, the photometer (used to determine if two illuminated surfaces were equally bright), and used it as the basis of a second instrument, the polarimeter (used to determine the relative intensities of the natural and polarized parts of light that is only partially polarized). In a photometer, if two separate light rays were directed to coincide with one another, the resulting ray would be partially polarized if the two original rays were not of equal intensity; if they were of equal intensity, natural light would appear. To determine the point at which both rays were of equal intensity, Wild redesigned a photometer so that it could adjust the polarized portions of the light (Figure 10). His photometer worked thus: Two parallel rays originated at *J′* and *J*; *J′* was horizontally polarized, while *J* was vertically polarized. A glass plate *D* inserted in the path of *J* altered *J* to polarize a part of its light horizontally. To detect the point at which *J′* and *J* neutralized one another (that is, when they had the same intensity and were polarized perpendicular to one another), Wild inserted in the path of *J* a crystal plate *K* and tour-maline *T*; for polarized light would exhibit interference lines in a crystal, whereas natural light would not. The two rays could be combined by the system of mirrors *S* and *S'*. With the glass plate *D* and the crystal in place, Wild determined the point at which the two rays neutralized one an-other—that is, when the intensities of both rays were equal—by rotating *D* until no interference lines appeared in *K*. In the absence of completely polarized light, the same procedure could be used in an apparatus that Wild called a polarimeter (Figure 11).[38]

By inserting the glass plate *D* and the crystal *K*, Wild made his instru-ment more sensitive than normal photometers, and hence his modifica-tions enabled him to detect equivalent optical intensities with greater precision than had been formerly achieved. His instrumental modifica-tions maximized the probability that the ideal conditions described in the optical laws pertinent to his investigation would be met. Other seminar students had mathematically analyzed physical imperfections in experi-mental apparatus or in the "environment" of the experiment, thereby producing correction factors that were appended to the mathematical formulas describing the experiment. Wild, in contrast, directly modified his apparatus so that the physical phenomenon under consideration was displayed as best it could be before any corrections were added to ideal formulas. He described this procedure as achieving the state wherein "the apparatus corresponds completely to its idea [*Idee*]."[39]

With Wild's investigation, we see more explicitly than in those of the 1840s how notions of precision were evolving in the seminar. Wild was the first seminar student to refer to precision explicitly in terms of decimal places. Noting that the human eye can generally differentiate a 1/64 difference in brightness and that exceptional eyes can detect a 1/100 to a 1/120 difference, he believed that "an accuracy of 1/100 is, however, not sufficient for scientific photometric investigations." He undertook his investigation in order to achieve a precision of 1/500 to 1/1000, or as he put it, "to three decimals." (A third-decimal-place precision placed Wild's measurements in line with the most accurate measurements of the eigh-teenth century.) His was also the first seminar investigation to use a "precision" instrument to attain a *precision measurement*. In previous stu-dent investigations, the results had been first adjusted for errors; they produced precise measurements but as *corrected results*. Although Wild acknowledged that in principle one had "to determine the value of the errors [in the apparatus] that cannot be eliminated and to correct the formulas accordingly,"[40] he pointed out that in the apparatus he had constructed, "constant errors," such as those resulting from the imperfect

38. Heinrich Wild, "Ueber ein neues Photometer und Polarimeter nebst einigen damit angestellten Beobachtungen," *AP* 99 (1856): 235–74.
  39. Ibid., p. 262.
  40. Ibid., pp. 235, 262, 253.

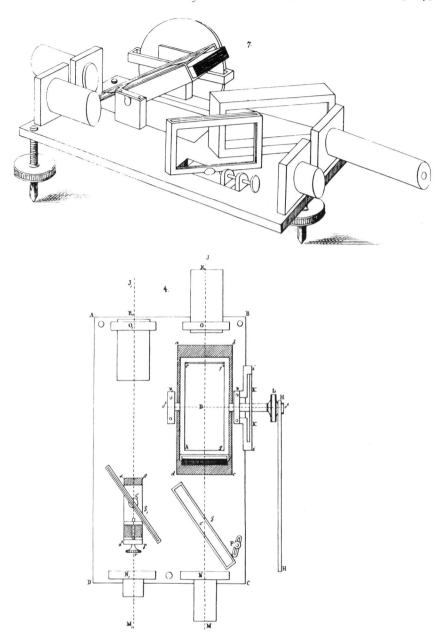

Figure 11. Wild's polarimeter. From Wild, "Ueber ein neues Photometer und Polarimeter," Table II, Figures 7 and 4.

parallel of the two light rays, were "insignificant" and that corrections for "variable errors"—that is, for the accidental errors of the investigation, computed by the method of least squares—did not affect the three decimal place accuracy.[41]

Wild's polarimeter undoubtedly was inspired by Neumann's interest in understanding the theory of optical intensities, but Wild did not draw explicit connections between his investigation and Neumann's. What his investigation demonstrated, however, was that even as the technical means for obtaining data improved, yielding data of greater precision, the sensitivity to the condition of data, which had been cultivated by the error analysis component of Neumann's seminar exercises, intensified. Wild, for instance, considered it necessary to distinguish more clearly data with different degrees of certainty even though all data were eventually combined so as to form the basis of a generalized result. He used his apparatus to measure the ratio of the optical intensity of two rays, one consisting of polarized sunlight and the other, natural sunlight, and presented his results as a graph (Figure 12). The x axis marked the angular distance of the sun from the north; the y axis represented the ratio of the two intensities multiplied by 100. Wild took measurements every 10° from 20° to 50°, at 85°, and then every 10° from 90° to 150°. He then connected the points from 20° to 50° in a smooth curve, as well as the points from 85° to 150°. The interpolation of points between those 10° intervals presented no problem for him, and he represented them graphically as if he had actually measured them. But for points above 150°, below 20°, and between 50° and 85°, he did not feel so certain. "The dotted parts of the curve," he remarked, "are supposed to signify the probable [*muthmaßlich*] course of those places where no measuring observations could be conducted."[42] His reason for representing those points as probable and the other interpolated points as certain seems to have been connected with the nature of the measurements he had taken. He knew that his measurements from 20° to 50° and from 85° to 150° were precise to three decimal places. But the range of 35° from 50° to 85° was apparently too large for him to impute any degree of certainty to his estimates. The same held for the range above 150° and below 20°, where he relied upon the quantitative observations of others; for while that data seemed to coincide with his, the fact that it was derived under conditions whose range of accuracy was not known was enough to convince Wild that it was qualitatively different from data he had taken directly with his own instrument.

Wild's hesitancy in interpolating and extrapolating data was not unique: Paul du Bois-Reymond also had been unwilling to describe a curve

41. Ibid., p. 262.
42. Ibid., p. 273.

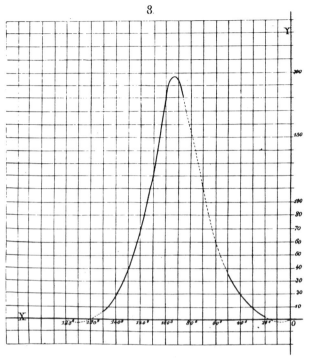

Figure 12. Wild's graph of optical intensities as a function of angular distance from the sun. From Wild, "Ueber ein neues Photometer und Polarimeter," Table II, Figure 8.

through interpolation.[43] Yet Wild's case, more than du Bois-Reymond's, points to the fact that as the degree of precision in measuring data increased, students were unwilling indiscriminately to combine precise data with data, including interpolated data, whose range of error was unknown, larger, or in some way questionable. His sensitivity to the differences among the sets of data at his disposal led to a particularly striking result in his graph. Error analysis of the type Neumann taught and he and others deployed, especially in astronomy, was intended in part as a means of combining measurements. But such measurements were generally "represented" in tabular form. No similar adaptation of error analysis to the graphical analysis of data was as yet in common practice. Wild, for instance, did not think of representing the probable errors of his observations (as well as of others) as error bars and then finding a "best fit"

43. P. du Bois-Reymond, "Experimentaluntersuchung über die Ausbreitung von Flüssigkeiten," p. 224.

curve. Instead, for him, data derived under different conditions were, in a sense, incomparable to one another or qualitatively different from one another *when represented graphically*. The graphical representation of data thus became a way to distinguish clearly—and at a glance—between data with different or unknown degrees of accuracy or precision. Merely listing two or more sets of data together in tabular form would have implied that they could be combined because all sets fell within certain tolerances of error.

Wild returned to the physical division of the seminar in 1856/57, when he was joined by nine other students. This group proved to be more mature than others because two postdoctoral students participated in what were now biweekly meetings of the seminar: H. Jacobson, a practicing physician and later *außerordentlicher* professor of medicine at Berlin; and Lothar Meyer, who had already taken a doctorate in medicine at Würzburg in 1854, had continued his studies in physics and chemistry with Kirchhoff and Robert Bunsen at Heidelberg, and then, at the suggestion of his brother Oskar Emil Meyer, registered for Neumann's lecture courses and seminar in 1856. During the summer semester of 1856, when the seminar students had to be divided into two sections, these postdoctoral candidates, like all Neumann's beginning students, achieved "fluency" in the application of mechanical principles. Students in the upper section were among the most experienced Neumann had taught to date: Wild, in his third and final year in the seminar; Quincke, who had participated in the seminar from 1853 to 1855 but spent 1855/56 studying under Neumann's student Kirchhoff at Heidelberg; and C. J. H. Lampe, in his third year in the seminar. For these more experienced students, Neumann designed problems in thermoelectricity, and he especially focused upon methods for determining the constants of thermal conductivity "with consideration of the electric current produced by heat." In executing their solutions, students learned Fourier's techniques for integrating partial differential equations. In his report, Neumann stressed less the mathematical component of the seminar exercises than how thermoelectricity affected the measurement of the electric current. To show students how to correct for thermoelectric disturbances, Neumann introduced the theory of the multiplicator in a new way, explaining how the instrument could be used to measure variable electric currents and developing the theory of measurements thus taken. Ultimately, this led to a discussion of the Peltier effect.

During the winter semester of 1856/57, when all students united into one section, Neumann continued to discuss the theory of heat by introducing complex theoretical principles through their application: equations for the propagation of heat through media were applied to the method of mixtures for the determination of specific heats; then to the propagation of heat in a system of discrete points and then in a wire; and finally, general equations for thermal propagation were developed for crystalline bodies and for fluids, where in addition the viscosity of the fluid was considered. While in all these cases, hypothetical considerations such as the nature of heat were disregarded, the development of theory through application nevertheless led to a consideration of more abstract physical situations and mathematical techniques. This time, Neumann ended the seminar with a discussion of spherical harmonics and the equations for the internal viscosity of liquids.[44]

In his last year in the seminar, Wild published a second investigation based on seminar exercises. It, too, emphasized what could be achieved by improving instruments and refining experimental protocol. His study concerned Neumann's method for determining galvanic polarization and transition resistance associated with a decomposition cell that had been added to a closed circuit (the decomposition cell increased the resistance and decreased the electromotive force of the entire circuit). Analytically and experimentally determining both the polarization and the transition resistance was difficult, Wild demonstrated, and so calculations of the transition resistance had been neglected in prior investigations. But, he argued, that omission introduced a significant error. Both Neumann's analytic determination of both variables and his measurements proved that the transition resistance existed. Wild concluded that "the error, which one has committed through the neglect of the transition resistance is thereby in no way small and makes all prior determinations of polarization inaccurate."[45]

Neumann's modifications of three instruments—the differential galvanometer, the rheostat, and the magnetic compass—were crucial for Wild's investigation. Wild used Neumann's combination of a differential galvanometer and a Wheatstone bridge to achieve greater precision. Wild did not abandon error and data analysis as the technical quality of his measuring instruments improved. The continued relevance of error analysis to an experiment was especially prominent in Wild's comparison of Neumann's magnetic compass with Wilhelm Weber's. Weber's, Wild explained, was a massive copper ring through which ran the current to be measured. For the compass to be accurate, a magnetic needle in the center had to have a length of at least one-fourth of the radius of the ring. Because the accuracy of Weber's instrument depended on what Wild considered bulky material conditions, he questioned its usefulness in taking refined observations but without citing its overall error, a criticism of questionable tact when referring to one of Germany's leaders in precision measurement. The real problem with Weber's instrument, Wild

44. "Bericht über die Leitung der physikalischen Abtheilung des math. physikalischen Seminars der Königl. Universität Königsberg während des Zeitraums von Ostern 1856 bis Ostern 1857," 28 August 1857, ZStA-M, Math.-phys. Seminar Königsberg, fols. 226–27.

45. Heinrich Wild, "Die Neumann'sche Methode zur Bestimmung der Polarisation und des Uebergangswiderstandes nebst einer Modifikation derselben," *Vierteljahrsschrift der Naturforschenden Gesellschaft in Zürich* 2 (1857): 213–43, on 232.

thought, was that it required a knowledge of errors that could not easily be determined quantitatively from Weber's construction. So Weber's "arrangement" for the compass, Wild concluded, "is suitable for ordinary observations [gewöhnliche Beobachtungen], but for accurate measurements [genaue Messungen], it is unsuitable." Thus, although both Neumann's and Weber's compasses produced precise measurements, Wild distinguished between the precision of their measurements on the basis of how well the instrumental construction lent itself to the analysis of errors. Neumann's principal modifications—to allow the needle to slide on the axis of the cylinder, to increase the amount of wire on the cylinder, and to protect the needle from the motion of the air by encasing it in a glass container—not only helped one to determine the errors of the instrument, Wild argued, but also increased its sensitivity.[46]

It is not accidental that an investigation like Wild's appeared in the 1850s but not earlier. With their emphasis on the construction of precision instruments, his experimental investigations reflected the new opportunity offered in Neumann's residential laboratory for refining instruments. The retention of the central role of error analysis meant that the essentials of the Besselian experiment had not been abandoned but instead had taken on added significance. The ease with which errors could be determined for more complex instruments became, in Wild's case, an indicator of the scientific value of the instrument and especially of the degree of precision of the measurements it yielded.

Error analysis of various types had been linked to measurement since antiquity. Ancient astronomers had a healthy mistrust of sensory observations, were aware of the approximate nature of their results, and had crude measures of the reliability of their data. They also used "error analysis" inconsistently and did not hesitate to dismiss out-of-the-ordinary observations. During the scientific revolution of the sixteenth and seventeenth centuries, more accurate measurements were possible, but "error analysis" was largely a qualitative affair. Micrometers transformed telescopes into measuring instruments and vastly improved the accuracy of measuring large distances. Galileo seems to have considered positive and negative deviations in measurements equally probable. At the Academia del Cimento, where the motto was "trial and error," standards of experimental protocol were taking shape. But the analysis of errors was still only a matter of talking about the crude instruments, inaccurate scales, and the effect of disturbing "environmental" conditions, such as air pressure, temperature, and humidity. With the exception of corrections for refraction on astronomical tables, errors were not rigorously and quantitatively determined. The quantification of experimental physics, outside of mechanics, began in the second half of the eighteenth century, when, in

46. Ibid., p. 235.

some areas of experimental physics, one could count on three-decimal-place precision and, with luck, four. The increased precision was due to more refined instruments and changing standards in the practice of physics, especially to the quantification of constant errors.[47]

Although the histories of error analysis, techniques of measurement, notions of precision, and improvements in instrumentation are sketchy and in desperate need of more detailed elaboration, especially in areas where they overlapped one another and where they affected the development of theory, it is nonetheless clear that the institutionalization of these techniques in the first half of the nineteenth century, especially in pedagogical settings like Neumann's seminar, was a crucial stage in their further refinement and in their acceptance as standard techniques in the practice of physics. But the physical seminar at Königsberg was more than a setting in which students learned to achieve the precision of the next decimal place. The techniques that came into play in seminar exercises were not only more exact and discriminating but also of greater variety than those used by the eighteenth-century quantifiers of physics. Neumann's incorporation of the method of least squares and of the analysis of accidental errors was an especially novel—and demanding—addition to the investigative enterprise in physics because it prompted students to go beyond the material conditions of an experiment and to examine the uncertainties that remained in the data itself.

In commenting on how notions of error were transmitted in antiquity, G. E. R. Lloyd noted that "a student must be *taught* what he *sees*."[48] Because error and data analysis so disciplined students in the practice of physics, bringing them to the point where they "possessed" what they had learned and focusing their energies in certain directions, it was the single most important component of Neumann's strategy for professional training. The adoption of rigorous error analysis altered the practice of physics by orienting the questions asked in a theoretical physical inves-

47. G. E. R. Lloyd, "Observational Error in Later Greek Science," in *Science and Speculation: Studies in Hellenistic Theory and Practice*, ed. Jonathan Barnes et al. (Cambridge: Cambridge University Press, 1982), pp. 128–64; Giora Hon, "Is There a Concept of Experimental Error in Greek Astronomy?" *BJHS* 22 (1989): 129–50; O. B. Sheynin, "Mathematical Treatment of Astronomical Observations: A Historical Essay," *AHES* 11 (1973): 97–126; "Examples of Experiments in Natural Philosophy Made in the Academy del Cimento [1667]," in W. E. Knowles Middleton, *The Experimenters: A Study of the Accademia del Cimento* (Baltimore: Johns Hopkins University Press, 1971), pp. 83–251; Giora Hon, "On Kepler's Awareness of the Problem of Experimental Error," *Annals of Science* 44 (1987): 545–91; Albert van Helden, *Measuring the Universe: Cosmic Dimensions from Aristarchus to Halley* (Chicago: University of Chicago Press, 1985); Maurice Daumas, "Precision of Measurement and Physical and Chemical Research in the Eighteenth Century," in *Scientific Change*, ed. A. C. Crombie (New York: Basic Books, 1963), pp. 418–30; J. L. Heilbron, *Elements of Early Modern Physics* (Berkeley and Los Angeles: University of California Press, 1982), pp. 65–70; Theodore Feldmann, "Applied Mathematics and the Quantification of Experimental Physics: The Example of Barometric Hypsometry," *HSPS* 15 (1985): 127–97.

48. Lloyd, "Observational Error in Later Greek Science," p. 133.

tigation not only toward technical matters in the execution of protocol but also toward epistemological matters in the assessment of measurements. Yet the emphasis on error and data analysis in the seminar also created blind spots in a student's field of vision by deflecting attention away from the deeper philosophical and conceptual elements in the study of physical phenomena.

Wild, whose work embodied the tenets of Neumann's teaching, had learned his lessons well. His case demonstrates that as both error and data analysis *and* the refinement of instruments became a part of seminar exercises, the gap between two forms of quantification—that of an exact experimental physics and that of a mathematical physics—widened even further. Closing the gap between experimental results and their analytic expression in a physical law—or between physical theory and its empirical confirmation—required increasingly sophisticated techniques and more refined judgments. Technically, graphical analysis seemed to be a perfectly acceptable and natural way to bridge experimental and mathematical forms of quantification; for in graphs, mathematical expression and experimental data intersected. But graphical analysis proved in Wild's case to be incompatible with the degree of certainty he could ascribe to his measurements. For Wild and, as we shall see, more so for other seminar students, it was not easy to use graphs, to make the transition from the graphical representation of experimental data to a mathematically based physical law, or to confirm a mathematically expressed theory by comparing its graph with that of empirical measurements. Wild and other students from Neumann's seminar viewed curves constructed from measurements as approximating not only theoretical values but also what they believed to be "more precise" experimental ones as well. Error analysis, refined instruments, and the desire to express physical relations in mathematical form made exactitude—the measure of achievement and the guarantor of degrees of certainty—an ever more difficult boundary condition to satisfy.

CHAPTER SEVEN

Interpolation
and Certainty

THE PURSUIT OF both the mathematical and the experimental dimensions of Neumann's seminar exercises restrained students from excessive mathematical abstraction and presumably from hypothetical speculation but still encouraged analytic sophistication. Yet Neumann's strategy in teaching theoretical physics was problematic. Although he had taught his students how to apply partial differential equations to increasingly refined situations in physics and how to reduce data, he seems not to have dwelt upon an equally important consideration—how to unite the mathematics of theory with the quantitative results of experiment—except by adjusting and refining the former and by increasing the precision and decreasing the errors of the latter. Trying to achieve these complementary goals simultaneously could be difficult if one adhered to the values and standards promoted by seminar exercises, because in order to create mathematical generalizations, one had to abandon the compulsion to achieve further experimental precision and work more closely with the conceptual foundation of theory itself. Student investigations by Otto Hagen, Georg Quincke, and others during the 1850s and 1860s demonstrate how complicated these matters of theoretical physics, measurement, and data analysis could become.

In the summer semester of 1857, Neumann divided the eleven students in the seminar into two sections when the "unequal standpoint" of the students meant that they could not all be taught together. After spending his first year of university studies in the mathematical seminar, the young student H. Kaul entered Neumann's division. Three students who had transferred from Berlin also took part. Two of them, O. E. J. Reichel and Franz Eduard Gehring, had already completed introductory work in the mathematical seminar; the third, E. G. H. Schindler, came to Königsberg "in order to take part in the seminar" after attending classes at Berlin for two years. Two experienced seminar students, Wilhelm Fuhrmann and Louis Saalschütz who had been in the mathematical division since, respectively, 1853 and 1854, participated in the physical division. All six had matriculated to study mathematics; they participated in the physical division partly to prepare for the *Lehrerberuf*. More dedicated to the study of physics itself in this year were Oskar Emil Meyer, Ludwig Minnigerode, and C. J. H. Lampe. Meyer, who had studied physiology with Carl Ludwig in Zurich, arrived at Königsberg in 1856 to learn physics with Neumann; Quincke, who had by then made his acquaintance, had taken the trip with him.[1]

Minnigerode was the only one who upon matriculation had declared physics as his major. Lampe, who had matriculated in mathematics and the natural sciences, participated so enthusiastically in the physical division that Richelot reported with regret that Lampe was leaning more toward physics than mathematics. In 1857, after four years in the mathematical seminar, Lampe joined Meyer in working on the philosophical faculty prize question, on the internal viscosity of fluids. No doubt these three students were a joy to Neumann. Yet in this year the seminar continued to attract others not in physics or mathematics. Two young professors of chemistry, Oskar Emil Meyer's brother Lothar and Leopold Pebal, sought postdoctoral training there. In fact, Lothar Meyer participated in both the elementary and advanced sections of Neumann's seminar for two years and published a brief note in physical chemistry on the influence of pressure on chemical affinity. Although his observations were made in Königsberg's chemical laboratory, Meyer's study focused on the physical aspects of the problem. He had taken the seminar to become acquainted with exacting investigative procedures, which he carried over to chemistry. In his paper on pressure and chemical affinity, he questioned the experimental results and protocol of previous investigations.[2]

In the summer semester of 1857, when Neumann lectured on magnetism and theoretical physics, students in the lower section of the physical seminar studied mechanics to learn how to apply calculus to physical problems. Neumann chose the theory of collisions, Newton's theory of resistance, and the ballistic pendulum as mathematical exemplars in physics. He used the pendulum again to demonstrate the influence of

        1. "Jahresbericht über die mathematische Abtheilung des mathematisch-physikalischen Seminars zu Königsberg in Pr.: Per Ostern 1857 bis Ostern 1858," 31 July 1858, and "Bericht über die Leitung der physikalischen Abtheilung des math. physikalischen Seminars der Kgl. Universität Königsberg während des Zeitraums von Ostern 1857 bis Ostern 1858," 27 August 1858, ZStA-M, Acta betr. das mathematisch-physikalische Seminar an der Universität zu Königsberg, Rep. 76Va, Sekt. 11, Tit. X, Nr. 25, Bd. I: 1834–61, fols. 241–44; Georg Quincke to Neumann, 23 March 1856, FNN 53.IIA: Briefe von Schülern.
        2. "Jahresbericht über die mathematische Abtheilung . . . , 1857/58," and "Bericht über die . . . physikalische Abtheilung . . . , 1857/58," fols. 241, 243; Lothar Meyer, "Ueber den Einfluss des Drucks auf die chemische Affinität," *AP* 104 (1858): 189–90.

disturbing factors in experiments. He also discussed relative motion and applied the concepts he developed to the motion of the earth. In the upper section, he put students to work on the prize question. He expected his students to design and execute their own trials to determine the coefficient of viscosity of fluids. Some were able to continue their investigations from the previous year, examining the propagation of heat in solid media, especially the production of thermal currents in unequally heated solid media. Often Neumann used geothermal examples and thus probably drew upon the geothermal measurements he and other students had taken. The purpose of these exercises was to calculate thermal constants. The analytic techniques taught in the physical and the mathematical divisions converged this semester when Richelot taught the principles of geodetic lines, showed how to apply elliptic and transcendental functions to them (leaving several original problems for his students to solve), and derived the relevant integrals. These assignments were relevant not only to the determination of isothermals in Neumann's seminar exercises but also to problems in geodesy and more generally to the reduction of data by the method of least squares.[3]

By the beginning of the winter semester of 1857/58, when Neumann offered courses on elasticity and optics, he was able to unite the sections of his division and discuss the distribution of heat in crystalline media and Gauss's capillary theory. Both topics he treated by using specific, concrete illustrations: thermal effects in crystals were determined by measuring changes in crystalline angles, while capillarity was studied by deriving the forces at work in an oil drop placed on alcohol and by determining changes in its shape; both were difficult, advanced topics.[4]

Emil Meyer, Minnigerode, Reichel, Kaul, Schindler, and Gehring returned to the physical seminar in 1858/59, when they were joined by five other students: three who had just matriculated in physics the summer semester of 1858—Friedrich Gustav Adolph Just, Karl Zöppritz, and F. J. Allemann; a student of mathematics and philosophy, G. Arnold; and Carl Pape, who had already taken a doctorate at Heidelberg University and was now continuing his studies in mathematical physics at Königsberg. Because it was now unnecessary to divide the seminar into two sections, Neumann built on his treatment of thermal phenomena in the summer semester by integrating theoretical and experimental issues more closely. He reviewed several methods for determining specific heats but developed the equations for the propagation of heat for only one, his own method of mixtures. Having introduced students to how heat propagated in crystalline media, he then asked them to generalize this case to the propagation of heat in any solid media, starting first with discrete masses

---

3. "Jahresbericht über die mathematische Abtheilung . . . , 1857/58," and "Bericht über die . . . physikalische Abtheilung . . . , 1857/58," fols. 241–44.
4. "Bericht über die . . . physikalische Abtheilung . . . , 1857/58," fols. 243–44.

and then developing the mathematical expressions for the distribution of temperature in a heated bar. More complex problems followed, including deriving mathematical equations for thermoelectric currents in a manner suitable for experimentally determining their constants. Just, Pape, and Zöppritz became so involved in the experimental part of this particular investigation that they stayed at Königsberg over the summer vacation to carry out their own trials on thermoelectricity. In the course of their investigation they developed a "theory" for the motion of a magnetic needle under the influence of a variable electric current. As in past years, some students were exempted from the seminar exercises and allowed to pursue their own investigations. This year Emil Meyer continued his independent investigation of the viscosity of fluids.[5] Such independent work was, however, still rare. In its third decade, the seminar still had not become an institute dedicated principally to the production of original research. At most, the students usually laid the foundations for such investigations and then, for various reasons, completed them elsewhere as doctoral dissertations or *Habilitationsschriften*.

Neumann must have felt confident that his students in the winter semester of 1858/59 had mastered the fundamentals of physics sufficiently to undertake simple measuring investigations in magnetism; for he assigned exercises in determining magnetic axes; assessing the effect of a magnet on a horizontal magnetic needle; and applying Gauss's method for measuring geomagnetism. Some students continued their investigation of thermal phenomena. Theoretical expressions for geothermal phenomena were also translated into forms suited for measuring operations, including the determination of the degree of solar radiation heating the surface of the earth. Students certified their formulas for isothermals by measuring the heat of the earth at various depths, by determining the thermal constants for Königsberg, and by making quantitative estimates of how much of the original heat of the earth was still on hand.[6] Whether they conducted the last investigation by extrapolating data from analytic formulas is not certain; although it would have been natural to do so, Neumann did not specifically mention extrapolation or interpolation techniques in his seminar reports.

In the second half of the 1850s then, Neumann continued to assign problems that were neither elementary nor original. None involved simple applications of the theories taught; his exercises were not merely "cookbook" problems. They were drawn from what he had on hand—including his geothermal station—but the most complicated steps were regularly left for students to derive on their own. The measuring observa-

---

5. "Bericht über die Leitung der physikalischen Abtheilung des math. physikalischen Seminars der Universität Königsberg während des Zeitraums von Ostern 1858 bis Ostern 1859," 3 October 1859, ZStA-M, Math.-phys. Seminar Königsberg, fols. 250–51.
6. Ibid.

tions that followed the mathematical exercises were built into his practical exercises, but his students had to determine how to carry them out largely on their own. Sometimes they saw a way to improve upon these techniques and so to begin an original investigation. But because his measuring exercises repeated what were by then well-formed techniques (e.g., Neumann's method of mixtures or Gauss's procedure for measuring geomagnetism), the chance of original investigations arising from them became less likely. For the most part, Neumann's exercises were now such that theory could be compared with experiment in straightforward ways because it was already securely grounded and its experimental limitations fairly well known. When such assurances were lacking or inadequate, however, his students encountered difficulties in modifying or creating mathematical expressions of theory on the basis of their measurements.

Otto Hagen, for instance, tried in his independent investigation on the absorption of light in crystals to analyze his results as Neumann had encouraged his students to do. Hagen, who had matriculated in 1855 and spent one year in the seminar, eventually completed his seminar investigation in Berlin, where he submitted it as a dissertation. Initially Neumann had suggested that Hagen examine how the coefficient of optical absorption changes with the wavelength of light by taking photometric measurements on pleochroic crystals. Hagen carried out his trials successfully. When it came time to analyze his data, however, he encountered difficulties. He first represented his results in tabular form and then tried to represent graphically the ratio of the absorption coefficients of ordinary and extraordinary polarized light as a function of wavelength by drawing curves from his data on three different crystals (Figure 13). He constructed his own graph paper, plotted his points as rather large solid circles, and then drew smooth curves between them. His measurements appeared sufficiently precise to justify the smooth curves, but it is not clear whether the curves representing his data were signs of his skill as an experimenter or an artifact of how he constructed his graph, whose scale was not refined enough to display the error in his measurements. Despite the regularity of his curves, Hagen did not take the next obvious step: deducing their equations. Rather than being a step on the path to further quantification, his graphical analysis merely remained an idealized pictorial representation of his data. Even the precision he claimed to have attained—1/5000 in his measurements of wavelength—failed to provide him with the confidence that he could construct a mathematical law from his curves. Instead, he was content with making "general conclusions" that merely reinforced the strictly *visual* purpose of his graphs: (1) that the functions represented were "continuous," (2) that the functions had a maximum and a minimum, and (3) that the functions "appear symmetrical."

Although a simple application of calculus—or even a general familiarity with the graphical representation of the most common functions—

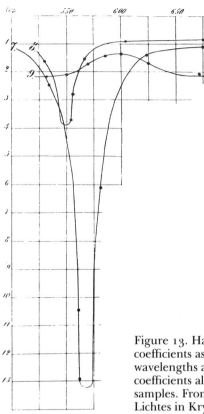

Figure 13. Hagen's graph representing absorption
coefficients as a function of wavelength, with
wavelengths along the x axis, the ratio of the absorption
coefficients along the y axis, and curves for three crystal
samples. From Hagen, "Ueber die Absorption des
Lichtes in Krystallen," Table III, Figures 7–9.

could have helped Hagen transform his "general conclusions" into a
mathematically exact relation, he made no attempt to construct one.
Instead, he firmly believed that to formulate a mathematical law, he
needed "more accurate" and "sharper" observations even though his
measurements were already accurate to between the third and fourth
decimal. Hagen had clearly made a choice here. Rather than use his
results to construct a mathematical law, if only an empirical one, which
could be further tested, he instead allowed the pursuit of certainty in
measurements to determine the next step in his investigation. Nevertheless, he still expressed the hope that what he had done would eventually
lead to a mathematical expression relating absorption coefficients to
wavelengths.[7] Unfortunately we will never know how Hagen might have
developed his ideas because he died a few years after the publication of his
investigation. His mathematical analysis of data, like Wild's, nonetheless

7. Otto Hagen, "Ueber die Absorption des Lichtes in Krystallen," *AP* 106 (1859): 33–
55.

reveals how the investigative strategies and values cultivated in Neumann's seminar shaped the decisions made in the course of an investigation.

Similar concerns for certainty arose in the investigations of Georg Hermann Quincke. Drawn to the study of physics in his penultimate year at the Friedrich Werder Gymnasium in Berlin, Quincke entered Berlin University in 1852 and eventually completed his dissertation there in 1858. Not all his university years were spent at Berlin despite his feeling that he was a "genuine Berliner."[8] Seeking out those from whom he thought he could learn the most, he studied for a few semesters at Heidelberg and for several under Neumann at Königsberg, where he was a member of the seminar during 1853/54, the summer semester of 1854, and the summer semester of 1856. Although his circle of teachers was wide, Quincke reserved a special place for Neumann. "Because you always have shown me so much love and interest," he wrote to Neumann from Heidelberg (where he had returned for the winter semester of 1854/55) on 9 November 1854, "I consider it my obligation to send you a written report and to tell you that I am very satisfied with my new field of activity. Perhaps I have not found [Heidelberg] as stimulating as Königsberg . . . but Bunsen did receive me with all the warmth I could have wanted." Once at Heidelberg, he was pleased to learn that Neumann's former student Kirchhoff had just joined the faculty. But when Kirchhoff first arrived, he offered a course on experimental physics, which Quincke did not want to attend. Quincke nevertheless became friends with Kirchhoff, who showed him the university's physical cabinet, which was equipped primarily with demonstration apparatus. While Quincke waited for Kirchhoff to offer more interesting courses, he independently studied Gauss's dioptric investigations, a topic Neumann had introduced to him in the seminar exercises.[9]

Although he profited from his tutelage under Bunsen and Kirchhoff, Quincke was thinking by the spring of 1856 of returning to Königsberg once again. His reasons were practical. In his spare time, he had studied, at Neumann's suggestion, the area of physics that was the foundation of Neumann's pedagogical program: mechanics. But even with Kirchhoff's occasional help, Quincke felt something lacking. "Now I have the feeling," he confided to Neumann, "that I no longer know enough mathematics." He thought first of returning to his home city of Berlin, but at the university Kummer was teaching a course on analytic geometry, which Quincke did not think would be of help. So with his father's permission, he planned to return to Königsberg by the middle of April even though he had not inquired into what Richelot would be teaching. Quincke

---

8. Walter König, "Georg Hermann Quinckes Leben und Schaffen," *Die Naturwissenschaften* 31 (1924): 621–27, on 622.

9. Quincke to Neumann, 9 November 1854, FNN 53.IIA: Briefe von Schülern.

hoped nevertheless "that I will be able then to hear [mathematical] lectures that will be useful for my profession."[10] At Königsberg, he continued to attend Neumann's cycle of lectures on mathematical physics,
sharpened his experimental skills, and became intensely interested in
capillarity. The strongest continuity in Quincke's physics education was
his experience at Königsberg.

Quincke's first two published papers stem from investigations undertaken at Heidelberg. He had initially been attracted to Bunsen's teaching
laboratory, where, as he reported to Neumann, "Bunsen is in the laboratory from 8 to 5 and supervises 40 students who work with him." From
Bunsen he "learned all methods" and occupied himself with "qualitative
investigations" in the laboratory such as the analysis of gases and chemi-
cals as well as techniques in glass blowing.[11] In his first published paper,
Quincke sought through quantitative chemical analysis to determine the
origin of red and gray gneiss. His analysis of data was at first curious.
Although he reported that his data, which he listed in tabular form, had
been computed according to the method of least squares, he did not
bother to specify the range within which they were accurate. The differences between the theoretical and experimental values of his investigation were large enough—some as high as 50 percent—to suggest that his
reduction of data was a formal step carrying little significance for the
interpretation of his results. Large errors were common in this type of
investigation, but Quincke did not use error analysis to suggest how the
experiment could be modified to reduce error. He also did not indicate
whether the error was large enough to consider the data within range of
theory or small enough to challenge the theoretical foundation of his
computations.[12]

Quincke's interpretation of the results of error and data analysis improved, however, once he turned to physics. His second published investigation was on the distribution of an electric current on a metal plate, a
topic that, although undertaken at Heidelberg, had its origins in earlier
seminar exercises at Königsberg and that was closely related to Kirchhoff's first seminar investigation there. Quincke's motivation for taking
up a problem already adequately solved by Kirchhoff was not only to
apply Kirchhoff's formulas for the lines of equal "tension" to novel cases
but also to improve upon the agreement between theory and experiment.
So although Quincke complained to Neumann that the study involved
"very large and unpleasant numerical calculations," he thought that "the
subject, upon which two students of yours have worked, is interesting
because it is a new proof for the mathematical theory of the electric
current."[13]

10. Quincke to Neumann, 23 March 1856, ibid.
11. Quincke to Neumann, 9 November 1854, 23 March 1856, ibid.
12. Georg Quincke, "Beitrag zur Kenntniss des rothen und grauen Gneisses des
Erzgebirges," *Annalen der Chemie und Pharmacie* 94 (1856): 232–40.
13. Quincke to Neumann, 23 March 1856, FNN 53.IIA: Briefe von Schülern.

"The following investigation," his paper began, "was undertaken in the laboratory of Professor Kirchhoff [at Heidelberg] with the intent of proving through experiment the theory of the distribution of an electric current on a metal plate given [by Kirchhoff]."[14] Noting that the theory had thus far only been proven for a circular metal disk to which two electrodes were attached at the periphery, Quincke measured the distribution of electricity on a square lead plate upon which electrodes moved across the diagonal and also on a circular disk of lead and copper soldered together along a diameter. He used the measuring techniques Kirchhoff had described earlier and used for finding the "curves of equal tension." Although Quincke drew upon Kirchhoff's theory and protocol, he detailed his experimental procedure and how he analyzed data in ways Kirchhoff had not. For example, Quincke compensated for more constant errors, including the effect of thermoelectricity generated by the heat in his hands. Most striking were Quincke's tabular and graphical representations of his data, in which he drew sharper distinctions between theoretical and experimental results than Kirchhoff had done.[15]

When Kirchhoff drew his equipotential lines, he represented only the "*approximate* position of the observed points" (Figure 9).[16] Although it is difficult to tell from Kirchhoff's graph just how "approximate" those points and lines were, it is telling that he supplied no x and y axis and, more important, gave no indication of the scale he was using. One suspects that, overall, Kirchhoff's illustration was drawn either from theory or from points judged, but not demonstrated, to be within the range of error of the experiment; for even though "points" were depicted on his graph, they were so carefully connected—and perfectly drawn—that the vertical axis divides what looks to be two sets of perfect, mirror-image smooth curves. So whereas Kirchhoff's tables of observed and calculated values readily showed the differences between experimental and theoretical results, his graphs did not.

Quincke, eleven years after Kirchhoff's publication, showed in his investigation another way to compare experimental and theoretical values for a related problem. For the square lead plate, Quincke calculated the potential U for seven points along the diagonal of the plate where $x = y$ in whole, even numbers from two to fourteen. Each coordinate pair corresponded to a particular value of U, which Quincke calculated from Kirchhoff's formulas. Then, from these seven values of U, Quincke found that "the values of y can be calculated through interpolation." The calculated values of y formed the basis of the theoretical equipotential curves (Kirchhoff's "curves of equal tension"), represented by solid lines marked

14. Georg Quincke, "Ueber die Verbreitung eines elektrischen Stromes in Metallplatten," *AP* 97 (1856): 382–96, on 382.

15. Ibid., Table IV, Plates 6 and 7.

16. Gustav Kirchhoff, "Ueber den Durchgang eines elektrischen Stromes durch eine Ebene, insbesondere durch eine kreisförmige" [1845], in *Gesammelte Abhandlungen* (Leipzig: Barth, 1882), pp. 1–17, on p. 12 (emphasis added).

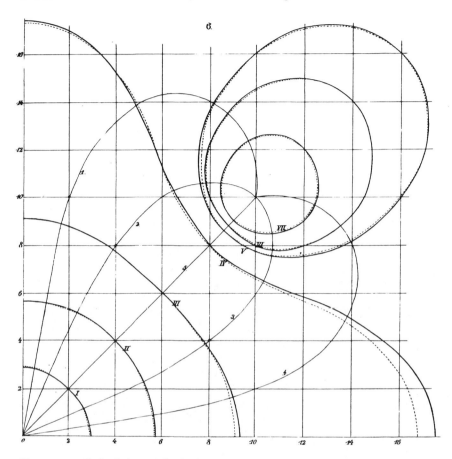

Figure 14. Quincke's graph depicting experimental (dotted) and theoretical (solid) equipotential lines of an electric current on a square lead plate. From Quincke, "Ueber die Verbreitung eines elektrischen Stromes," Table IV, Plate 6.

with roman numerals in Figure 14. Quincke represented the measured equipotential curves by dotted lines on the same graph. Perpendicular to the calculated (theoretical) equipotential lines, he drew "theoretical curves of electric current"; these are marked with arabic numerals in Figure 14. He noted that these curves allowed the values of x or y to be computed "through interpolation." Quincke repeated the same procedure for the lead and copper disk, with one difference: he tried to refine his measurements by reading the scale on his galvanometer through a telescope, thereby reducing the chances that he would introduce disturbances (e.g., the heat of his hands) into the environment of the experiment. (In Figure 15, theoretical equipotential curves for this disk appear as solid lines designated by roman numerals; measured equipotential curves, by

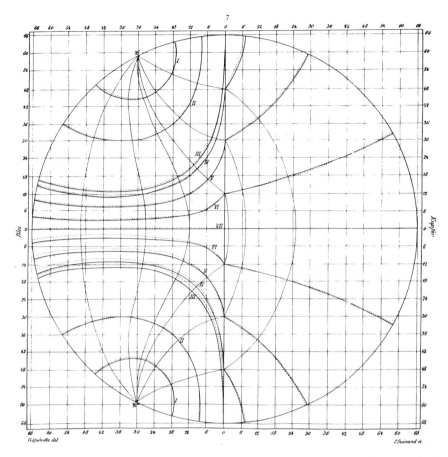

Figure 15. Quincke's graph depicting experimental (dotted) and theoretical (solid) equipotential lines of an electric current on a circular plate of lead (left side) and copper (right side). From Quincke, "Ueber die Verbreitung eines elektrischen Stromes," Table IV, Plate 7.

dotted lines; and "theoretical curves of electric current," by the lighter solid lines that intersect the theoretical and measured equipotential curves at right angles.)[17]

For both the lead plate and the lead and copper disk, Quincke's graphs displayed the differences between theory and experiment as readily as did his tables, thus improving upon Kirchhoff's graphical representation of similar data. Quincke was also clearer on how those curves were constructed. Unlike previous students, he was explicit about the formulas he

17. Quincke, "Ueber die Verbreitung eines elektrischen Stromes," Table IV, Plates 6 and 7.

used for interpolation: Kirchhoff's theoretical formulas for the potential or for the current intensity. By establishing points in this way, Quincke acknowledged that interpolation could be accomplished mathematically from theoretical considerations (such as formulas) as well as mechanically from measured values (as when measured points were joined to create the most probable smooth curve). He raised no doubts about using interpolation for either his theoretical or his experimental curves. But he also failed to comment in detail on how closely his theoretical calculations and experimental values were to one another or on how reliable his experimental values actually were and whether their error was such as to make them fall within range of the theoretical values. Undoubtedly the faith he placed in his techniques stemmed in large part from the agreement with Kirchhoff's results that he had achieved. Yet like Hagen, Quincke did not use his curves as a source for the further mathematical analysis of his data. His graphs, then, were visual representations of his data which substituted for tables, showed nothing new, and demonstrated pictorially the approximate difference between theory and experiment.

Quincke appeared to be comfortable using interpolation in this study, but he was not always so confident. In his next investigation he expressed doubts about the validity of using interpolation as a technical aid in comparing theoretical and experimental values over nonmeasured ranges in the data. In 1856 when he began determinations of the capillary constant of mercury, a topic that had captured his interest at Königsberg and that became his doctoral dissertation at Berlin in 1858, he kept Neumann informed of his progress. After spending the summer semester of 1856 in the seminar, he wrote to Neumann from Berlin on 3 October 1856. His apparatus was a special concern. He was satisfied with his microscope (it had 20x power), but his work was held up "because the various parts of the [experimental] apparatus have to be made by different people," so he was "unable to assemble everything" at once. Nevertheless, the material advantages Berlin offered for carrying out this delicate experiment convinced Quincke he should remain there. Magnus, he told Neumann, was "friendly with his equipment" and had even "recommended" that Quincke "examine the shadow of the [mercury] drop . . . and take measurements of this shadow" so that he would have a better idea of the drop's exact boundary. Yet equipment and suggestions were not all Quincke needed to continue; for his purpose in writing to Neumann was to find out the best way to measure the drop's configuration. Moreover, the theoretical side of the problem was proving troublesome: "I am still much behind in the integration of the differential equations of the mercury drop," he confessed. He had sought out the help of Neumann's recent doctoral student, Alfred Clebsch, who was then working at a *Realschule* in Berlin, and was grateful for Clebsch's assistance; but recently, he explained, "Clebsch has been sick, and therefore my study of elliptic functions has been neglected."[18]

18. Quincke to Neumann, 3 October 1856, FNN 53.IIA: Briefe von Schülern.

Quincke, who did not find it easy to obtain the kind of precise numerical data Neumann looked for, did not contact Neumann again for almost five months. He wrote on 28 February 1857 only to assure Neumann that he was "still occupied with the matter." Once again he related his problems and the shortcomings of his investigation. Now he was concerned about Johann Poggendorff's suggestion that mercury oxidizes in air. If such were true, he explained to Neumann, "then it is very natural that the observations do not agree with theory, and especially that the [capillary] angle that is made with the surface is variable. . . . I must therefore first investigate, if the mercury is really oxidized, and [I] want to make the following measurements." Unfortunately for Quincke, those measurements "left much to be desired" in accuracy; he admitted that an accuracy of "more than 0.001 mm can hardly be guaranteed."[19]

These concerns were merely a preamble to Quincke's frustration in not being able to live up to Neumann's standards in this investigation. Aware that Neumann might criticize his experimental procedure, Quincke challenged him to do better: "If you had made the measurements yourself, then you would have been convinced of how many small things stand in the way, things that do not belong at all to the investigation. Such small matters take up a lot of time in the experiment." Quincke was particularly troubled that earlier measurements completed at Königsberg did not yield consistent results.[20] He did not represent himself to Neumann as a failure but as a student trying to come to terms with the standards of the master. So he asked Neumann to sympathize with his difficulties and to bear with him, emphasizing that he did not want to appear ungrateful for all the interest and love Neumann had shown. Quincke's problem was that he could not fully satisfy the standards of precision he had been taught.

Especially irksome to Quincke was the lack of agreement among the many theories of capillary phenomena as well as between the theoretical and experimental values of capillary constants. For him, a satisfying resolution to the problem of which capillary theory was correct could result from a better agreement between theory and experiment than had been achieved to date. So he mentioned to Neumann the limits within which his own measurements were accurate. He tried to understand variations among his values and to achieve greater agreement and consistency but was stymied. He reported to Neumann that the Berlin chemist Eilhard Mitscherlich did not consider the deviations in Quincke's measurements of the drop's surface to be due to Quincke's *method* of measurement. Instead, it was Mitscherlich's opinion, Quincke reported, that as soon as "I applied formulas other than pure interpolation formulas, I no longer would have had pure observations, but the mathematics could have and would have contributed something false." By "formulas other than pure interpolation formulas" Mitscherlich (and Quincke) must have

19. Quincke to Neumann, 28 February 1857, ibid.
20. Ibid.

meant interpolation formulas derived from a theory of capillarity, because Quincke also told Neumann that Mitscherlich did not believe in
Laplace's capillary theory and that he himself "had already found much to
doubt strongly in current theories of capillary phenomena."[21] Earlier,
when Quincke had had the benefit of Kirchhoff's already confirmed
theoretical investigation of "curves of equal tension," he had viewed
interpolation based on theoretical formulas as an acceptable mathematical technique for analyzing data. Now, confronted with data that appeared inconsistent and convinced that existing capillary theories had
shortcomings, Quincke no longer considered it appropriate to use a
theoretical formula to interpolate values.

By the time Quincke finally published his results in 1858, he had
systematically worked through several different analytic methods of
interpolation—the "pure interpolation formulas" he had mentioned to
Neumann—as crucial steps in interpreting his data. In his investigation
he tried various ways of calculating and measuring the capillary constant
of mercury from the configuration of a mercury drop on a plane surface.
Assuming the mercury drop to be a surface of rotation (Figure 16), he
first tried to determine the angle Θ made between the tangent at a point
on the surface of the drop and the x axis from "a simple method of
interpolation," a quadratic equation, $y = Ax + Bx^2$, whose constants could
be determined from observation. Accurate measurements were paramount here, so he equipped his apparatus with microscopes, one of which
had an ocular micrometer (Figure 17). To assure that the plate upon
which the mercury drop sat was perfectly horizontal, he adapted the
method of coincidence observations Neumann had detailed in his lectures on theoretical physics. After "countless measurements," Quincke
computed the constants A and B by the method of least squares.

But Quincke did not get consistent results because, as he thought, of
"the inaccuracy of the method applied." So he tried something else. Aware
that capillary phenomena were "easily affected by unnoticeable causes,"
he applied another method, suggested by Neumann, which also treated
the drop as a surface of rotation, but as a surface that slowly changed its
configuration over time. Unfortunately, Neumann's method, although
more refined and sophisticated mathematically, was not, Quincke found,
"a more accurate interpolation method." It was only after working
through these interpolation methods and achieving no satisfying results
that he finally turned to the construction of the experiment itself, devising
a method for determining Θ directly from observation, with measurements depending on the coincidence of two reflected images of the drop.
But he still did not achieve a consistent value; his measurements showed
"irregular" changes, which he attributed to the "accidental shaking of the
building or travel on the street." Only an observatory, he concluded, would

21. Ibid.

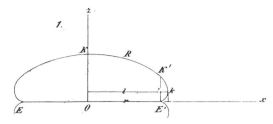

Figure 16. Quincke's schematic drawing of a mercury drop, used to calculate its capillary constant. From Quincke, "Ueber die Capillaritätsconstanten des Quecksilbers," Table I, Figure 1.

provide the conditions necessary for maintaining an equilibrium surface. In the end, he concluded that Θ changed so quickly in the course of any one experimental trial that its accurate determination was not possible. The changes in the drop, he believed, stemmed not from either the oxidation of the mercury or changes in temperature but rather from the simple fact that an equilibrium surface could not be maintained long enough to take measurements. "If it is not a pleasant, satisfying conclusion

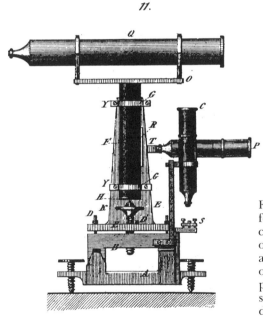

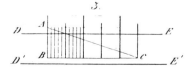

Figure 17. Quincke's apparatus for determining capillary constants, where Q is an observation microscope and C is a microscope equipped with an ocular micrometer. The principle of the latter is sketched below; 0.01 mm divisions are marked on the scale, but one can measure 0.0001 mm. Like Wild's polarimeter, this is a precision instrument. From Quincke, "Ueber die Capillaritätsconstanten des Quecksilbers," Table I, Figures 11 and 3.

to an investigation that one must say [that] the agreement of theory and observation is not to be demonstrated," he reported, "then I hope through the publication of these trials to have saved others time and effort which otherwise would have been wasted." That he did. Like the first student paper to have issued from the seminar, Brix's, Quincke's revealed nothing new but detailed the limits to what was known. The first person Quincke thanked at the end of his article was Neumann, who had suggested key mathematical steps in the investigation.[22]

Hagen's and Quincke's investigations exemplify how the techniques cultivated in seminar exercises limited the range of questions that could be asked—and answered—when those techniques dominated experimental protocol. Students sometimes found themselves in a conundrum when they simultaneously pursued precision in experimental results and certainty in mathematical description. Interpolation—a bridge between experimental data and its mathematical expression—was especially problematic. Neumann cautioned students about the tentative nature of interpolated formulas, and so they viewed the method skeptically. Hagen never mentioned interpolation in his investigation, but he must have used it to construct his curves. That he could not go beyond his graphs to express mathematically the relations depicted in them is puzzling but not inexplicable or unexpected. The "conclusions" he drew from his graphs suggest that he used them only to express qualitative relations visually. To have taken more seriously the ability of graphs to assist in quantitative data analysis would have meant overlooking the fact that the more numerous and unobserved interpolated values had not been ascertained directly and hence were for all practical purposes unknown "invisibles." When Hagen said he wanted sharper and more accurate measurements, he was partly acknowledging the tentative nature of interpolated values to which no accuracy could as yet be ascribed.

Equally revealing of the perceived lack of certainty when using interpolation and graphical analysis are Quincke's electrical and capillary investigations. He could represent his data pictorially in graphs without complication when he had Kirchhoff's already proven theoretical formulas for equipotential lines to guide him. The dotted lines, representing his measured values, were no doubt constructed in part by referring to the theoretical curves. But no similar clear-cut theoretical guideline was available for his capillary investigation. Doubting Laplace's theory as well as other theories and put on guard by Mitscherlich on the issue of interpolating on the basis of theoretical formulas, Quincke went through several analytic or "pure" interpolation formulas before giving up his attempt to bring theory and experiment into agreement.

Hagen's and Quincke's investigations suggest that students could reach

22. Georg Quincke, "Ueber die Capillaritätsconstanten des Quecksilbers," *AP* 105 (1858): 1–48, esp. pp. 10, 26–33, 47–48.

an impasse when using interpolation and the graphical representation of data. Interpolated values and measured values were not easily distinguishable in graphs, and hence graphs made it appear as if all values possessed the same degree of precision, even though interpolated values were not derived from experiment but from some analytic formula. Without knowing the degree of precision that could be assigned to interpolated values, one could not legitimately perform the one function that error analysis facilitated: combining observations. The drive for certainty thus inhibited the use of interpolation and graphs, impeded the formulation of mathematical laws, and made the confirmation of theoretical expressions more difficult.

The research values cultivated in seminar exercises, especially the certainty that was associated with error analysis, were not the only causes of doubt concerning interpolated values. Another was a lack of understanding about what, exactly, graphs represented. The graphical representation of empirical data had only recently been introduced into scientific investigations and then merely as pictorial representations of numerical tables, which had been the principal means of presenting data in the eighteenth century. In the eighteenth century, according to Laura Tilling, graphs were rare, were generally not viewed as a step in establishing functional relationships, and "usually arose unintentionally out of the method of measurement used." Conclusions (one could hardly call them laws) drawn from graphs were considered approximations, and there was in general no consideration of experimental errors on graphs. The striking example of graphical analysis that Tilling found in the work of J. H. Lambert during the 1760s and 1770s demonstrates one way in which graphs were first viewed without considering error analysis. For Lambert, graphs gave "facts as cannot be observed but which nevertheless are needed." In contrast to Mitscherlich, Lambert considered it advantageous to have a theoretical law in hand before constructing a graph from data because "if the rule relating the ordinates and abscissae are known, then we may proceed [in constructing the graph] far more methodically."[23]

Lambert's case, however, was unusual. Graphical analysis was not revived until the first decades of the century in Great Britain, and at a much later time (we still do not know precisely when) on the Continent. Even when graphs did appear in Continental literature, they were accompanied by a tabular representation and generally were not used for the analysis of data because investigators were more comfortable working from numerical tables. At Königsberg, graphs were used in one of two ways, neither of them intended for processing experimental results. One of Neumann's early achievements was a method for illustrating crystal structure by using projective geometry.[24] His "graphical method," how-

---

23. Laura Tilling, "Early Experimental Graphs," *BJHS* 8 (1975): 193–213, on 194, 205.
24. Neumann, "Beiträge zur Krystallonomie" [1823], in *GW* 3:175–322.

ever, depicted qualitative features of a crystal and was not intended for analyzing empirical data, even though it incorporated the generalized results of empirical investigations. As student notebooks from both his lecture courses and seminar amply confirm, Neumann also used graphs to illustrate the functions representing theoretical relations. Thus to the extent that Neumann or his students did use graphs for experimental concerns, especially as an instrument of data analysis, they were deploying a relatively new technique whose usefulness was circumscribed by the value they placed not only on precision measurement but also on accuracy as determined by the closeness of theoretical calculations and experimental results.

The infrequency of more imaginative uses of graphs was also the result of the omission or relative unimportance in Prussian mathematics instruction of both descriptive geometry and the graphical analysis of functions until later in the century. The seminar at Königsberg was no exception. Richelot assigned geometry a foundational role in his curriculum, but he always thought primarily in terms of analytic geometry. To a certain extent analytic geometry what to Richelot was mechanics was to Neumann: a subject that could serve as a foundational course not because it imparted key concepts but because it conveyed exemplars of the investigative techniques needed for original research. But Richelot never thought of geometry as the only foundation of his curriculum and so did not teach it as frequently for introductory students as Neumann did mechanics. When two students, fresh out of the gymnasium, entered his seminar in 1856/57, Richelot sent one of them to a special exercise session conducted by Eduard Luther, director of the Königsberg observatory. As Sohncke had done years before, Luther offered elementary lectures and exercises—this year on analysis and integral calculus—"so that the gaps in the student's knowledge could be filled." Richelot also sent two transfer students from Berlin to these elementary sessions and later remarked that they "would not have been in the position to reach their goal, namely to understand the lectures in mathematical physics, the theory of elliptic functions, and analytical mechanics if they had not acquired at that time the necessary skills in integration." In the following year, when elementary topics had to be taught again because "most participants were not properly trained," Richelot assigned problems from the theory of determinants.[25]

Richelot did teach analytic geometry most frequently for beginning students but for special reasons. As he explained in 1856, he had

allowed by way of exception two students, who had just graduated from the Altstädtisches Gymnasium in Königsberg, to take part in the seminar at their

25. "Bericht über die mathematische Abtheilung des mathematisch-physikalischen Seminars in Königsberg für die Zeit von Ostern 1856 bis Ostern 1857," 1 July 1857, ZStA-M, Math.-phys. Seminar Königsberg, fols. 228–31, on fol. 229; "Jahresbericht über die mathematische Abtheilung . . . , 1857/58," ibid., fols. 241–42.

request. In this school, which has for a long time noticeably distinguished itself among all gymnasiums through an especially successful mathematical curriculum, both students acquired a good and certain knowledge [of mathematics]. One student, H. Kaul, whose exceptional talent was made known to me many years ago and who since then has been supervised by me, acquired unusual dexterity in the solution of geometric problems. I cannot pass up this opportunity without making remarks on the importance of this means of instruction [i.e., solving geometric problems] for the school. It is chiefly suitable not only for sharpening judgment but also for acquiring in particular a rather important independent mental skill for every future profession. If, therefore, at gymnasiums in other provinces, less consideration should be given to this part of mathematical instruction, if especially in the *Abitur* the knowledge of geometric systems should be valued more than mental dexterity, then in my opinion the true nature of the discipline [of mathematics] will remain misunderstood.

Even in the advanced sections of the mathematical seminar, Richelot sometimes found it necessary to teach analytic geometry, as in the summer semester of 1856, when, even though Quincke had returned to the seminar to learn advanced mathematical techniques, Richelot taught analytic geometry because of the inadequate preparation of the remainder of his students. (Not until the winter semester of 1856/57, after Quincke had departed, did Richelot cover the more complex topics that Quincke had returned to Königsberg to learn: series and their convergence; definite integrals; quadratic, algebraic, and transcendental equations; and topics from the works of Lagrange and Jacobi.) Again in 1858/59, Richelot assigned problems from analytic geometry to his introductory students.[26]

Richelot was an attentive teacher, perhaps more so than Neumann. In his introductory sections Richelot probed students to discover "those points in [his] lectures that they did not completely understand." By 1857/58 he was enhancing the dialogues he had made a part of his classes by holding a "discussion section at definite hours" on days between scheduled seminar meetings. In this way, "I was able to achieve," he boasted, "a very enthusiastic and successful participation [of the students]."[27] Richelot held steadfast to the conviction that students learn mathematics and its techniques systematically. When the future mathematician Paul Gordan, for instance, registered for his seminar in 1858/59, Richelot judged his preparation poor, even though Gordan had already studied mathematics for one and a half years at Berlin and Breslau. Richelot implied that Gordan would have to start his study of mathematics all over again, this time systematically. Another student, who had studied mathe-

---

26. "Bericht über die mathematische Abtheilung . . . , 1856/57," fols. 228–31, on fol. 228; "Jahresbericht über die mathematische Abtheilung des mathematisch-physikalischen Seminars der Universität Königsberg während der Zeit von Ostern 1858 bis Ostern 1859," 22 September 1859, ibid., fols. 252–54.

27. "Jahresbericht über die mathematische Abtheilung . . . , 1857/58," fols. 241–42.

matics first at Königsberg and then at Berlin, returned to Königsberg "in order to engage in a systematic study of the material in the seminar," but he was soon asked to leave when Richelot found that his preparation, interest, and diligence were too far below that of other students.[28] Yet although Richelot was more sympathetic to beginning students and more agreeable to teaching geometry than Jacobi had been (although Jacobi had taught descriptive geometry at least once in the seminar[29]), Richelot's teaching did not help Neumann's students in a key area where they needed assistance: graphical analysis.

Although less than half Neumann's seminar students took Richelot's seminar, most took one or more of his lecture courses. Generally Richelot's instruction was immensely helpful to students learning quantitative techniques in physics. Richelot even presented in class his own thoughts on the foundation of the method of least squares. But the prominence of analytic geometry in Richelot's seminar, in particular its pivotal position as an introductory course, failed to contribute to the higher levels of quantification involved in Neumann's seminar exercises, especially for the interpretation of experimental data through the graphical comparison of theoretical and experimental results. Combining graphical representation and data analysis necessitated assigning visual ways of thinking—as used in diagrams, graphs, and drawings—parity with the more abstract and analytic ways of thinking required to analyze data in tabular form. Although some seminar students recognized on their own how important visual thinking could be in a physical investigation (as had Quincke, who did not understand some previous works on capillarity just because they lacked diagrams), their attempts to think visually about mathematical relations among data were rarely sustained long enough to create quantitative laws of physics from graphs.[30]

The problems Neumann's students had with interpolation and the graphical analysis of data were not entirely Richelot's fault. Descriptive geometry and the graphical analysis of functions were not taught in Prussian secondary schools until much later in the century—in many places, not until the 1880s.[31] Early in his career the mathematician Felix Klein observed that "mathematical physicists continually ignore the ad-

28. "Jahresbericht über die mathematische Abtheilung . . . , 1858/59," fols. 252–54, on fol. 253.

29. "Bericht über das Königsberger mathematisch-physikalische Seminar im Jahre 1840: Mathematische Abtheilung," 8 January 1842, ibid., fol. 89.

30. Richard Henke reported Richelot's interest in least squares in *Ueber die Methode der kleinsten Quadrate*, Inaugural-Dissertation (Dresden: B. G. Teubner, 1868), p. 37; Quincke, "Ueber die Capillaritätsconstanten des Quecksilbers," p. 7.

31. Rudolf Schimmack, *Die Entwicklung der mathematischen Unterrichtsreform in Deutschland*, Abhandlungen über den mathematischen Unterricht in Deutschland, ed. Felix Klein, vol. III.1 (Leipzig: B. G. Teubner, 1911), pp. 3–13; Heide Inhetveen, *Die Reform des gymnasialen Mathematikunterrichts zwischen 1890 und 1914* (Bad Heilbrunn: Klinkhardt, 1976), p. 192. On the difficulties involved in integrating applied mathematics, including descriptive geometry, into German mathematical curricula see Paul Stächel, "Angewandte

vantages that a merely modest development of projective thinking could provide in many cases." According to David E. Rowe, the most likely place to find courses in descriptive geometry was in Germany's *Technische Hochschulen* before Klein began teaching it at Leipzig University in the 1880s.[32] Even in the 1890s, physicists were known to complain that graphical methods were underappreciated and inadequately utilized in physics teaching and research.[33] Thus curricular preferences in mathematics teaching helps explain why graphical analysis and the interpolation it entailed were not only problematic but also misunderstood and mistrusted by several of Neumann's seminar students.

Whereas graphs create a data continuum over which one can make a continuous comparison between theory and experiment, the techniques of data analysis Neumann's students customarily employed, such as the method of least squares, dealt with discrete data points. Comparing points over a continuum was possible only if one accepted that interpolated values had the same certainty as measured points. But Neumann was skeptical of interpolation and, although he used it, did not especially advocate it. Yet it was interpolation and graphical analysis that brought together experimental data and theoretical calculations, that united measurement with other forms of quantification. Closing the gap between members of these pairs was not easy for seminar students who lacked not only training in the functional analysis of data but also suitable analogous cases upon which to model their attempts to create quantitative laws from graphs.

In retrospect it seems obvious that graphical representation and data analysis must be used together to produce exemplars of a mathematical physics. But for Neumann's students, that combination was troublesome and difficult. The problems they encountered were ironically the result of the cultivation of an awareness that both certainty and precision were always circumscribed, and defined in terms of, the analysis of errors. It seems not surprising that in the few published investigations that issued from the seminar through the 1850s, there was a marked preference either for error analysis or for mathematical expression, but not for both. Kirchhoff used sophisticated mathematical techniques such as elliptic transcendentals but represented his data in tabular form. His graphs were in part guesses, idealized representations of what he thought equipotential lines should look like. Brix concentrated his efforts elsewhere. In an effort to understand the relationship between error and measurement, he tried to eliminate error altogether in order to find "truth." Du Bois-

Mathematik und Physik an den deutschen Universitäten," *JDMV* 13 (1904): 313–41, esp. 324–25.

32. Quoted in David E. Rowe, "Klein, Hilbert, and the Göttingen Mathematical Tradition," *Osiris* 5 (1989): 186–213, on 190.

33. E. Wiedemann, "Die Wechselbeziehungen zwischen dem physikalischen Hochschulunterricht und dem physikalischen Unterricht an höheren Lehranstalten," *ZMNU* 26 (1895): 127–40, on 131.

Reymond's paper, while not as closely related to the activity of the seminar, deserves mention; for in his case, linear interpolation was problematic because it would have masked precisely the curvature he wished to
find. While it was therefore natural for him to eschew it, it is telling that he
made no effort to use other more sensitive methods of interpolation. Wild
also found interpolation problematic; for he represented long stretches
of interpolated and extrapolated points differently from those he had
measured or about which he was more confident. Although influenced by
Mitscherlich's criticism that interpolation formulas introduced something
"false" into consideration, Quincke in the end included interpolation as
one possible route toward a mathematical law (which he did not succeed
in constructing). Hagen freely interpolated points but did not formulate a
mathematical expression for want of more accurate measurements.
Hence, those students who worked from data augmented by interpolation stopped short of a mathematical law, even though they may have
been only steps away from one. In contrast to the British scientist William
Whewell, who in 1840 argued that through graphical analysis "are obtained data which are *more true than* the individual *facts themselves*,"[34] for
Neumann's students, the points comprising graphs lacked certainty.

    In later seminar investigations, beginning in the 1860s, a few students
did use interpolation as a mathematical tool, but primarily in areas such as
atmospheric or earth temperatures, where interpolation and even graphical analysis were considered reliable for constructing empirical formulas,
in part because of the deeper understanding of the operation of thermometric instruments as the century wore on. In his determination of the
thermal absorption coefficient of the atmosphere, Oskar Frölich interpolated to calculate temperatures for locations where he did not have actual
measurements. He also considered interpolation suitable for creating a
smooth curve out of a jagged, rough one constructed from straight lines
connecting data points. Emil Meyer interpolated to obtain intermediate
values and to reduce all his observations to the same temperature. But he
refused to calculate interpolation formulas from some questionable data,
believing that the resulting "numerical value would turn out to be too
uncertain." Saalschütz, too, expanded his geothermal measurements
through interpolation. As was common practice in meteorology, he interpolated the temperature of Königsberg from those of Danzig and Tilsit
for periods in which he had no temperatures from Königsberg.[35]

    In other areas of physics, however, students continued to doubt the

---

34. Quoted in Tilling, "Early Experimental Graphs," p. 209.

35. Oskar Frölich, *Ueber den Einfluss der Absorption der Sonnenwärme in der Atmosphäre auf
die Temperatur der Erde*, Inaugural-Dissertation (Königsberg: Dalkowski, 1868), pp. 7, 21–22;
O. E. Meyer, "Ueber die Reibung der Flüssigkeiten," *AP* 113 (1861): 55–86, 193–238, 383–
425, on 408; Louis Saalschütz, "Ueber die Wärmeveränderungen in den höheren Erdschichten unter dem Einflusse des nicht-periodischen Temperaturwechsels an der Oberfläche," *AN* 56 (1862): 1–44, 161–206, 273–98.

reliability of interpolated points. Their doubts were in part a reflection of Neumann's own: Albert Wangerin remembered that when Neumann introduced the theory of circular and elliptical polarization in 1864/65, he warned that its meaning was unclear, so "the investigation had the character of an interpolation"; in other words, of being nothing more than an empirical formula.[36] When former seminar students did represent measured and interpolated points in the same way in graphs, their graphs continued to serve as little more than visual representations of data. When Emil Meyer sketched the viscosity constant of salt solutions as a function of the concentration of each solution (Figure 18), using dotted and dashed lines to signify measurements taken for different substances. "All such obtained curves," he noted, "have the appearance of an algebraic curve of the second degree." He did not use the curves as a basis for constructing an empirical equation that he could compare with a theoretical one but instead compared theory and experiment rigorously only for points he had actually measured. Although he constructed a theoretical equation, he thought it only "very probable" and claimed that the patterns in his graphs were "on theoretical grounds also very probable."[37]

Anton Müttrich treated the graphical representation of data similarly. Suspecting that the optical constants of crystals changed with temperature just as their optical axes did, he wanted not only to determine the value of optical constants at certain temperatures "but also to ascertain *according to which law* the constants changed with varying temperature." Using the precise techniques he learned in seminar exercises, he obtained measurements at various temperatures of the optical axis and the coefficient of refraction in crystals, as well as the propagation velocity of light in the crystal. To achieve an "overview" of his results, he "drew the results *as curves*" in the final section of his paper. In separate graphs he plotted optical constants as a function of temperature (Figure 19). That several of his "curves" were not curves at all, but broken lines drawn between his corrected measurements, seemed not to concern him. So whereas he was able to incorporate the results of the calculation of constant errors into his graphs, he did not use the accidental errors of his data to assist in the curve-fitting process. Data analysis could thus make it difficult to deal with interpolation in more imaginative, and perhaps more successful, ways. When Müttrich derived empirical formulas, he did so from his measured points, not from his graphs.[38]

36. Albert Wangerin, *Franz Neumann und sein Wirken als Forscher und Lehrer* (Braunschweig: F. Vieweg & Sohn, 1907), pp. 106–7.
37. Meyer, "Ueber die Reibung der Flüssigkeiten," p. 405.
38. A. G. Müttrich, "Bestimmung des Krystallsystems und der optischen Constanten des weinsteinsauren Kali-Natrons; Einfluss der Temperatur auf die optischen Constanten desselben und Bestimmung der Brechungsquotienten des Rüböls und destillirten Wassers bei verschiedenen Temperaturen," *AP* 121 (1864): 193–238, 398–430, on 193–94, 430 (emphasis added).

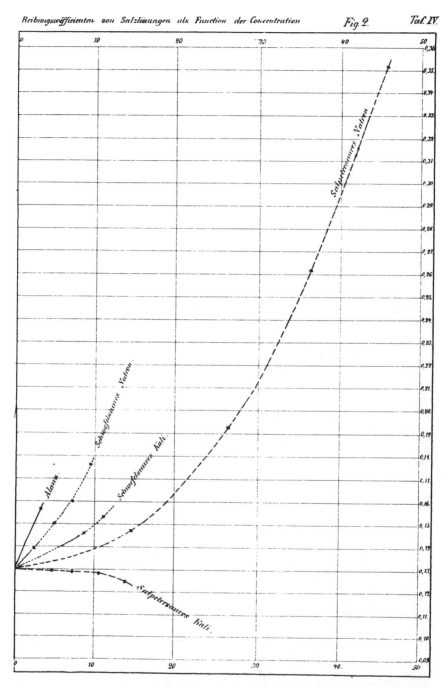

Figure 18. Meyer's graph depicting the viscosity constant of salt solutions as a function of concentration. From Meyer, "Ueber die Reibung der Flüssigkeiten."

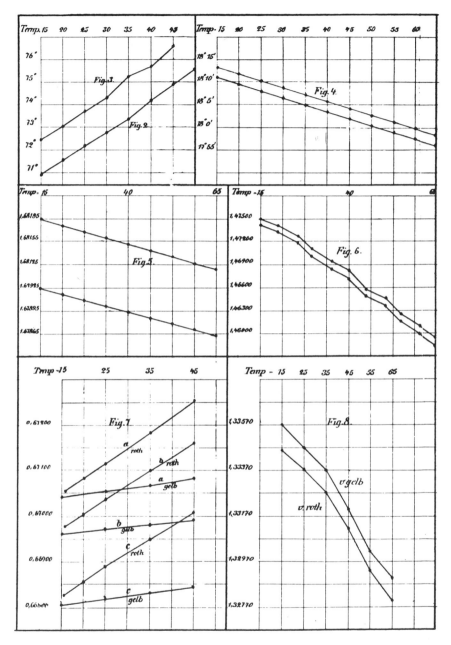

Figure 19. Müttrich's graphs of optical constants as a function of temperature. Figures 2–4 depict optical axes; 5, 6, and 8, the coefficient of refraction; and 7, propagation velocity. From Müttrich, "Bestimmung des Krystallsystems und der optischen Constanten des weinsteinsauren Kali-Natrons," Table IV.

The history of science has repeatedly shown that scientists do not always use to their fullest potential the analytic and mathematical tools available to them. Scientific tools, like scientific concepts, are often found in sets internally united by the shared values and objectives of their elements. The factors limiting the range of investigative techniques and the domain in which they are deployed are often embedded in professional and social contexts. That Neumann's seminar students did not fully exploit interpolation and the graphical analysis of data is understandable only partly in terms of their mathematical education and the value seminar exercises placed on error analysis. It is also a consequence of how he presumably taught the method of least squares. Even though Neumann knew of several of Bessel's discussions on the method of least squares, including Bessel's 1838 essay on the distribution of observational errors, it is unlikely that he developed the probabilistic basis of the method to any great extent when teaching it in the seminar. In the classroom it would have been easier, more convenient, and of greater utility for him to teach the method merely as an analytic tool for assessing the error that remained after the investigator believed all constant errors had been taken care of.

If Neumann did in fact ignore the probabilistic considerations inherent in the method of least squares, some of the advantages of least squares, particularly for the construction of graphs, would have gone unnoticed. Although both he and his students used the method to determine the constants of empirical equations and to compute the range of error in data, Neumann seems not to have distinguished between various kinds of deviations. Nor did he or his students always express accuracy in the same way. Although his students computed accidental errors, they did not consistently express their results in a form (e.g., ± 0.001 or 0.05%) that could have aided graphical representation: into error bars that marked the range of tolerance within which a smooth curve could be drawn. Even when a range of accuracy was specified, it was not translated into visual terms on graphs. The nature of his students' graphs also indicates that they were not taught that smooth curves could be constructed by making sure that the sum of the squares of the deviations above the line equaled the sum of the squares of the deviations below the line.

To point out these limitations—to identify what Neumann did not teach or did not realize could be done with the method of least squares—is a sobering reminder that possession of a technique does not necessarily create the conditions for its deployment in the full range of circumstances with which it is later identified. The particular way in which Neumann taught the method of least squares; the state of mathematical education, especially in geometry, graphical analysis, and probability; the purpose of the pictorial presentation of data in physics; the professional cultivation of values for determining the certainty and reliability of data—all these factors and more shaped the context within which Neumann's seminar students practiced and viewed interpolation and the graphical analysis of data. For the students, these methods were still problematic avenues to the formulation of mathematical laws in physics because they were ridden with uncertainty. Familiar with similar situations in astronomy and well acquainted with mathematical techniques, they had the training but not always a reason to push certain analytic tools to their limits. What was ironic about their use of error analysis was that it was a form of mathematization or quantification that did not always enlighten; for it sometimes focused attention on the determination of the unknown to the exclusion of the mathematical expression of the known. Equating certainty with the diminution of error made these students very cautious about the mathematical construction of physical laws, especially from graphs.

Jerome Ravetz has claimed that teachers exercise "judgments of value" in curriculum construction.[39] Judgments of value in turn shape the intellectual outlook and the investigative practices of students, particularly in their fledgling researches. The types of questions his students raised, in particular about the graphical representation of data and especially about interpolation, indicate that it was still a *theorist's* perspective that was cultivated in the seminar, one that had emerged from Neumann's adaptation of the Besselian experiment. Wouldn't more data increase precision and so contribute to a more convincing conclusion? Can't some of the disturbances still be removed from the data? How can interpolated and extrapolated intervals on graphs be endowed with greater certainty? These were questions not of the mathematician or experimentalist but of the still-inquisitive theorist. That Neumann's students did not deduce mathematical equations from their graphs (even equations based on tables they considered little more than empirical formulas) is less significant than their recognition that, to attain reliable formulas, they had first to take other theoretical steps suggested by the analysis of errors if they were to achieve certainty.

It was in fact their views on the meaning of data that most strongly distinguished Neumann's students from those trained under two other leading physics teachers, Gustav Magnus at Berlin and, at a later time, Friedrich Kohlrausch at Göttingen. Magnus's alleged unsympathetic view toward "mathematical theories and elaborate calculations"[40] is not, on the surface, evident in the first three physical investigations (published between 1845 and 1847) undertaken in his Berlin laboratory. In several respects these investigations—on the density of ice at various temperatures, the determination of the temperature and conductivity of solid media, and the cohesion of liquids—were strikingly similar to those un-

39. Jerome Ravetz, *Scientific Knowledge and Its Social Problems* (New York: Oxford University Press, 1971), p. 225.

40. J. T. Merz, *A History of European Scientific Thought in the Nineteenth Century*, 2 vols. (rpt. New York: Dover, 1965), 1:205n.

dertaken at Königsberg.[41] Concerned largely with theory confirmation, they demonstrated that earlier investigations, mostly performed by French scientists, did not properly account for experimental conditions. Although "significant progress" had been made in "the mathematical theory of thermal phenomena through the analytic investigations of Fourier, Poisson, etc.," Christian Langberg thought that these theories had only "restricted influence" upon knowledge of thermal phenomena because they had been "only little" confirmed and proved through trials, there being a lack "of accurate methods to determine the temperature changes in solid media." Regarding the pattern of temperature readings produced in a solid bar with a thermal source at one end, which C. M. Despretz claimed to have shown was in agreement with theory, Langberg wrote, "It still remains undetermined if the observational deviation is to be ascribed to an erroneous method of observation or a shortcoming of the theory."[42] Laplace, Gauss, and Poisson had performed experiments to confirm their capillary theories, but C. Brunner similarly doubted if their experiments had achieved the accuracy required for a rigorous confirmation.[43] So to ground these theories, Langberg and Brunner deployed several techniques of an exact experimental physics, including compensating for constant errors and applying the method of least squares to some of their data.

A closer look at their investigations reveals, however, that although they deployed techniques similar to those used in Neumann's seminar, they used them less rigorously and for different objectives. Even though they had numerical measures of the constant errors of their experiments, they estimated these errors from the known material constraints of their instruments rather than from analytic formulas rigorously derived. They compensated for constant errors as the French had done, by varying the conditions of the experiment, and even accepted the method of repetition as a way to acquire data if the results fell within the observational error of the experiment. Their investigations were instrument oriented in that they sought greater precision primarily through improved instruments and measuring methods and even expressed accuracy in terms of the quality of their instruments. They explicitly rejected going back to theory after the experiment was performed; Brunner wrote, "It is not my purpose to derive the formulas for capillary phenomena" because this was something that had to be undertaken by "great mathematicians." Most important, they had an experimentalist's view of their data; for if the actual measured data became "too copious" to deal with in a rigorous quantitative fashion (e.g., through the method of least squares), they used

41. C. Brunner, "Versuche über die Dichtigkeit des Eises bei verschiedenen Temperaturen," *AP* 64 (1845): 113–24; Christian Langberg, "Ueber die Bestimmung der Temperatur und Wärmeleitung fester Körper," *AP* 66 (1845): 1–30; C. Brunner, "Untersuchung über die Cohäsion der Flüssigkeiten," *AP* 70 (1847): 481–529.

42. Langberg, "Die Bestimmung der Temperatur," pp. 1, 3.

43. Brunner, "Die Cohäsion der Flüssigkeiten," pp. 482, 487.

graphical interpolation to produce more tractable data. Probably because they accepted data Neumann's students would have considered less certain, Magnus's students seemed more amenable to considering hypothetical constructions in physical investigations. For example, unlike Neumann's students, they sometimes delved below the level of the phenomena and more easily considered events at the molecular level.[44]

Even sharper differences in the handling of data, especially in the use of interpolation, are evident in the investigations by Kohlrausch's students. Of the seven Kohlrausch directed in the 1870s, the last decade in which Neumann directed his seminar, only two dealt in sustained ways with theory; most concerned the quantitative determination of constants, the refinement of experimental procedures, and the derivation of empirical relations.[45] As for Magnus's students, for Kohlrausch's the preferred way to compensate for uncertainties in data was through technical means, including the variation of experimental procedure. Yet even though the determination of constant errors was an essential step in such investigations, it was not necessarily one that had to be done quantitatively, much less analytically. So, for example, when the values for electrical conductivity obtained using a constant current differed from values based on alternating currents, Johann Tollinger admitted that there was "no reason" to seek the cause of the difference in the possible "polarization of the alternating current or . . . in an incomplete validity of Ohm's law" because other factors in the environment of the experiment had not been excluded. Tollinger, however, considered it adequate to give only estimates of the quantitative values of these errors, which he judged to be only 0.1 percent, making the values for electrical conductivity "not essentially different" from one another.[46] Sometimes Kohlrausch's students also used interpolation to obtain estimates of constant errors, but for a "quantitative determination" of error that had "certainty," interpolation was insufficient.[47]

In general Kohlrausch's students expressed less concern than Neu-

44. Brunner, "Die Dichtigkeit des Eises," pp. 116, 118; Langberg, "Die Bestimmung der Temperatur," pp. 3, 4, 6, 10, 29–30; Brunner, "Die Cohäsion der Flüssigkeiten," pp. 482–83, 489, 507, 529, on pp. 483, 507.

45. In chronological order they are Heinrich Schneebeli, "Bestimmung der horizontalen Componente des Erdmagnetismus auf chemischem Wege," *AP* 144 (1871): 640–43; Eduard Riecke, "Beiträge zur Kenntniss der Magnetisirung des weichen Eisens," *AP* 149 (1873): 433–74; Johann Tollinger, "Bestimmung der elektrischen Leitungsfähigkeit von Flüssigkeiten mit constantem Strome," *AP* 1 (1877): 510–16; V. Strouhal, "Ueber eine besondere Art der Tonerregung," *AP* 5 (1878): 216–51; W. Kohlrausch, "Ueber die experimentelle Bestimmung von Lichtgeschwindigkeiten in Krystallen," *AP* 6 (1879): 86–115; idem, "Zweite Mittheilung: Schiefe Schnitte in zweiaxigen Krystallen," *AP* 7 (1879): 427–35; Carl Barus, "Die thermoelectrische Stellung und das electrische Leitungsvermögen des Stahls in ihrer Abhängigkeit von der Härtung," *AP* 7 (1879): 383–414; August Schleiermacher, "Ueber die auf einem benetzten Körper verdichtete Flüssigkeitsmenge," *AP* 8 (1879): 52–83.

46. Tollinger, "Bestimmung der elektrischen Leitungsfähigkeit," p. 515.

47. W. Kohlrausch, "Zweite Mittheilung," p. 435; Strouhal, "Ueber eine besondere Art der Tonerregung," pp. 244, 245.

mann's over the epistemological status of their data. They achieved "entirely satisfying results" that they believed were "precise" even though they employed the one technique Neumann's students often found either wanting or entirely unacceptable: interpolation. To compare sets of data taken under different experimental trials, Kohlrausch's students often interpolated from a graphical representation of their measured points, all the while acknowledging that graphical interpolation could introduce additional errors.[48] They sometimes used graphical analysis so that the "various relations, as they are given in calculation, can be judged more easily and in clearer ways" than possible through a tabular listing of data, and they believed the empirical relations or laws thus produced "with great probability" to be valid.[49] Although, like Magnus's students, Kohlrausch's more readily considered hypothetical assumptions, especially concerning molecular theory, like Magnus's, too, they generally did not return to theory to offer modifications after their experimental investigation and even argued for the superiority of experiment for examining issues where theory was unsettled (e.g., when dealing with molecular forces). Both Eduard Riecke in his investigation of magnetization and Wilhelm Kohlrausch in his experimental determination of the velocity of light in crystals using Fresnel's laws could easily have modified theoretical expressions at the end of their investigations, but they did not, focusing instead on more experimental objectives.[50] Thus whereas Neumann's seminar cultivated a theorist's perspective toward data, the laboratories of Kohlrausch and Magnus promoted an experimentalist's point of view wherein the calculational advantages of interpolated data were valued more than the greater degree of epistemological certainty that might be gained by increasing the number of readings overall.

The more intense emphasis in Neumann's seminar on data actually measured accounts for the similar views of several of his former students on the illegitimacy of interpolation in mathematical physics and on the questionable certainty of interpolated values. Although their views on these and related matters were sufficiently alike and distinctive for others to consider Neumann's former students a "school," they were neither universal nor permanent features of the emerging discipline of theoretical physics. By the beginning of the twentieth century, physicists moved easily between the tabular and graphical forms of data presentation. Eventually, tables disappeared from printed scientific papers in physics, surviving only in the private laboratory notebook. Graphs became essential as aids to quantification, especially as analytic tools for distinguishing

48. E.g., as in Tollinger, "Bestimmung der elektrischen Leitungsfähigkeit," pp. 512–13.

49. Strouhal, "Ueber eine besondere Art der Tonerregung," pp. 240, 241; see also ibid., p. 228, and Barus, "Die thermoelectrische Stellung . . . des Stahls," pp. 409, 411, 413.

50. Riecke, "Beiträge zur Kenntniss der Magnetisirung des weichen Eisens," pp. 471–74; Schleiermacher, "Ueber die auf einem benetzten Körper verdichtete Flüssigkeitsmenge," p. 53; W. Kohlrausch, "Experimentelle Bestimmung."

between and deciding among competing theories. Sometimes the graphical representation of data played a central role in the resolution of conflict, as in the early stages of the debate over the laws of black body radiation.[51] The epistemological value of interpolation was viewed more favorably as graphs were used more frequently; for mathematical equations constructed from graphs were considered ideal constructions approximating real conditions rather than exhaustive descriptions of the conditions under which a phenomenon became manifest in an artificial laboratory setting. The cautious attitude of Neumann's students toward the graphical analysis of data thus formed a bridge between an era when it and other techniques of quantification were not used at all and one when theoretical physicists without hestitation deployed interpolation, graphs, and several other modes of representing and enhancing measurements and processing data without necessarily conducting additional trials.

51. E.g., as in Otto Lummer and Ernst Pringsheim, "Die Vertheilung der Energie im Spectrum des schwarzen Körpers," *Verhandlungen der Deutschen Physikalischen Gesellschaft* 1 (1899): 23–41.

CHAPTER EIGHT

Error over Truth

The numerical calculations and mathematical approximations that had engulfed Neumann's seminar students were later remembered by some of them with fondness and even excitement, as if performing them had been an initiation rite of sorts. When in 1879 Neumann sent
his recently reprinted work on spherical harmonics to the Tübingen
chemist Lothar Meyer, Meyer reveled in "old friendly memories" of his
postgraduate days in the seminar, which he had attended with his brother
Emil Meyer, now a physicist. Lothar remembered "vividly"

> the beautiful time twenty years ago when we listened to your lectures in the
> auditorium of the Albertina, which was not elegant but to which we were
> comfortably accustomed. We had to exert ourselves vigorously afterward in
> order to work through everything at home and to derive calculations only
> hinted at in class. I can still see my brother sitting next to me in our room on
> Löbenicht Street as we calculated the numerical coefficients of spherical
> harmonics and mutually confirmed our results. Although a long time has
> since gone by, [this manner of teaching] appears [now] to me to be the
> opposite of how my students want to be taught and actually must be taught.

Meyer, who was by then involved in shaping the discipline of physical
chemistry, concluded his letter by saying he was glad he had been educated differently from other chemists.[1]

Another former seminar student, Johann Pernet, recalled that although Neumann was "very mild in his judgment of the performance of
beginners," he "still demanded an energetic participation in the seminar."
When Pernet and Neumann's last doctoral student Woldemar Voigt shud-

---

1. Lothar Meyer to Neumann, 25 January 1879, rpt. in LN, p. 460.

dered one day when they realized how much work would have to be done to complete a seminar assignment, Neumann reminded them that "only if one works can one achieve something." So both Pernet and Voigt subjected themselves to the discipline of completing weekly seminar assignments, which were reviewed and evaluated by Neumann. Pernet recalled that Neumann was particularly attentive to how students had worked out their solutions, and if a more elegant or simpler solution could be found, he showed them how to do it. According to Pernet, Neumann began on vacations the measuring observations he later assigned to his students. The results he obtained, which he shared with his students, were intended less to indicate their own success than to inspire them to surpass the accuracy Neumann had achieved. Neumann asked Pernet, for instance, to check the scale on a thermometer to find out why it was not behaving properly. So involved did Pernet become in that exercise that he later made the calibration of thermometers the center of his professional research agenda.[2]

The circumstances in which Neumann pursued this way of teaching physics were quite different from the setting in which his pedagogical strategy had first taken shape. During the last decade and a half of the seminar, 1860 to 1876, the natural sciences assumed a position of prominence in German culture, especially in the educational policies of the German states. Physical laboratories began to appear,[3] and practical exercises similar to Neumann's became more common and eventually were recognized as a required part of all physics instruction.[4] The student clientele of physics courses broadened in 1871 when *Realgymnasium* graduates were granted the privilege of entering Prussian universities to study modern languages, mathematics, and the natural sciences. In 1866 the Prussian state teaching examination was revised once again, not only assigning the natural sciences a more defined role but also incorporating several aspects of the Königsberg philosophy of teacher training.

What made the 1866 Prussian state teaching examination so different was the addition of the requirement of a substantial written work—for all practical purposes equivalent to a doctoral dissertation—as a part of the examination. Not only did this requirement enhance the scholarly attributes of secondary school teachers; it also encouraged more refined disciplinary specialization than had been possible in the past. The requirements in mathematics, which incorporated several of Richelot's recommendations, stipulated that the candidate show he was a "trained mathematician" in higher geometry, higher analysis, and analytical mechanics and, at Richelot's insistence, that he could conduct his "own

2. J. Hann, "Nekrolog [Franz Neumann]," *Almanach der Kaiserlichen Akademie der Wissenschaften zu Wien* 64 (1896): 271–80, on 278, 279.

3. David Cahan, "The Institutional Revolution in German Physics, 1865–1914," *HSPS* 15 (1985): 1–65.

4. On practical exercises in various German physical seminars, see below, Chap. 10.

investigation in these areas with success." To teach physics to the upper forms, a candidate had to have "knowledge of the theories of mathematical physics and its methods, along with a precise understanding of physical instruments and exercises in their use." In addition to demonstrating that he could teach mathematics, a candidate who wanted to teach physics also had to be able to teach mathematical geography and astronomy. If a teaching candidate had attended a university seminar, he had to say so on his vitae. Largely at Richelot's insistence, Bonn seminar graduates lost the privilege of exemption from the natural sciences portion of the examination. Between 1839 and 1867 there had been only one examiner for all the natural sciences on each regional examination board, but after 1867 more specialized examiners were added for each discipline or pairs of related disciplines (e.g., chemistry and mineralogy, botany and zoology, etc.).[5]

Despite these and other improvements on the examination, however, Richelot was not completely satisfied and strongly resisted the retention of older, less scholarly parts, especially those that ran counter to enhanced disciplinary specialization. He objected to the broad scope of the general education requirements—which covered religion, philosophy, pedagogy, history, geography, and languages—as well as to the requirement that those who wanted to teach mathematics and physics also had to show a general competency in the natural sciences. His drive for even greater specialization came to a halt when the ministry instructed the state teaching commission in Königsberg to remind Richelot that evidence of a candidate's possession of "practical teaching skill" was in the state's best interests.[6]

We know less about how Neumann reacted to the 1866 examination, but he must have been pleased. The requirement of a substantial written work fulfilled his earlier hope that physics would be treated like mathematics on the examination. But of greater significance for his pedagogical program was the examination's explicit promotion of *mathematical* physics. Teaching candidates who wanted to teach physics not only had to know mathematics but had to be trained in mathematical physics. Even candidates interested primarily in teaching mathematics would be predisposed toward teaching a more mathematically oriented physics. Furthermore, the requirement that a candidate have a precise knowledge of school instruments and student exercises using them reinforced the relevance of Neumann's practical physics to teacher training, but it did not necessarily

5. "Reglement für die Prüfungen der Candidaten des höheren Schulamts, 12. Dezember 1866," *Centralblatt für die gesammte Unterrichtsverwaltung in Preußen* 8 (1867): 13–35; "Ausführung des neuen Reglements für die Prüfungen der Candidaten des höheren Schulamts," ibid., p. 209.

6. Kultusministerium to the Director of the Wissenschaftliche Prüfungskommission in Königsberg, 31 July 1869, rpt. in Gert Schubring, *Die Entstehung des Mathematiklehrerberufs im 19. Jahrhundert* (Basel: Beltz, 1983), p. 289.

reinforce his emphasis on the analysis of both accidental and constant errors.[7]

A scientific culture was more noticeable in Königsberg after 1860. Although the new university building completed in 1862 did not include the physical laboratory Neumann hoped for, several scientific institutes were renovated or recipients of budgetary increases in the 1860s.[8] Together with Moser, Neumann received additional funds for instruction in 1862. After the natural sciences seminar was canceled in 1852, Moser received all its physical equipment. Like Neumann, he then conducted laboratory exercises in his own home but considered this arrangement provisional until a physical laboratory could be built. On the eve of completion of the new university building, Moser pressed the university for support for laboratory instruction; the university curator then transmitted his request to the ministry, arguing especially for the needs of the medical students in Moser's courses. Because Neumann was already using his own house for laboratory instruction, the curator requested that both Neumann and Moser receive 200 taler each. In early 1862, when an agreement was finally made, Neumann and Moser actually received substantially more: Neuman received 200 taler for instruments plus 160 taler for their maintenance; Moser, a 300 taler rent subsidy, 200 taler for instruments, and 140 taler for their maintenance. Because Neumann still had at his disposal 158 taler for the physical cabinet, 100 for the mineralogical collection, and part of the 350 taler budget for the mathematico-physical seminar, the ministry's new award made the budgets of Neumann and Moser approximately equal.[9]

Yet even as a scientific culture became more prominent and the facilities for scientific instruction more richly endowed, the number of students concentrating in mathematics or the natural sciences failed to increase proportionate with the growth in the student population. During the early 1850s, the Königsberg philosophical faculty had only about 50 students, about 20 percent of whom took Neumann's seminar, and that ratio later proved unsustainable. Although between the summer semester of 1859 and the winter semester of 1875/76, the number of students in

7. The Kultusministerium knew in advance that this requirement would be difficult to satisfy, but it hoped nonetheless to encourage candidates to become familiar with instruments; see the ministry's letter of 24 December 1866 to the state examination boards: *Reglement für die Prüfungen der Kandidaten des höheren Schulamts vom 12. Dezember 1866* (Berlin: C. Heymann, 1880), p. 20. Another edict had to be issued on 20 July 1876: "Königl. preußische Ministerial-Verordnung, betreffend die Anleitung künftiger Lehrer der Physik zur Bekanntschaft mit den physikalischen Lehrmitteln und im Experimentieren während der Studienzeit," *ZMNU* 8 (1877): 186.

8. Hans Prutz, *Die Königliche Albertus-Universität zu Königsberg i. Pr. im neunzehnten Jahrhundert* (Königsberg: Hartung, 1894), p. 297.

9. Ibid.; Königsberg University Curator to Kultusministerium, 13 December 1861, ZStA-M, Acta betr. eines Seminars für die naturwissenschaftlichen Studien auf der Universität zu Königsberg, Rep. 76Va, Sekt. 11, Tit. X, Abt. X, Nr. 21, [1827–76], fols. 228–31; Kultusministerium to Königsberg University Curator, 7 October 1863, ibid., fol. 244.

the philosophical faculty increased from 79 to 223 and the number of students who upon matriculation declared mathematics, physics, and the natural sciences also increased, from 27 to 58, the proportion had actually declined, from 34 to 26 percent. Over the same period, the total number of registrants in Neumann's lecture courses fluctuated but did not increase overall. Although enrollment information for Neumann's seminar is scarce after 1871/72, it appears that participation did not grow proportionally either to the number of students declaring the natural sciences or mathematics as a major or to increasing enrollments in the philosophical faculty. In the winter semester of 1860/61, when there were 102 students in the philosophical faculty, there were only 12 in Neumann's seminar. For the last semester in which we have seminar enrollment figures, the winter semester of 1874/75, the philosophical faculty had 197 students; the mathematico-physical seminar, only 9.[10]

These are only very rough comparisons, however, because there is no way of estimating the number of auditors in Neumann's seminar. First in 1869/70 Neumann remarked that "a great number of students regularly attended the seminar, but I have not listed their names because they did not solve the assigned problems." Again the next year he remarked that "a great number of students regularly participated in the recitation without solving the assigned problems."[11] Although Richelot had admitted auditors decades earlier than Neumann without altering the operation of his seminar, auditors could weaken Neumann's seminar because they did not have to complete the written assignments that served as preparation for the measuring operations that were at the heart of the seminar's activity.

Yet even without a real increase in enrollment and—even more striking—without significant changes in curriculum (which drew primarily upon old problems and techniques rather than absorbing new ones from recent advances in physics, especially thermodynamics), Neumann produced more doctorates in the 1860s—seven—than he had in all the other years he had been teaching (from Brix in 1841 to Clebsch in 1854 he had only produced five doctoral students). Publications resulting from seminar investigations also increased, although the number remained modest. Six of the seven doctoral students published all or part of their investigations in scientific journals.[12] The seventh, Friedrich Gustav Adolph Just,

10. Königsberg Universität, *Amtliches Verzeichniss des Personals und der Studirenden*, 1857–76.

11. Draft reports, physical division, 1869/70 and 1870/71, FNN 48: Seminar Angelegenheiten.

12. O. E. Meyer, "Ueber die Reibung der Flüssigkeiten: Theoretischer Theil," *JRAM* 59 (1861): 229–303; 62 (1863): 201–14; idem, "Ueber die Reibung der Flüssigkeiten," *AP* 113 (1861): 55–86, 193–238, 383–425; Louis Saalschütz, "Ueber die Wärmeveränderungen in den höheren Erdschichten unter dem Einflusse des nicht-periodischen Temperaturwechsels an der Oberfläche," *AN* 56 (1862): 1–44, 161–206, 273–98; Anton Müttrich, "Bestimmung des Krystallsystems und der optischen Constanten des weinsteinsauren Kali-Natrons; Einfluss der Temperatur auf die optischen Constanten desselben und Bestimmung

who became a secondary school teacher, later published part of his doctoral dissertation in his school's annual program.[13] In addition, two students, Carl Pape and Karl Zöppritz, both of whom began and almost finished investigations in the seminar which they later used as *Habilitationsschriften*, published their findings.[14] Another student, Ludwig Bernhard Minnigerode, submitted his seminar investigation as his dissertation at Göttingen in 1861. When he published parts of it in the 1880s, connections to the seminar exercises were no longer evident.[15] In contrast, between 1870 and 1876, when he ceased teaching in the seminar, Neumann directed only one doctoral dissertation, that of Woldemar Voigt, who later reworked his investigation for publication in the *Annalen der Physik* while he taught secondary school.[16] Overall, the number of doctorates and publications from Neumann's seminar remained modest in comparison to other prominent centers for learning physics, such as Berlin.[17]

Neumann was fortunate in the summer semester of 1859; for only older students enrolled in the seminar, enabling him to hold only one section and to discuss more difficult examples from his lecture course on the electric current from the semester before. He directed his students' efforts in particular at understanding the motion of electricity in a plane and in space when different conductors or various configurations of conductors were present. Almost apologetically, he pointed in his report to the "general" nature of these exercises, as if he were addressing readers who would have preferred to have heard that his students had accomplished considerably more. He added, however, that these "general" in-

---

der Brechungsquotienten des Rüböls und destillirten Wassers bei verschiedenen Temperaturen," *AP* 121 (1864): 193–238, 398–430; Albert Wangerin, "Die Theorie der Newton'schen Farbenringe," *AP* 131 (1867): 497–523; Oskar Frölich, "Zur Theorie der Erdtemperatur," *AP* 140 (1870): 647–52; Karl Von der Mühll, "Ueber die Reflexion und Brechung des Lichtes an der Grenze unkrystallinischer Medien," *MA* 5 (1872): 471–556. That Neumann's *draft* reports between 1860 and 1875 do not mention these and other independent investigations is an indication of how incompletely they render the seminar's operation.

13. Friedrich Just, "Geschichte der Theorie des Regenbogens," *Programm*, Städtisches Gymnasium Marienburg, 1862/63.

14. C. Pape, "Ueber die specifische Wärme wasserfreier und wasserhaltiger schwefelsaurer Salze," *AP* 120 (1863): 337–84, 579–99; K. Zöppritz, "Theorie der Querschwingungen schwerer Stäbe," *AP* 128 (1866): 139–56.

15. Ludwig Bernhard Minnigerode, "Untersuchungen über die Symmetrieverhältnisse und die Elasticität der Krystalle," *Nachrichten von der Kgl. Gesellschaft der Wissenschaften zu Göttingen* (1884): 195–226, 374–84, 488–92; idem, "Ueber Wärmeleitung in Krystallen," *Neues Jahrbuch für Mineralogie, Geologie und Palaeontologie* (1886): 7–13; idem, "Ueber die Symmetrieverhältnisse der Krystalle," ibid., (1894): 92–97.

16. Woldemar Voigt, "Bestimmung der Elasticitätsconstanten des Steinsalzes," *AP* E7 (1876): 1–52, 177–214.

17. For example, over seventy published investigations by almost thirty individuals issued from Gustav Magnus's laboratory at Berlin between 1843 and 1870; see "Wissenschaftliche Untersuchungen aus Magnus' Privat-Laboratorium" and "Wissenschaftliche Untersuchungen aus dem physikalischen Laboratorium," *Berichte der Deutschen Chemischen Gesellschaft zu Berlin* 3 (1870): 1099–1101.

vestigations "had to be carried through in special cases to the point where methods could be derived for proving results empirically."[18]

In the winter semester of 1859/60, in contrast, when "the unequal strengths of the new members necessitated the separation [of the students] into two sections," Neumann divided his energies between older students, who first covered the mathematical theory of the distribution of magnetism and then applied their findings to the calculation of such variables as magnetic inclination, and younger students, who were required to learn the "mechanical parts of physics." That in his report he merely listed the topics covered by these younger students—motion in a resisting medium, the ballistic pendulum, the moment of inertia, the determination of the earth's shape through an analysis of centrifugal force and of the attraction of an ellipsoid, and so on—was indicative of the standard form his teaching of mechanics had achieved. Still not a course on mechanics as we know it, it included lessons in measurement, especially the use of the pendulum to measure small intervals of time. Perhaps aware that his standardization of instruction in theoretical physics might evoke the misunderstanding that students were being "spoon-fed" material, he added that he always covered these subjects "only to the point where I have stimulated enough interest in the issue at hand so that the students themselves can achieve the solution." Lacking original student investigations from this semester, Neumann sent the ministry instead the recently completed work of two former students, Saalschütz and Schindler. Altogether he reported teaching nine students in the seminar; but two others were listed in his draft report, the chemistry student H. B. Rathke and the mathematics student L. O. Bock. Although three of his students—Just, Minnigerode, and Zöppritz—committed themselves upon matriculation to the study of physics, a fulfillment of Neumann's earlier hope that his seminar created institutional conditions conducive to the study of physics, he neglected to mention that fact to the ministry.[19]

During the summer semester of 1859, students were unable to enjoy Richelot's tutelage in mathematics because he had taken a medical leave of absence. He intended, however, "not to deprive seminar students of this means of instruction for a long time," so he gave students with sufficient background in mathematics problems to complete in his absence. He also spoke to them about their work both before and after his leave. For Richelot, the engagement in learning that seminar exercises offered was too valuable to omit, too routine to ignore. When he returned in the winter semester of 1859/60, he, like Neumann, had to divide his

18. "Bericht über die Leitung der physikalischen Abtheilung des mathematisch-physikalischen Seminars zu Königsberg während des Zeitraums von Ostern 1859 bis Ostern 1860," 20 August 1860, ZStA-M, Acta betr. das mathematisch-physikalische Seminar an der Universität zu Königsberg, Rep. 76Va, Sekt. 11, Tit. X, Nr. 25, Bd. I: 1834–61, fols. 259–60, on fol. 259.

19. Ibid., fol. 260.

students into two sections. He remained sensitive to the needs of students not ready for introductory seminar exercises. He remarked that even though two new seminar students each had one year of university-level mathematics, they were unequally prepared for the seminar. While one had achieved "very good" knowledge, the other "had rather good school knowledge [but] had barely acquired the first elements of higher analysis." He worked closely with two other students who had just entered the university in the winter semester of 1859/60 and who were not yet prepared to take part fully in the seminar. "At their request," he reported, "I gave them more problems and occasionally spoke with them about their work."[20] That semester, Richelot covered integration, differentiation, series, and hypergeometric series with special problems taken from Gauss's work.

So in 1859/60 both divisions of the seminar operated routinely, and Neumann's and Richelot's reports were appropriately unexceptional. Over the next sixteen years Neumann's annual reports were little more than lists of the topics of his seminar exercises.[21] In only a few respects did his exercises differ from those offered in previous decades. Topics in electricity appeared less frequently. From time to time he taught students how to derive the general differential equations for the distribution of electricity on various surfaces, as Kirchhoff and Quincke had done earlier in their publications; these exercises often employed the use of elliptic functions. He handled topics in magnetism by returning to the work of Gauss and Weber, in which he found the opportunity to develop mathematical equations and to teach students how to use measuring instruments. Students learned how to determine the moment of inertia of a bifilar magnetometer and to measure the absolute intensity of the earth's magnetism, as Gauss had done. Using Weber's theory of magnetic inclination, they learned how to determine the effect of an induced differential current on a multiplicator needle and then to measure electrical resistance. Neumann required students to derive the equations for the distribution of magnetism in pure iron of various shapes.

He handled other topics also without significantly changing his approach. In most of his theoretical and practical exercises in heat and light, he employed crystals to illustrate general principles. Neumann always started with the conduction of heat in discrete solid bodies, then turned to a continuous series of finitely many bodies, and finally applied the derived formulas to continuous crystalline bodies or to the conduction of heat in the earth. Avoiding hypotheses on the constitution of matter, he assumed

20. "Jahresbericht über die mathematische Abtheilung des physikalisch-mathematischen Seminars in Königsberg für die Zeit Ostern 1859 bis Ostern 1860," 25 September 1860, ibid., fols. 261–64.

21. Draft reports, physical division, 1860 to 1876 (with the exception of the summer semesters of 1868, 1872, and 1874; the winter semester of 1872/73; and the academic year 1875/76), FNN 48: Seminar Angelegenheiten.

instead that certain phenomena such as thermal conduction could be investigated by creating an abstract image of a body: as constituted of finitely many differential volume elements. Innovative work in thermodynamics, such as that by Rudolph Clausius and James Clerk Maxwell on the kinetic theory of gases, never became a permanent part of seminar exercises. Neumann derived the laws of optics first for crystalline, and then for uncrystalline, media. He considered reflection, refraction, double refraction, and the theory of lenses to be elementary topics and preferred to assume knowledge of them, but frequently he could not. Optical polarization was an especially important topic in the 1860s. Students learned how to calculate and observe the intensity of polarized rays, the polarization angle, and the plane of polarization. To improve his students' assessment of the accuracy of their own observations, he gave them his own measurements of polarization angles as well as those of another investigator for comparison. They examined the rotation of a plane of polarization as well as new topics: circular polarization, elliptical polarization, metallic reflection, and dispersion. Here the mechanical equations for elastic media and small vibrations were central; for in the case of isotropic media Neumann explained optical phenomena by assuming the existence of an optical ether whose elasticity modulus varied. He brought together several of his courses on physics when he discussed the effect of electricity and magnetism on the rotation of a plane of polarization; here the mechanical equations of small vibrations were used to explain the laws of metallic reflection and refraction, as well as those for crystalline surfaces.

Neumann's course on theoretical physics remained the anchor of his curriculum. The pendulum was still central to his exposition of certain principles such as the moment of inertia. Now he found the pendulum suitable for illustrating the conservation of energy. He covered the equations of elasticity used in his optical exercises; he computed the living force of a system of elastic masses; and he derived equations of motion out of the principle of virtual velocity. Hydrodynamic considerations, such as the internal viscosity of fluids and the finer points of capillary phenomena, also figured prominently in seminar exercises. When treating capillarity Neumann often began with Laplace's treatment based on molecular forces, but he preferred to deal with capillarity in terms of potential theory, thus providing his students with the most generalized treatment available.

More important than these topics was the foundation that theoretical physics still provided for his practical exercises. Neumann continued to be steadfast in his conviction that "these [mathematical and theoretical] investigations had to be carried through to the point where the results directly provide the order of observations that one of the seminar students will carry out."[22] He thereby insisted that it is from mathematical

22. Ibid., 1861/62.

theory that students learn to construct experimental protocols. When they learned how to derive the equations for the motion of heat in a sphere, for example, it was for the purpose of understanding the flow of heat in the earth and for measuring geotemperatures. Sometimes these "observations" began with a theoretical analysis of instruments, as Neumann suggested for the prisms of Nichol and Fresnel, before making any measurements. Often parts of experimental protocol, especially measuring operations, led back to theory, as when Gauss's method for measuring geomagnetism was used to develop notions of the magnetic potential. At other times, measurements provided the means to understand instruments; when, for example, Neumann asked his students to determine the capillary constant of mercury, it was for the purpose of understanding its influence upon the depression of mercury in a barometer, and hence upon barometric measurements. Sometimes a topic was treated solely through techniques of measurement, as in 1867/68, when the theory of magnetism was discussed. In spite of the central role of instruments in these practical exercises, students did not modify them materially except to achieve more precise results. True to the original precepts of the Besselian experiment, the procedural and material aspects of experiment were cast in theoretical and analytic terms.

Albert Wangerin, who attended the seminar during the mid-1860s, recalled that in 1863/64, Neumann worked through the analytical theory of heat in such a way that the formulas led to several practical applications, including the theory and calibration of thermometric instruments. During the summer semester he derived formulas for temperature distribution in a sphere in such a way that the constants of equations could be determined through measurement. He then developed equations for more complicated cases, especially ones that took into account the effect of cooling. The derivation of equations for patterns in geotemperatures culminated Neumann's coverage of theory in this seminar. He performed hardly any theoretical derivations in the winter semester, when he directed students to turn their full efforts to practical considerations. They examined and applied the method of mixtures; considered how the electric multiplicator could be used as a thermometric instrument; and derived equations for the motion of heat in a bar, once again for the purpose of understanding patterns in geotemperatures, taking into account the effect of solar heating. Neumann followed a similar strategy when teaching magnetism in 1865/66. After covering Gauss's method for determining the intensity of geomagnetism, he asked students to improve upon the method of bifilar suspension and Weber's method for measuring integral currents.[23]

Introductory seminar students fastidiously copied Neumann's derivations in notebooks. Paul Peters audited the seminar in the summer semes-

---

23. Ibid., 1863/64; Albert Wangerin, *Franz Neumann und sein Wirken als Forscher und Lehrer* (Braunschweig: F. Vieweg & Sohn, 1907), pp. 163–64, 172–73.

ter of 1870, when Neumann drew his exercises from his course on the theory of light, offered in the previous semester. Neumann presented the wave theory of light either by working entirely from the optical properties of crystalline matter or by assuming the existence of an optical ether, the pendulumlike vibrations of which were presumed to produce optical phenomena; in his lecture course from the winter semester of 1869/70, he had taken the first approach, which was simpler conceptually. In the next semester's seminar, he began with the geometric derivation of the path of light waves, first in isotropic media and then in crystals. His analytic derivations were ordered so as to lead to measuring operations. At every step he asked questions, urging students to draw conclusions on their own. At what angle does a light ray falling on a crystal become polarized? How does polarization take place in a crystalline medium? Neumann presented general conclusions—in this case the definition of the angle of polarization—only after students solved these and similar problems. Then he gave them the measurements of Seebeck and Brewster on polarization angles and asked them to compare these experimental values with those they themselves had calculated from theory.[24]

Peters's 1872/73 winter semester seminar notes bristle with the fine details of the derivations Neumann discussed. We learn that Neumann altered the parameters of problems from time to time so that students could adapt mathematical descriptions to different physical conditions. For example, in the derivation of capillary equations, one normally assumes that the capillary tube is vertical; but as an exercise, Neumann asked his students to derive the equation for the case when the tube is at an angle. Capillary theory gave him the opportunity to discuss more complex ways of handling the figure a liquid drop makes on a plane surface, such as treating the drop as a surface of rotation and deriving its structure on an arbitrary surface. Technical operations thus dominated the seminar. Neumann did less to develop or discuss the hypothetical assumptions incorporated into certain theories. For instance, although he generally began his discussion of capillarity with Laplace's theory, which assumed short-range intermolecular forces, he did not carry through the physical assumptions because they would not lead to the kind of measuring operations he wanted his students to perform. "The sphere within which these [capillary] forces are active is assumed to be very small," he told his students, "so small that they cannot be measured microscopically." Throughout his discussion he drew diagrams on the blackboard to illustrate the geometry of the problem, as Peters's overview of the seminar illustrates (Figure 20).[25]

24. FNN 23.A: Ausgewählte Kapitel der Optik, gelesen von Prof. Dr. Neumann, Wintersemester 69/70. Paul Peters; FNN 23.B: Physikalisches Seminar, geleitet von Prof. Dr. Neumann, Sommersemester 1870. In the following year Neumann taught optics once again, but this time began with the properties of the optical ether; FNN 23.C: Theorie des Lichtes, gelesen von Prof. Dr. Neumann, Wintersemester 1871/72, Paul Peters. See also draft reports, physical division, 1869/70, 1871/72, FNN 48: Seminar Angelegenheiten.

25. FNN 17: [1.] Theorie der Capillarität, gelesen von Prof. Dr. Neumann, Nachschrift,

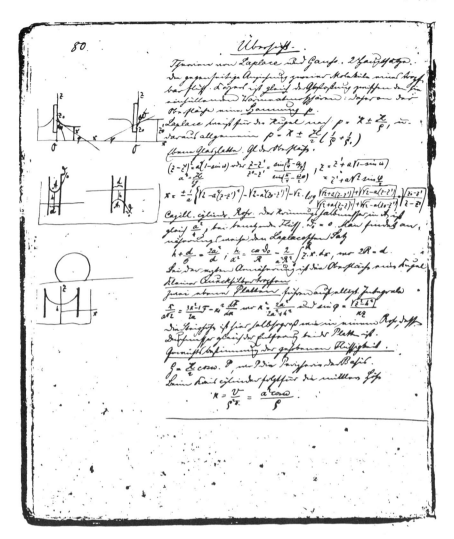

Figure 20. Page from Paul Peters's seminar notes on capillary theory, winter semester 1872/73, where he begins his overview of the theories of Laplace and Gauss. From FNN 17: [1.] Theorie der Capillarität [and 2.] Physikalisches Seminar, with permission to reproduce granted by UBG-Hs.

In the winter semester of 1869/70, Peters had attended Neumann's seminar on the theory of heat. The equations for the mechanical theory of heat being less settled than those for capillarity or the wave theory of

Wintersemester 72/73, Paul Peters [and 2.] Physikalisches Seminar, gelesen und geleitet von Prof. Dr. Neumann, Wintersemester 1872/73, Paul Peters. In 1857, when Neumann had also taught capillarity, he had approached the subject differently, by developing his equations from potential theory; cf. FNN 16: Capillaritätstheorie, F.N.'s Vorlesungen und Seminare von 1857: Heft von O. E. Meyer.

light, Neumann chose not to begin that seminar with theoretical topics but "with consideration of the experimental determination of constants." "This determination," he continued, "is among the most difficult [, first,] because *heat* is not itself constant but is given up by an object continuously and, second, [because] the resources of the theory cannot be used satisfyingly." He began with a discussion of specific heats, which were constant within certain intervals. He introduced Lavoisier's calorimetry experiments; the method of cooling and Pouillet's pyrheliometer; and his own method of mixtures, with its characteristic apparatus.[26] Especially when discussing the last two methods, he relied upon Fourier's differential equations for heat flow. Neumann made certain that at each step the equations had a recognizable physical meaning, and from time to time he represented them graphically. He left the final step of integrating the equations, not a trivial operation, for students to complete at home. Although he incorporated topics from thermodynamics, he devoted most of his attention to the experiments of Joule and Regnault and to the experimental determination of the velocity of sound and not to the theoretical work of König, Clausius, and Carnot (although their investigations were mentioned). He talked about the physical assumptions at the foundation of thermodynamics, including the redefinition of pressure as the sum of the collisions of small elastic spheres, but he did not examine these assumptions critically (especially that of perfect elasticity) or consider in detail the mathematics involved (including statistical considerations). He referred to work by a variety of terms (*Arbeitsvorrath*, *Wirkungsfactor*, *Wirkungsgröße*), but in the end, he seemed somewhat indecisive about its physical meaning, remarking that work "probably consists of living force and internal force." Peters' notes reveal that in class, Neumann concentrated mostly on deriving thermal constants from Fourier's equations.[27]

Peters recorded that Neumann made frequent reference to Adolph Wüllner's *Lehrbuch der Experimentalphysik*, especially when discussing the theory of heat. He urged students to read especially Wüllner's treatment of the thermometer, the method of cooling, and his account of Neumann's method of mixtures. That Neumann chose Wüllner's text as supplementary reading is somewhat curious; for Wüllner demonstrated a refined sense for *taking* measurements but not for *interpreting* them. Wüllner believed the "proof" of a physical law was to be found in measurements precisely taken, but he did not suggest just how that precision was to be achieved and assessed. Unlike Neumann, he did not stress error analysis, including the method of least squares, as a way of evaluting the reliability of measurements. Instead, he believed that "how great the error of observation can be in individual cases, how closely the observed

26. FNN 10: Wärmetheorie, physikalisches Seminar geleitet von Prof. Dr. Neumann, Wintersemester 1869/70, Paul Peters, p. 1.
27. Ibid., p. 67.

value can correspond to the real one—that cannot generally be determined. It depends on, other conditions being equal, the difficulty of the measurements."[28] Such words may have elicited a sigh of relief from Neumann's students, but their pleasure could only have been momentary. Wüllner merely exemplified different attitudes regarding experiment, especially the deployment of rigorous mathematical techniques in error analysis (including the method of least squares), that had emerged in German physics by the 1870s.

As the published versions of Neumann's lecture courses reflect his teaching only partially, so do these student notes taken in his seminar. The epitome of Neumann's seminar program remained practical exercises in measuring operations; all other work was directed toward that task. Neumann consistently mentioned the students' measuring operations in his seminar reports. During the winter semester of 1869/70, for example, he covered the theory of heat in the classroom sessions of the seminar, but what he reported to the ministry was that students had derived equations for the propagation of heat and for internal and external conductivity for the purpose of applying them to measuring operations. Significantly, he did not mention to the ministry how, if at all, his coverage of the theory of heat had changed in response to recent advances in thermodynamics.[29] Evidence from the late 1860s and early 1870s suggests that Neumann's seminar exercises were not keeping pace with recent developments in physics.

Neumann had long maintained that instruments and the experimental procedures governing measurement were not fully understood until all sources of error were eradicated or mathematically accounted for, but in his seminar reports after 1860 he mentioned error analysis less frequently. Error analysis nonetheless remained a critical part of his curriculum, as the progression of seminar topics and exercises testifies. Yet his excessive concern for the computation of error was no longer necessary because opinions were not what they had been. In the 1830s when Neumann had first introduced error analysis, he did so not just as a way of reducing data but as one aspect of confirming mathematically expressed theories in physics and of finding the limits within which theory was valid. By the 1860s the work of not only Neumann but also Weber, Kirchhoff, Clausius, and others had so extended mathematical methods in physics that their use was no longer so controversial. Mathematical physics was also more strongly institutionalized in the 1860s; although full professorial positions in it had not yet appeared, it was taught at most Prussian universities. The increased use of mathematics in physics also

28. Adolph Wüllner, *Lehrbuch der Experimentalphysik*, 2d ed., 3 vols. (Leipzig: B. G. Teubner, 1870–71), 1:11; see also 3:273 for Wüllner's discussion of Neumann's method of mixtures.

29. Draft report, physical division, 1869/70, FNN 48: Seminar Angelegenheiten.

created subtle distinctions among practitioners of mathematical physics. At least three different groups can be identified: those who used physical problems as a vehicle for developing mathematical techniques; those who were fundamentally interested in the theoretical or conceptual content of mathematical expressions; and those who still thought mathematical expressions of physical phenomena could only rightfully be understood in the context of the experimental conditions that demonstrated the lawful relations contained in them.

Thirty years after Neumann introduced his course on theoretical physics, error analysis could thus comfortably be confined to experimental investigations and to the analysis of constant errors within them without harm to the integrity of mathematical expressions of physical theory. That is not, however, what Neumann chose to do. Instead, as mathematical physics gained ground both intellectually and institutionally, he retained and refined the pedagogical exemplars he had developed over the past thirty years. As we shall see in the investigations that originated in the seminar after 1860, error analysis remained a prominent investigative technique. It may even have been enhanced in seminar exercises during the 1860s and 1870s as a consequence of Neumann's keen interest in techniques for making refined measurements, especially the measurement of small time intervals and time-dependent variables.

Vividly demonstrating the pivotal role of Neumann's practical physics in training students were the doctoral dissertations written under him and the *Habilitationsschriften* that stemmed from seminar exercises between 1860 and 1874. Emil Meyer, Louis Saalschütz, Fritz Just, Anton Müttrich, Albert Wangerin, Karl Von der Mühll, Oskar Frölich, and Woldemar Voigt completed doctoral dissertations under Neumann between 1860 and 1874. Most of these students had spent two years in the seminar and presumably had taken the full cycle of Neumann's lecture courses. Their investigations covered diverse topics: the viscosity of fluids, the thermal dependency of optical constants in crystals, Newton's rings, reflection and refraction, geothermal theories and observations, and the elasticity constants of rock salt. In addition, Neumann's former students Carl Pape and Karl Zöppritz developed their seminar investigations into *Habilitationsschriften* on, respectively, specific heats and transverse elastic vibrations. Although most of these investigations were linked in some way to Neumann's research interests, they were inspired less by Neumann's own investigations than by his seminar exercises. It is not a coherent research program that links these investigations but a common investigative strategy that emphasized primarily, although not exclusively, the eradication of error rather than the discovery of new truths or the construction of comprehensive theories.

In 1857, Emil Meyer began to work on the prize question of the philosophical faculty (proposed by Neumann): determining the coefficient of viscosity of fluids using Coulomb's methods. He repeated Coulomb's experiments in order to explain them through "theoretical" considerations, which meant conducting an exhaustive analysis of the sources of error in them. Meyer began by working out the theory of the experimental apparatus, a modification of Coulomb's: a thin circular disk suspended at its center by a wire and immersed in a liquid-filled container. The disk could rotate about its vertical axis, and its angular displacement gave a measure of the resistance of the fluid. Meyer's theoretical analysis of his instrument and his experimental protocol proved stronger than his observations. Despite his unsatisfying measurements, he was awarded the prize on 18 January 1858. He then set out to improve the experimental part of his investigation for his doctoral dissertation. Throughout his trials Meyer held steadfast to the definition of "theory" used in seminar exercises: "theory" meant principally, although not exclusively, the mathematical description of what actually happened during the experimental trial; the identification of constants that could be determined through measurement; and the isolation and, where appropriate, the analytic expression of the causes of errors afflicting the experiment.

When he completed his trials in 1860, Meyer's theoretical equations still had not led to the empirical results he wanted. Deciding to separate the theoretical and experimental parts of his investigation, he wrote to Neumann on 6 February 1861 that he was sending the theoretical part "essentially unchanged" from what he had done at Königsberg to Borchardt's (Crelle's) *Journal für die reine und angewandte Mathematik*. The imperative to measure shaped the course of his otherwise mathematical analysis. Working from Newton's assumption that the viscosity between two fluid layers was independent of pressure but proportional to the difference in the velocities of each layer and to their contact surface, he developed equations suited for measuring the viscosity. He even compared his approximation methods with his experimental errors and found that his equations were refined enough. Meyer concluded that Coulomb's method was adequate for determining the coefficient of viscosity because the accuracy of the constants so obtained was no less than what could be gotten by other methods.[30]

But when Meyer wrote to Neumann, he revealed that the experimental part of his investigation was still plagued by so many problems that he was not yet ready to report publicly on it. Part of his experiment involved determining the viscosity of mixtures of salt solutions, but he had "little success" with his calculations because the observations "unfortunately did not have the accuracy that is required in order to reach a certain conclu-

---

30. O. E. Meyer, "Ueber die Reibung der Flüssigkeiten: Theoretischer Theil," pp. 229, 272, 280; O. E. Meyer to Neumann, 6 February 1861, FNN 53.IIA: Briefe von Schülern. See the excellent treatment of Meyer's work in Stefan L. Wolff, "Die Rolle von Reibung und Wärmeleitung in der Entwicklung der kinetischen Gastheorie" (Ph.D. diss., Ludwig-Maximilians-Universität, Munich, 1988), pp. 56–76.

sion" and because he could not determine if the solutions had been chemically transformed when mixed. Apologetically he asked Neumann to assist him by making parallel measurements in Königsberg with which he could compare his own. Another part of his investigation involved examining the properties of the wire by which the disk was suspended. He explained to Neumann that insofar as his own measurements of the internal resistances of the various wires were concerned, "scarcely the first [decimal place] is certain. I doubt, therefore, that the observations performed will be suitable for proving the lawful dependence of the resistance of a wire upon its radius." Referring to E. Wiedemann's work on the magnetization of metal wires, Meyer decided to examine how resistance changed with magnetic state.[31]

Yet despite the complex and still-unfinished state of the experimental part of his investigation, Meyer decided to present his results to the Berlin Physical Society, where his audience included secondary school teachers as well as academic physicists. His lecture did not go well. He later told Neumann that he "was not well understood because in mentioning partial differential equations, I assumed too much mathematics on the part of the audience. The mistake was in any case mine because an abbreviated lecture that rests on mathematics is difficult to follow." Although he was discouraged by the reaction to his investigation, his lecture was not a complete failure. Brix, who had heard the presentation, assured Meyer that at least his thermometer calibrations were correct.[32]

By August 1861, Meyer had reworked his investigation sufficiently and had data from enough trials—some of them performed by Neumann—to send it to the *Annalen*. For this report he made several stylistic changes to address a new readership: he reformulated his equations to make physical considerations more prominent; he rewrote the introduction to accommodate those "unaccustomed to mathematical considerations and who do not read Crelle's journal"; and he repeated much of what was in his earlier paper with a "completeness of detail" that he himself found unpleasant but which he felt served his readers well. The reformulation was apparently worthwhile because he told Neumann that a young physiologist in Berlin understood this presentation but not his earlier one.[33] What Meyer did not do, however, was to abandon the dominant strategy in seminar exercises that focused on the calculation of error as the culmination of a physical investigation.

From a theoretical perspective, Meyer's investigative strategy had matured. Now, rather than giving Coulomb's method a thorough mathematical treatment as he had in Borchardt's *Journal*, he attempted "to give through concrete ideas a view of the procedure" that he had earlier

31. O. E. Meyer to Neumann, 6 February 1861, FNN 53.IIA: Briefe von Schülern.
32. Ibid.
33. O. E. Meyer to Neumann, 6 August 1861, ibid.

worked out "in numbers and drawings." He wanted at first to understand more clearly the behavior of the particles in the layers of fluid closest to the horizontally rotating disk. Although he claimed that the "law of viscosity" could only be found through observation, he in fact used hypotheses and analogies to guide his derivation. He viewed, for instance, the coefficients of external and internal viscosity as analogous to external and internal thermal conductivity in Fourier's theory of heat. Meyer's "hypothesis about the form of the unknown law" was Newton's, that the viscosity of a fluid was proportional both to the difference in the velocity of two fluid layers moving in the same direction and to the size of the contact surface, but independent of pressure and the absolute values of velocities. To find the "fundamental law of viscosity" Meyer advocated using a hypotheticodeductive method, calculating "in advance the results of the experiment on the basis the hypothesis" and then comparing "the result of this calculation with trials actually performed."[34]

The detail with which Meyer worked out his mathematical equations was intended less as a means of exploring the conceptual foundation of the theory than as a means of avoiding what he viewed as the principal shortcomings of other methods. The apparent simplicity of other experimental methods, he argued, was "illusory" because certain conditions of the experiment "were still very far from a mathematical determination." Since the mathematics did not adequately represent the experiment, he concluded that "one can therefore doubt if in the methods performed until now, the conditions under which the mathematical theory is most rigorously correct were actually filled." His own goal was to obtain those conditions that provided "the greatest possible sharpness of observation." So he held up the publication of his results until he could account quantitatively for several errors, including those from Gauss's method for determining the moment of inertia (it could not be applied, Meyer found, to apparatus with a small moment of inertia); from a larger than suspected effect of the resistance of the air; and from incomplete knowledge of the elasticity of the wire.[35]

Meyer's corrections for constant errors, which dictated the course of his investigation, all sharpened the mathematical description of the theory of his experiment. They included corrections for air resistance (drawn in part from Bessel's seconds pendulum investigation), for the moment of inertia of his apparatus, for the finite thickness of his disk, and so on. He improved upon Coulomb's protocol by observing the oscillations of the disk with a telescope equipped with cross hairs (Coulomb had used his unaided eye) and computed the errors produced in the course of reading the angles marked on the disk. Almost all his measurements were reduced by using the method of least squares. All his trials were conducted with an

34. O. E. Meyer, "Ueber die Reibung der Flüssigkeiten," pp. 61, 67.
35. Ibid., pp. 59–60, 56.

apparatus, designed by Neumann, from the physical cabinet at Königsberg. Despite the care with which he conducted his experiment, the
deviations between theoretical and experimental values were large
enough for him to reconsider the conditions of the experiment. He found
that at the beginning of his observations, the motion had not gone on long
enough "to be able to justify the approximations made in the final formula of the theory." So he performed observations on several different
disks "to prove the law through sharper observations and more rigorous
calculations." Like Bessel, Meyer wanted to make his data as accurate as it
could be by freeing his results from the conditions of the experiment, but
he could not. His results, he found, depended on the radii of the disks he
used, and so his theoretical equations were valid only for small disks.
Although the precision of his data was inconsistent—it generally varied
between the third and fifth decimal place—he thought that he could do
no more. The remaining deviations, he claimed, could not be explained
any further by the "probable errors of observation" but instead originated
"in the approximations introduced in the derivation of the method from
Coulomb's theory"; for these approximations led to constants whose
values were too large. Meyer was satisfied, nonetheless, because "the
assumptions of the theory [were] filled to the limits of the accuracy of the
observations."[36]

Meyer generally considered "theory" insofar as it related to experimental protocol. But ultimately his investigation concerned the nature of
intermolecular forces. So when measurements produced a "surprising"
result—"that the viscosities of air and of water are not significantly different from one another"—he had to shift from one definition of "theory" to another. The "surprising" result he obtained could be explained
not by referring back to the errors afflicting the experiment but instead by
calling upon very recent molecular theories in thermodynamics. Meyer
believed that his result did not agree with "the usual hypothesis about the
molecular constitution of matter, according to which the molecule lies in a
fixed position and deserts [this position] only through external forces." So
he turned instead to the "Bernoulli-Clausius" view on the constitution of
gases, which assumed that molecules not only vibrated but moved along a
"free path" until a collision occurred. By adopting Clausius's view that
viscosity could be explained as the transfer of velocity from a rapidly
moving layer of gas to a slower moving layer, Meyer found that his
experimental results were "not in the least shocking."[37]

Despite its significance, Meyer's discussion of "pure" theory was but a
brief interlude in a study far more technically oriented. Although the
kinetic theory of gases later became a more prominent thread in his
investigations, for the time being "theory" remained for him the analysis

36. Ibid., pp. 220, 226, 420, 235.
37. Ibid., pp. 386, 386–87.

of experimental protocol. The conclusions he drew from his data, for example, did not at all concern intermolecular forces. Nor were his conclusions precise quantitative statements that further elucidated the physical process taking place in contiguous fluid layers in motion with different velocities. His results were instead a confirmation of the advantages his method had over others. Most of his conclusions issued from his refined error analysis: the decrease in internal viscosity with increasing temperature; the significant viscous effect of air; and the fact that under certain conditions, Gauss's method for determining the moment of inertia could lead to significant errors. Meyer concluded that his and Coulomb's observations confirmed "the theory of this method" and that his experiment was a "new proof" that external viscosity was proportional to the difference in the velocities of the layers of the fluid (liquid or gas), and internal viscosity to their differential quotient.[38]

The brief excursion into physical theory proper in Meyer's investigation was a more prominent feature of only two other doctoral dissertations that Neumann directed between 1860 and 1874. One was written by Karl Von der Mühll, who had been a student of Meyer's at Breslau in 1863, when Meyer advised him to study with Neumann in Königsberg. Von der Mühll appeared "capable," Meyer told Neumann, and probably had "significant talents and great scientific eagerness."[39] By 1866, Von der Mühll had finished his doctoral dissertation on the derivation of dynamical equations for reflection and refraction of light at the boundary of two transparent uncrystalline media. He hoped that the exact form of these equations would help him decide between Fresnel's and Neumann's theories because observations thus far had not supported one theory decisively. Originally Von der Mühll had concluded that the deviation between theoretical and experimental values could not be considered in terms of the "errors of observation" because even though observations agreed "very well" with, say, Cauchy's formula, "that alone can hardly speak for the correctness of the formula." When he published his findings six years later, he tried to find "if and how [Fresnel's] formulas can be brought into agreement with the results of observation." After reviewing some empirical data, he concluded that "small deviations from Fresnel's formula" occurred in nature. Although he claimed to be seeking not mere numerical agreement between theory and experiment but a more fundamental refinement of theory (particularly of Fresnel's and Neumann's assumptions concerning the elasticity of the ether), Von der Mühll was held back by the incomplete knowledge of certain constants, such as the ratio of ether densities in two transparent media. He did not discuss his quantitative results in detail and, it is significant, did not follow through with the kind of experimental analysis Meyer had done. Von der Mühll's

38. Ibid., p. 424.
39. O. E. Meyer to Neumann, 20 February 1863, FNN 53.IIA: Briefe von Schülern.

study, although rooted in the kinds of questions asked and operations undertaken in the seminar, leaned more toward the kind of mathematical physics then taking shape in the German universities, a mathematical physics that tended not to incorporate the techniques of an exact experimental physics.[40]

The second doctoral dissertation that strongly considered "pure" theory was written by Oskar Frölich in 1868, but unlike Von der Mühll, Frölich was far more attentive to experimental considerations. Before he arrived at Königsberg in the fall of 1865 "to study physics under the direction of Professor F. E. Neumann," Frölich had attended Bern University, where he had taken courses with Neumann's former student Heinrich Wild; had four semesters of practical laboratory training, three in chemistry and one in physics; and had served as Wild's laboratory assistant for two semesters. Wild told Neumann he hoped "that this young man, if he concentrates more than he did here, [would become] a capable physicist under your direction. He does not lack the ability." For five semesters Frölich took Neumann's lecture courses and participated in the seminar, and he "occupied [himself] in private under [Neumann's] auspices" with experimental and theoretical matters in the theory of heat. An investigation from 1867/68 on the absorption of solar heat in the atmosphere and its effect on geotemperatures became his doctoral dissertation in 1868. By using temperatures taken at Königsberg and Brussels, and by examining the conditions under which the Königsberg measurements had been taken, he was able to prove that the absorption of solar heat in the atmosphere was constant, as Poisson had presumed but which Frölich considered unproved because Poisson's formulas were not suited for practical calculations and Poisson's observations contained what Frölich thought were "improbable values." Although he continued to examine the theories of Poisson and others, his objective was not to refine theory but to create equations best suited for practical calculations, in particular for measuring the variables (such as the absorption coefficient of the atmosphere) that Poisson either had assumed were constant or had not determined at all.[41]

The determination of error guided most of the remaining investigations directed partly or entirely by Neumann during the last decade and a half of the seminar. So much did the concern for error set the course of investigations that it inhibited less rigorous (more theoretical) thinking and sometimes even led to poignant realizations. Louis Saalschütz, faithful member of the seminar for two years while he attended Königsberg

40. Von der Mühll, "Ueber die Reflexion und Brechung des Lichtes," pp. 503, 505, 549; the first two parts constituted Von der Mühll's dissertation.

41. Oskar Frölich, *Ueber den Einfluss der Absorption der Sonnenwärme in der Atmosphäre auf die Temperatur der Erde*, Inaugural-Dissertation (Königsberg: Dalkowski, 1868), p. 29 (vitae); Heinrich Wild to Neumann, 15 January 1866, FNN 53.IIA: Briefe von Schülern; Frölich, "Zur Theorie der Erdtemperatur."

between 1854 and 1859, finished his doctorate under Neumann in 1861. Seeking to find out how nonperiodic changes in the temperature of the earth affected the temperature of various layers of the atmosphere, he exercised discrimination in accepting measurements. He threw out one set, taken at a depth of 5 feet, because the error introduced by an unreliable null point on the thermometer had not been taken into consideration. But he kept Neumann's observations, taken at a depth of 24 feet, from the late 1830s because they "carried the stamp of accuracy and reliability that one would expect given the caution and care with which they were taken."[42] In a similar fashion, an accurate determination of errors guided Müttrich's dissertation on the thermal dependency of optical constants in 1863, which he had begun many years earlier in the seminar. Presumably it was his concern for error that prevented him from using his graphs more productively, too.[43]

After studying at Halle during 1862/63, Albert Wangerin matriculated at Königsberg in the winter semester of 1863/64, when he registered for Neumann's courses on theoretical physics and topics in mathematical physics and for Richelot's course on elliptic functions. He also immediately attended both divisions of the seminar. Between then and 1866, when he received his doctoral degree under Neumann, Wangerin took one or two courses a semester with both Neumann and Richelot in addition to attending the physical seminar for four semesters and the mathematical seminar for five. By the time he finished his university study, he had covered almost the full cycle of Neumann's lecture courses. Two weeks after defending his dissertation on 16 March 1866, Wangerin took the state teaching examination, a testament to Richelot's belief that with proper university training, additional preparation for the examination was unnecessary.[44]

Wangerin considered himself chiefly a mathematician, not a physicist, but when a seminar investigation on Newton's rings (for which he used an apparatus belonging to Neumann) suggested novel results, he turned it into his doctoral dissertation. The novelty in Wangerin's investigation was the result of considering smaller analytic terms in the formula for the rings which had been neglected in prior investigations and whose effect had not been empirically determined. In Wangerin's view, the explanations for Newton's rings given thus far were only "first approximations," approximations for which "one can decide nothing about the degree of approximation." He believed that "in a complete theory" one had "to give an accurate accounting of the neglected terms." By taking into account all

42. Saalschütz, "Ueber die Wärmeveränderungen," p. 3.
43. Müttrich, "Bestimmung des Krystallsystems."
44. Albert Wangerin, Lebenslauf, Darms. Samml. Sig. H(13) 1899; Wilhelm Lorey, *Das Studium der Mathematik an den deutschen Universitäten seit Anfang des 19. Jahrhunderts* (Leipzig: B. G. Teubner, 1916), p. 95.

terms that could be measured, his theory predicted that the rings were not circular but had an eccentricity that in principle could be observed. But his "exact measurements" of the rings were plagued by numerous errors that he could neither eliminate through his protocol nor account for mathematically. As a partial correction, he took as many measurements as he possibly could "in order to draw a somewhat certain conclusion." Although he was interested in reducing the range of uncertainty in his observations, he did not discard outlying ones but instead reported both the smallest and largest observed values, their average, and the probable error of the average computed by the method of least squares. The probable error was an extremely important figure in Wangerin's investigation because it indicated to him not only what sets of measurements were unreliable altogether but also whether he could regard his more refined theory as confirmed.[45]

But his sophisticated mathematical analysis of errors in the experiment compelled him to circumscribe his conclusion with caution. "And so I believe," Wangerin wrote, "that my measurements made an eccentricity of the rings in the plane of incidence at least experimentally probable. I scarcely believe that a greater accuracy in the individual measurements can be attained. A more rigorous experimental proof could at best only be shown through a still greater number of observations." Unfortunately, he could not undertake the next logical step in an investigation of this type: making more precise measurements. He explained that he had "been prevented from conducting these [observations] because the necessary apparatus does not stand at my disposal in my current position [at a secondary school]. Therefore I had to restrict myself to communicating the measurements I made a year ago at Königsberg."[46] The demands of precision thus exposed to Wangerin the gap between a world where experimental protocol could in principle be endlessly refined and one where the practice of science was constrained by material and professional resources. He acutely, and regrettably, sensed the distance between them.

Even in the last doctoral dissertation Neumann supervised, Woldemar Voigt's on the elasticity constants of rock salt, a painstaking analysis of error shaped the contours of the initial investigation and the course of revisions leading to publication. After graduating from Leipzig's St. Thomas Gymnasium in 1867, Voigt entered Leipzig University, where until the summer semester of 1870 he completed a wide-ranging, but for the most part basic, curriculum in mathematics and the natural sciences, including physics, chemistry, and mineralogy. Until his final year he had studied only experimental physics and, after taking physical exercises in the summer semester of 1868, served as assistant in those exercises for the

45. Wangerin, "Die Theorie der Newton'schen Farbenringe," pp. 497, 518, 522.
46. Ibid., p. 523.

following four semesters. In 1869/70 he finally began to take advanced courses such as analytical mechanics, higher optics, and potential theory with two of Neumann's former seminar students, Carl Neumann and Karl Von der Mühll. Voigt had not taken mathematics until his second year at Leipzig, when he began with analytic geometry and differential and integral calculus. Not until he was registered in advanced mathematics courses—such as infinitesimal calculus—did he attend the more difficult, mathematically oriented physics courses taught by Neumann and Von der Mühll.[47]

Voigt's university study was interrupted by the Franco-Prussian war in 1870/71. He had heard about Neumann's program from Von der Mühll and Carl Neumann, and he also had made the acquaintance of two other former seminar students then teaching secondary school in Leipzig, Johannes Eduard Böttcher and Gustav Louis Baumgarten, who both urged him to continue his studies at Königsberg. Baumgarten told Neumann that Voigt wanted to pursue mathematics and physics "undisturbed." Voigt left for Königsberg in June 1871 even though he had just returned from the war and the semester was almost over. From the winter semester of 1871/72 to the winter semester of 1873/74 he attended only mathematics and physics courses at Königsberg. In his first semester he took Richelot's course on ultraelliptic functions, Neumann's courses on optics and topics in mathematical physics, and both divisions of the seminar. Neumann did not teach in the summer semester of 1872, so Voigt attended only the mathematical seminar and Richelot's course on variational calculus.[48]

Voigt quickly immersed himself in learning Neumann's way of doing physics by the beginning of his second year at Königsberg. In the month before the winter semester of 1872/73 began, from 24 September to 17 October 1872, he mastered on his own the investigation that, almost forty years earlier, had shaped Neumann's pedagogical physics: Bessel's seconds pendulum investigation. Voigt's notes on Bessel's pendulum reveal that he was interested primarily in what Neumann considered important in the investigation; for Voigt worked out in detail only those sections of Bessel's extensive appendixes which concerned the constant errors of the experiment, especially the hydrostatic and hydrodynamic influences of air on the pendulum's period. He also studied Bessel's methods of calculation, especially Bessel's techniques of integration.[49]

The following semester, however, proved only partially rewarding.

---

47. "Von dem Universitäts-Gerichte zu Leipzig," WVN 1; Voigt's vitae, FNN 53.IIA: Briefe von Schülern; Woldemar Voigt, *Untersuchung der Elasticitätsverhältnisse des Steinsalzes*, Inaugural-Dissertation (Leipzig: Pöschel,1874), p. 56 (vitae).

48. G. L. Baumgarten to Neumann, 28 June 1871, FNN 53.IIB: Briefe anderer Schüler; "Anmeldungsbuch [Königsberg University]," WVN 1.

49. WVN 6: Ausarbeitung einiger wichtiger Capitel aus Bessel's Untersuchungen über die Länge des einfachen Secundenpendels, 24/9–17/10 1872.

Voigt signed up for Richelot's course on variational calculus and, probably as a result of his study of errors in the seconds pendulum investigation, also for a course on the method of least squares taught by the Königsberg astronomer Eduard Luther. But Voigt never completed the courses he began with Neumann that semester; they were on the electric current, topics in mathematical physics, and the physical seminar. During his final year at Königsberg, the summer semester of 1873 and the winter semester of 1873/74, he took with Richelot mechanics and dynamics and one semester of the mathematical seminar and with Neumann the induced electric current, elasticity, two courses on topics in mathematical physics, and two semesters of the physical seminar. He received his doctoral degree in March 1874. Throughout his life Voigt retained a deep and abiding admiration and warmth for Neumann, who deeply influenced his career and investigative style.[50]

Voigt undertook his doctoral dissertation in the spirit of Neumann's earlier use of the properties of crystals and minerals as a means of examining physical phenomena. Attempts to understand more clearly the "causal nexus" between the structure of crystals on the one hand and the facts of light, sound, heat, electricity, and pressure on the other were hampered, Voigt thought, by "the lack of exact methods of observation and correspondingly reliable [and] useful numerical data." He sought to close this gap and especially to determine "exact numerical values" for elasticity relations that in his view were represented only by questionable empirical formulas. In particular he wanted to measure the elasticity constants of rock salt "with consideration of all possible sources of error" and to bring these and other measurements together under "a single point of view."[51]

The course of his investigation, however, was shaped less by the desire to create a unitary theory than by the imperative to achieve precise results. Using an apparatus that had been designed and constructed by Neumann and deployed by other seminar students (Figure 21), Voigt measured the elasticity of rock salt in the Königsberg observatory, the best location for avoiding disturbances. When the winter's chill made it impossible to use the observatory any longer, Neumann offered Voigt use of his physics auditorium in the main university building when classes were not held. Voigt modified the instrument, accounted for all sources of error, and in general used one sequence of observations after the next to determine still finer sources of error. He had much the same attitude toward numerical results that Bessel had: they were not really useful without

50. "Anmeldungsbuch [Königsberg University]," WVN 1. Neumann's courses for the winter semester of 1872/73 had not been canceled. Voigt's private letters to his family indicate that his personal and intellectual bond to Neumann was stronger than his letters to Neumann suggest (W. Voigt family correspondence, held by Maria Voigt, Göttingen).
51. Voigt, *Untersuchung der Elasticitätsverhältnisse*, pp. v, vi, viii.

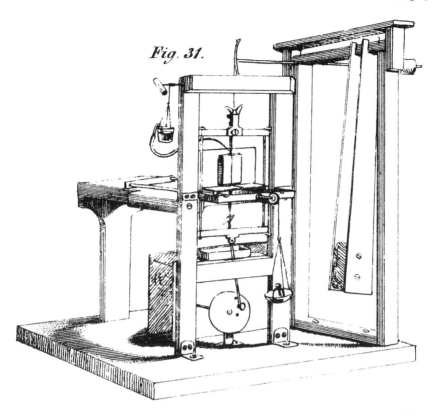

Fig. 31.

Figure 21. Neumann's apparatus for measuring elasticity constants. From Voigt, *Untersuchung der Elasticitätsverhältnisse*, Table III, Figure 31, courtesy of Library of Congress Photoduplication Services.

information on the conditions under which they had been made and the corrections that compensated for the errors afflicting them.[52]

Building on the work of two former seminar students, Leonhard Sohncke and G. L. Baumgarten, Voigt perfected an experimental procedure for measuring the elasticity constants of rock salt. In the course of his investigation, he, like other seminar students, had difficulties in interpreting his numerical data. Sometime in 1873 he had completed two, and possibly three, sets of observations but was having trouble determining the elasticity modulus of his samples from the second set.[53] By the time he defended his dissertation on 21 March 1874, he had not solved all the problems in his investigation, but he had taken several important steps in

52. Ibid., pp. vi, viii, 5–7.
53. W. Voigt to Neumann, n. d. [1873], FNN 53.IIA: Briefe von Schülern.

constructing an exact experimental procedure, in analyzing data, and in assessing errors, especially in determining the compensations one had to make for the permanent deformation a sample suffered when bent. Like several essays stemming from the seminar, Voigt's reported exhaustively on his protocol and error analysis.

Voigt devoted the final quarter of his dissertation to what had proved over the years to be the most indecisive aspect of seminar investigations: the graphical representation of numerical results. Believing that graphs conveyed results with "better clarity" than tabular data listing, he nonetheless called his graphs "curve tables" (Figure 22), as if they were a hybrid form that retained all the properties of tables, in particular their entirely empirical foundation. He could claim to have "attained rather the same certainty" with graphs as he would have "through the direct communication of numerical values" only because he did not go substantially beyond his measured points by drawing smooth curves. Yet in other ways he admitted that the graphical representation was both less and more than would have been rendered in tables: less because his graphs omitted observations that did not have "the desired degree of accuracy," more because a graph inevitably included points not measured. Although he used his graphs to draw general qualitative conclusions, Voigt found them less useful for determining precise mathematical formulas and constants governing elasticity relations, and so for these he worked more closely with his numerical data.[54] The specific conclusions his graphs yielded were not as important, though, as the way the graphs were constructed.

In the graphs shown in Figure 22, for example, Voigt depicted the effect of stress on his sample. His circled points were calculated from the formula on *PW33¾°I*; he connected them by a "curve" drawn with dotted lines he claimed were "arbitrary." Like several other seminar students, he did not draw curves through his measured data points but connected points with straight lines. Although some of his "curves" gave a direct visual comparison between his empirical formula and the points he had actually measured, he did not explain in detail why one deviated from the other, nor did he explain how the deviations could be accounted for in the construction of a physical law. (At one point, on *PW33¾°I*, he used an exclamation point to signify his surprise over the deviation). Although Voigt resisted relying heavily on graphs for rigorous mathematical generalizations, he did use them for constructing qualitative conclusions, especially when comparing results from different samples. He probably could not have drawn such general statements from his tables because to be able to conclude, for instance, that two different samples had the same general elasticity pattern was something that could only be done visually from graphs. Yet, in the end, his qualitative generalizations from graphs were no more refined than those of former seminar students decades earlier.[55]

54. Voigt, *Untersuchung der Elasticitätsverhältnisse*, p. 35.
55. Ibid., pp. 35–41.

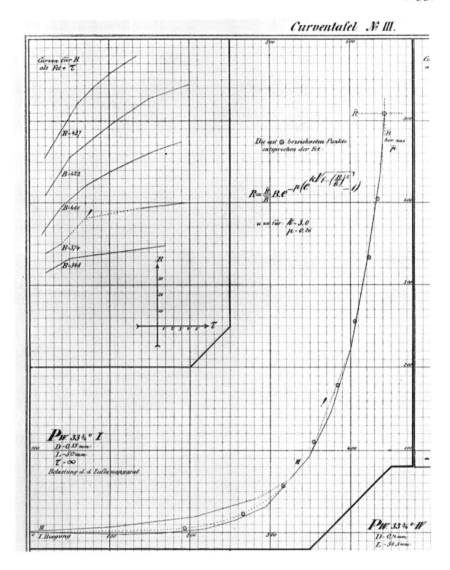

Figure 22. Voigt's "curve tables." From Voigt, *Untersuchung der Elasticitätsverhältnisse*, Curventafel III, courtesy of Library of Congress Photoduplication Services.

It is significant that when Voigt revised his investigation for publication, he relied less on graphical representation and more on numerical determination of certain elasticity constants. Still dissatisfied, he took up the problem again not nine days after the defense of his dissertation, when he began teaching secondary school in Leipzig. What Neumann had taught,

Voigt claimed, created a path for him.[56] If Voigt's reworking of his doctoral dissertation can be taken as the next set of markings on his investigative trail, then he defined that path in a curious way. Rather than following up on the suggestive physical implications of his investigation—as he put it, the causal dependency of thermal, optical, electrical, and other phenomena upon crystalline structure—or on the interpretation of his graphs, he instead concentrated on sharpening the precision of his data. Although his understanding of what happened in the experiment sharpened as well (he was able to describe the deformation more accurately and to show that Navier and Poisson had been wrong in assuming that a crystal could return to its natural state after stress had been applied), he singled out for special emphasis his protocol and his analysis of errors.

Even though he had not yet completed his move to Leipzig, he hoped as soon as possible "to devote a greater part of the day to the investigation." His teaching responsibilities, his marriage in July 1874, and the perfectionist spirit he brought to his measuring operations held up his revisions. But by November 1874 he became deeply involved again in making measurements and perfecting his protocol. Still he was not confident enough to bring the investigation to completion without asking Neumann if he had had any further thoughts on the matter. Voigt was at this time trying to understand how the torsion angle of the rock salt varied with the length of the sample; for if the relationship were linear, then his calculations would be simplified considerably. By December he realized that more detailed experimentation would be necessary in order to replace the approximate values he had for the torsion angle with more precise numbers. So he decided to perfect his method of observation, hoping to complete his experiment during Christmas vacation. A year later, though, he was still working on his calculations and had not overcome all the problems remaining in his experimental procedure. Thankful for Neumann's help but sensing he should have accomplished more on his own by then, Voigt apologized to Neumann for the slow pace of his observations, claiming they were time consuming.[57]

On 2 October 1875, after being named *Privatdozent* at Leipzig, Voigt was called to Königsberg as *Außerordinarius*. Thus his investigation of elasticity constants of rock salt (which appeared in the 1876 *Annalen*) marked not his first publication as a school teacher but the beginning of his career as an academic physicist. Deviating significantly from his doctoral dissertation, it still exhibited strong connections to the seminar and the style of work it promoted; to Neumann; and to other seminar students, especially Baumgarten. His continued pursuit of error not only enabled him to throw into sharper relief his finding that in some cases

56. Voigt to Neumann, 30 March 1874, FNN 53.IIA: Briefe von Schülern.

57. Voigt to Neumann, 30 March 1874, 11 June 1874, 12 November 1874, 16 December 1874, [?] December 1875, 22 December 1875, ibid.

rock salt did not return to its original state and that a permanent deformation could occur but also to revise significantly the values for certain elasticity constants. Voigt once again described meticulously his analysis of errors. He mentioned his reasons for neglecting errors of less than 1/1000 and for using the method of least squares. Driven to reduce the inconsistencies that had plagued the data in his dissertation, he went even further. Even though he had obtained reasonable agreement between values of elasticity constants predicted by Neumann's theory and his own observations, he still thought his results left "much to be desired." He found that the surface of the sample deteriorated in the course of successive measurements, adding to the errors of the experiment. He took into account the error that arose in reading a scale from behind the cross hairs of his microscope. So involved did he become in detailing the sources of error and his compensations for them that he become self-conscious. Not too far into his two-part paper, he explained, as he had done in his dissertation, that he had "communicated the preceding countless discussions of error in spite of their forbidding appearance in order to show that no precaution is neglected in freeing the final result from disturbing factors and, at the same time, to prove that the final result is of the greatest importance."[58] Voigt concluded that regular crystals do not behave elastically like uncrystalline media because the dilation and torsion coefficients are directionally dependent in crystals.

Like earlier seminar students, Voigt had to pull away from his data in order to make progress in his investigation. In contrast to his dissertation, where he used interpolation (but in a primitive way) to construct his graphs, now he relied on interpolation to construct a law that expressed the effect of stress on a crystal. He also used graphs differently. This time, the section devoted to the graphical representation of results was considerably shorter. Whereas he had previously depicted his results as approximate functions in two dimensions, now he represented the coefficients of dilation and torsion as surfaces in three dimensions and supplied the analytic formulas for both (Figure 23). Now maxima and minima could be easily determined either graphically or analytically. Indeed, it appears that because these "graphs" were so perfectly drawn and because Voigt gave no extended discussion of their reliability, they must have been constructed from his formulas and not from his data. In this sense they functioned as Voigt said they did: as an "illustration" (*eine anschauliche Vorstellung*) of his results.[59] Voigt used graphical representation only to draw rough qualitative conclusions; these graphs were pictures, no more and no less.

Voigt's experiences and investigations testify to the persistence of cer-

58. Kultusminister Falk to Voigt, 2 October 1875, WVN 1; Voigt, "Bestimmung der Elasticitätsconstanten," p. 21; cf. Voigt, *Untersuchung der Elasticitätsverhältnisse*, p. 7.
59. Voigt, "Bestimmung der Elasticitätsconstanten," pp. 205, 212.

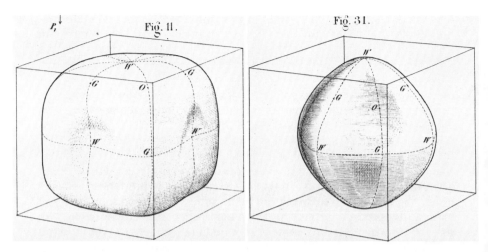

Figure 23. Voigt's representation of the coefficients of dilation (Fig. 11) and torsion (Fig. 31). From Voigt, "Bestimmung der Elasticitätsconstanten," Table I and Table II, courtesy of Library of Congress Photoduplication Services.

tain investigative strategies and research values in Neumann's seminar, especially to error analysis and the epistemological consequences it held for the interpretation of data. He represents the culmination of Neumann's pedagogical physics, vividly exemplifying the active role Bessel's seconds pendulum investigation still played in shaping the research techniques and values of Neumann's students almost fifty years after that investigation had been completed. Years later, Voigt continued to argue that in his measurement of gravity by the seconds pendulum, "Bessel gave an exemplary model for the union of the virtuoso art of observation with penetrating analysis."[60] But at the same time there is a sense of tragicalness in Voigt's adaptation of Neumann's style because problems in that style evident almost from the beginning had not yet been surmounted.

That doctoral students should strongly feel the influence of a mentor's way of doing things is to be expected. But Neumann's way of training students in the investigative techniques of physics also had a profound effect on students who worked less directly under his guidance. Carl Pape, for example, completed two years of postdoctoral work in the seminar and then left for Berlin, where he hoped to become *Privatdozent*. Despite the fact that the Berlin statutes stipulated that to get the *venia legendi* (the right to teach), the unanimous vote of the entire faculty was necessary, many young aspiring academics nevertheless sought to begin their careers there. Having been assured by Berlin officials that nothing stood in the way of his *Habilitation* as long as he completed the legal requirements (chiefly, a research investigation), Pape set to work in February 1861 on

60. Voigt, *Physikalische Forschung und Lehre in Deutschland während der letzten hundert Jahre* (Göttingen: Dieterich, 1912), p. 5.

the combustion of gunpowder, an investigation he had begun under Bunsen in Heidelberg. A new investigation, he argued to Neumann, "would have taken too long to finish." He was probably right. Even though this investigation "was very tiresome and time consuming owing to many numerical calculations," he was able to complete it in eight weeks.[61]

While he waited to see whether he would be able to teach at Berlin, Pape returned to an investigation on the specific heats of salt solutions that he had begun in the seminar at Königsberg; he planned to have the Berlin faculty consider this investigation as well. Eager to complete his calculations as quickly as possible, he was nevertheless held back by his inability to make certain measurements. Because his concentrations varied very little, the errors in his measurements were too large for him to discern a pattern in his data. "Unfortunately, the trials are not giving me the expected results," he told Neumann. "The conclusions that I want to draw out of them are not there." Undeterred, he set out to determine "the specific heats of a great number of salts which until now were not known." He was able to report to Neumann that "an interesting result appeared to me in the experiment: that the specific heats of salts with high water content change quickly with temperature [and increase] with temperature." Thus satisfied, he thanked Neumann for his friendliness in Königsberg.[62] Continuing his investigation of specific heats turned out to be worthwhile, for when Pape's hopes to teach at Berlin were not realized (partly because of internecine struggles at the university), he submitted it as his 1862 Göttingen *Habilitationsschrift*.

Pape chose to focus his investigation of the specific heat of salts on a comparison of the methods used by Neumann and the leading French experimentalist, Henri Victor Regnault, whose measurements of thermal constants were considered exemplary.[63] In Pape's view, the basis of Regnault's experiments had not been subject to the kind of scrutiny for which Neumann's seminar was known. Neumann's method of mixtures for determining specific heats guaranteed that all parts of the heated body reached the same temperature. Pape naturally used this method as the basis of his own investigation but corrected the mathematical formula for errors and modified Neumann's apparatus slightly to reduce heat loss to the surrounding air. Achieving results similar to Neumann's, Pape now compared Neumann's values to Regnault's. The difference between the two was regular: Regnault's measurements were consistently higher than Neumann's by 0.6–15.9 percent. "One must . . . assume," he concluded, "that either Regnault's or Neumann's method contains a constant error upon which the regular deviation depends."[64] Already convinced that

61. C. Pape to Neumann, 6 June 1861, 53.IIA: Briefe von Schülern.
62. Ibid.
63. Pape, "Ueber die specifische Wärme."
64. Ibid., p. 379.

Neumann's method accounted for all possible errors, he examined the conditions under which Regnault's measurements were taken. Noting that one error in Regnault's investigation consistently changed the value of specific heats, Pape reviewed Regnault's method, the construction of his apparatus, and the execution of his trials. He found that Regnault had been unable to guarantee, during the duration of his measurements, a constant temperature for either the substance or the liquid into which it had been plunged. Although Pape identified several procedural errors, he found the most problematic parts of Regnault's investigation to be in how closely mathematical assumptions represented experimental conditions. According to Pape, Regnault had "ascertained the correction [formula] through trials performed for this purpose in advance, represented [the correction] by an interpolation formula, and added the derived value to [the results of] every trial." Pape did not object to Regnault's procedure, but he did point out that "one had to assume that Regnault determined the interpolation formula out of a sufficiently large number of trials and was careful to produce exactly the same external conditions in all remaining trials."[65] Pape found, however, that the conditions of Regnault's correction might not be fulfilled in all of Regnault's trials.

Pape repeatedly demonstrated that Regnault had represented corrections by interpolation formulas "which [allowed] the correction to be determined out of two trials set up under similar conditions."[66] He suspected that Regnault could not really guarantee the constancy of experimental conditions. Without that guarantee, he deemed Regnault's corrections, and hence his results, unreliable. Having thus loosened the connections between the experimental and mathematical parts of Regnault's investigation, he then argued that "if one considers the various sources of error to which Regnault's observations are subject and their significant influence, then it appears doubtful if they correspond to the degree of accuracy which was hitherto attributed to them and if the result obtained . . . warrants the far reaching conclusions that Regnault drew out of them."[67] By so closely scrutinizing (along the lines of what had been done in seminar exercises) the relationship between theory and experiment in Regnault's investigation, Pape was able to cast doubt on Regnault's conclusion. Pape's examination of Regnault's experimental protocol seemed to open up a path to an alternative formulation of the theory of specific heats or to a deeper examination of the conceptual foundations of thermal phenomena. He did not, however, pursue either alternative and was content instead that his analysis of Regnault's errors constituted an investigation in itself.

In their fledgling investigations, Neumann's students seemed to derive

65. Ibid., pp. 581, 582.
66. Ibid., p. 593.
67. Ibid., p. 599.

a good part of their professional identity as practicing physicists from their ability to identify errors and assess data critically. For Voigt, Pape, and others who pursued physics in this way, successes were not very easy to come by, as their experiences demonstrate. Other former seminar students were unable, for various reasons, to follow so closely and execute so successfully the investigative strategies they had learned. Wilhelm Fuhrmann had one of the longest records of attendance in the seminar. From 1853 to 1858 he attended the mathematical division, and from 1854 to 1858, the physical division. He spent the next year as a private teacher, returning to the seminar for 1859/60. From all appearances, he should have been well trained, not to say confident. But shy and self-effacing, he reluctantly sought Neumann's advice in completing his seminar exercise on determining the attraction of small magnetic balls over a large distance. Hoping to be able to work undisturbed on the problem, he instead found himself without a place to work, without energy, without a "satisfying thought," and without a conclusion. Worse yet, "without coming into contact with colleagues, without other stimulation, and from time to time a despairing hermit," he wrote, "I now find it difficult to do that which at an earlier time would have been easier for me to do. Now it is the case that I no longer have my main work time, the morning." He had accomplished very little but wanted to write to Neumann "in order not to wander in the dark for too long." Fuhrmann persisted, but three months later he reported, "It is impossible for me to solve or to handle the problem in the way expressed by you, to make it possible to bring the result to the required practical application. I have reached this conclusion after many fruitless attempts, after which I saw that I was not making any progress."[68] Eager and willing to work but frustrated when he could not execute the problem so as to include a practical component, Fuhrmann never published his investigation. It was probably acceptable without measurements and the accompanying error analysis, but in his eyes, it did not measure up to seminar standards. Finding his goal—measuring observations—outside his reach, Fuhrmann expressed to Neumann the pathos of the professional training offered in the seminar.

Fuhrmann's experience was not unique. Practical measuring exercises and error analysis were the foundations of Neumann's pedagogical physics; without them, students knew that their work was below par, if not unprofessional. Measuring exercises and the analysis of errors were time consuming, and students often had to choose between a protracted investigation that included them or one of lesser quality that did not. Fritz Just, who attended the seminar from 1858 to 1861, when he began to work on the theory of the rainbow, wrote to Neumann asking if a simple comparison of theory and experiment, *without* his own observations, would

68. W. Fuhrmann to Neumann, 22 February 1859, 20 May 1859, FNN 53.IIB: Briefe anderer Schüler.

suffice for his doctorate at Königsberg. He explained that "to make my own observations is at the moment impossible owing to the lack of a locale and apparatus, although I find myself very strongly in need of conducting such observations." Time was clearly a factor; his position required that he "obtain the doctorate as soon as possible." Even though he held "promotion at [Königsberg] to be more honorable (in our subject at least) than at any other university," he nonetheless considered submitting his work to the Breslau faculty if Neumann would not accept it without his own observations, and he asked Neumann's opinion. He did get his doctorate in 1862 under Neumann, but whether he was required to conduct his own observations is unknown; for he only published a part of his investigation, its historical introduction, in a gymnasium program.[69]

Measurements and errors also beleaguered Karl Zöppritz, who, after spending from 1858 through 1861 in the seminar, began work on transverse elastic vibrations. Zöppritz's decision to study mathematics and physics troubled his family. The eldest son of a Darmstadt businessman, Zöppritz had attended a higher trade school in Darmstadt and then, with the intention of becoming a practical chemist, entered Heidelberg in 1856 to study with Bunsen. He also took courses with Kekulé, Hesse, and Kirchhoff. After taking Kirchhoff's course on theoretical physics, however, Zöppritz decided to pursue an academic career in mathematics and physics, subjects in which he had had private lessons after leaving trade school. "In consequence of this decision," he later wrote in his vitae, "I transferred to Königsberg in order to enter the famous mathematico-physical school under Neumann and Richelot." In addition to attending the seminar for six semesters, he attended Neumann's lecture courses on mineralogy, theoretical physics, the theory of light, the electric current, induction, potential theory, elasticity, and heat. Finished, he thought, with his course work, he began work on transverse elastic vibrations in 1861.[70]

That he was still working on calculations in late 1862 troubled his mother, who thought he was taking too long. She wondered if he had the ability to finish, and she asked Neumann if it would be better for her son to "be promoted at another university with a smaller investigation." Zöppritz's problem, he soon found out, was not his ability but a deficient background in mathematics; he had not registered for Richelot's section of the seminar. So even though at times he found his investigation "intractable," he persisted, hoping that by taking several courses between 1862 and 1864 with Richelot (differential equations, statics, dynamics, elliptic functions) and Luther (approximation techniques and the method of least squares), he would be able to complete his investigation. Still, he remained despondent over his work, frustrated by such matters as the "various approximation methods" and the "complications of algebraic

69. F. Just to Neumann, 24 June 1862, 2 July 1862, ibid.; Just, "Geschichte der Theorien des Regenbogens."

70. Zöppritz's vitae [July 1865], UAT 126/787.

calculations." He was particularly concerned about the "boring work" that took "many weeks" to compute a transcendental equation whose solution had to be found in order to calculate the elasticity constant out of the observed frequency of vibration. He succeeded, but then he had to turn to his observations, whose calculations, he estimated, would take him at least 168 days, or five and a half months, to complete.[71]

At this point, his doubts about his abilities, fears of overwork, and parental pressure (not to say the power of their suggestion) prevailed; reluctantly Zöppritz made a "quick" decision to leave Königsberg, a decision made all the more "difficult," he claimed, because he had not been able to talk to Neumann about it. Ashamed and embarrassed that he did not have the time to complete the calculations, the practical work, and the analysis of errors he had been taught were the core of a physical investigation, he went to Heidelberg, where he knew he could fulfill the requirements for a doctorate much more easily (Heidelberg did not require printed dissertations), even though it had been his intention all along to be promoted at Königsberg. He was nonetheless grateful to Neumann, so much so that he thought words could only "approximate the thanks and the respect" he felt. In the end, Zöppritz thought the best way to show his gratitude would be "to help promote the many directions of scholarship opened up by [Neumann]."[72]

Despite his difficulties, Zöppritz thought enough of his seminar investigation to use part of it for his *Habilitationsschrift*, which he hoped he could present at Heidelberg, or perhaps at Marburg. He preferred Heidelberg because the study of mathematics and the natural sciences had been strongly promoted there, whereas at Marburg he suspected there might only be a few students interested in mathematical physics. Neither prospect materialized, so in 1865 he presented his revised and deepened seminar investigation as his *Habilitationsschrift* to the Tübingen philosophical faculty. He remarked in its preface that although his theory had been completed three years earlier, it had taken him time to perform his calculations, to integrate the differential equations, and to construct a suitable approximation method for the measurements that would be made in experimental trials. Eventually, he published only the theoretical part of his investigation, developing the mathematical side of the problem and leaving out, for the time being, the observations and most of the calculations upon which the investigation was based. But Zöppritz was not ashamed to admit on the vitae he submitted to the Tübingen faculty that he had worked on transverse elastic vibrations since 1861.[73]

In their failures as well as in their successes, Zöppritz and the others

---

71. Frau M. Zöppritz to Neumann, 15 December 1862, FNN 53.IIA: Briefe von Schülern; K. Zöppritz to Neumann, 30 August 1864, ibid.

72. K. Zöppritz to Neumann, 30 August 1864, ibid.

73. K. Zöppritz, *Theorie der Querschwingungen eines elastischen, am Ende belasteten Stabs* [Habilitationsschrift] (Tübingen: Laupp, 1865); Zöppritz, "Theorie der Querschwingungen schwerer Stäbe"; Zöppritz's vitae [July 1865], UAT 126/787.

who worked on seminar investigations in the seminar's final decades demonstrate the impact and persistence of a set of related investigative techniques, all directed at the eradication of error through a rigorous analysis of data and of experimental instruments, conditions, and protocol. The most prominent characteristic of those investigations was what might be called a predilection for error over truth.[74] Eliminating or compensating for errors remained a central component of how Neumann taught theoretical physics. His students, in their fledgling investigations, translated their training into practice. Even as instruments improved during the 1860s and 1870s—reducing the range of errors afflicting data and generally producing more precise data—Neumann not only continued to emphasize the importance of calculating constant errors but also continued to teach the method of least squares as a means of expressing in quantitative terms the last remaining residue of error. As Bessel had maintained years earlier, Neumann and his seminar students did not believe that redesigning an instrument could reduce all errors. In their view, it could be futile to pursue the material perfection of instruments over the accurate mathematical determination of errors. "Owing to the nature of thermometer observations," Pape remarked, for example, "a greater accuracy than 1/100 part of the value sought is, at present, not to be attained."[75]

So great was the emphasis on the technical aspects of a physical investigation—such as correction formulas, error analysis, instrument construction, and reduction of measurements—that the execution of these techniques, methodically learned in seminar exercises, shaped the protocol of investigations and the format of published papers. Neumann's students extensively examined and compared the methods used by others; questioned if ideal conditions were met or modified in experiment; amended formulas with corrections; scrutinized instruments; and detailed their execution of observations.[76] There was no room for pure theory here. Even where the first step in a protocol was the derivation of a theoretical expression, the derivation was accomplished with an eye toward the clear expression of the variables to be measured.[77] From time to time, their published papers indicated that students kept laboratory notebooks in which they logged their observations. They studied these, not expecting to reach novel theoretical formulations on the basis of their

74. The emphasis on "error" was not peculiar to Neumann's physical seminar. Anthony Grafton, who has studied classroom practices in German philology seminars, has found in them a "preference for error over truth." I have borrowed Grafton's apt phrase in this passage and for the title of this chapter. Anthony Grafton, "Polyhistor into *Philolog*: Notes on the Transformation of German Classical Scholarship, 1780–1850," *History of Universities* 3 (1983): 159–92, on 181.

75. Pape, "Ueber die specifische Wärme," p. 378.

76. As, for example, in the investigations of Pape (ibid.) and Meyer ("Ueber die Reibung der Flüssigkeiten").

77. For example, Pape ("Ueber die specifische Wärme," p. 341), derived "the expression for specific heat as it is used in the calculation of observations."

data but to account for deviations in measurements and experimental conditions. From his records, Müttrich noted that "observations [of optical constants] conducted on different days deviated more from one another than could be explained by the errors of observation," so he was led "to consider these deviations as a consequence of the various temperatures under which the individual observations were conducted."[78]

So powerful, in fact, was the imperative to conduct an investigation with flawless technique that students were inhibited from rushing into print and felt compelled to apologize for omissions and imperfections. Often, as for Wangerin and Fuhrmann, their apologies were signs of the disparity between their image of the practice of physics and the reality of their workaday world. Rarely was there an apology for not sufficiently considering conceptual foundations or venturing further into the realm of mathematically expressed theory. Hence, these students tended not to render theory abstract but to acknowledge, as Meyer did, that a theory and its assumptions were confirmed only "to the limits of the accuracy of the observations."[79]

Seminar exercises inculcated what seems to be a curious paradox: theory was scrutinized and sharpened, but it was also skirted. For seminar students, theory remained something considered, if at all, only in conjunction with measurement. One could formulate a hypothesis on the basis of theory, but in the end, theory and experimental results had to agree in order to assert that the hypothesis "is fulfilled in nature."[80] Theory never functioned alone without consideration of the conditions under which it would become manifest. Even the corrections added to a mathematical formula were considered a "control of theory."[81] As the comparison between theory and experiment sharpened in these student papers, so did the difference between the two. Their investigations thereby contributed to the emergence of a distinctly theoretical physics in Prussia less by making contributions directly to theory than by setting standards for what was and was not a well-grounded theory. Their collective contribution to physics was thereby not insignificant; for theories pass, while these techniques, in contrast, show remarkable longevity (albeit not in theoretical physics). Neumann's seminar students may have overemphasized the need to measure constantly, but they showed by so doing how mathematical expressions in physics, and sometimes theory itself, not only gain a firm foundation in reality, but achieve exactitude. Their penchant for scrutinizing but skirting theory is therefore curious only from our perspective of theory (and even its history) as something primarily conceptual, contributing to a worldview or a unified set of prin-

---

78. Müttrich, "Bestimmung des Krystallsystems," p. 193.
79. Meyer, "Ueber die Reibung der Flüssigkeiten," p. 235.
80. Ibid., p. 83.
81. Ibid., p. 75.

ciples, and relatively divorced from experiment. Neumann saw it otherwise: "theoretical" meant the practical considerations one must bring to
bear on the confirmation of mathematical relations in physics.

That sense of "theoretical" dominated seminar instruction where a
technically competent, quantifiable, and accurate physics exerted a
powerful influence over how Neumann's students perceived the practice
of physics. "Work" in physics was, as it had been in seminar exercises, the
meticulous execution of technique. The mathematical techniques he introduced in his lectures and then reinforced through seminar exercises—
partial differential equations, spherical harmonics, elliptic and transcendental functions, and so on—were difficult to apply to physical phenomena, not just because the mathematics itself was difficult but because,
for each experimental situation, one had not only to decide which terms
were relevant but to frame the mathematical expression so that its constants and variables could in fact be measured: hence the importance of
approximation techniques and the determination of relevant terms in
series and other formulas. But first, in order to decide which terms were
relevant and which to cast aside, one had to have a good idea of the range
of error in the experiment, because the only terms one could delete from
a mathematical equation describing physical phenomena were those
smaller than the error afflicting the experiment. Reduce the error, and
terms cast off earlier could become important and might even modify a
theory, a fact Neumann's students knew from their own investigations.

Neumann's lecture courses were well known for the comprehensive
coverage of physical theory that they provided. They unified the separate
branches of physics by spreading common analytic techniques throughout its domain. Potential theory and elasticity theory were especially
important in this regard because they provided ways of treating different
classes of phenomena in mathematically analogous ways. Neumann's
lecture courses also promoted the critical comparison of alternative theories, which often resulted in a preference for Neumann's because it was
"simpler" and "more direct."[82] What his courses and especially his seminar exercises did little to promote was the use of hypotheses, which were
shown to guide physical thinking much less effectively than controlled
measuring observations.

In publishing Neumann's lecture courses, his students emphasized—to
the extent they could—the "pure" theoretical element in them. With the

---

82. As was said, for instance, about Neumann's theory of capillarity, which he introduced after he compared the capillary theories of Gauss and Laplace (F. Neumann,
Vorlesungen über die Theorie der Capillarität, ed. A. Wangerin [Leipzig: B. G. Teubner, 1894], p.
226). So, too, in his lectures on electricity, Neumann is reputed to have made the comment
that "when it is a matter of measuring the induced current, the potential law [one resem-
bling his own] leads immediately to the simplest expression, while in the application of
Weber's law, special conditions are considered which aggravate the calculations" (F. Neumann, *Vorlesungen über elektrische Ströme*, ed. K. Von der Mühll [Leipzig: B. G. Teubner,
1884], p. 308).

exception of the published version of his lecture course on electricity, and to a lesser extent his course on the theory of light, his students systematically suppressed the practical element of his pedagogical program, even in the course in which the practical element was most strategic and prominent: his course on mechanics as an introduction to theoretical physics. In contrast to the published versions of his lectures, notes taken in class show a consistent pattern: Neumann began with theory, turned to measurements for analytic refinements in the theory, and then returned to theory, now more accurately expressed.[83] Although making measurements was not strictly a part of the lecture courses, students were nevertheless shown *how* to make measurements in preparation for seminar exercises. The essence of Neumann's instructional program was always found in its practical elements.

There are many reasons why Neumann's lecture courses as well as his seminar exercises were not as conceptually innovative after 1850, the most obvious being his retreat from active research. Yet as important as stimulating research is to good teaching, his manner of instruction never depended entirely on his research interests. What the retreat from research seems to have meant is that he was no longer as closely in touch with the field as he had been. The critical perspective he sought to cultivate through a comparison of theories would have been well served by considering recent theories, but instead he compared the same theories (for example, the capillary theories of Gauss or Laplace or the theory of the induced electric current by himself or Weber) over and over again until they became paradigms in his curriculum.

A second, perhaps stronger, reason for the absence of innovation had to do with his student clientele. As we shall see, the bulk of Neumann's students were teaching candidates. Although candidates for the Prussian teaching examination after 1866 had to demonstrate that they could perform original research, that research did not necessarily have to be at the forefront of the field. It thus sufficed for Neumann to continue to teach the skills of research through paradigmatic examples drawn up earlier and to assign students small, unsolved problems in them rather than radically new, unsolved problems.

In addition to teaching candidates, the seminar attracted students with a more "professional" interest in physics, including postdoctoral students in physics, mathematics, chemistry, and medicine. Although the number of these more serious students was never as high as those who intended to teach secondary school, they were nonetheless an important component of his clientele because it was this group that Neumann purported to want to reach. But in what way? Neumann was never able to rely on *Privat-*

83. As, for example, in FNN 6: Vorlesungen über Wärmelehre: Heinrich Wild [Wintersemester 1854/55]. Compare Wild's notes to the more rationalized and theoretical presentation of Neumann's lectures on heat by Carl Neumann in FNN 5: Carl Neumann's Manuscript über die mechanische Wärmetheorie von F.N.

dozenten or even Moser to take care of elementary courses in physics. By the 1860s the largest portion of Moser's clientele came from the medical faculty, and he instructed these students with consideration "that they must complete their study of the natural sciences before they visit the clinic."[84] So the courses he offered them were designed as self-contained units, what we would call survey courses today. Moser's program thus never developed the hierarchical ordering or the conceptual variety that Neumann's had, nor was it even an adequate introduction to what Neumann wanted to teach. The particular way in which the student clientele in physics divided itself at Königsberg meant not only that Neumann had the luxury of teaching advanced students but also that he had the responsibility of introducing them to physics. It is thus understandable that Neumann viewed the function of an *ordentlicher* professor in part in terms of the "systematic instruction" he had to offer.[85]

Yet despite the need to return periodically to introductory issues, Neumann did occasionally incorporate new material. But by and large this new material touched upon the practical rather than the conceptual aspects of physics. For example, although Neumann ceased to publish after 1850, he continued to observe and measure and, having developed an interest in the mechanical theory of heat, wanted to refine thermometric procedures. His extensive notes on the topic indicate that he was trying to unite theoretical and experimental considerations by considering the work of Robert Mayer and Rudolph Clausius in conjuction with the experimental determinations of James Prescott Joule and with measurements taken with refined pyrheliometers and thermometers. As early as 1853 he is reported to have developed a method for determining the mechanical equivalent of heat. His lecture course on heat from 1854/55 reveals that he approached the subject through measuring operations (two-thirds of the course concerned the experimental determination of temperature and thermal constants) and reserved the mechanical theory for the final section.[86]

It was the few practical and technical innovations of Neumann's research that made their way into his seminar exercises. Between 1859 and 1862 he perfected a method for determining the internal and external thermal conductivity of a substance. On 4 March 1862 his former seminar student Rudolph Radau asked him to explain his method in more detail, mentioning in passing that the difference between Neumann's and Ångström's results appeared "very great."[87] Ångström's method for de-

84. Königsberg University Curator to Kultusministerium, 13 December 1861, ZStA-M, Naturw. Seminar Königsberg, fols. 228–31, on fol. 228.

85. [Neumann's recommendations concerning the replacement of the *außerordentlicher* professor of mathematics at Königsberg], 1856, rpt. in LN, pp. 440–41, on p. 441.

86. FNN 4: F.N.'s Aufzeichnungen zu Wärmelehre; FNN 6: Vorlesungen über Wärmelehre, Heinrich Wild [Wintersemester 1854/55].

87. R. Radau to Neumann, 4 March 1862, FNN 53.IIA: Briefe von Schülern.

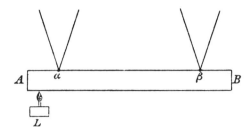

Figure 24. Neumann's
apparatus for measuring the
thermal conductivity of metals.
From *GW* 2:140.

termining the thermal conductivity of metals involved measuring the
periodic temperature changes in a bar, one end of which was kept at the
temperature of its surroundings while the other was alternately cooled
and heated.

Neumann's method, in contrast, used a metal bar (*AB* in Figure 24)
heated at one end by a flame, *L*, and to which two thermal circuits were
connected at α and β.[88] A differential multiplicator was linked to the two
circuits in such a way that both the sum and the difference of the currents
in them could be measured. Neumann was able to calculate the internal
conductivity of the bar from the difference of the currents; the external
conductivity from the sum. Among the advantages were that only two
constants had to be determined in comparison to Ångström's seven; that
time-dependent temperatures could be measured; that "no kind of deformation [of the bar] takes place which cannot be rigorously accounted for
in the calculation"; and that "by means of this method the absolute value
of the internal and external conductivity can be simultaneously determined." He reported his results to Radau but could not explain why they
differed from Ångström's. Neumann knew his method would not produce rigorous results if conductivity varied with temperature because, as
he put it, "the mathematical theory of heat in relation to the dependency
of conductivity upon temperature is still *entirely incomplete*—and perhaps
will remain incomplete for a long time."[89]

With two new alternative measuring techniques in hand, Neumann
introduced them into his seminar exercises in 1866/67. After developing
Fourier's equations in the first semester, he turned in the second semester
not only to techniques for measuring time-dependent temperatures but
also to Ångström's method for determining conductivity, which he had
students duplicate and assess.[90] So even when revising his curriculum,

88. Neumann, "Ueber eine neue Methode zur Bestimmung der innern und äussern
Wärmeleitungsfähigkeit" [letter to R. Radau, 10 March 1862], in *GW* 2:137–42; draft in
FNN 53.IIA: Briefe von Schülern; original in Darms. Samml. Sig. Flc(2) 1841.
89. Neumann, "Ueber eine neue Methode," p. 142. Radau published Neumann's
method as "Experiences sür la conductibilité calorifique des solides," *Annales de chemie et de
physique* 66 (1862): 183–87; rpt. in *GW* 2:143–46. See also "An den Neumann'schen Untersuchungen sich anschliessende Beobachtungen von Radau," in *GW* 2:146–59.
90. Draft report, physical division, 1866/67, FNN 48: Seminar Angelegenheiten.

Neumann preferred to integrate newer and more rigorous methods for measuring and determining errors rather than to introduce conceptual or theoretical innovations.

Given that his seminar exercises emphasized the practical elements of physics and that Neumann refined his curriculum primarily in terms of them, it is not at all surprising that his students, immersed for the first time in the study of physics, tended to conduct their investigations in ways that sought greater degrees of technical perfection. Success in the seminar was certainly measured by the accuracy of observations, but it was also gauged by finesse in technical performance. A student did not necessarily have to solve a problem, just work his way through it. As Neumann explained to the ministry in 1870/71, "Naturally the members of the seminar did not often obtain the solution to the suggested problem, but I believe [that I] prepared them better in this way than through idle endeavors."[91] By emphasizing technical mastery over conceptualization, he promoted the pursuit of means over ends, of technical competence over creative ideas or grand syntheses. His method of instruction might well be seen as a part of what Max Weber later termed rationalization, the movement toward increasing clarity and calculation, which Weber thought was accompanied by a disenchantment or loss of meaning or purpose. It might also be viewed in terms of what Jean-François Lyotard has called the emphasis upon "performativity" in modern science.[92] But to view his students' investigations as merely *mechanically* executing the techniques of physics would be to miss the point Frederic L. Holmes has noted, that technical refinements require *creative* imagination no less than conceptual innovation does.[93]

After 1870 the seminar endured interruptions. Owing to the Franco-Prussian war, Neumann, in the winter semester of 1870/71, had to offer exercises in theoretical physics because "the older members of the seminar were called to the field." In the last years of the seminar, Neumann, now over seventy (Figure 25), fell ill frequently. He now taught only one course a year, in the winter semester. The pace of activity in the seminar slowed as well. For 1873/74 he submitted only a brief, one-page report to the ministry, not mentioning how many students participated. During the summer semester, students worked on standard themes in capillary theory, such as motion caused by capillary forces and the angle made by a liquid in a narrow tube. In the winter semester Neumann chose familiar problems from hydrodynamics; he had his students describe the motion

91. Draft report, physical division, 1870/71, ibid.

92. Max Weber, "Science as a Vocation," in *From Max Weber: Essays in Sociology*, ed. H. H. Gerth and C. Wright Mills (New York: Oxford University Press, 1946), pp. 129–56; Jean-François Lyotard, *The Postmodern Condition: A Report on Knowledge*, Theory and History of Literature vol. 10 (Minneapolis: University of Minnesota Press, 1984), esp. pp. 41–52.

93. Frederic L. Holmes, "Carl Voit and the Quantitative Tradition in Biology," in *Transformation and Tradition in the Sciences: Essays in Honor of I. Bernard Cohen*, ed. E. Mendelsohn (Cambridge: Cambridge University Press, 1984), pp. 455–70, on pp. 455, 467.

Figure 25. Neumann in his study. From LN, courtesy of
Library of Congress Photoduplication Services.

of a fluid, taking into account internal viscosity.[94] He did not hold seminar
exercises in the summer semester of 1874.

The last available seminar report, remarkably brief, covered the winter
semester of 1874/75. Nine students participated—eight from mathematics and one from philology; all thus appear to have taken the seminar in
preparation for the state teaching examination. Most were new students;
for Neumann lectured concurrently on theoretical physics. In the seminar, he first conducted exercises on the theory of heat and developed a
"method for comparing thermometric instruments with consideration of
the variations in their dimensions." Students developed the formulas for
the motion of heat between discrete masses for the determination of
latent heat; measured thermal constants and reduced data; and learned

94. Draft reports, physical division, 1870/71, 1873/74, FNN 48: Seminar Angelegenheiten.

how to use Pouillet's pyrheliometer. Then Neumann returned to the foundations that had initially made the seminar possible—in this, his last recording of the seminar's exercises: "The work of the seminar members in the second half of the winter semester was bound closely to my lecture course on theoretical physics. Thereby I intended to make familiar through simple applications the ideas and methods developed in the lectures. At the same time, these applications were supposed to amplify further many points which were only touched upon in my lectures." Seminar members derived a complete theory of Atwood's machine, taking into consideration all disturbing conditions (such as roll), and then developed the theory of vertical motion in a resisting medium by following Newton's derivations. Students discussed related issues such as the ballistic pendulum. Neumann supplied "old observations" so that they could calculate pendulum velocity and compute the effect of air resistance. Methods for measuring small intervals of time with an electric current followed, and "finally the barometer was treated and a complete theory of the motion resulting from air pressure, when the barometer hangs on the arm of a scale, was developed."[95] All these exercises were simple, but it was just that simplicity that made them, over the years, convenient vehicles for teaching more sophisticated techniques in observation, measurement, and the mathematical techniques of physics.

In 1875 when Richelot died, Neumann lost a devoted colleague who, as his partner in the seminar, had for so long taught physics students some of the mathematical skills they needed to understand Neumann's lectures. Especially for students whose mathematical abilities were lacking, Richelot's lectures and exercises were important adjuncts. One student, Heinrich Martin Weber, remembered Richelot's lectures as "not systematically worked out, . . . not easy to understand, and [demanding] a great deal of extra work by the student" but "packed with material." Weber's recollection may have been singular; for Richelot worked hard, as his seminar reports showed, to construct a foundation for the study of mathematics. Weber also remembered how easy it was to talk to Richelot.[96] Richelot's sensitivity to the needs of students, in mathematics as well as physics, made Neumann's section of the seminar stronger; Neumann's success in teaching a once novel mathematical physics was due partly to their cooperative efforts. Despite his loss, Neumann directed the seminar for one more year, 1875/76, but he did not submit a report. Anticipating retirement, he had begun negotiations in late 1874 to bring back to Königsberg as director of the seminar his last doctoral student, Woldemar Voigt.[97]

When asked by university officials in April 1876 why a physical labora-

95. Draft report, physical division, 1874/75, ibid.
96. Weber's recollections are partially reprinted in Lorey, *Das Studium der Mathematik*, pp. 97–98, on p. 97.
97. Voigt to Neumann, 16 December 1874, FNN 53.IIA: Briefe von Schülern.

tory was necessary at Königsberg, Neumann drew upon his experience in teaching physics for fifty years, forty-two of them in the seminar. His conviction that a physical laboratory was a place where primarily critical acts of judgment were taught and executed remained unchanged. He especially cited the need for the academic physicist to be able to perform the techniques of practical physics so that he could "confirm" the investigations and observations of his colleagues and "create an independent judgment about the value of and limits to the reliability of their results." But it was as a center for training (*Ausbildung*) in physics that Neumann considered a physical laboratory especially important. In a draft of his statement on the necessity of a physical laboratory, he explained that instruction in physics

> must satisfy a two-fold purpose. First, it must furnish the knowledge and views in physics which belong to the general education [*allgemeine Bildung*] of those who want to devote themselves to one of the various natural sciences—chemistry, mineralogy, botany, and so on—or who want to study medicine. [Second,] instruction [in physics] must take care of those who want to become physicists. The first task can be accomplished through the lectures on experimental physics. These lectures form at the same time the foundation for a more thorough study of physics. But they are not suitable for those who want to be trained as teachers of physics at institutions of higher learning, or for physicists by calling. The lectures on theoretical physics are designed for these people. Here the exact mathematical connections of phenomena, which experimental physics links loosely together, are developed as far as science has succeeded in doing until now.[98]

In this one paragraph Neumann brought together several of the complementary issues with which he had grappled over the past fifty years: theory and experiment; the role of mathematics in physics instruction; student clienteles; teacher training and professional instruction in physics.

But, he went on to say, one could not be content with offering only lecture courses in experimental and theoretical physics. Instruction in physics had, in his view, an "essential component": practical training (*Ausbildung*). Physics instruction, he believed, was "incomplete" without learning "the handling of apparatus, the art of observation, methods for measuring, methods for deriving numerical elements out of phenomena, and so on," all of which required a physical laboratory. Although he had been able to offer practical exercises in his seminar, he knew then, as he had known earlier, that physics instruction at Königsberg was not all it could have been. Lacking a laboratory, he had advised several of his students to complete their investigations elsewhere if they could afford to

98. Draft recommendation found in FNN 61.7: Kampf um das Laboratorium. Cf. Neumann, "Nothwendigkeit eines physikalischen Laboratoriums," in LN, pp. 455–56.

do so.[99] But even without a laboratory, as we have seen, Neumann had succeeded in impressing upon his students the central importance of practical techniques. Theirs never was an armchair physics. In learning physics his students became acquainted with the workaday world of that science.

Among the remarks Neumann made in his recommendation, the most significant concerned *"Physiker von Beruf,"* physicists by calling. By pairing instruction in theoretical physics with the shaping of practicing physicists, Neumann acknowledged the pedagogical, professional, and intellectual superiority of theoretical over experimental physics. Theoretical physics at Königsberg had by then come to mean the combination of his set of lecture courses, which spanned the branches of physics, expressed the phenomena in them mathematically, and unified the field of physics, with the set of techniques cultivated in his seminar exercises, which gave quantitative expression to the errors in an experiment, set the limits within which theory was confirmed, and generally made possible the critical assessment of theory in quantitative terms.

Neumann used the phrase *"Physiker von Beruf"* only in the draft of his recommendation. In the final version he claimed that theoretical physics was for "those who wanted to be trained as physicists [*die sich zu Physikern ausbilden wollen*]."[100] He must originally have been thinking of *Beruf* in older terms, as an inner, personal calling. Had he thought of *Beruf* in modern terms, as profession, there really would have been no need to reframe his ideas in terms of *Ausbildung*, or training, because it was incorporated in the notion of profession. The replacement of *Beruf* by *Ausbildung* points, I believe, to the continuing vacillation and uncertainty in his own mind about whether he had in fact accomplished his original goal of creating a training program that replaced inclination and talent as a way of regenerating the ranks of the discipline. In principle he seems to have accomplished his goal; for his lecture courses and his seminar exercises constituted precisely the training program he had initially sought. In practice, however, the results—the number of this seminar students who practiced physics as a livelihood—were less than he had perhaps hoped for.

The sociologist Max Weber later wrote that an academic *Beruf* was characterized by the execution of increasingly precise scholarly techniques.[101] In that sense, Neumann's seminar contributed to the professionalization of physics through its emphasis on techniques of quantification, especially those that assessed errors. Through his work and his teaching, he contributed significantly to the institutionalization in Prussia of a mathematical physics built in part on French style and methods, but

99. Ibid.
100. Neumann to [Königsberg University Curator], c. April–May 1876, rpt. in Wangerin, *Forscher und Lehrer*, pp. 183–85, on p. 185.
101. Weber, "Science as a Vocation."

from a technical and epistemological perspective Neumann's pedagogical physics surpassed that of the French. The quantitative techniques associated with error and data analysis, which were either not considered or incompletely used in French investigations, proved crucial to legitimating and refining French results, to confirming and extending theory, and especially to training students.

Despite the successes of Kirchhoff, Von der Mühll, and Frölich in conducting investigations that were largely theoretically and mathematically oriented, Neumann's exercises had been drawn less from a French way of doing physics than from Bessel's way of analyzing an experiment. His students directed their efforts at quantification less at shaping new theoretical ideas and expressions that might be fulfilled under ideal circumstances than at closing the gap between what could be expressed mathematically and what was actually realizable in practice under laboratory conditions. The construction of the laws of nature was thus a matter of balancing, in quantitative terms, the ideal and the real. For his students, the proper object of a physical investigation was located somewhere between perfect abstract expressions and imperfect experimental conditions. To practice physics in their terms was a matter of deploying the mathematical and experimental techniques that could temper mathematical expressions and refine experimental protocols. Demonstrating that one had been "called" to physics was thus a matter of reckoning again and again the "impurities" in experiment as well as in theory by quantifying error, thereby gaining higher and higher degrees of both precision and accuracy.

Whereas in 1838/39 Neumann had created his course on mechanics as an introduction to theoretical physics in order to teach beginning students the mathematical and measuring methods of physics, by the time of his retirement in 1876 he identified theoretical physics with almost the entirety of his pedagogical program. In the interim he had introduced the practical elements of his course on mechanics into other lecture courses. This tactic proved decisive in shaping theoretical physics for a pedagogical setting. It was not just that Neumann's lecture courses synthesized ideas in physics, thus expanding the conceptual boundaries of theory in physics pedagogy. It was that the stability and even the impact of his pedagogical model rested largely on the certainty and exactitude believed inherent in the methods of the Besselian experiment. Neumann's unplanned and unanticipated introductory course thus became the template for his pedagogical physics, the repository of the techniques defining the labor of science for his students, and most important, a source of the values guiding the search for scientific truth.

PART II

PRACTICE

CHAPTER NINE

The Workaday World
of *Physiklehrer*

SECONDARY SCHOOL TEACHING was the most popular career choice among students who attended Neumann's seminar. As Table 2 illustrates, sixty-four students (42.7%) remained in secondary schools. Another nineteen (7.9%) taught secondary school at some time during their careers (Table 3). The fluctuation over time of the ratio of temporary to permanent teachers roughly follows the expansion and contraction of occupational opportunities in Prussian science. During the 1830s, when teaching was just about the only profession open to students trained in physics and mathematics, almost all students who began their careers in teaching remained there. By the 1850s, however, when Prussia had a surplus of teachers, temporary teachers outnumbered permanent ones as those who started at the lower ranks in secondary schools failed to obtain permanent positions. The proportionally greater number of teachers from 1860 to 1876 reflects both an enrollment expansion and growth of secondary schools. Despite the greater need for mathematics and science teachers in *Realschulen* over the years in which the seminar was in existence and especially after 1859, when *Realschulen* achieved parity with gymnasiums, former Königsberg seminar students are not overly represented in them; gymnasium teachers outnumber *Realschullehrer* by a ratio of about 3 to 2.[1]

---

1. The sixty-four permanent secondary school teachers were Baumgarten, Behr, Besch, Bock, Böttcher, Brandis, Czwalina, E. F. J. T. Ebel, Ellinger, Feyerabendt, Fleischer, Fritsch, Fuhrmann, R. H. H. Hagen, Haub, Haveland, Heideprim, Heinemann, Hermes, Hoffmann, Hossenfelder, E. L. H. F. Hübner, O. H. J. E. Hübner, Hutt, Jänsch, Just, Kade, Kiessling, Kleiber, Kostka, Krueger, Lampe, Lange, Lautsch, Matthias, K. O. Meyer, Michelis, Milinowski, Mischpeter, Momber, Mörstein, Nicolai, Noske, Oelsnitz, J. P. G. Peters, Pietzker, Powel, Radicke, Reuter, Rumler, P. Sanio, T. Sanio, J. A. H. von Schäwen,

Physics as a Calling

Table 2. Professions of Neumann's seminar students

	Number	Field	Number
		A. University	
Ordinarius	19	Physics	8
		Mathematics	7
		Chemistry	3
		Geophysics	1
Außerordinarius	1	Mathematics	1
Privatdozent	1	Mathematics	1
Total	21 (14.0%)		
		B. *Technische Hochschule*	
Ordinarius	5	Physics	2
		Mathematics	3
Privatdozent	1	Mathematics	1
Assistant, Physical Institute	1	Physics	1
Total	7 (4.7%)		
		C. Vocational school (all types)	
Director	2	Mathematics	1
		Unknown	1
Professor	2	Physics	2
Lehrer	1	Physics	1
Total	5 (3.3%)		
		D. Precision industry, engineering, statistics	
All occupations	6	Physics	5
		Physics and Astronomy	1
Total	6 (4.0%)		
		E. Astronomy	
Observatory director (nonacademic)	2	Astronomy	9
Observatory assistant (3 academic, 1 nonacademic)	4		
Astronomer (1 academic, 2 nonacademic)	3		
Total	9 (6.0%)		
		F. Medicine	
Außerordinarius	1	Physiology	1
Practicing physician	1	Physics	1
Total	2 (1.3%)		
		G. Secondary schools	
Director/rector	7		
Professor	36		

Table 2. Continued

	Number	Field	Number
Oberlehrer	11		
Lehrer	9		
Mittelschullehrer	1		
Total	64 (42.7%)		
		H. Other teaching/research	
Professor, Swiss foundation	1		
Private teacher	3		
Archaeologist	1		
Total	5 (3.3%)		
		I. Other	
Agriculture	2		
Commerce	2		
Land surveyor	1		
Librarian	1		
Music	2		
Postal service	1		
Early deaths	6		
	15 (10%)		
		J. Unknown	
Total	16 (10.7%)		
TOTAL	150 (100%)		

School teaching generally began with a probationary year during which a candidate worked as an unpaid instructor, followed by a temporary appointment as *wissenschaftlicher Hilfslehrer* (assistant teacher). Market conditions dictated when a teacher became *ordentlicher Lehrer* (ordinary teacher), in which position he generally taught several of the lower forms. Teachers of the two upper forms, *Secunda* and *Prima*, were called *Oberlehrer*; they constituted the upper echelon. *Oberlehrer* who distinguished themselves through the publication of scholarly works received the title *Professor*, which held the same civil service ranking as an *außerordentlicher*

P. C. A. von Schäwen, Scheeffer, Schindler, Schlicht, Schönemann, J. H. C. E. Schumann, J. H. E. Schumann, Thalmann, Wegner, Wiesing, Wittrien. The nineteen students who were temporarily employed as secondary school teachers were Albrecht, Amsler-Laffon, Brix, Clebsch, Dorn, Friedrich, Frölich, Hesse, Müttrich, C. Neumann, Reichel, Schinz, Schoch, Schroeder, Sohncke, Thiesen, Voigt, Wangerin, Wild. There may have been others who taught in secondary schools. G. Arnold, who attended Neumann's seminar between 1858 and 1860, sought Neumann's advice for the written part of the state teaching examination (FNN 52: Lose Blätter), but it is not known if he ever took the examination and taught.

Table 3. Secondary school teachers from Neumann's seminar

| | Decade of departure from seminar | | | | | |
	1830s	1840s	1850s	1860s	1870s	Total
Temporary	1	5	6	5	2	19
Permanent	8	7	2	29	18	64
Total	9	12	8	34	20	83

Professor at a university. The highest position in a secondary school was that of director or rector, for which scholarly qualifications were important; after the 1850s it increasingly became a political appointment. Almost two-thirds of the permanent teachers among Neumann's former seminar students advanced to the highest professional levels (professor and director or rector), where they were rewarded for their scholarly achievement with reduced class hours and the opportunity to teach physics and mathematics in the upper forms. Career advancement in administration was easier in *Realschulen* than in gymnasiums. Although teachers in gymnasiums outnumbered those in *Realschulen,* six of the seven who attained the position of rector or director worked in *Realschulen.* (Directors at gymnasiums were generally humanists, most frequently philologists.) Rank determined salary and class hours, both of which varied by location despite attempts at standardization. Directors generally had 14–16 class hours; *Oberlehrer,* 20–22; and ordinary teachers, 22–24. When class size increased after the 1860s, often to over thirty students, and longer hours became more commonplace, it was more difficult for teachers to contribute to scholarship. Hence using teaching as a steppingstone to another career became less likely as the century wore on.[2]

Although Neumann had sensed the tensions that could arise between scholarship and teaching in the secondary school, he nonetheless cultivated close ties to the secondary school community in Königsberg and helped his students obtain positions, continue their research, and maintain contacts with a scholarly community. Other academic scientists were not as favorably disposed. When, in 1822, Bessel's student August Rosenberger was forced to seek a teaching position for economic reasons, Bessel thought that the position "would distance [him] entirely from his studies in astronomy." After Bessel's son Wilhelm spent 1835/36 in the seminar, Bessel noticed that Wilhelm knew "so much [pure mathematics] that he could be one of those admirable individuals around here who preen as *Oberlehrer*" and who had very high opinions of themselves. In his worry that Paul du Bois-Reymond would become a teacher, Helmholtz too

---

2. R. Bölling, *Sozialgeschichte der deutschen Lehrer: Ein Ueberblick von 1800 bis zur Gegenwart* (Göttingen: Vandenhoeck & Ruprecht, 1983), esp. pp. 30–35.

expressed his reservations about the profession. Yet other professional opportunities were not so abundant that students of mathematics and physics could neglect the state teaching examination without suffering the consequences. Carl Pape, who was not successful in habilitating at Berlin in 1861, decided a year later to pursue teaching rather than "the uncertain position of *Privatdozent*." But he had not taken the teaching examination, so he asked Neumann both for a letter testifying to his ability to teach physics and for help in arranging to take the examination at a later date. Pape managed in the meantime to habilitate at Göttingen instead, but he viewed school teaching much as others had: as a "secure position."[3]

Teaching School Physics

The predominance of secondary school teaching among the professions of former Königsberg seminar students upholds the traditional historical interpretation of seminars—that they existed primarily to train teachers—but it also raises questions about the significance of seminar exercises and investigations for teaching candidates who would not teach physics at the level they had learned it and who would have limited opportunity to conduct advanced research. To be sure, a secondary school teacher did not have to view seminar participation in such narrowly utilitarian terms; for teaching could lead to other professions where specialized knowledge was more in demand, as it did for the nineteen seminar students who taught secondary school for anywhere from one semester to ten years. Of these, eventually six entered university positions, seven entered trade or technical schools, five engaged in professions that in some way related to precision measurement or that required techniques of quantification, and one when into commerce.[4] But their

3. F. W. Bessel to Kultusministerium, 26 December 1822, ZStA-M, Acta betr. die Sternwarte bei der Universität zu Königsberg, Rep. 76Va, Sekt. 11, Tit. X Nr. 16, Bd. II: 1814–25, fol. 143; F. W. Bessel to K. E. von Baer, 10 December 1836, von Baer Nachlaß, UBGi-Hs. Despite Helmholtz's admonishments against becoming a teacher, Paul du Bois-Reymond taught secondary school in Berlin until 1865, when he finally became *Privatdozent* in mathematics at Heidelberg (Paul du Bois-Reymond, vitae, Personalakt, UAF). Carl Pape to Neumann, 9 January 1862, 26 March 1862, 5 April 1862, and 24 September 1863, FNN 53.IIA: Briefe von Schülern. On the perceived security of secondary school teaching see, e.g.; A. Amsler and F. Rudio, "Jakob Amsler-Laffon," *Vierteljahrsschrift der Naturforschenden Gesellschaft in Zürich* 57 (1912): 1–17, on 2; and J. Böttcher to Neumann, 29 April 1872, FNN 53.IIB: Briefe anderer Schüler. Among the students who relied on Neumann's contacts with the secondary school community was R. Radau; see R. Radau to Neumann, 21 October 1870, 7 November 1870, FNN 53.IIA: Briefe von Schülern.

4. C. Neumann taught secondary school while holding a university teaching position; Clebsch, Dorn, Hesse, Voigt, and Wangerin moved from teaching into academic positions. Albrecht, Müttrich, Reichel, Schinz, Schoch, Schroeder, and Sohncke obtained positions at technical high schools or vocational schools. Amsler-Laffon, Brix, Frölich, Thiesen, and Wild went into precision industries and related concerns. Friedrich went into commercial pursuits.

mobility depended in part on the fact that most of them (16 out of 19) had a doctorate, whereas only one-fourth (16 out of 64) of the permanent teachers did.

In spite of the conceptual distance between the school and the seminar, advanced training in physics was of great value to permanent teachers; for it gave them the resources with which to shape their workaday world. The seminar's emphasis on an exact experimental physics, including error analysis, proved especially influential in helping these teachers to enhance existing and emerging pedagogical practices in secondary schools. Independent of any effort on their part, measuring exercises (not necessarily with error analysis) were assuming a more prominent role in Prussian secondary schools.[5] It was also primarily an exact experimental physics that enabled these teachers to sustain identities as physics practitioners.

My stress on the importance of the seminar experience for the professional lives of teachers trained at Königsberg is not intended as an argument in favor of educational determinism. It is meant to acknowledge the crucial role of education in delineating for prospective science teachers the fluid boundaries of what is possible both pedagogically and intellectually in their professional lives. After all, these teachers identified scholarly instruction with the process of becoming a teacher. As Johann Pernet wrote to Neumann in 1868, it was "at the advice of Herr Dr. H[einrich] Wild" that he had "decided to study theoretical physics under your experienced guidance in order to be trained as a teacher of physics."[6] Others did the same, including Gustav Louis Baumgarten and Johannes Böttcher, both of whom had studied with Karl Von der Mühll and Carl Neumann in Leipzig, and Ernst Schroeder, who had studied with L. O. Hesse, Robert Bunsen, and Gustav Kirchhoff at Heidelberg. Not all, however, were so attracted to physics; some took the seminar to prepare for teaching in the upper forms of secondary schools, or even to enhance their teaching of natural history in the lower forms.

Sometimes a student did become more deeply engaged in physics than he had intended. Louis Saalschütz, for instance, had been interested in the exact sciences since *Prima*, but initially he directed his efforts primarily at learning mathematics at the university. After spending some time in Neumann's seminar he found himself drawn by an "inner inclination" to study mathematical physics. Still he struggled between his old love and his new attraction, because he later claimed that it was "half against his will" that the physics portion of his *Oberlehrerarbeit*, on geothermal measure-

5. Kathryn M. Olesko, "Physics Instruction in Prussian Secondary Schools before 1859," *Osiris* 5 (1989): 94–120.

6. Johann Pernet to Neumann, 4 October 1868, FNN 53.IIA: Briefe von Schülern. On the historical study of *Physiklehrer* see Kathryn M. Olesko, "The Mental World of *Physiklehrer*: Subject and Method in History of Mentalities," *Recherches en didactique des mathématiques* 6, nos. 2/3 (1985): 347–62.

Table 4. Secondary school teachers identifying with physics

	Semesters in physics division			
	1–2	3–4	5+	Total
Number of permanent teachers	27	21	16	64
Number who published in physics	5	9	10	24
Percentage of the total	18.5	42.9	62.5	37.5

ments taken at Neumann's makeshift geothermal station, "proved suitable as a doctoral dissertation," completed under Neumann.[7] What proportion of these *Physiklehrer* had chosen physics as their specialty beforehand and what proportion were called to it as a result of their seminar experience cannot be determined with certainty. Even though most students enhanced their qualifications as teachers of physics by attending the seminar, more than one-third of the permanent teachers also published in physics (Table 4); many of them had drawn upon Neumann's seminar exercises as a source of problems for their original investigations. For these teachers, the seminar experience thus helped in maintaining professional identities as physicists. Yet among the 24 who identified with physics through research, there was no greater preference for working in *Realschulen*, where physics courses were more numerous, than among the entire population of secondary school teachers. Of these 24 teachers, 15 taught primarily in gymnasiums; 9, in *Realschulen*.

Neumann's seminar trained teachers mainly for a rather small geographic area, in and around East Prussia. His connections with the secondary school community in Königsberg proved especially useful in helping his students begin their careers at one of the seven schools in that city: the Löbenichtsche Schule and the Realgymnasium auf der Burg, both semiclassical schools, and the Friedrichs Collegium, Altstädtisches Gymnasium, Kneiphöfisches Gymnasium, Königliches Wilhelms-Gymnasium, and Königliches Waisenhauses-Progymnasium. Of the 64 permanent teachers who issued from the seminar, 28 taught at Königsberg schools at some time during their careers; 19 of those 28 remained permanently. Of the 19 temporary teachers, 5 also taught in Königsberg. Former seminar students especially enhanced the progressive science curricula of the Friedrichs Collegium, the Realgymnasium auf der Burg, the Löbenichtsche Schule, and the Altstädtisches Gymnasium. At the Altstädtisches Gymnasium, for instance, Anton Müttrich's father, Johann August Müttrich (who had studied under Bessel) had established a tradition of advanced and rigorous courses in physics and mathematics. One of Neumann's first seminar students, J. H. C. E. Schumann, joined Müttrich in 1844. After Müttrich's death in 1858, his son Anton, also Neumann's

7. Louis Saalschütz to unknown, 4 November 1910, Darms. Samml. Sig. H 8 (1870).

student, took his place and taught there until 1866. Neumann's former student Albert Momber taught there from 1867 to 1876, as did Louis Hübner from 1876 to 1883. The influx of new blood reoriented the school's *Abitur*, which after 1858 included more physics questions in the mathematics section.[8] Their common educational background undoubtedly helped these teachers work effectively together, especially in coordinating physics and mathematics classes. Together they also formed an important support network for seminar students in their probationary period of teaching.

"Every young man who feels called to scholarship," Max Weber wrote, "has to realize clearly that the task before him has a double aspect. He must qualify not only as a scholar but also as a teacher. And the two do not at all coincide."[9] Weber's remark was intended for academics, but with a slight change it applies equally well to school teachers: they had to qualify not only as teachers but also as scholars. Although not all school teachers aspired to an academic life or even to a career punctuated by research, research was unavoidable in their professional lives. It defined their preparation for teaching, and it was essential for professional advancement, even though their daily performance consisted entirely in teaching physics in simpler ways for a few hours and in teaching other subjects, not always the sciences and mathematics, for quite a bit more. In the Prussian gymnasium curriculum, the number of hours per week devoted to all natural science instruction in all forms decreased in the first half of the century, from 20 hours in 1812 to 14 in 1856. In 1882 it rose to 18, but physics was still taught for 2 hours per week in both years of *Prima* and one (sometimes 2) in both years of *Secunda*. Out of an average weekly teaching load of 25 hours, a *Physiklehrer* taught physics for 6 at most. But class time designated for physics could be used for other sciences such as natural history, anthropology, and later in the century, physiology. Science instruction was much more intense in *Realschulen*, where physics classes were over twice as long: 3–4 hours per week in *Prima* and *Secunda* and 2 hours on average in *Tertia*.[10]

So the first boundary that defined the workaday world of these teachers was created by the state's regulation of class hours in the secondary school curriculum, and more specifically by the distribution of hours between the sciences. The experiences of these teachers varied widely; often official duties crowded out the few hours that might have been available for

8. *Programm*, Altstädtisches Gymnasium, Königsberg, 1885/86.

9. Max Weber, "Science as a Vocation," in *From Max Weber: Essays in Sociology*, ed. H. H. Gerth and C. Wright Mills (New York: Oxford University Press, 1946), pp. 129–56, on p. 133.

10. Max Nath, *Lehrpläne und Prüfungs-Ordnungen in höheren Schulwesen Preußens seit Einführung des Abiturienten-Examens* (Berlin: Pormetter, 1900), pp. 99–104; Peter Lundgreen, *Sozialgeschichte der deutschen Schule im Ueberblicke*, Teil 1, *1770–1918* (Göttingen: Vandenhoeck & Ruprecht, 1980), pp. 324–32.

research. Lampe, at the Danzig Städtisches Gymnasium in the 1870s, was responsible for two hours of physics instruction in *Prima* and upper and lower *Secunda* but also taught fourteen additional hours of mathematics, arithmetic, and natural history to the lower forms. As had been customary at many Prussian schools for some time, mechanics and mathematical geography were taught in *Prima*; Lampe alternated these subjects with optics and the wave theory of light. In *Secunda* he taught heat, galvanism, general physics, static electricity, and magnetism. In contrast, Ellinger, at a Tilsit *Realgymnasium* in 1849, taught primarily mathematics, also natural history from time to time. He did not teach physics until 1870, and then only temporarily.[11]

Across the city from Ellinger, Schindler began working at a gymnasium in 1860 after having spent his probationary year at the Friedrichs Collegium in Königsberg. Within a year he was promoted to *ordentlicher Lehrer* with a salary of 450 taler, and in this position for only several weeks in 1861, he became *Oberlehrer* because of a colleague's death. Schindler now had the responsibility for teaching all physics and upper mathematics courses, which totaled twenty-seven class hours per week. Within a short time two more hours were added to his teaching schedule. Despite the necessary reduction in office hours that came with the increase in class hours, *Prima* students, he reported, still insisted on consuming his office hours. Overwhelmed and overworked, he felt "physically and intellectually incapable" of extending his teaching any further. He considered himself to be in a "very dangerous position" because, as a young teacher, his "teaching energy can thereby be used up in a few years." Rather than becoming discouraged, Schindler turned the situation to his advantage. A year later he reported to Neumann the *Tertia* classes in natural history had been added to his schedule. Instruction in natural history, Schindler believed, was the "holy obligation of every physics teacher in the gymnasium"; for in reaching the lower forms, he shaped a "ready and interested public" for his physics courses. He found astronomy and physical geognosy especially suited for introducing *Tertia* students to the study of science because these sciences, he thought, were more intuitive. He prepared for his physics classes enthusiastically, constructing instruments with materials obtained from the Tilsit Polytechnic Society, where he was an active member. He also took his students on class trips so that they might see for themselves the way in which physical principles were manifested in daily life. *Secunda* students listened to a local organ; *Prima* students visited a telegraph station and a paper factory, where they learned about the operation of the steam engine, air pump, hydraulic press, and a centrifugal machine.[12]

11. *Programme*, Städtisches Gymnasium zu Danzig, 1869–72; *Programme*, Realgymnasium zu Tilsit, 1849/50, 1869/70.

12. H. Schindler to Neumann, 9 June 1861, 10 July 1861, 2 November 1862, FNN 53.IIB: Briefe anderer Schüler.

Teaching physics as if he were at a *Realschule* rather than a gymnasium, Schindler must have infected his students with an enthusiasm for what they were learning. Whereas a year earlier he had misgivings about the time he spent with students during his office hours, now he stressed to Neumann how important it was to talk to students so that he might achieve "success" with a few. But he also knew that his teaching took its toll on his research. In his last letter to Neumann he confided that "when I think, however, of my further scientific concerns . . . then I am lacking nothing but everything here: the time . . . ; the scientific apparatus . . . ; and above all, colleagues in my subject with whom I can speak about difficult points." Although there were two teachers at the Tilsit Realschule whose interests overlapped his own, Schindler found them too busy for collegial discussion.[13] Eventually he moved, first to Elbing and then to Brandenburg, Frankfurt, and finally to Berlin, where, while employed at the Joachimsthal Gymnasium, he operated a meteorological station. Undoubtedly he drew his students into it, showing them how to take barometric and thermometric measurements and teaching physics through measuring operations as Neumann had done.

It was not easy to give physics instruction the time and preparation some teachers would have liked. Eduard Hübner was fortunate in bringing physics lessons to *Tertia* at the Memel Gymnasium in 1874, but Gustav Louis Baumgarten found it difficult to survive professionally at the Nicolai Gymnasium in Leipzig. Primarily responsible for natural history and geography, Baumgarten had forty-seven or forty-eight students in each of his four classes; he did not find it easy to teach "such masses." His close friend Johannes Böttcher, also a former seminar student, who worked at a Leipzig *Realschule*, had, Baumgarten reported, seven mathematics courses (at all levels) with no more than thirty-five students per course. Although Baumgarten was pleased by the "tight discipline" in his school, he nonetheless believed that his research suffered with such heavy teaching. Optimistic, he expected "an improvement in the future." Still at the Nicolai Gymnasium in 1873, he found that he had been able to complete investigations begun in the seminar, but he felt "overburdened" because a fourth mathematics teacher had not been hired. "The 36 hours [of teaching] per week completely absorb my time and energy," he told Neumann. Not even on vacations could he do research, so much was he in need of rest.[14]

Luckily, by December 1873, Baumgarten reported that the Saxon ministry had offered him a position at a new Dresden gymnasium "with very good terms." Not only was he asked to teach mathematics and physics in the upper classes and thereby to become an *Oberlehrer*, he was given what

13. Schindler to Neumann, 2 November 1862, ibid.
14. *Programm*, Memel Gymnasium, 1874/75; G. L. Baumgarten to Neumann, 28 June 1871, 31 December 1873, FNN 53.IIB: Briefe anderer Schüler.

every *Physiklehrer* wanted: his own physical cabinet. With only eighteen hours per week of instruction, he felt encouraged about his research, the theoretical part of which he had "made little progress on" during 1873.[15] Baumgarten did manage to achieve a satisfying balance between teaching and research, so much so that when he left Leipzig, he was praised for both his scholarly and pedagogical talents. His replacement was Neumann's last doctoral student, Woldemar Voigt, who remained there less than a year, having been called as *außerordentlicher* professor of theoretical physics to Königsberg. The director of the gymnasium admitted that Voigt was moving to a "working environment" more satisfying than a secondary school.[16]

Capturing the professional lives of these teachers in greater detail would involve observing them in action in the classroom. But unlike the therapeutic activity of physicians, which has been admirably shown by John Harley Warner to be accessible through hospital patient records and physicians' case histories, the study of secondary school classroom practices can draw upon no correspondingly rich source that depicts a day of teaching in fine detail.[17] Textbooks and official curricula can at best approximate the intellectual texture of a year-long physics course. In Prussia from the late 1830s to the 1870s, the recommended physics curriculum followed closely what Lampe had taught and included primarily heat, electricity, galvanism, magnetism, and other subjects that could be treated without mathematics in *Secunda*; *Prima* included mechanics, optics, and other mathematically oriented topics such as mathematical geography and astronomy. As areas of physics became quantified, mathematical topics from *Secunda* moved into *Prima* until finally, by the 1880s, sections of mechanics were moved down to *Secunda*, where they could be used to introduce students to the more difficult mathematical problems in physics. Recommended curricula never carried great weight, however, and teachers could offer topics in physics that suited their own inclinations and the needs of their students.

Textbooks could be more influential in determining what was taught, but with them, too, local variations occurred. In Prussia between the 1820s and the 1840s, E. G. Fischer's *Lehrbuch der mechanischen Naturlehre* and F. Kries's *Lehrbuch der Naturlehre für Anfänger* were popular. Fischer emphasized the use of instruments and laid the basis for understanding physical phenomena mathematically; Kries, who had been educated in Heyne's philological seminar at Göttingen, stressed the contributions of physics to formal *Bildung*. From the 1840s to the 1860s, the most popular textbooks were H. A. Brettner's *Leitfaden für Unterricht in Physik* in gym-

15. Baumgarten to Neumann, 31 December 1873, FNN 53.IIB: Briefe anderer Schüler.
16. *Programm*, Nicolai Gymnasium zu Leipzig, 1875/76.
17. John Harley Warner, *The Therapeutic Perspective: Medical Practice, Knowledge, and Identity in America, 1820–1885* (Cambridge, Mass.: Harvard University Press, 1986).

nasiums and J. Heussi's *Die Experimental-physik, methodisch dargestellt* in *Realschulen*. Although Brettner's textbook promoted an abstract understanding of physics based on mathematics and diagrams and Heussi's stressed a more intuitive understanding of physical phenomena based on experiment, both were often used in the classroom in ways that belied their orientation. By the late 1870s, when classroom experiments were more in evidence, the ministry recommended E. Jochmann's *Grundriß der Experimentalphysik*.[18]

The fine structure of the daily classroom activity of *Physiklehrer* is distorted if not made completely inaccessible by textbooks and official curricula. One gets little insight, for instance, into how Königsberg-trained *Physiklehrer* utilized quantification of all types in physics lectures; this lack makes it especially difficult to draw correlations between their professional training in the seminar and their teaching. Even if such information were available, it is unlikely that the practices of former seminar students were distinctive enough to have influenced significantly the character of secondary school physics instruction. The period from the 1830s through the 1860s saw the emergence of "chalk physics" characterized by the enhanced role of mathematics in physics instruction, a process that had been aided by textbooks such as Fischer's. One indication that the mathematical orientation of physics classes had indeed changed was the widespread appearance of problem sets. Initially these were viewed as a way to re-create mentally the creative process of discovery (including experimental discovery), but as more books devoted exclusively to physics problems (and their solutions) for students appeared, problem sets evolved into routine exercises in the execution of the numerical, mathematical, and experimental techniques of physics.[19]

When Königsberg-trained teachers entered their first teaching positions in the late 1830s and early 1840s, problem sets and homework were gaining ground in popularity and on their way to becoming standard in physics instruction. Composing, assigning, and correcting problems were important parts of the workaday activity of *Physiklehrer*. A mathematical approach to physics was especially important in *Prima* classes, where in addition to mechanics, two other subjects taught from the 1830s to the 1870s demanded a strong mathematical treatment: mathematical geography and astronomy. Not only did learning both mean having a familiarity with sophisticated mathematical techniques in geometry and trigonome-

18. Olesko, "Physics Instruction," pp. 110–12.

19. On problem sets, see Carl Gützlaff, "Bemerkungen über den mathematischen Unterricht auf Gymnasien," *Programm*, Kgl. Gymnasium zu Marienwerder, 1843/44, p. 11; "Physikalische Aufgaben," *ZPCU* 1 (1887): 110–11; C. T. Anger, "Grundzüge der neueren astronomischen Beobachtungskunst," *Programm*, Städtisches Gymnasium zu Danzig, 1846/47; Friedrich C. Kries, *Sammlung physikalischer Aufgaben nebst ihrer Auflösung* (Jena: Frommann, 1843). School programs began to mention problem sets in the physical sciences during the 1840s; see, e.g., *Programm*, Stadtschule zu Schwedt, 1842/43, and *Programm*, Petri-Schule zu Danzig, 1846/47.

try; it also meant becoming acquainted with apparatus such as the telescope and with measuring instruments such as goniometers, chronometers, and surveying instruments. How much quantification occurred in physics classes varied greatly, however, with location. In *Realschulen*, where one might have expected a strong mathematical approach, it was not until 1859 that graduates were required to develop natural laws mathematically and to express qualitative relationships in quantitative terms. And in both *Realschulen* and gymnasiums, it was not until the last two decades of the century that two of the most useful pedagogical aides for the mathematical analysis of physical relationships—graphical analysis and the geometric and algebraic methods upon which it was based—were widely introduced into secondary school instruction. Related techniques in diagraming and drawing, useful for rendering the construction of instruments and the progression of operations in experimental protocol, were introduced about the same time.[20]

So while Neumann's student von Mörstein claimed to have offered his physics courses at the Königliches Wilhelms-Gymnasium in Königsberg from "the present mathematical perspective," he could not have meant much more than deriving and expressing physical laws with simple calculus, and he probably ended up doing less than that. The same holds for Paul Peters's claim that at the same gymnasium in 1883/84, his *Prima* class in physics emphasized "the most difficult sections of mechanics." Highly unusual was the success of Louis Hübner in teaching his own elegant mathematical derivations of orbital deviations in planetry motions, which he incorporated into his physics course at the Memel Gymnasium in 1874/75.[21]

It was the distinctive feature of Neumann's seminar instruction—its promotion of an exact experimental physics emphasizing error and data analysis—that proved to have the greatest impact on how his students taught physics in secondary schools. But even here his former students merely enhanced existing trends rather than shape new ones, because measurement had secured its role in the secondary school curriculum by the late 1850s, and by a decade later, physical laboratories began to appear in Prussian secondary schools.[22] From the 1830s to well beyond

20. See, e.g., *Programm*, Elbing Gymnasium, 1831/32; *Programm*, Städtisches Gymnasium zu Danzig, 1836/37. On *Realschulen* see R. Tagmann, "Zur Realschulfrage. Zweiter Artikel: Die Unterrichts- und Prüfungs-Ordnung vom 6. Oktober 1859," *Programm*, Realgymnasium zu Tilsit, 1859/60. On graphs and drawings in science instruction see Heide Inhetveen, *Die Reform des gymnasialen Mathematikunterrichts zwischen 1890 und 1914* (Bad Heilbrunn: Klinkhardt, 1976), p. 192; E. Wiedemann, "Die Wechselbeziehungen zwischen dem physikalischen Hochschulunterricht und dem physikalischen Unterricht an höheren Lehranstalten," *ZMNU* 26 (1895): 127–40, on 131; Otto Ohmann, "Ueber die Anwendung der zeichnenden Methode im naturwissenschaftlichen Unterricht des Gymnasiums," *Programm*, Humboldt Gymnasium zu Berlin, 1898/99, p. 21.

21. *Programme*, Kgl. Wilhelms Gymnasium zu Königsberg, 1874/75–1876/77, 1883/84; *Programm*, Memel Gymnasium, 1874/75.

22. Olesko, "Physics Instruction."

the end of the century, physical cabinets at secondary schools gave *Physik-lehrer* the best opportunity to tailor their physics courses to their own interests. We learn about them from annual school programs that often reported on additions to cabinet holdings. Apparatus and instruments, both for teaching and research, were the material expression of the workaday world of the *Physiklehrer*. In the hands of several Königsberg-trained teachers, these resources provided the opportunity to go beyond the standard fare of demonstration experiments and measuring exercises and even to introduce their students to the role of error analysis in the practice of physics.

While a member of Neumann's seminar when it first opened in the winter semester of 1834/35, Theodor Schönemann had worked on the balance and weighing techniques, including corrections for constant errors. What were at first routine seminar exercises became for Schöne-mann later in his career a way to practice physics; for although most of his publications were in mathematics—an achievement for which he received the rank of professor at the Brandenburg Gymnasium—he turned by the early 1850s to two related investigations: measuring procedures for weight and water pressure. He developed a method for using small weighbridges to make precision measurements of momentary pressures. Discontent with the custom in physics instruction of using the hypo-theticodeductive method as an exemplar of scientific practice, he now found an alternative approach in his technical achievements. He published his results on weighbridges so as to publicize their pedagogical importance. He believed that "the purpose of introductory instruction in physics" was to involve students "as directly as possible" in experimenta-tion. His own weighbridge, he argued, was exactly the kind of instrument that students should learn to use. He called it an "indispensable instru-ment in the physical cabinet," and to make sure that *Physiklehrer* were clear on how he intended that it be used, he showed how to compute its experimental errors.[23] Although Schönemann's hopes were rather op-timistic (his weighbridge was a refined and delicate instrument not espe-cially suited for introductory students who knew little about physics), the priority he placed on exact experiment demonstrates the lingering influ-ence of the primacy of error analysis in learning what physics was all about.

At the Leipzig Realgymnasium, Johannes Böttcher explored how deeply one could actually engage students in the quantitative aspects of experimentation. Before he arrived in 1871, the school had acquired a superb instrument collection that included many measuring instruments. The physics classes of *Prima* and *Secunda*, in which the theory of modern machines such as the steam engine and the telegraph were discussed in

23. Theodor Schönemann, "Ueber den Gebrauch empfindlicher kleiner Brückenwaagen für physikalische Zwecke," *Archiv der Mathematik und Physik* 24 (1855): 264–85, on 265.

the context of an otherwise ordinary physics curriculum, were year after year "illustrated by countless experiments." Lacking a designated laboratory, the school turned for assistance in 1865 to the local Polytechnic Society, which permitted Realgymnasium faculty to use one of its laboratories, a room equipped with an experimental table (complete with gas jets), and two storage rooms with cabinets for equipment. School officials were pleased that "the room corresponds so closely to its purpose."[24]

At first Böttcher taught only mathematics, but by 1875 he assumed responsibility for physics in *Tertia* and lower *Prima*. Rather than accept the standard curriculum for *Prima*, he discussed the "significance of experiment," emphasizing the distinction between "what is observed" and "what is concluded." In the meantime, he worked on a classic Königsberg topic, the circular pendulum, the results of which he published the following year in his school's program. For his classes he focused on simple problems, such as free fall, believing that the contrast between the simplicity of the observation and the complexity of the required analytic treatment (both theoretically and experimentally) would stimulate critical thinking in his students. In an 1884 essay on how students could observe the path of the sun in the sky, he revealed that he had grappled with the problem of understanding observation in his private classes (indicating that some students, at least, were motivated to go beyond classroom material) before he introduced similar matters in mathematical geography into his *Prima* classes. To the question of "what could all students find in the sky themselves, without calendars, textbooks, and artful instruments," he answered: the shadow cast on a flat surface by an object (somewhat like a shadow cast by a sundial). Böttcher directed his students' attention to the symmetric curve marking the progression of the shadow. The task he assigned them was "to examine accurately the configuration of the curve." Invoking Kirchhoff's statement that the purpose of physics was to describe physical phenomena as accurately as possible, Böttcher urged his students not to be content with the "shadow curve" but to "strive to describe" the process of the daily motion of the sun at first simply and then completely. He admitted that the assignment was not easy, but because his students knew descriptive geometry, he thought they "had fun" completing it. The important point about this exercise, though, was that Böttcher was directing his students to quantify physical phenomena not by analogy or other theoretical means but by beginning with independent observations that required critical scrutiny.[25]

Among the most well known *Physiklehrer* in the second half of the nineteenth century was Karl Johann Hermann Kiessling. Between 1858

24. *Programme*, Realgymnasium zu Leipzig, 1862/63–1870/71, quotations from 1863/64 and 1865/66.

25. Ibid., 1875/76, p. 46; Johannes Eduard Böttcher, "Die Bewegung eines Kreis-Pendels auf rotirender Pendel-Ebene," *Programm*, Realgymnasium zu Leipzig, 1876/77; idem, "Beobachtung des Sonnenlaufs durch die Schüler," *ZMNU* 16 (1885): 161–80.

and 1863 he studied at Göttingen, Halle, and Königsberg. He spent his last four semesters of training between 1861 and 1863 at Königsberg in Neumann's seminar. Before he took the state teaching examination in Berlin in 1864, he assisted in teaching the upper classes at the Kneiphö-fisches Gymnasium in Königsberg. He then spent his probationary year at the Joachimsthal Gymnasium in Berlin (where he joined the Berlin Physical Society and worked in Magnus's laboratory), was called to teach in Schleswig-Hölstein in 1867/68, entered military service, and then was appointed to the Gelehrtenschule des Johanneums in Hamburg (a classical gymnasium) in 1870, only to be drafted into the Franco-Prussian war. He finally began teaching there in 1871, remaining at the school until he retired in 1902, when he moved to Marburg, where he could maintain contact with that city's scientific community.[26]

One of the reasons Kiessling had been called to Hamburg was that the *Prima* class had grown to over forty students, an enrollment the director considered too high. Kiessling was fortunate in having at his disposal newly granted funds for physical equipment. His purchases were so extensive as to indicate that there must have been little equipment present before his arrival. Most of what he chose was for demonstration experiments, but he included several sensitive apparatus and instruments: balances, apparatus for determining specific gravity, barometers, and apparatus for measuring the constants of electrical, magnetic, and optical phenomena. In the following years, Kiessling added to the collection; taught a very mathematically oriented physics course; and even integrated topics from physics into his *Tertia* science classes. When a colleague, also responsible for science instruction, retired in 1876, Kiessling, who had by then published three scientific papers in school programs and one in the *Annalen*, was given the title Professor.[27]

Undoubtedly buoyed by his new position as the senior science teacher, Kiessling introduced a novelty into physics instruction in 1876. Student participation in learning physics through problem sets and measuring exercises was by then not unusual. What Kiessling did was to take student instruction one step further and introduce optional exercise sessions outside normal classroom hours: two hours a week in physics and two in chemistry for students in *Prima* and upper *Secunda* who had an interest in the natural sciences. The chemistry session provided a "systematic course in the fundamentals of chemistry as well as analytic exercises," while in the physics session, students gained experience in the "practical handling of the most important measuring instruments in physics as far as is possible with the presently still very incomplete apparatus on hand in the

26. Kiessling's obituary is found in *Programm*, Gelehrtenschule des Johanneums (Hamburg), 1905/6, pp. 2–4.
27. Ibid., 1868/69–1876/77.

instrument collection." Registration, although optional, was twenty-five in 1876/77.[28]

In the following year Kiessling's regular classroom instruction in physics turned more decisively toward measuring operations, while in the physics exercise session he taught the experimental determination of physical constants, exercises in the use of the microscope, and polarization experiments. So important had this exercise session become (in 1877/78 twenty-four students registered for it) that in 1878/79 he handed over his *Prima* and *Secunda* physics courses to an *Oberlehrer* and devoted all his attention to his exercise sessions, in which thirteen students had registered for the summer semester, fourteen for the winter semester. Kiessling later taught *Prima* and *Secunda* again, but the exercise sessions stayed intact. In 1881, when the gymnasium was pressed for space, the physical cabinet moved to Kiessling's residence. In 1883 the *Oberlehrer* who worked with Kiessling taught students the methods for taking observational errors into account.[29]

For the remainder of the century, Kiessling continued to be an outstanding teacher; to promote student participation in experimentation and measuring exercises (he considered experimentation to be the center of gravity of physics instruction); to articulate for other *Physiklehrer* the materials needed for physics instruction; and to argue for exercise sessions like the ones he had directed in Hamburg. He even pressed for the introduction into physics instruction of the concept of a function and its graphical representation, and for showing students the difference between mathematical constants (such as e, pi, etc.), which were "defined with refined accuracy," and physical constants, "the accuracy of which depends upon the size of the observational errors." Reflecting on the manner in which he had been instructed in Neumann's seminar, Kiessling maintained that it was "scarcely justified" to begin physics instruction with those parts of mechanics that dealt with the properties of matter because so many hypotheses entered into consideration. Instead he believed that the richest introduction to physics was through the repetition of measuring exercises and independent observations.[30] Kiessling thus not only taught several methods of an exact experimental physics but did so in a way that reflected the values cultivated in Neumann's seminar.

The parallels between the teaching styles of Schönemann, Böttcher, and Kiessling on the one hand, and Neumann's on the other, suggest that these former seminar students drew upon their educational experiences

28. Ibid., 1876/77, pp. 6–7, 19.

29. Ibid., 1877/78–1884/85.

30. J. Kiessling, "Physik," in *Handbuch der Erziehungs- und Unterrichtslehre für höhere Schulen*, ed. K. A. Baumeister, Bd. 4.2, Sekt. X (Munich: Beck, 1898), pp. 1–73, esp. pp. 4, 8, 24–25, 26–30, 32.

in creating their workaday world. The abundant documentary evidence on the teaching practices of these three is more a peculiarity of the interests of their school directors, who saw fit to write about them in school programs, than it is of the singularity of their achievements. They were not alone in revamping physical cabinets and instruction according to their image of the practice of physics. At the Danzig Städtisches Gymnasium, Lampe, who determined the pressure in the water lines of Danzig, had at his disposal several pieces of demonstration equipment as well as measuring instruments such as a small balance with an attachment for determining specific gravity, and a manometer, none of which were standard equipment in gymnasiums. By the late 1870s he had also acquired a rheostat, a galvanometer, an Edison phonograph, an interference prism, and a host of other items. At the Rastenburg Gymnasium in the 1870s, Hermann von Schäwen similarly added apparatus to that school's physical collection. During his first year there, Schäwen, who was then investigating the direction of vibration of ether particles in polarized light, ordered for the collection not only such standard equipment as a water balance and a differential thermometer but also several instruments related to his optical investigations, including a stereoscope, prisms, crystals, and an apparatus for performing double refraction experiments. After teaching for six years in Altmark, Hermann Wiesing moved in 1871 to the Nordhausen Gymnasium, where he immersed himself in building up the school's physical cabinet. A year after his arrival he remodeled an unused room into a laboratory for physical exercises. The school not having the funds to pay for the materials, he successfully solicited support from local patrons. For six years he built up the physical cabinet and upgraded laboratory instruction. These achievements undoubtedly contributed to his next professional offer, to be director of the Nordhausen Realschule, which had a long and rich tradition of demonstration and laboratory exercises and where both quantitative and qualitative elements were illustrated in experiments.[31]

Measuring exercises and especially error analysis occupied less time than did lectures in physics courses, and hence they were never as important as more traditional methods of teaching. Schönemann, Böttcher, and Kiessling did not have the opportunity to influence their students to the degree that Neumann had. Still, they considered accurate measuring, error analysis, and the epistemological values associated with them important enough to include in their courses. Measuring exercises and error analysis constituted a skill-based reform of physics instruction that stressed a practicotechnical mastery and competence rather than a conceptual or philosophical one. Significantly, their relevance to physics

31. *Programme*, Städtisches Gymnasium zu Danzig, 1871/72–1879/80; *Programme*, Rastenburg Gymnasium, 1870/71–1875/76; *Programme*, Nordhausen Gymnasium, 1872/73–1877/78.

instruction, as well as their claim to scientific truth, were challenged at the end of the century.[32]

Error, Experiment, and Professional Identity

Improvements in physical cabinets and creative use of local resources helped *Physiklehrer* to engage in research and to mold a workaday world more in harmony with their training. But the initial formulation of problems worth investigating as well as the techniques used to solve them were drawn largely from seminar experiences and even from continued contact and correspondence with Neumann. Yet the Königsberg connection, so important in helping to maintain an identity as physics practitioners, was also a continuing source of frustration. Neumann had presented error and data analysis as essential instruments of persuasion in a scientific argument. Outside an academic setting, though, measurements and error analysis were often difficult to carry out. They were also difficult to apply; often they were not even relevant to the problem at hand. When addressing a disciplinary community in a scholarly journal, *Physiklehrer* took Neumann's lessons so seriously that they applied these rigorous techniques to problems whose solutions could have been achieved with much less effort and not as much concern for epistemological certainty. But in the principal organ of publication available to *Physiklehrer*, annual school programs, error analysis was rarely used to convince a more diverse audience of the reliability of results.

The omission of measurements and error analysis from school programs was in part a matter of expediency rather than a lack of initiative. The scholarly essays accompanying school programs since 1824 in Prussia were often written by the school's youngest teachers, who were fresh from the university and burdened by their first year of teaching. Some of Neumann's former seminar students, however, tried to hold out before publishing their investigations so that they could include measurements and error analysis; often they could not. Gustav Kade's first and only publication in physics, an expository essay on refraction, without measurements, which appeared in 1842, was drawn from seminar exercises from the late 1830s on the eye, the optical properties of crystals, and Goethe's theory of color. In the decade between 1864 when Eduard Hutt entered the seminar and 1874 when he finally published his study on methods for determining magnetic inclination, Hutt occupied himself with gathering and reducing his data. Writing in 1866 to ask Neumann to review in particular his methods for determining the magnitude of the disturbances produced by pure iron and an electric current, Hutt viewed his approach as the proper way to strive "for the greatest possible gen-

---

32. Olesko, "Physics Instruction," p. 120; see also below, Chap. 10.

erality of treatment." But under pressure to produce an essay for his school's program and unable to complete this investigation on time, he published instead in 1868 a mathematical study on the quadrature of an elasticity surface. Interested in the direction of vibration of ether particles in polarized light, Hermann von Schäwen attempted to synthesize the ideas of Neumann and several former seminar students, including Carl Neumann, Karl Von der Mühll, and Georg Quincke, but considered his study inconclusive without additional data on other media. He nevertheless published his incomplete results in his school's program in 1873.[33]

When asked for suggestions on topics suitable for the written part of a state's teaching examination, Neumann directed his students to investigations requiring exact measurements. Before taking the state teaching examination, but while employed as a teaching candidate at a gymnasium in Lyck, Adalbart Powel asked Neumann in 1875 for a problem in physics that would be appropriate for the teaching examination, adding that if he had his choice, he would like a problem in electricity.[34] In 1874 when Oskar Ruppel landed a position as assistant to another former seminar student, Leonhard Sohncke, at the physical cabinet of the Karlsruhe *Technische Hochschule*, he told Neumann that he had not given up "the intention of taking the teaching examination before the East Prussian testing commission [in Königsberg]" because now he had the time to work on the written part of the examination and, more important, because "the physical cabinet here stands completely at my disposal and contains several really good optical instruments. Besides, Herr Professor Sohncke will gladly purchase those [instruments] I need. Then I have at my disposal the library of the physical cabinet, which among other things contains a complete run of the *Annalen*, and the library of the technical high school. Consequently it would be entirely possible for me to have a theme whose execution required observations." So he asked Neumann "to direct me to a theme from physics for the written part of the examination," explaining, however, that the construction of the physical cabinet did not always guarantee the stability of instruments and that magnetic measurements (should these be required) would not be feasible in the winter. Five months later, Ruppel wrote back to say that Sohncke asked him to measure the force of cohesion along various axes of crystals, a topic he hoped Neumann would approve for the state's teaching examination.[35]

For teachers who were unable to produce their own data, Neumann was

33. Gustav Kade, "Ueber die Veränderungen des Orts und der Gestalt durch einfache Brechung," *Programm*, Realschule zu Meseritz, 1841/42; Eduard Hutt, "Die Quadratur der parallelen Oberfläche der Elastizitätsoberfläche," *Programm*, Realschule zu Tilsit, 1867/68; idem, "Die Bestimmung der magnetischen Neigung," *Programm*, Gymnasium zu Brandenburg, 1873/74; E. Hutt to Neumann, 19 February 1866, FNN 53.IIB: Briefe anderer Schüler; Hermann von Schäwen, "Ueber die Schwingungsrichtung der Aethertheilchen im polarisirten Licht," *Programm*, Königliches Gymnasium zu Rastenburg, 1872/73, p. 20.

34. F. A. Powel to Neumann, 14 January 1875, FNN 53.IIB: Briefe anderer Schüler.

35. F. O. Ruppel to Neumann, 13 December 1874, 19 May 1875, ibid.

a ready source. From 1864 to 1869, G. Rumler attended Neumann's lecture courses, and for five semesters between 1865 and 1868, he attended the seminar, where he claimed he had "learned how to make use of knowledge." In May 1877, while *Physiklehrer* at the Friedrichs Gymnasium in Gumbinnen, he wrote Neumann that he had "unfortunately denied myself the scientific bond which has tied me to you, as all of your students are firmly drawn to you, because the entirely new working environment which I entered as well as the care of my family absorbed my time and energy so that only a little time remained for the further progress of my study which was begun at the university. Not until now have I succeeded in being able to devote a part of my time to this study." Under Neumann, Rumler had examined thermal conduction in solid media and had been awarded one of the seminar's stipends for his investigation on the thermal conductivity of a cylindrical bar with an initial stationary temperature. For the written part of his teaching examination he also examined methods for determining the internal conductivity of solids. Now, he told Neumann, he "would like to use a part of this work for a paper for the program of this royal gymnasium at which I have been a teacher of physics and mathematics for six days." But the investigation, we find, was not entirely of his own making. The theoretical and mathematical foundation was Neumann's; it had not yet appeared in print, so Rumler asked permission to publish it. Lacking measurements of his own, he used data Neumann had given him in 1869. Rumler published "his" study quickly, in the fall of 1877, and thus with Neumann's help was kept "on track" at his secondary school.[36]

To be sure, not all physical investigations by former seminar students required measurements and data analysis or showed ties to the type of work done in Neumann's seminar. The investigations both of R. H. H. Hagen in chemistry and of Hugo Fritsch on the mechanical theory of heat bear little resemblance to seminar exercises.[37] School publications of former students who had been drawn more to Richelot's mathematical division reveal the close alignment of the exercises they had undertaken in both divisions, the near fusion of mathematics and physics in their education, and the importance they placed on the role of mathematical analogies in the development of physical theory. P. C. A. von Schäwen, Johannes Böttcher, and Albert Kostka identified primarily with mathematics, but each of them in their publications incorporated problems from Neumann's seminar, including ones on the circular pendulum and on ellipsoid equilibrium figures in a rotating fluid. It was Richelot who

36. G. O. E. Rumler to Neumann, 26 May 1877, 27 September 1877, ibid.

37. R. H. H. Hagen, "Ueber die Constitution der Aepfelsäure, ihre Salze und über das Verhalten der letzteren in höherer Temperatur," *Annalen der Chemie und Pharmacie* 38 (1841): 257–78; Hugo Fritsch, "Ueber die Erregung der Electricität durch Druck und Reibung," *AP* 5 (1878): 143–44; idem, "Lässt sich die Anwendung der lebendigen Kraft in der mechanischen Wärmetheorie rechtfertigen?" *AP* 153 (1874): 306–15.

had introduced Albert Momber to Poisson's study on the distribution of electricity on two conducting spheres, but when Momber expanded on it, he drew upon material from Neumann's lectures, Kirchhoff's work, and Carl Neumann's discussion of the distribution of temperature in a spherically bounded area. Not only did Momber thereby unify several mathematical and physical considerations; he also drew analogies between heat and electricity by using potential theory.[38]

Königsberg-trained *Physiklehrer* published less on the kinds of physics problems that arose in Richelot's seminar than on exact experimental ones of the type represented by Neumann's seminar exercises. Even so, there was one crucial difference between seminar exercises and their investigations: seminar exercises always dealt with experimentation on mostly theoretical matters, whereas the later investigations also dealt with practical problems drawn from everyday life. Böttcher used his mathematical talents to work out formulas for the optimal regional density of railroad networks and for the reliability of one formula over another. After years of publishing only in mathematics, Schönemann constructed and theoretically analyzed small weighbridges whose measurements, he believed, agreed "so accurately with the formula" he had developed that he thought one could not doubt the reliability of his analytic corrections and technical refinements. Emil Schinz used error analysis to demonstrate the effect of wind on telegraph signals; examined the behavior of locomotive steam using the equations of thermodynamics; and even described the operation of a new church clock in Bern—including the conditions impeding its motion—by using the theory of the pendulum's motion as he had learned it in the seminar. In a commentary on mild Novembers in Danzig, Albert Momber criticized temperatures thus far on record because "they were taken by different observers, sometimes with defective thermometers" and were therefore difficult to analyze and combine. Momber also reported on meteorological advances owing to a closer examination of measuring procedures; on how electrical units were determined by measurement; and on how the graphical analysis of sunrise and sunset in the new mid-European time zone (introduced in 1893) might be used for practical purposes, such as determining the hours of a school day. Finally, in a lecture on Fahrenheit, Momber focused on the history of improvements in the fixed points and scales of thermometers. Several former seminar students also produced theoretical analyses of instru-

38. P. C. A. von Schäwen to Neumann, 17 January 1882, FNN 53.IIB: Briefe anderer Schüler; Böttcher, "Die Bewegung eines Kreis-Pendels"; C. F. A. Kostka, "Ueber die Auffindung der ellipsoidischen Gleichgewichtsfiguren einer homogene, um eine feste Axe rotirenden Flüssigkeitsmasse," *Monatsberichte der Akademie der Wissenschaften in Berlin* (1870): 116–25; Albert Momber, "Ein Beitrag zu den Lösungen des Poisson'schen Problems: 'Ueber die Vertheilung der Elektrizität auf zwei leitenden Kugeln,'" *Programm*, Altstädtisches Gymnasium, Königsberg, 1871/72.

ments. In 1846, Schinz showed how Nörrenberg's polarization apparatus could produce more accurate crystallographic measurements. A decade later, intrigued by Amsler's polarplanimeter (of which more below), Schinz worked out its mathematical theory. J. H. C. E. Schumann, in his last publication, from 1862, explained the principles of the magnetic compass as constructed by Neumann. In each of these cases, *Physiklehrer* extended the exacting methods of Neumann's pedagogical physics to matters of practical concern and thus extended the domain over which exactitude ruled.[39]

Error analysis was an especially strategic and necessary tool where epistemological certainty was required, and *Physiklehrer* had to establish the trustworthiness of their results: in the measurement of constants, a classical experiment in the style of seminar exercises often directed at the modification of theory; in studies designed to establish the superiority of one analytic or theoretical expression over another; and in the determination of the practical relevance of results. These were all special kinds of investigations in that none were undertaken merely to demonstrate a natural phenomenon, whether common or rare, or to reproduce one in the laboratory under constrained conditions. Highly unusual and irregular among the investigations these *Physiklehrer* undertook were the experiments of Kiessling, who despite his preference for exact experiment in teaching, practiced physics otherwise. After publishing a mathematical study of the moment of inertia in 1866, he examined sound waves produced by a tuning fork, a study he completed in Magnus's Berlin laboratory in 1867. Kiessling reviewed prior experiments on the topic, concluded they were incomplete, and hoped "to be able to observe and measure interference phenomena accurately." Despite his reference to accuracy, his treatment lacked error analysis or any other rigorous quan-

---

39. J. E. Böttcher, "Mass für die Dichte des Eisenbahn-Netzes," *Geographische Zeitschrift* 6 (1900): 635–39; T. Schönemann, "Experimental-Untersuchung über den Druck, welchen das Wasser während des Fliessens auf seine eigenen Theile ausübt . . . ," *Monatsberichte der Akademie der Wissenschaften in Berlin* (1858): 273–80; idem, "Ueber den Druck im fliessenden Wasser [Nachtrag]," ibid. (1861) 1136–46, on 1143; Emil Schinz, "Die durch Blasen erzeugten Aspirationserscheinungen," *Mitteilungen der Naturforschenden Gesellschaft in Bern* (1859): 105–16; idem, "Die Aufhängung der Kirchenglocken," ibid. (1864): 33–62; idem, "Ueber den Einfluss des Windes auf die Richtung der Signal-Scheiben," ibid. (1864): 65–76; Albert Momber, "Daniel Gabriel Fahrenheit," *Schriften der Naturforschenden Gesellschaft in Danzig* 7.3 (1889): 108–39; idem, "Graphische Darstellung der Zeiten des Auf- und Untergangs der Sonne für Danzig nach mitteleuropäischer Zeit," ibid. 8.3/4 (1894): 261–62; idem, "Das allgemeine Windsystem der Erde," ibid. 9.1 (1895): xx–xxxiii; idem, "Vorführung einschlägiger Apparate und Experimente über die elektrischen Maasseinheiten," ibid. 10.1 (1899): xxvii–xxix; idem, "Ueber milde November in Danzig," ibid. 10.2/3 (1901): xxxviii–xxxix; Emil Schinz, "Modifikation von Nörrenberg's Polarisationsapparate," *Verhandlungen der Schweizerischen Naturforschenden Gesellschaft* (1846): 38–39; idem, "Ueber das Polar-Planimeter von Prof. Amsler in Schaffhausen," *Mitteilungen der Naturforschenden Gesellschaft in Bern* (1857): 153–76; J. H. C. E. Schumann, "Eine neue Tangentenboussole," *Programm*, Altstädtisches Gymnasium, Königsberg, 1861/62.

tification. Two more school program essays followed, including one from 1874 on the refraction of light in the eye, in which he applied mathematical treatments by Gauss and Carl Neumann.[40]

Kiessling's research was sporadic until he became intrigued by unusual diffraction phenomena in the night sky over northern Europe during the winter of 1883/84. Rather than construct an experimental protocol that enabled him to study diffraction phenomena directly as they occurred in nature, Kiessling created a controlled laboratory experiment in which he reproduced diffraction patterns in a chamber by regulating its air, water vapor, and dust contents. The problem with Kiessling's experiment was that, while it may have succeeded in *imitating* nature, Kiessling could not know for certain if it actually *duplicated* nature because its conditions were so idealized that Kiessling called them "artificial." By so rigidly controlling dependent or disturbing factors rather than calculating the effect of those actually present, Kiessling's experimental style ran counter to the thrust of that promoted by seminar exercises, where certainty was tied to the analysis of disturbing conditions operating even in the constraints of an experiment.[41]

Unfortunately, even in the most common case where error analysis was applied—the determination of physical constants—students were unable to overlook uncertainties sufficiently to derive reliable empirical relations from data. Gustav Louis Baumgarten started measurements of elasticity coefficients of crystalline media in the seminar in 1870 but took four years to complete it. Before entering the seminar he had intended to study chemistry after graduating from the *Höhere Gewerbeschule* in Leipzig, but he switched to physics and mathematics after taking classes with Neumann's student Karl Von der Mühll at Leipzig University. Months after leaving Königsberg in 1870 he told Neumann that in spite of his best intentions he had gotten nowhere. In his first teaching position at Leipzig's Nicolai Gymnasium in 1871, he continued working on his investigation but was concerned about taking enough controlled and certain measurements so as "to arrive at a passable interpolation formula" from his data. He completed enough of his investigation to submit it to the Leipzig faculty as his doctoral dissertation in 1872 but thought that still more conditions had to be examined before it could be published. In the meantime Neumann assisted him by sending him formulas for calculating the elasticity of crystals and an apparatus for measuring elasticity.

40. K. J. H. Kiessling, "Ueber die Schallinterferenz einer Stimmgabel," *AP* 130 (1867): 177–206, on 177; idem, "Die Brechung der Lichtstrahlen im Auge," *Programm*, Gelehrtenschule des Johanneums (Hamburg), 1873/74.

41. K. J. H. Kiessling, "Einfluss fremder Beimischungen in der atmosphärischen Luft auf Nebelbildung und Diffraktions-Erscheinungen," *MZ* 1 (1884): 34; idem, "Diffraktions-erscheinungen in künstlich erzeugten Nebel," ibid., p. 33; idem, "Ueber den Einfluss künstlich erzeugter Nebel auf direktes Sonnenlicht," ibid., pp. 117–26; idem, "Nebenglüh-Apparat," *Abhandlungen der Naturwissenschaftlichen Verein in Hamburg* 7 (1884): 1–8; idem, "Zur Erklärung der ringförmigen Gegen-Dämmerungen," *MZ* 2 (1885): 70–72; 230–32.

Baumgarten found both difficult to handle, especially when his mind was on his upcoming transfer from Leipzig to Dresden, where a new school was under construction. "Better days for my scientific work will come," he told Neumann. By 1874 he did indeed complete his investigation, publishing it first in a school program and then in the *Annalen*. In the *Annalen* version, he acknowledged that he had been stimulated by Neumann's lectures to examine how one could determine the elasticity coefficients of crystalline media and that his experiment had been "extraordinarily simplified" by Neumann's apparatus, which was constructed in part from ideas contained in Bessel's unpublished papers.[42]

Reliable measurements of parameters such as surface area, width, length, density, and weight of a crystal were difficult to obtain without well-prepared samples and precise measuring techniques, but Baumgarten was able to obviate prior difficulties sufficiently to present some of his results in graphical form. Nevertheless he placed little confidence in his graphs for want of additional measurements and of an independent theoretical derivation of the functional relations they represented. His desire for a theoretical derivation was especially significant; for he reaffirmed the sense among all seminar students that the mathematical representation of physical relations could not be grounded solely in empirical formulas but had to emerge from some theoretical system. Like other former students, Baumgarten drew only qualitative conclusions from the immediate visual features of his graphs. About the dependency of elasticity upon the weight of the sample, he merely observed that elasticity was not linearly dependent upon weight (Figure 26). Without the dimensions of the x and y axes, the magnitude of his measurements, and especially some indication of the extent to which his constructed curve deviated from a straight line, no other independent quantitative conclusion could be drawn from his graphical "results." The overall conclusion of his paper was a negative one: that the linear differential equations of elasticity were invalid for crystalline media. Baumgarten hoped to solve the exact form of the revised equations by taking additional measurements, but a projected second study never appeared.[43]

Baumgarten's tactic of using precise measurements to criticize and possibly to amend existing theories was a common one in seminar exercises and proved especially popular among *Physiklehrer* with access to the necessary instruments and apparatus. Leonhard Sohncke, son of Neumann's and Jacobi's first codirector of the seminar, Ludwig Adolf Sohncke, attended Neumann's seminar for three semesters in 1863 and

42. Baumgarten to Neumann, 18 September 1870, 28 June 1871 (quote), 31 December 1873 (quote), FNN 53.IIB: Briefe anderer Schüler; G. L. Baumgarten, "Die Elasticität von Kalkspathstäbchen," *AP* 152 (1874): 369–97. Baumgarten submitted this article, with a new title page, as his doctoral dissertation to the Leipzig philosophical faculty: *Die Elasticität von Kalkspathstäbchen*, Inaugural-Dissertation (Berlin: Schade, 1874).

43. Baumgarten, "Die Elasticität von Kalkspathstäbchen."

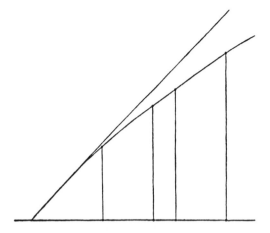

Figure 26. Baumgarten's graph depicting elasticity (*y* axis) as a function of the weight of the samples (*x* axis). The upper diagonal line represents the theoretical linear dependency; the curved line below it, with verticals to the *x* axis, is constructed from Baumgarten's measurements. From Baumgarten, "Die Elasticität von Kalkspathstäbchen," p. 397.

1864 while teaching at the Friedrichs Collegium in Königsberg, where he was responsible for teaching twenty-two hours a week. First through Eduard Heine and Carl Neumann at Halle and then through Richelot at Königsberg, Sohncke was drawn to work in mathematics. But influenced by Neumann, he turned to theoretical physics, applying the seminar's investigative strategy to problems in crystallography. In his first paper, a theoretical discussion of crystallographic systems of classification, he raised objections about abstract mathematical systems that had little connection to crystals themselves (much as Neumann had argued some forty years earlier) and about using apparatus whose theoretical basis had not been explained. In 1867 he used error analysis as a way to cast doubt on the proposition that measurements of the refraction of light by a star could tell "if and with what velocity" the star moved with respect to the earth, a proposition that ran counter to the Doppler effect. Demonstrating from data at hand that the value of its error was close to the difference sought, Sohncke concluded that the observations were "not in the state to prove a new view . . . in a convincing and irrefutable way." Sohncke left open the possibility that the mathematical theory of the effect the light source's motion had on refraction might itself not be refined enough for dealing with the data, but he considered any challenge to the Doppler effect as currently understood to be premature. A major study on the cohesion of rock salt, undertaken with an apparatus borrowed from Neumann and set up in the corner of his Königsberg apartment, gained for Sohncke his entré into an academic career, the *venia legendi* at Königsberg. In his study he explained that his accurate measurements would have been "to the highest degree difficult" had Neumann not placed at his disposal the necessary instruments and advised him on the analysis of his measurements.[44]

44. Leonhard Sohncke, "Die Gruppirung der Molecüle in den Krystallen: Eine theoretische Ableitung der Krystallsysteme und ihrer Unterabtheilungen," *AP* 132 (1867): 75–

Jakob Amsler's commitment to the techniques of an exact experimental physics predated his years as a secondary school teacher in Schaffhausen between 1852 and 1858. After attending Neumann's seminar he became an assistant at an observatory in Geneva in 1848 and in the following year, *Privatdozent* at Zurich, where he taught mathematics and branches of mathematical physics, including the mathematical theory of heat, optics, and acoustics. His investigations up through 1852 concerned the analytic equations used in the determination of specific heats, more precisely, the use of the elastic vibrations of solid media at constant volume to determine specific heats. He concluded after reviewing various methods for determining the coefficient of elasticity that the range of values these methods yielded could not be accounted for by the heat produced by compression. He therefore set out "to develop a rigorous formula" and, departing from his earlier predominantly analytic and theoretical style, to account for previous deviations by scrutinizing experimental conditions, especially to find out whether a constant temperature had been maintained throughout the experiment. Convinced there was a better way to determine the specific heats of solids, Amsler returned in his second paper to potential theory and extended to the study of heat, by analogy, equations he had earlier used for magnetism. In a third and final investigation, he developed more fully a mathematical theory for determining specific heats along the lines of seminar exercises, with an eye toward the measurements that would have to be performed. He held the amenability of his formulas to measurement to be especially important; for he believed that those who had been solely concerned with the mathematical theory of heat conduction had not taken into account the ways their equations had to be modified for actual experimental conditions. He pointed out that Wilhelm Weber's formula, for example, was "incorrect" because it did not consider the error cooling introduced while observations were being made. In contrast, Amsler sought an analytic equation corresponding completely to the idea (*Idee*) of the experiment, and although he did not then perform any observations, he was satisfied that his final formula "appear[ed] especially suited to compare with observation."[45]

For a not entirely convincing reason—that he needed time to learn more mathematics—Amsler left academics in 1852 to become a secondary school teacher. What happened, in fact, was exactly the opposite; for while teaching, he became drawn more to the practical side of physics. He

---

106; idem, "Ueber den Einfluss der Bewegung der Lichtquelle auf die Brechung," *AP* 132 (1867): 279–91, on 279, 284; idem, "Ueber die Cohäsion des Steinsalzes in krystallographisch verschiedenen Richtungen," *AP* 137 (1869): 177–200.

45. Jakob Amsler, "Ueber die Anwendung von Schwingungsbeobachtungen zur Bestimmung der specifischen Wärme fester Körper bei konstantem Volumen," *Mitteilungen der Naturforschenden Gesellschaft in Zürich* 2 (1850): 241–48; idem, "Zur Theorie der Anziehung und der Wärme," *JRAM* 42 (1852): 316–26; idem, "Ueber die Gesetze der Wärmeleitung im Innern fester Körper . . . ," ibid., pp. 327–47, on p. 346.

later recalled his seminar experience of making his own instruments using the seminar's lathe, plane workbench, and vise and often "reflected on just how this primitive arrangement, which provided problems [to be solved] step by step and promoted the spirit of discovery, would prove useful to him."[46] In 1854 he invented the polarplanimeter, a precision instrument used to measure the area inside a plane curve but which also had applications in both mathematics and physics, where it could be used to determine the coefficients of a Fourier series as well as moments of inertia. What is of interest in the mathematical theory he gave of this device (which was written with the needs of engineers in mind) is the extent to which he dwelt on error analysis. "Every measurement," he explained, "is to be viewed as faulty or even as uncertain. One has to differentiate between absolute and relative errors. The absolute error is the difference between the true value of a quantity and what is found through measurement. The ratio of this difference to the measured value is the relative error. The measure of accuracy is the relative error. One can just as well differentiate between absolute and relative uncertainty." Through concrete examples Amsler demonstrated how instruments such as the planimeter could not eradicate either error and that hence all measurements were partially uncertain. The relevance of error analysis, Amsler believed, for engineers and "practical men" was simple and direct: errors in measurements, especially of land, had economic implications, and therein was to be found the principal advantage of his invention. The polarplanimeter was more economical because measurements could be executed more quickly, with greater accuracy, and hence at considerable savings.[47] Concurrent with his invention of the polarplanimeter, Amsler opened a shop for the construction of precision instruments. He left secondary school teaching four years later.

These *Physiklehrer* had a strong professional stake in error analysis when writing for a professional community attuned to the meaning of precision: it established the trustworthiness of their results and their credibility as practitioners of their craft. They would have deployed error analysis independent of the economic benefits of greater precision. Consequently, the distinction between precise results and practically useful results was not always maintained, justifying exhaustive analyses of errors for situations that may not otherwise have required them. Thus error analysis played a central role in C. J. H. Lampe's study of the Danzig water lines, conducted between 1869 and 1872.

Lampe had left Königsberg in 1858 after attending Neumann's seminar for four years and Richelot's for five. In Neumann's seminar he

46. A. Amsler and Rudio, "Jakob Amsler-Laffon," p. 2. After his marriage in 1854, Amsler hyphenated his name.

47. Jakob Amsler-Laffon, "Ueber die mechanische Bestimmung des Flächeninhalts, der statischen Momente und der Trägheitsmomente ebener Figuren, insbesondere über einen neuen Planimeter," *Vierteljahrsschrift der Naturforschenden Gesellschaft in Zürich* 1 (1856): 41–70, 101–39, on 126.

worked on the internal viscosity of fluids. He had been motivated to investigate the motion of spheres in fluid media when Neumann proposed the topic as a prize question in 1857 and required students to go back to Coulomb's original experiments to improve upon them. Lampe's investigation won the competition. He continued to develop his investigation on the written part of the state's teaching examination, which he completed in 1858. The momentum of his research was broken when he began to teach at a Danzig gymnasium, however, because

> I lacked here all experimental material so that I could not think of applying in practice the theoretical results I had achieved. I then sought, as far as my official responsibilities allowed, first to perfect the approximate theory of oscillating disks, which I had developed earlier. I made calculations according to the more accurate formula. I developed my earlier observations, made in Königsberg, of the coefficient of viscosity of water and several salt solutions. When I finished this work I communicated it with the developments concerning the sphere to the Danzig Natural Sciences Society with the intention of publishing both in its journal.

Lampe's efforts to reach beyond his workaday world met with success; for the Danzig Natural Sciences Society gave him the funds to construct the apparatus he needed. Unfortunately, construction took over a year, and when Lampe finally had the opportunity to work with the apparatus, he found that it did not increase the accuracy of his observations. In the meantime, Oskar Emil Meyer claimed to have done the same experiment, and done it better; so Lampe canceled his plans for publication. When five years went by and Meyer's work had not appeared, Lampe decided to publish the theory upon which his experimental investigation was based. It appeared in the program for the Danzig Gymnasium in 1866. He considered his theory to have more than practical value "because it contains a new example for the application of spherical harmonics, which enter into so many problems in mathematical physics."[48]

Lampe succeeded in sustaining his research by adapting to the resources of the city of Danzig. In 1869 he began to examine public water lines, pressing into service the analytic skills he had learned in the seminar. His investigation appears not to have been commissioned by the city officials because he made a point of explaining why his data was sporadic and sparse: he had to rely on the good will of the city's technicians, taking measurements only when the occasion suited them. It is not clear from his final published study if he had been motivated primarily by the abstract problem of how to describe the pressure of water in large tubes analytically or by the practical issue of showing Danzig officials the advantages

---

48. C. J. H. Lampe, "Ueber die Bewegung einer Kugel, welche in einer reibenden Flüssigkeit um einen senkrechten Durchmesser als feste Axe rotierend schwingt," *Programm*, Städtisches Gymnasium zu Danzig, 1865/66, p. 2.

of using accurate manometers to monitor water usage. The epistemologi-cal issues he brought to bear upon his manometric data suggests, how-ever, that Lampe found his investigation suited to the kinds measure-ments and analyses a classic Königsberg investigation required.

Lampe devoted a good portion of his essay to analytic expressions for the pressure and velocity of liquids in tubes. He was dissatisfied with existing theories for several reasons: the lack of agreement between the-ory and experiment; the inadequate consideration given to tubes having larger diameters and to the errors larger diameters might introduce; and the absence of any theoretical discussion of certain variables. Seeking to formulate a more general expression that was valid for tubes with large diameters, he conducted laboratory measurements of his own but in-creased the ratio of the length of the tube to the diameter by 10^3 (so as to be able to use existing formulas at least approximately) and took his velocity measurements with a "somewhat higher degree of accuracy." He based his measurements on formulas worked out by Heinrich Jacobson, another of Neumann's former seminar students, because Jacobson rested "the execu-tion of his very valuable experimental investigation on the rigorous theory of F. E. Neumann, which [Jacobson] published with Neumann's ap-proval." About Jacobson's "countless exact measurements," Lampe con-cluded that "a greater accuracy can scarcely be expected." By comparison, Lampe claimed, other studies "can only be considered interpolation for-mulas, not as true expressions" of natural laws, because they lacked both precise data and a firm theoretical foundation. Driven to achieve an accurate agreement between theory and experiment, he subjected his data and his manometers to a rigorous error analysis and, on the basis of his results, modified his theoretical formula.[49]

But when it came time to apply his theoretical results to the pressure and velocity of water lines in Danzig, Lampe had to alter some of the rigorous standards he had earlier used as instruments of criticism, espe-cially as they concerned the epistemological certainty of measurements or theoretical expressions. With his limited opportunity to make measure-ments, Lampe did not have enough data to confirm his theory, and moreoever the data he did have were taken across two miles of a water line over three years' time. How then could he combine data from different "trials"? Although he applied the method of least squares to his data, he apparently did not consider his results sufficient in themselves for com-bining his observations. So he also justified combining his measurements on the grounds that they agreed "among one another and with the values calculated from the interpolation formula that had been derived from them." Moreover, some of Lampe's measurements were never more than

49. C. J. H. Lampe, "Allgemeine Bemerkungen über die Bewegung des Wassers in Röhren, nebst Messungen von Druck und Geschwindigkeit an der ca. 450 Fuss langen neuen Danziger Wasserleitung," *Schriften der Naturforschenden Gesellschaft in Danzig* 3 (1872): 1–72, on 3, 4, 22, 45.

"provisional" because they had been made quickly with inadequate manometers. Not surprisingly, troublesome differences between his observed and calculated values still remained, even after he computed the
constant and accidental errors of the experiment. Because the remaining
differences were, according to Lampe, "so irregular," he claimed that
"without a doubt they could be viewed as errors of observation," rather
than as a shortcoming of his experimental design or protocol. To attain
practical results, he employed interpolation once again, this time to find
the average velocity of the water flowing through the Danzig lines. Satisfied that subsequent daily manometric measurements in the water line
agreed with his calculated results, he suggested to the city council that
they make use of a manometer to monitor water usage, which they did.[50]

Lampe's attempt to deal with the water lines of Danzig in terms of exact
experimental conditions was unusual among investigations stemming
from seminar exercises, yet he viewed his problem as suited for the same
kinds of exacting techniques he had used in his seminar investigation on
the internal viscosity of fluids. So much did he consider those techniques
essential that he strained them by making ad hoc additions (i.e., the
comparison with interpolated values) of questionable epistemological
soundness, just to achieve what he thought would be certain results. His
case is instructive. The function of exact experiment in the lives of these
Physiklehrer transcended its role in solving scientific problems. It legitimated their identities as practitioners of physics and secured both the
trustworthiness of their results and the credibility of their arguments.
Investigations without measurements and error analysis were often considered incomplete, suited perhaps for publication in a school program
but not necessarily in a scholarly journal. Significantly, Königsberg-
trained *Physiklehrer* tried to avoid hasty publication, but the pressures of
school customs and careers sometimes meant suspending standards for
the sake of expediency. It is remarkable that despite the limited means
available to *Physiklehrer*, they produced investigations that were in every
respect like those issuing from seminar exercises and completed under
more favorable circumstances.

In more substantial investigations, error and data analysis functioned
as instruments of persuasion, used to establish the incontestability of their
results. Baumgarten, Sohncke, and Amsler used error analysis to criticize
theoretical expressions; Amsler and Lampe, to argue for the correctness
and hence superiority of their practical results. The power of error analysis in scientific problems is easier to understand than its force in dealing
with other issues. For nonscientific problems it was less that circumstances
demanded error analysis than that these *Physiklehrer* had a stake in using
it. The consequences of applying error analysis to practical problems are
worth mentioning; for they are signs of how the meaning of exact experi-

50. Ibid., pp. 1, 57.

ment changed in the workaday world of *Physiklehrer*. Baumgarten, who applied error analysis to a classical scientific problem, was skeptical of interpolated results; Lampe, who applied it to a practical one, was not. What had been a coherent set of techniques and values in seminar exercises could thus break down, changing the epistemological meaning of data. What is most striking, however, is that in their application of error analysis to practical problems—Amsler to the polarplanimeter and Lampe to the Danzig water lines—these *Physiklehrer* were acting as carriers of a culture of precision and accuracy, helping to objectify the world. The epistemological significance of precise data in seminar exercises inevitably became an economic one when precision was demanded outside the laboratory. When the water lines of Danzig became a laboratory, the issue was not merely whether and how one could measure the pressure and velocity of water but what value there was in achieving greater accuracy.

The Geothermal Station at Königsberg

The geothermal station that Neumann had first used in the late 1830s for both official and pedagogical purposes remained an important institutional setting in which former seminar students who became secondary school teachers in Königsberg could practice physics as they had learned it. For four decades Neumann either allowed students to appropriate the geothermal data he had taken in the late 1830s or assigned them "duty" at the geothermal station, where they took measurements of their own. His students used the data for projects ranging from introductory seminar exercises to independent investigations. Because the data had been carefully taken—with a high degree of control under natural conditions— they were suited for a variety of purposes: to understand climatic patterns, to assess older (mostly French) theories of solar radiation and thermal conductivity, and finally, to analyze errors and instruments.

The first student to analyze Neumann's data was the *Physiklehrer* J. H. C. E. Schumann, one of the original seminar members. Schumann used the data in the late 1830s to confirm Joseph Fourier's formulas for thermal conductivity, an investigation that required an extensive analysis of errors and led to suggested modifications of Fourier's theory. Schumann's study circulated among seminar students for decades, but he never published his results. When he became *Physiklehrer* at the *Höhere Bürgerschule* in Elbing in 1841, he instead began to analyze local temperature patterns. When he published these results in the *Annalen der Physik*, he included an analysis of his errors, which were crucial to the outcome of his study. Although he described his thermometer as "error free" and thought he had produced readings accurate to 0.1°C, Schumann in fact found after a careful examination that the "probable error" of his measurements, ±0.219°C, exceeded what he had believed to be the total constant error of his measurements.[51]

After Schumann, the analysis of geothermal measurements proved enduring as a seminar exercise that often led to published articles. W. Schoch left the seminar in 1856 to take up a position at a gymnasium in Schaffhausen, where he tried to determine how the average annual temperature was a function of longitude and latitude by using the mathematical techniques Neumann had developed in 1838 as an alternative to the method of least squares. In 1862, Louis Saalschütz became the first to publish an analysis of Neumann's earlier measurements, work he based in part on Schumann's manuscript. Oskar Frölich produced what those familiar with the geothermal station considered to be the best detailed theoretical discussion of geothermal measurements taken at Königsberg and elsewhere in his 1868 doctoral dissertation, the epitome of which was published in the *Annalen* in 1870.[52]

With the exception of Frölich's dissertation, however, most of the work of Neumann's students on geothermal measurements had contributed little to an understanding of why such measurements continued to be important in physics, although they had obvious empirical value in meteorology. An empirical equation based on Fourier's theory, adapted to local conditions, might require confirmation, but it was the type of confirmation that did little to amplify or refine theory further. Studies of geothermal measurements by former students who became secondary school teachers—Schumann, Schoch, and as we shall see, Ernst Dorn and Emil Mischpeter as well—cultivated and expanded certain mathematical and numerical techniques, including error analysis, more than they explained the theoretical significance of the measurements themselves. Schumann, for example, had used his data (and some of Neumann's formulas) as the basis for an empirical formula and for a graph depicting a "temperature curve." He even determined what he called the "normal wave" of temperature patterns by introducing a new method for calculating the average temperature for any single month. Unlike similar methods, his preserved the relative natural positions of maxima and mimima. Although he used crude graphical methods (straight lines joining together the maxima

51. J. H. C. E. Schumann, "Die Temperaturverhältnisse von Elbing in Preußen," *AP* 68 (1846): 575–82, on 575, 578. Schumann's study appeared first in *Programm*, Höhere Bürgerschule zu Elbing, 1842/43. Through his suggestive questions concerning internal and surface geotemperatures, Fourier had invited the comparison between his theory and exact observations; see Joseph Fourier, *The Analytical Theory of Heat*, trans. A. Freeman (Cambridge: Cambridge University Press, 1878), pp. 4, 9, 12–13, 20–21.

52. W. Schoch to Neumann, [c. 1855 or 1856], FNN 53.IIA: Briefe von Schülern; Louis Saalschütz, "Ueber die Wärmeveränderungen in den höheren Erdschichten unter dem Einflusse des nicht-periodischen Temperaturwechsels an der Oberfläche," *AN* 56 (1862): 1–44, 161–206, 273–78; Oskar Frölich, *Ueber den Einfluss der Absorption der Sonnenwärme in der Atmosphäre auf die Temperatur der Erde*, Inaugural-Dissertation (Königsberg: Dalkowski, 1868); idem, "Zur Theorie der Erdtemperatur," *AP* 140 (1870): 647–52.

and minima for each month) to approximate his "temperature curve," the extreme and mean values depicted in his curve agreed well enough with his measurements for Schumann to consider his curve "accurate." Schoch also based his analysis of geothermal measurements on formulas developed by Neumann, but his broader goal of relating temperature to geographic location required "knowledge of the temperature at places on the earth's surface where observations were lacking," so he determined those values through interpolation formulas, which he carefully explained in a letter to Neumann.[53]

Motivated by advances in meteorology, by what he believed to be the growing importance of geothermal measurements, and by his successes in using geothermal measurements in seminar exercises, Neumann argued for the establishment of a formal geothermal station at Königsberg. With the support of the director of the university's botanical garden (where the station was to be moved from Neumann's garden at home) and with the financial assistance of several local scientific societies, especially the Königsberg Physico-economic Society, the station was built in 1872 at a cost of 380 taler, of which 230 taler was used for thermometers. The formal incorporation of the geothermal station, although it occurred almost forty years after the station's actual inception, enhanced the station's professional role at Königsberg, making it a stronger institutional setting for gathering data, for training neophyte scientific practitioners, and for legitimizing data reduction as an essential activity in the practice of physics. Neumann's former student Ernst Dorn, who had attended the seminar from 1866 to 1869 and had just returned to Königsberg to teach secondary school after spending a year at a Berlin gymnasium, was named director. Dorn identified his first task as determining the reliability of the station's instruments, and soon after assuming the directorship he published what he considered to be the definitive study of the theory and experimental calibration of the station's thermometers (Figure 27).[54]

Dorn's 1872 discussion of the geothermal measurements taken at Königsberg was devoted to a technical matter: improving the theoretical and practical elements of Neumann's method for calibrating thermometers. At Neumann's suggestion, Dorn explored how interpolation could improve the accuracy of calibration. Employing Neumann's own observation journal, Dorn averaged the dozen or so measurements taken at each thermometer of the station, interpolated between the nth and the $(n + 2)$ average, and then checked to see if the value of the interpolated point

53. Schumann, "Die Temperaturverhältnisse von Elbing," pp. 575, 579, 580; Schoch to Neumann, [c. 1855 or 1856], FNN 53.IIA: Briefe von Schülern.

54. Ernst Dorn, "Die Station zur Messung von Erdtemperaturen zu Königsberg i. Pr. und die Berechtigung der dabei verwandten Thermometer," *SPGK* 13 (1872): 37–88, 159–60. Dorn's observations continue in volumes 15–18, 20, and 23. On the history of the station see, in addition to Dorn, Paul Volkmann, "Franz Neumann als Experimentator," *Verhandlungen der Deutschen Physikalischen Gesellschaft* 12 (1910): 776–87, on 781.

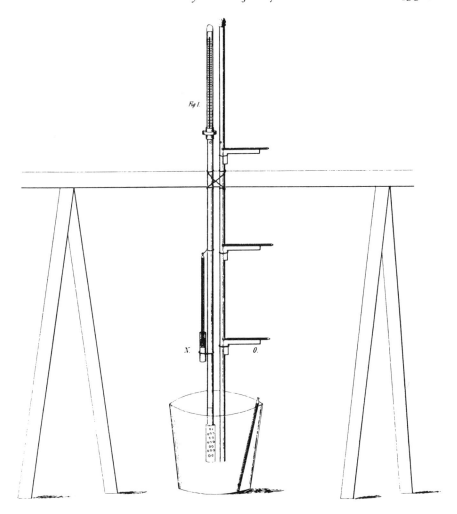

Figure 27. A Königsberg geothermometer. From Dorn, "Die Station zur Messung von Erdtemperaturen," Figure I.

corresponded reasonably well to the $(n + 1)$ average temperature. He considered his technique of interpolation to be novel and important enough to warrant a separate explanation in a footnote (Figure 28). The difference between the interpolated value and the average of the measurements was crucial because it indicated the error introduced by interpolation. Dorn knew he was taking extraordinary measures in defining the error produced by interpolation, but he persisted. "It appears to me necessary," he wrote, "to make the presentation in such a way that complete insight into the accuracy of the results can be achieved and that

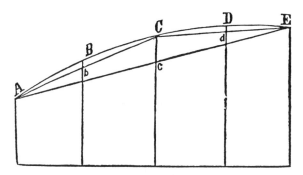

Figure 28. Dorn's method of interpolation for thermometer calibration. The *x* axis represents divisions on the thermometer scale; the *y* axis, temperatures assigned to those divisions. Dorn's method determines the error introduced by interpolation and helps determine if interpolation can be used for calibration. Points A, C, and E are average temperature readings, at three points on the scale, joined to form a curve. To interpolate between A and E, draw line $\overline{AE}$; the difference between the observed and calculated values at C is Cc, which is also the error introduced by interpolation. Similarly, interpolating between A and C and be-tween C and E produces errors $\overline{Bb}$ and $\overline{Dd}$, which are both less than $\overline{Cc}$. Dorn, like Baumgarten (Figure 26), did not think it necessary to provide scales for his *x* and *y* axes. From Dorn, "Die Station zur Messung von Erdtemperaturen," p. 55n.

everyone, who wants at some time to use the observations, may be able to judge their reliability for themselves." For Dorn, interpreting the measurements was a matter of understanding the instruments that produced them and the quantitative techniques that reduced them rather than a matter of the theory that gave rise to them. So preoccupied was he with determining the accuracy of his measurements that he later published an appendix to his paper in which he detailed the "environment" of his "experiment": the physical features of the geothermal station (e.g., its exact geographic location and height above sea level) and other features of his experimental protocol (e.g., the times at which he took his measurements). In the course of his study, he used as research assistants some of Neumann's current seminar students: L. Hübner, J. P. G. Peters, and E. Scheeffer.[55]

Dorn's study, with its emphasis on technique and accuracy, laid the foundation for and set the tone of a twenty year program in geothermal measurements at Königsberg, which included several of his reports as well as those of subsequent directors. During that time, Frölich's doctoral dissertation continued to be the only theoretical discussion of the relevance of those measurements. Although Frölich had believed he needed more data, he was confident that he had confirmed aspects of Poisson's

55. Dorn, "Die Station zur Messung von Erdtemperaturen," pp. 55n, 37 (quote), 159–60 (app.).

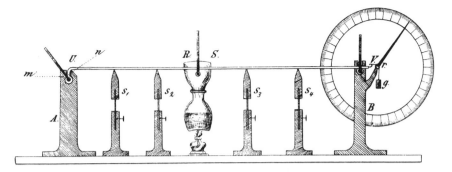

Figure 29. Mischpeter's apparatus for measuring thermal expansion. From Mischpeter, "Ueber die Ausdehnung ungleichmässig erwärmter Körper," Figure 3, by permission of Staatsbibliothek Preußischer Kulturbesitz, Berlin.

theory of solar radiation; specifically, that the atmosphere absorbed a constant amount of solar radiation, that the amount of solar radiation reflected on the earth's surface was the same for each angle of incidence, and that the internal thermal conductivity of the earth had not changed over time.[56]

Emil Mischpeter, who spent five semesters in Neumann's seminar in the late 1860s and had finished his doctorate under Richelot in 1874, replaced Dorn as director of the geothermal station in 1886. After he began his teaching career at the Realschule auf der Burg in Königsberg in 1869, Mischpeter reconsidered his devotion to mathematics. Two years after he received his doctorate he was working in physics and on the kinds of problems the seminar exercises had cultivated. His first publication in physics was on the thermal expansion of an unequally heated object. After developing the mathematical equations for the conductivity of heat in a cylinder whose radius was small in comparison to its length, Mischpeter measured the thermal expansion of several bodies using an instrument he had designed and constructed (Figure 29). A bar, *UV*, was fixed to two immovable stands, *A* and *B*, so that the temperature of the ends of the bar could be measured. (The bar was supported at four points, S_1 to S_4 to prevent bending.) A glass container, heated and filled with a liquid, acted as a heat source; its temperature could also be measured. The end of the bar located at *V* had a small hook that, when expanded, pushed a wire (attached at *r*), which was counterbalanced by a weight *g*. A needle attached at *r* deflected whenever the bar expanded. Mischpeter's formulas related the degree of deflection to the amount of thermal expansion.[57]

56. Frölich, *Ueber den Einfluss der Absorption der Sonnenwärme.*

57. Emil Mischpeter, "Ueber die Ausdehnung ungleichmässig erwärmter Körper," *Programm*, Realschule auf der Burg, Königsberg, 1875/76.

Mischpeter's experiment was not a difficult one theoretically. Probably because he published his investigation in his school's annual program, though, he did not subject his data to an exhaustive analysis. His audience would not have needed to know how he compensated for the temperature of various parts of his apparatus, including the air surrounding it; how he corrected his measurements; or how he combined observations from various trials. Nevertheless, its conceptual simplicity invited the kind of error analysis and comparison between theory and experiment that was common in seminar exercises. His protocol was sensitive enough for members of the Königsberg Physico-economic Society to notice it, primarily because it had potential practical applications to the kinds of corrections needed for thermometric observations. After Mischpeter accepted their offer to direct the station, he continued to teach secondary school while taking geothermal measurements and publishing six reports on them between 1886 and 1893.[58]

By 1889, Mischpeter had compiled several years of measurements but had paid little attention to experimental conditions (except to note that one of Dorn's thermometers had broken); nor did he discuss the value of his measurements or the use to which they might be put. Sensing the growing concern at the Königsberg Physico-economic Society that these measurements might not be worth the effort and financial support given to make them, Paul Volkmann drew up a two-page article on the scientific significance of the geothermal station. Volkmann was one of the most ardent supporters of Neumann's way of doing physics, having attended his lectures in the mid-1870s, taken his doctorate under Woldemar Voigt at Königsberg, and being now the *außerordentlicher* professor of theoretical physics at Königsberg. Volkmann first mentioned the obvious: that the data were useful in meteorology and geology, especially for understanding the relation between temperature and climate and in determining the age of the earth.[59]

The real significance of the station for Volkmann lay much closer to home, especially in the opportunity it offered to analyze data. Frölich's dissertation, Volkmann pointed out, had provided key reasons for establishing the geothermal station; observations made since then could be used to calculate the most important variable in Frölich's study, the absorption coefficient of the atmosphere. Still, several in the society raised concerns about measurement reliability. Volkmann admitted that "one will always be able to find fault with the value of observations that are made under externally complicated circumstances," but he explained that

58. Emil Mischpeter, "Beobachtungen der Station zur Messung der Temperatur der Erde in verschiedenen Tiefen im botanischen Garten zu Königsberg in Preußen," *SPGK* 27 (1886): 9–32; 28 (1887): 1–26; 29 (1888): 1–26; 30 (1889): 1–26; 31 (1890): 33–58; 34 (1893): 62–76.

59. Paul Volkmann, "Ueber die wissenschaftliche Bedeutung von Erdthermometerstationen," *SPGK* (Sitzungsberichte) 31 (1890): 3–4.

"as soon as one abandons the laboratory," as one did when one worked in the station, "one is no longer master of all conditions." But that did not necessarily mean that the geothermal measurements were worthless, because, as Volkmann continued, "the value of the station itself can be judged quite sufficiently through theoretical considerations." Criticisms leveled against the station were thus, in Volkmann's view, not only directed at its existence but also at a particular way of doing physics. To accept Volkmann's defense of measurements taken under "complicated circumstances" was tantamount to recognizing the value of error analysis—the "theoretical consideration" to which Volkmann referred—in transforming data into a useful and reliable form.[60]

Volkmann's defense was motivated by the society's recent creation of a seven-member commission charged with evaluating and determining the future of the geothermal station. The commission included Volkmann and two of Neumann's former seminar students, Carl Pape, now professor of experimental physics at Königsberg, and the *Physiklehrer* Mischpeter. Although, when the commission first met on 9 December 1889, it acknowledged that the Königsberg geothermal station was one of the most important worldwide, it persisted in viewing the purpose of the station primarily in workmanlike, technical ways—somewhat along the lines of seminar exercises. The commission listed among the station's first functions the "comparison of the theory of thermal conductivity with observations," but the object of such a comparison was less theory confirmation than theory adjustment for real, empirical meteorological conditions. The commission considered "a discussion of the reasons for the deviation between calculation and observation"—hence a discussion of errors—to be an equally important task. The final two functions it cited also involved workmanlike aspects of scientific practice: recording the earth's temperature at various depths and determining the thermal absorption coefficient of the atmosphere according to Frölich's method.[61]

Yet when the commission met two days later, on 11 December 1889, the physical interpretation of the geothermal measurements dominated their discussion. The mathematician Ferdinand Lindemann presented an evaluation of Frölich's dissertation. Lindemann did not see a firm connection between Frölich's theory and the measurements taken at the station because, as he pointed out, Frölich relied primarily upon atmospheric temperatures taken in conjunction with the technically more involved geothermal ones. Lindemann also raised questions about whether it would even be possible to interpret some of Frölich's formulas in physical terms because they contained imaginary values. At first the commission reacted cautiously to Lindemann's criticisms, agreeing only to examine the mea-

60. Ibid., p. 4.
61. Julius Franz, "Bericht der Kommission zur Beratung über das fernere Schicksal der . . . Bodenthermometerstation," ibid., pp. 4–6, on p. 4.

surements made during 1890 from a "theoretical" perspective that included an analysis of the thermometers. But about a week later it decided to take more decisive action. On 19 December 1889 the commission released to the public a prize question worth 300 marks: "The Society would like as comprehensive as possible a theoretical evaluation of the geothermal measurements made at Königsberg, especially to understand the thermal conductivity of the earth and the causes of it, and directs [contributors] especially to the preliminary work done by O. Frölich in his dissertation." Replies were due 1 February 1891.[62]

With both a commission and a prize question in place, the stage was set for a trial of sorts. But what was on trial—the geothermal station, Frölich's dissertation, or a particular way of practicing physics that had been a part of the seminar and then of the professional lives of *Physiklehrer* from Schumann to Mischpeter? The wording of the prize question, with its emphasis on the *physical* understanding of the measurements, suggested that Lindemann's views held sway over the commission and that conceptually oriented replies would be favored over technical or workmanlike ones. But there was an ambiguity in asking for a *theoretical* evaluation. To former seminar students, a theoretical evaluation consisted primarily of applying the techniques of Neumann's practical physics, especially error analysis, and not necessarily of scrutinizing physical causes, definitions, or theoretical formulations in general. The wording of the prize question, however, leaned toward the latter. What the commission soon found out was that the outcome of the "trial" hinged as much on how the term *theoretical* was understood as on the norms that the commission itself, acting as both judge and jury, applied to the case. A further complication was that in its reprinting of the prize question, the *Annalen der Physik* replaced the word *theoretical* with the word *wissenschaftlich*, which carried similar ambiguous meanings for the respondents and for the commission.[63]

After the competition was over, the commission reported on two entries. Ernst Leyst, an administrator at the meteorological observatory in St. Petersburg, authored one of them. He seized the opportunity to make something of the *Annalen* faux pas by commenting on the limitations of a theoretical treatment of geothermal measurements. Leyst argued that a purely "mathematico-physical" treatment of the Königsberg geothermal measurements was inappropriate. He claimed that a "theoretical" treatment was also out of the question because the mathematical theory of heat was itself not "satisfyingly constructed," being at this time based "only on an ideal case, which perhaps is realized in the laboratory, but not in free nature." Only "with hestitation," he believed, could the theory be applied

62. Ibid., pp. 5–6.
63. The commission took note of the error in the *Annalen*; see F. Lindemann et al., "Bericht zur Beurtheilung der einzelnen Preisarbeiten," ibid., pp. 33–38, on p. 34.

to the Königsberg geothermal measurements. He instead argued for a geophysical analysis, one that took into account the quantitative impact meteorological phenomena had on geotemperatures. "Not until then can one give a solution to the problem of the distribution of heat in the earth," he emphasized. "If all the causes are correctly recognized and determined, then their effect upon the temperature of the earth is given." Leyst's causal analysis was in essence an analysis of disturbing conditions, or of what under laboratory conditions gave rise to the constant errors of an experiment. He spoke of this "causal analysis" in the same way Neumann's students spoke about the analysis of constant errors: that it was essential because, without it, the theory would be "incomplete," incorporating merely "ingenious speculation" and "deficient ideas." He went further, identifying a genuinely theoretical treatment with what occurs under ideal laboratory conditions and juxtaposing it to what he believed to be a more comprehensive scientific (*wissenschaftlich*) treatment based on physical causes.[64]

Leyst's position posed a dilemma. Although the commission was sympathetic to the issue of taking disturbing factors into account, it disagreed with his view that Frölich's dissertation was "completely inadmissible" and that Dorn's protocol and calibrations were inadequate. Furthermore, to the commission, answering the prize question was a matter "of determining *numerically* what the mathematical theory of thermal conductivity may or may not tell about the actual conditions of the earth." The commission in fact had to take this position in order to salvage the station and its measurements. Yet even though it did not consider Leyst's critique of the mathematical theory of thermal conductivity adequate, it was impressed enough by his examination of what affected the data to award him an unplanned second prize. Even Paul Volkmann was sufficiently moved by Leyst's analysis to go back to Neumann, now ninety-three, to ask him if it were true, as Leyst had assumed, that the thermometers used in the 1830s did not have protective coverings; Neumann confirmed that they did not. (Volkmann later questioned whether covered thermometers would have given a "true picture of the temperature distribution in the surrounding area" and so reaffirmed the primarily data-oriented, rather than instrument-oriented, approach to experiment taken by Neumann and his students.)[65]

First prize went to Adolf Schmidt, a *Physiklehrer* with no connection to Königsberg. Schmidt, like Leyst, recognized that the problem was dichotomous, having both physical and meteorological components, and he

64. Ernst Leyst, "Untersuchungen über die Bodentemperatur in Königsberg in Pr.," *SPGK* 33 (1892): 1–67, on 64, 2; see also Lindemann et al., "Bericht der . . . Preisarbeiten," pp. 34–35.

65. Lindemann et. al., "Bericht der . . . Preisarbeiten," pp. 34 (emphasis added), 35; Paul Volkmann, "Beiträge zur Wertschätzung der Königsberger Erdthermometer-Station, 1872–92," *SPGK* (Sitzungsberichte) 34 (1893): 54–61, on 55, 57.

too made physical issues central; but unlike Leyst, he defined the "essence of the physical problem" as determining a law and calculating its numerical constants. He had at his disposal decades of measurements from the geothermal station which, because they had been obtained by what he called an "unusually exact method," justified "an accuracy in numerical calculations which in most cases would be pedantic." But how should one achieve that accuracy? Despite the number of measurements, there was no way for Schmidt to avoid interpolation as a means of assembling a denser series of numbers that would provide the foundation for what he believed would be a more accurate mathematical law. Alas, like Neumann's seminar students, he found interpolation problematic. At first he found purely mathematical (parabolic) interpolation unsatisfactory. Interpolating on the basis of a theoretical formula might work. Schmidt, however, wanted to evaluate his data independently of theoretical formulas, even though he knew that he could correct a theoretical formula for disturbances stemming from both laboratory and natural conditions and then use this corrected formula as an interpolation formula. Yet he viewed even a theory-based interpolation formula as problematic because in nature "every condition is not in the meantime fulfilled, and therefore interpolating according to a given [corrected formula] . . . is only approximately correct." Not all conditions could be examined or fulfilled, he furthermore admitted, so the theory-based interpolation formula itself remained hypothetical. He also acknowledged that it might be deceptive to incorporate a multitude of corrections into the formula; for accuracy would not necessarily increase. The effort to create such an interpolation formula might, he suggested, have a negative impact on the investigation overall.[66]

For Schmidt, simple analytic interpolation yielded mainly mathematical data devoid of physical significance, whereas a more complex, theory-based interpolation formula produced data tinged with hypothetical components. On this issue of interpolation Schmidt must have appealed to the sentiments of the commission because he had acknowledged that interpolation was an inadequate foundation for a mathematical law, even though he admitted that from interpolation one could get "at least a qualitatively correct picture of the true situation." What Schmidt was in essence arguing was that interpolation could only be used as a stopgap tactic when additional measured data was unavailable. In his view, the only way to sharpen the theoretical understanding of the Königsberg geothermal measurements was by expanding the empirical foundation of the problem: first, by making modifications in the thermometers and by taking climatic conditions into account and, second, by increasing the number of readings from thrice daily to hourly. Only by increasing the

66. Adolf Schmidt, "Theoretische Verwertung der Königsberger Bodentemperatur-Beobachtungen," *SPGK* 31 (1890): 97–168, on 100, 125, 132.

readings, he argued, could one construct a mathematical law and compute its constants.[67]

Naturally the commission was pleased by Schmidt's recommendation for the geothermal station. He told them what they wanted to hear and in the language they wanted to hear it. Years of observations did not have to be sacrificed. The commission considered Schmidt's approach superior to Leyst's because Schmidt had worked the problem through "theoretically," as the prize question had stipulated and as the commission had defined it. To the commission, Schmidt's investigation had "great insight and rigor"; for it had given "a clear picture of the difficulties and scope of the problem." But in fact the "picture" Schmidt had drawn was one that depicted less the physical meaning of the data than the analytic and experimental steps necessary to find a "lawful" relation residing in the measurements. In the commission's view, Schmidt's suggestions for improving the accuracy of the data—especially by taking more measurements—led to "the next task of the Society." Schmidt's study legitimated the past and future operation of the Königsberg geothermal station and publicly reinforced a style of physical research practiced by Neumann's seminar students, including the local *Physiklehrer* trained under him.[68]

A recommendation to make more measurements without having a firm idea of what their physical relevance would be was not what university officials and other members of the society wanted to hear. Despite the best efforts of the commission, the geothermal station was dismantled in 1892, ending a research project begun by Neumann and sustained by his students and local *Physiklehrer* for over fifty years.[69] While several considerations, including financial ones, contributed to the decision, it was the outcome of the society's prize competition that made clear the workmanlike nature of the station's operations and the thin conceptual foundation upon which it had been built. Former seminar students who worked at the station were guided by empirical and technical considerations. They did not view the theory of geothermal measurements in physical terms or even causal ones, nor did they view the experimental trials abstractly, in terms of ideal laboratory conditions. For them, a "theoretical evaluation" meant finding the constraints upon data that limited its accuracy. Going beyond the data itself once again proved problematic. And ironically, once again mathematical sophistication did not facilitate

67. Ibid., pp. 138 (quote), 156–57.
68. Lindemann et al., "Bericht der . . . Preisarbeiten," pp. 35, 36.
69. Criticisms of the geothermal station are found in *SPGK* (Sitzungsberichte) 30 (1889): 67; 31 (1890): 49; 32 (1891): 68–69. The dismantling of the geothermal station is reported in ibid. 33 (1892): 60–61. The Königsberg astronomer Julius Franz wrote the last report on the station's data: "Die täglichen Schwankungen der Temperatur im Erdboden: Nach der Bodenthermometer-Station der Physikalisch-ökonomischen Gesellschaft," *SPGK* 36 (1895): 51–66. Cf. Paul Volkmann, *Erkenntnistheoretische Grundzüge der Naturwissenschaften und ihre Beziehungen zum Geistesleben der Gegenwart* (Leipzig: B. G. Teubner, 1910), pp. 158–63, 183–88.

using interpolation. The geothermal station made former seminar students aware of the choices they faced: either use interpolation and get on with the business of formulating or confirming a law, or become engaged in the endless technical refinements and calculations that delayed and eventually obstructed a deeper conceptual understanding of the physical processes at work.

Balancing Teaching and Research

Jerome Ravetz has argued that much of scientific training and practice is craftsmanlike, involving the transmission and execution of skills.[70] His view aptly describes Neumann's seminar and the subsequent activity of the *Physiklehrer* who attended it. These *Physiklehrer* were extraordinarily determined to use the techniques they had learned. They engaged in the practice of physics largely through the execution of technique rather than the development of ideas. To create an identity of themselves as practitioners of physics, they had to reach beyond their workaday world. Physical cabinets of secondary schools, laboratories of local scientific societies, a geothermal station, and even the water lines of Danzig were the settings of their scientific investigations. Often lacking contact, though, with a broader scientific community and hence unable to participate in the kinds of dialogue that established consensus on scientific matters, these *Physiklehrer* publicly legitimated their findings by deploying techniques acknowledged as rigorous and as guaranteeing certainty. The limited research they were able to perform continued the commitment among Neumann's students to means over ends, to technical execution over conceptual refinement.

With the obstacles standing in their way, it is remarkable that so many former seminar students who became teachers were able to complete any research at all. The transition to the workaday world of secondary school teaching (especially the unavoidable pressures of the first years), the imperative to publish as soon as possible (if only in a *Schulprogramm*), and the increase in family obligations that customarily accompanied economic independence were some of the distractions that broke the sustained concentration and effort required to pursue research. Yet among several Königsberg-trained *Physiklehrer*, the drive to do research remained. If some of their investigations failed to measure up to standards Neumann set and other former students in more fortunate working conditions met, works by *Physiklehrer* satisfied, nonetheless, the aspiration that teaching be a scholarly profession. By identifying with the practice of physics they sustained the values acquired in the seminar.

70. Jerome Ravetz, *Scientific Knowledge and Its Social Problems* (New York: Oxford University Press, 1971), pp. 75–77, 85, 99–101.

Their lives, however, were filled with tensions that were in some respects greater than those in academics. Professors as well as *Physiklehrer* had to achieve a balance between teaching and research. But to some, being a successful *Lehrer* implied the loss of the style and substance of scholarship. In 1873, Emil Meyer told Neumann that Heinrich Schröter had recommended Anton Müttrich, then an *Oberlehrer* at the Johannes Gymnasium in Breslau, for a physics position at Breslau but that he "would agree [with Schröter] with joy if I were not concerned that Müttrich's long school activity [fifteen years] has already made him unsuitable." "Müttrich is a man too faithful to his duty," he added, meaning that Müttrich was no longer suited to become an academic.[71] But Neumann viewed matters otherwise. In a letter of recommendation for the mathematician Rudolf Lipschitz, who had attended some of his courses, Neumann pointed out that prior to his appointment at Bonn, Lipschitz taught mathematics for about eighteen months in the upper forms of a gymnasium in Elbing. "This [mathematical] instruction," Neumann wrote, "without a doubt will prove useful for his academic activity."[72]

Reality fell somewhere between the elitist views of Meyer, who had never taught secondary school, and the optimistic views of Neumann, who taught primarily teaching candidates. The workaday world of the secondary school offered no easy choices between the two extremes. In 1849 after teaching for only a few years at the *Cantonschule* in Aarau, Emil Schinz concluded that "the instruction that a natural sciences teacher receives in a seminar is too general, too high, and penetrates too little into life." The problem, of course, was not the seminar but his professional life. Discontent with his position by 1855, he told Neumann that he wanted to move to Bern, where he hoped he would find an academic position. "This is for me such an important goal," he wrote, citing the better scientific environment. Neumann recommended him, praising the "scope and fundamental nature" of his knowledge in mathematical physics and the "many and beautiful seminar works" that had earned prizes. Neumann agreed, though, that Schinz's secondary school position was confining because it "left unused a treasure of many-sided and fundamental knowledge and quieted a strength which . . . if transferred to an academic setting, would be enhanced."[73] Schinz did not get the academic position in Bern, but did take a position at a Bern *Realschule* from which he could draw upon Bern's scientific life.

Wilhelm Schoch's position was similar. Only a year after he left the seminar in 1856 and had assumed a position at a gymnasium in Schaff-

---

71. Emil Meyer to Neumann, 10 May 1873, FNN 53.IIA: Briefe von Schülern.

72. Neumann to C. Kappeler, 1861, FNN 53.IB: Anfragen wegen Berufungen.

73. Emil Schinz, "Vorträge über den naturwissenschaftlichen Unterricht in Volkschulen," *Verhandlungen der Schweizerischen Naturforschenden Gesellschaft* (1849): 50–58, on 52; Emil Schinz to Neumann, 1 May 1855, FNN 53.IIB: Briefe anderer Schüler; Neumann's recommendation for Schinz, c. 1855, partially rpt. in LN, p. 370.

hausen, he wrote Neumann that he was working as a teacher "but not in the manner that I desire and that is suited to my studies. My students are 12–15 years old, so that I only have to teach the beginnings of the subjects represented by me. These subjects are arithmetic, geometry, physics, chemistry, and geography. At first I had natural history rather than geography, but someone else took it over." Although among his students were those "who want to devote themselves to a scholarly calling," Schoch wanted to find another position with better working conditions so that he could teach more "scientific subjects"—those associated with the upper forms, such as higher levels of physics. Moreover, his present position, he explained to Neumann, allowed little time for work on his dissertation. As Schinz viewed the scientific circle in Bern as a more hospitable professional environment, so Schoch viewed the one in Basel. Schoch ended up teaching at the technical school in Winterthur, where he could have some contact with a scientific culture.[74]

The positions that Schoch and Schinz had in Swiss schools were somewhat less scholarly oriented than those in German states, especially Prussia. It was also more common in Switzerland than in the German states for university teachers, even full professors, to teach at local secondary schools. Ernst Schroeder tried to hold down two positions in Zurich, as a secondary school teacher and as a *Privatdozent*, between 1864 and 1868. But the "doubled position" just did not work, he later wrote in his autobiography, so he returned to his home in Baden, where he taught secondary school. While professor in Bern during the 1860s, Heinrich Wild considered it worthwhile to teach four hours a week at a gymnasium but later found that he had to give it up because it detracted from his research. When Carl Neumann was called to Basel as *ordentlicher* professor of mathematics in 1863, he was asked to teach six hours a week at the local *Pädagogium*, a task he gladly accepted. In fact when he received a call to Breslau in 1864, Neumann told officials that he would decline the offer if he could teach physics to the upper forms in the *Pädagogium*. Neumann was granted his request, not only because the university considered it in their best interests to keep him but also because they considered his teaching "profitable for instruction in the *Pädagogium*": it would help students make the transition to university physics instruction more easily. Although all did not work out as Carl Neumann had planned—he did not feel successful teaching secondary school and the dual position became onerous—and he later left for an *ordentlicher* position in Tübingen, he continued to believe that bridging school and university instruction was worthwhile. He therefore recommended Paul du Bois-Reymond for the Basel position because of his extensive secondary school experience.[75]

74. Schoch to Neumann, 1 June 1857, FNN 53.IIA: Briefe von Schülern.
75. J. Lüroth, "Nekrolog auf Ernst Schroeder," *JDMV* 12 (1903): 249–65, on 251–52; H. Wild to Neumann, 21 November 1864, FNN 53.IIA: Briefe von Schülern; also P. Merian to Erziehungscollegium, 23 May 1863; Universitäts-Curatel to Erziehungscollegium, 16

If their workaday world did not offer the professional identity that *Physiklehrer* sought, other settings did. Connections to the scientific community were important to them. Writing to Franz Neumann from Berlin in 1867, Albert Wangerin explained how a scholarly environment sustained research:

> Above all else the scientific stimulation holds me here; this would be lacking in a small provincial city. If this stimulation is not as direct as the stimulation I had the luck to enjoy in Königsberg with you and Herr Professor Richelot, then it nevertheless is strong enough to stir an interest in science continuously. Here it is the Berlin Physical Society and the journal circle associated with it that makes it easy for me to become acquainted to a certain degree with new discoveries in science. I have derived much stimulation for my own scientific work from the society.

Wangerin immersed himself in the activity of the society and even published reviews of recent publications in optics in its *Fortschritte der Physik*, although Neumann had advised him not to, perhaps believing it would drain too much of his time and energy. Wangerin did not continue reviewing, but he did remain active in the society during the ten years he was a secondary school teacher, as did other former Königsberg students such as Hutt, Brix, and Quincke.[76]

When a local scientific society was not available, former students had to sustain themselves on memories of their days in Königsberg. On a trip to Paris and other scientific centers in Europe, Emil Schinz wrote to Neumann about the memories he had of Königsberg, memories "so deeply impressed upon [his] soul" that he would not forget them. Baumgarten and Böttcher, living together in Leipzig, nurtured their memories of Königsberg by displaying pictures of Neumann and Richelot, "who look down from the wall upon the struggles of two small schoolmasters." Without access to many books, Ernst Bardey, a private teacher, turned to the lecture notes he had taken in Neumann's classes for intellectual stimulation. What the workaday world could not provide, renewed memories of educational experiences did.[77]

But maintaining linkages to the seminar produced ambiguous results. Drawing upon the seminar's investigative techniques and working on problems suggested by Neumann enabled *Physiklehrer* to sustain identities as scientific practitioners and to do so in a way that gave their work an air of credibility. The mental world carved out by those techniques could,

---

March 1864; Carl Neumann to Basel University administration, 4 February 1865, 14 February 1865; and Universitäts-Curatel to Erziehungscollegium, 10 February 1865; Universitäts-Curatel to Erziehungscollegium, 17 February 1865; all in SAKBSt, CC27.

76. A. Wangerin to Neumann, 19 October 1867, FNN 53.IIA: Briefe von Schülern.

77. Schinz to Neumann, 21 January 1844; Baumgarten to Neumann, 28 June 1871; Ernst Bardey to Neumann, 10 February 1853; all in FNN 53.IIB: Briefe anderer Schüler.

however, prove confining. Wangerin, for instance, may have relished his contact with the Berlin Physical Society, but he was burdened by carrying several teaching positions at once. "Unfortunately," he wrote to Neumann, "now I lack the time for peaceful and undisturbed work and for following in detail my ideas or those stemming from your lectures and seminar. I hope that when I have a permanent position and do not have to seek part-time positions, I will have more time for my work. For the time being I must restrict myself to science *au fait* and to collecting new facts. With great longing I therefore look back upon my time in Königsberg when I could devote myself to scientific work alone." Yet even had Wangerin had time to follow up on his ideas, their technical and material requirements probably could not have been satisfied by most secondary schools.[78]

Normally in the history of science, marginal individuals such as these *Physiklehrer* go unmentioned. They are what recent social history has called the "little people," who offer us the opportunity to view life from another perspective, from the ground up rather than from the top down. There are compelling reasons to examine the professional world and work of these teachers. Anthony Grafton has looked closely at a similar group of marginal scholars, the eighteenth-century polyhistors, the predecessors of philologists. Grafton detected "a certain pathos" in their work. They listed books, discussed them superficially, but never showed that they really understood them. "We find it easier to laugh at the polyhistors," Grafton noted, "than to take them seriously." Yet he also found that their work, despite its shortcomings as scholarship, enabled them to expand the range of topics in history instruction and so to enrich their courses.[79]

The same could be said of research performed by *Physiklehrer*. *Physiklehrer* returned to the past, to their seminar exercises, for the resources necessary to conduct research. They selectively deployed the techniques taught in the seminar, especially error analysis, using them when they needed to validate their results. Readers of the *Annalen* needed to know the reliability of their data; readers of *Schulprogramme* did not. But before we dismiss their quixotic adventures in trying to transform the secondary school into a quasi-academic environment or criticize their duplicity in adopting rigorous techniques of analysis when addressing one audience but not another, we should consider what their training brought to their professional lives as teachers, especially in the classroom, and to the collective mentality they shared. Schönemann's sensitivity to measurements, Böttcher's concern for the analysis of data, and Kiessling's exercise sessions would not have been possible without the benefit of instruction in

78. A. Wangerin to Neumann, 19 October 1867, FNN 53.IIA: Briefe von Schülern.
79. Anthony Grafton, "The World of Polyhistors: Humanism and Encyclopedism," *Central European History* 13 (1985): 31–47, on 40, 41.

research methods. It was not always possible, however, to integrate teaching and research satisfyingly. What Max Weber observed in 1919 could be applied to these *Physiklehrer*: that it was "hard for modern man, and especially for the younger generation . . . to measure up to a workaday existence."[80] A central tension in the lives of *Physiklehrer* sprung from the need to balance a workaday world where as teachers their intellectual authority was unquestioned and a disciplinary world where as researchers their professional credibility was gauged in part by their ability to quantify what they did not know.

80. Weber, "Science as a Vocation," p. 149.

CHAPTER TEN

The Ethos
of Exactitude

AMONG FORMER SEMINAR STUDENTS who worked at universities, *Technische Hochschulen*, and vocational schools or in precision industries, engineering, astronomy, or medicine, the techniques of an exact experimental physics were used in problems drawn from physics, chemistry, thermometry, meteorology, and astronomy as well as from forestry and physiology. Former students often used these techniques to make theory more rigorous, as Neumann had taught them. But for every adaptation of these techniques, there were modifications. More difficult problems in physics meant a deeper consideration of philosophical issues (molecular hypotheses in particular) that Neumann's physics instruction had excluded. The need for practical applications sometimes meant abandoning precision for results that were merely tractable. Technical advances in instrumentation helped some overcome barriers to using interpolation and graphical analysis. In these and in other ways, the relation between theory and experiment changed, altering the requirements for the empirical foundation of theory and thus modifying the standards for scientific truth. Neumann's particular combination of exact experiment and exact theory proved unsustainable in practice as well as in physics instruction as the century drew to a close.

Despite the waning importance of exact experiment in theoretical physics, several of Neumann's former students continued to be guided, in various ways and to various degrees, by the precepts of an exact investigative style. Their works exhibit what Jerome Ravetz has called the "craft work of science," which is "influenced by the institutional and social context in which it is conducted."[1] The focal point for understanding the

---

1. Jerome Ravetz, *Scientific Knowledge and Its Social Problems* (New York: Oxford University Press, 1971), p. 101.

"craft" these students practiced is an ethos of exactitude, which originated in the Besselian experiment as taught in the seminar.

The ethos of exactitude can be viewed as an instance of what Max Weber called an "ideal type": the synthesis of the distinctive and interrelated features of a certain kind of behavior (i.e., exact experimental practice) coupled with the values that sustained that behavior.[2] The defining feature of exactitude was the quantification of error, especially of accidental errors by the method of least squares, chosen over either the idealization of laboratory conditions or the perfection of apparatus as the preferred experimental strategy. The quantification of error helped to create exact theory, if one believed that theory could not be understood merely abstractly but had to account for the perturbing factors that affected its manifestation in nature or under laboratory conditions. This made a theoretical analysis of experimental conditions just as important as an experimental demonstration of analytic expressions. The demands of exactitude upon theory were in fact great, constraining subsequent mathematization by requiring that all analytic expressions be experimentally demonstrable by exact means and pass the stringent tests of error analysis.

The ethos of exactitude embodied a particular conception of scientific truth. Exactitude set up such high standards for certainty that it severely restricted the range of admissible data by casting doubt upon the reliability of certain techniques for processing data, such as graphical analysis, interpolation, and empirical formulas (what Neumann frequently referred to as interpolation formulas). The highest degree of truth was thought to lie in more numerous and precise measured data, but those data in themselves could not be the foundation of a mathematical expression of physical phenomena (that being merely an interpolation formula) because the limits within which the data were reliable had to be known and because all analytic expressions had to have solid theoretical underpinnings. Although exactitude denied the possibility of achieving in science the certainty of mathematical truth (errors always meant uncertainties), a mathematical conception of scientific truth nonetheless remained an unstated goal, competing with but not eliminating the more approximate, even probabilistic, conception of truth that seemed inherent in the methods of probability calculus used in the method of least squares.

Naturally no one former seminar student instantiated the ethos of exactitude in all its respects, but every former student who practiced experiment displayed enough traits of the ethos to justify treating it as a historical entity that carried deep meaning to those exemplifying even a part of it. Moreover, like Weber's ideal types, whose behavioral forms could persist in the absence of the original values that gave them meaning, the ethos of exactitude also evolved into something independent of

---

2. For an example of his ideal types see Max Weber, *The Protestant Ethic and the Spirit of Capitalism*, trans. T. Parsons (New York: Scribner's, 1958 [orig. publ. 1904–5]).

exact experimental practice in the physical sciences. The ethos of exactitude became a way of thinking identified with Neumann's seminar.

This chapter draws upon former seminar students who were advanced professionally (represented in groups A to F on Table 2). They are only one-third of all identified seminar students, but they constitute the most professionally active population. Among the 50 treated in this chapter, 32 obtained the *venia legendi* but only 22 (69%) remained in academic positions (21 in philosophical faculties and 1 in a medical faculty). The attrition rate in academics was thus roughly comparable to that in secondary school teaching, where of 83 teachers, 64 (77%) remained. It is also comparable to professions in astronomy, where 8 of the 13 who worked in observatories remained in them. Attrition rates in other institutions or professions was higher. Of the 14 who worked in *Technische Hochschulen*, 7 remained; and of the 13 who taught at vocational schools, 4 remained. Thirteen former students began their careers as assistants at physical institutes; all but one are known to have advanced to higher positions. The most professionally stable populations, showing no attrition, were the 6 who entered precision industries, engineering, and statistics and the 2 who were physicians.

Among the 13 students who took their doctorates under Neumann at Königsberg (Brix, Schinz, Ebel, Kirchhoff, Clebsch, Meyer, Saalschütz, Just, Müttrich, Wangerin, Von der Mühll, Frölich, and Voigt), 8 received the *venia legendi*. Two of these, Ebel and Just, died shortly after leaving the university (Just had taught briefly at a gymnasium). No doctoral student made a career in a secondary school, although 9 taught at one. Among the 11 who had sustained careers, three practiced primarily mathematics mixed with some mathematical physics: Clebsch and Wangerin, who eventually became *Ordinarien*, and Saalschütz, who became a nonsalaried *Außerordinarius*. The remaining 8 practiced physics: 4 as *Ordinarien* at universities, 1 as *Ordinarius* at a forestry academy, 1 as *Privatdozent* in mathematics, and 2 in precision industries. A greater percentage of higher-level professional occupations are represented among Neumann's doctoral students than in the entire population of his seminar students.

Exactitude in Practice

More than half of the former seminar students in groups A–F practiced exact experiment (Table 5). It was especially the rigorous determination of error and the subsequent analysis of data that so often determined the course of their experimental investigations, shaped judgments concerning the nature of experiment, and influenced standards of scientific explanation and truth. Although the nature of the problems they treated varied greatly, these former students all chose a refined understanding of experimental conditions—rather than exploratory experimentation or

Table 5. The practice of exact experiment in physics

Group	Total	Exact experiment	Percent
A. University	21	12	57
B. *Technische Hochschule*	7	4	57
C. Vocational school	5	1	20
D. Precision industry, etc.	6	5	83
E. Astronomy	9	4	44
F. Medicine	2	2	100
Total	50	28	56

Note: The students represented are, in *A* (sustained practice): E. Dorn, R. von Eötvös, G. Kirchhoff, O. E. Meyer, G. Quincke, W. Voigt, A. Wangerin, K. Zöppritz; in *A* (practice in early career): L. Minnigerode, C. Pape, L. Saalschütz, K. Von der Mühll; in *B*: F. O. Ruppel, L. Sohncke, H. Weber, E. Schinz; in *C*: A. Müttrich (in addition, W. Schoch's letters indicate he would have practiced exact experiment if able); in *D* (Theoretical examination of instruments and errors relative to a physical investigation): J. Amsler-Laffon, W. Brix, O. Frölich, J. Pernet, M. Thiesen; in *E* (Theoretical examination of instruments and errors relative to a physical investigation): F. E. Kayser, E. Mägis, R. Radau, H. Wild; and in *F*: H. Jacobson, C. Kohn-Akin. No other feature of a scientific style is so widely shared in groups A–F. Other approaches to physics were less broadly represented: 3 students practiced a strictly mathematical physics and 2 a physics of an unknown type. Of the others who did not practice exact experiment in physics, 8 were in mathematics or statistics, 3 in chemistry, 5 in astronomy, and 1 in an unknown field. Although the elements of an exact experimental physics overlapped with investigative practices in astronomy and were carried over into chemistry, I have not included the chemists or several of the astronomers in my counting because they did not necessarily do work relevant to physics.

theoretical speculation—as the preferred way to close the gap between theory and experiment. Hence Heinrich Wild, who after attending the seminar went to Kirchhoff's laboratory in 1858 "to set up several measuring observations" on the diffusion of salts, began by trying to find out why the theoretical assumption "that in the first moments when diffusion begins, traces of salt are already spread throughout all the water" did not "correspond to reality" by looking more deeply at the disturbances affecting his experiment so that he might modify the theory accordingly.[3] Leonhard Sohncke, aware that others might consider his investigation of the influence of temperature on the rotation of the optical axes of certain crystals superfluous because other exact studies already existed, pointed out that the small difference between theory and experiment that still remained had "stimulated" him "to examine the subject further," which he did by analyzing the errors of the experiment. Sohncke's interpretation of the accidental errors proved crucial because—in the absence of a more detailed calculation of the experiment's constant errors, including the uncertainty in the temperature readings—he considered the accidental errors an unreliable guide to the accuracy of the observations.[4]

3. Heinrich Wild to Neumann, 1 February 1858, FNN 53.IIA: Briefe von Schülern.
4. L. Sohncke, "Ueber den Einfluss der Temperatur auf das optische Drehvermögen des Quarzes und des chlorsauren Natrons," *AP* 3 (1878): 516–31, on 518.

Upon leaving Königsberg, Gustav Kirchhoff continued his purely theoretical examination of the laws of electrostatics and electrodynamics,
seeking to derive Ohm's laws from more general considerations, to create
a unified theory of electricity, and to clarify the notion of potential in
electricity. But he wrote to Neumann with a greater sense of excitement
and even pleasure about his examination of Poisson's elasticity equations
as they applied to the production of "tone-figures" by a vibrating circular
disk. Kirchhoff found two problems with them: the approximations Poisson had made in determining the solution of his equations did not satisfy
all boundary conditions, and more important, theory and experiment did
not agree. It was especially the latter that intrigued Kirchhoff because it
made "the comparison between the two to be so much more interesting."
About his work on this problem, Kirchhoff wrote: "I love it." He told
Neumann, "What made the anticipated measurements all the more interesting to me is the fact that I believed [myself] to have found a probable
explanation for the difference between theory and observation" in an
error introduced into the experiment when an ideal condition (the un-
constrained vibration of the disk) was assumed. On this basis, Kirchhoff
amended Poisson's theory. Neumann especially liked to hear of this kind
of investigation: later when recommending Kirchhoff for a position at
Heidelberg, Neumann chose to mention first that his work was "in part
the discovery of errors which are linked to a name like Poisson," only then
adding that it was also "in part new methods and new standpoints for
physics."[5]

Refining or confirming theory through error analysis meant that ideal
cases and abstractions were tactics of questionable epistemological soundness in experimentation. When Leonhard Sohncke and Albert Wangerin
collaborated on interference phenomena in thin films in the 1880s, a rival
theory complete with confirming measurements seemed to invalidate
theirs. But the detailed "theoretical" analysis that Sohncke and Wangerin
then produced gave such a complete description of the protocol, apparatus, and data of their experiment that they concluded that the formulas in
the rival theory had to be "thrown away" because they held only for the
ideal and completely unrealizable case in which the thickness of the
parallel plate on which the optical lens was placed was zero. Their own
observations, by contrast, "were not set up to confirm a finished theory";
for most of them "preceded the completion of the theory" and provided
the incentive to work the theory out according to the conditions under
which the interference phenomena actually became manifest. Sohncke
and Wangerin were really arguing not only against the supposition of

5. G. Kirchhoff to Neumann, 29 February 1848 (quote), 13 October 1848, 2 April 1849
(quote), FNN 53.IIA: Briefe von Schülern; Neumann to R. Bunsen, 20 March 1854, GLA
235/3135. See Gustav Kirchhoff, "Ueber die Schwingungen einer kreisförmigen elastischen
Scheibe" [1850], in his *Gesammelte Abhandlungen* (Leipzig: J. A. Barth, 1882), pp. 279–85
(hereafter cited as *Ges. Abh.*).

ideal conditions in an experiment but also against the predilection to view abstract theories as "naturally" "more correct." Ideal cases, they were saying, could be viewed as limits to how nature manifests itself in experiment but not as the foundation upon which theory could be based.[6]

Yet ideal conditions could not be completely avoided. It was impossible to account exhaustively for all conditions, so some assumptions inevitably had to be made. In addition, new theories in physics, especially the kinetic theory of gases, made it necessary to consider both hypotheses about the molecular behavior of matter and assumptions about the conditions (and hence errors) operating at the molecular level in an experiment. Hence the incorporation of ideal conditions became a way to test beliefs about the theoretical and hypothetical assumptions one made about the real world. In investigations by Neumann's former seminar students, however, what could be achieved by using ideal conditions was often compared to results attained by using the classical methods associated with seminar exercises, especially error analysis. This was the strategy Oskar Emil Meyer deployed in his lifelong studies on the internal viscosity of fluids, which he began by combining a theoretical consideration of kinetic theories of matter with experimental methods derived from Bessel's pendulum experiment.

After realizing that his initial measurements for the internal viscosity of gases were "too large" and hence "quite erroneous," Meyer redesigned his trials, obtained better results, and then set out in February 1863 "to determine the dependency of viscosity on the density and water content of the air." Both James Clerk Maxwell's assumption that the coefficient of viscosity depended in some way on the number of collisions in the gas and Josef Stefan's recent criticisms of Meyer's work—that he introduced errors by not considering the rotation of the molecules—forced Meyer to revise his earlier approach, even though he considered it "an advantage that I did not at all concern myself with this useless rotation" and "not worth the effort to answer Stefan in public." By the following summer Meyer thought he had achieved satisfying results confirming "the Clausius-Maxwell theory" that the viscous constant was independent of the density of the gas *provided* the density was not below a certain level. If not, Meyer argued, then "the theory is no longer suitable" because certain errors in the experiment became too large to achieve discriminating results.[7]

When he published his results in early 1865, Meyer considered the hypothesis of molecular collisions more directly but did so by determining experimentally, "*through measurements, in what way the constant of the internal*

---

6. L. Sohncke and A. Wangerin, "Ueber Interferenzerscheinungen an dünnen, insbesondere keilförmigen Blättchen," *AP* 20 (1883): 177–227, 391–425, on 193, 192, 391, 419.

7. O. E. Meyer to Neumann, 20 February 1863, 16 July 1864, FNN 53.IIA: Briefe von Schülern.

viscosity of air depends on the pressure and temperature of the air." He thought that his observations were good enough "to prove the reliability" of the hypothesis of molecular collisions because he was so confident of the method he used to obtain his measurements. He determined the viscosity of air by using pendulum experiments that involved an exhaustive analysis of errors, including an improvement of Bessel's novel hydrodynamic correction and other crucial corrections that, he reported, Neumann had "rescued from forgetfulness" and passed on to him in the course of his experiment. Thus Bessel's pendulum experiment provided the means for Meyer to test Maxwell's theory, including its crucial hypothesis of molecular collisions. Although Meyer's results supported the hypothesis of molecular collisions, he considered himself "far away from viewing the confirmation of the theory as proof for the truth of the hypothesis." He had two important reservations: he believed he had confirmed the theory only approximately because he considered it "only rigorously correct for ideal gases," and he thought it was "absolutely essential" to prove the theory and the hypothesis by another means, "namely through thermal observations."[8]

In his second study on the viscous constants of gases, Meyer explained in greater detail why he needed more measurements. He argued that "a natural law is proven not through preconceived theoretical views, also not through its agreement with *individual* observations, but only through an agreement confirmed by *all* various methods of observation through which its correctness can be proven." He was clearly setting up excessive standards for theory confirmation and reducing his chances of successfully completing his investigation. In his new trials he turned to another common exemplar found in seminar exercises, capillary action, and looked at the behavior of gases in capillary tubes in order to measure the viscous constants of gases in another way. He considered these measurements to be a "control" of his earlier ones. Meyer's treatment of experimental errors was not enough, however, to make his results "completely certain" because he believed he needed more observations in order to justify using Poiseuille's law for gases. Although the formulas he used corresponded "with very great approximation" to reality, he again viewed his results as only approximately confirming the empirical relations in Maxwell's theory: that the viscosity of a gas is independent of its pressure but increases with temperature. For want of additional measurements, Meyer did not speculate on how the coefficient of viscosity was related to temperature, and for the time being he left aside a discussion of the theory's hypothesis of molecular collisions.[9]

8. O. E. Meyer, "Ueber die innere Reibung der Gase. Erste Abhandlung: Ueber den Einfluss der Luft auf Pendelschwingungen," *AP* 125 (1865): 177–209, 401–20, 564–99, on 179–80, 180, 196, 180, 598. See also the excellent treatment of Meyer by Stefan L. Wolff, "Die Rolle von Reibung und Wärmeleitung in der Entwicklung der kinetischen Gastheorie" (Ph.D. diss., Ludwig-Maximilians-Universität, Munich, 1988), pp. 56–76.

9. O. E. Meyer, "Ueber die innere Reibung der Gase. Zweite Abhandlung: Ueber die

Conclusive results were slow to come by over the next few years. Error analysis, Meyer's principal tool for assessing experimental "proofs" of the empirical relations in the kinetic theory of gases, repeatedly forestalled the end of his investigation by setting up excessive standards for data certainty. Although Meyer argued for producing data by several means, he did not consider numerical agreement between sets of data sufficient for claiming the proof of a theory. His data on the viscosity of air agreed with data produced by Balfour Stewart and P. G. Tait and even by Maxwell, but Meyer trusted none of these results completely because, he believed, no investigator had accounted for all experimental errors. Meyer considered the errors introduced by the single wire in Maxwell's apparatus large enough to warrant using bifilar suspension. He repeated Maxwell's method with his own apparatus because "the repetition of the experiment with another apparatus guaranteed the best certainty against constant errors still on hand," even though Meyer's method of calculation gave only the square root of the internal viscosity of the air and thus doubled the constant errors present. In the end, Meyer skeptically concluded that Maxwell's data did not confirm the conclusion that the coefficient of viscosity was proportional to absolute temperature.[10]

Meyer hoped that he would "soon" return to the investigation, but first he took up two corrections whose further refinement, he thought, would help him to establish his results with greater certainty: Bessel's hydrodynamic correction for the motion of the pendulum in air and a more detailed proof of the validity of Poiseuille's law for gases in capillary tubes. For almost five years he conducted experiments with a pendulum suspended from Breslau University's highest staircase in order to distinguish the disturbances caused by two different kinds of air resistance: internal resistance and the viscosity of the air, a distinction Bessel had not made. Similar corrections, according to Meyer, had thus far been handled only through an interpolation formula and were "without theoretical grounding and of small practical value." Yet after fifteen years of additional experimentation, he was still unable to express all the disturbances affecting air viscosity measurements in quantitative terms. Not until 1887 did Meyer begin to use idealizations to overcome the calculational and technical difficulties arising in his experiments. Neglecting a part of the viscosity of air, an "assumption" he thought "not completely filled in reality" but which he considered "conceivable," he finally concluded that "the newly calculated numbers agree among one another a great deal better than the [ones] earlier found, so that my observations now satisfyingly confirm the fact that the coefficient of viscosity of air, as demonstrated by Maxwell's

Strömung der Gase durch Capillarröhren," *AP* 127 (1865): 253–81, 353–82, on 254, 255, 380, 257.

10. O. E. Meyer, "Zur Erklärung der Versuche von B. Stewart und P. G. Tait über die Erwärmung rotirender Scheiben im Vacuum," *AP* 135 (1868): 285–93; idem, "Ueber die innere Reibung der Gase. Dritte Abhandlung: Ueber Maxwell's Methode zur Bestimmung der Luftreibung," *AP* 143 (1871): 14–26, on 16.

theory, is independent of pressure and density." Meyer still thought, however, that his figures could demonstrate a higher degree of internal consistency and that the deviations among them might disappear if additional errors were accounted for. So once again he returned to a more rigorous analytic determination of the conditions affecting the measurement of the internal viscosity of the air.[11]

Meyer's investigations clearly were not about the kinetic theory of gases in the same way Rudolph Clausius's were. Yet only at a very superficial level could one say that Meyer's experiments were about measuring constants; for much more was at stake. Measuring viscous constants was a means toward several ends: to account for all errors, not by estimating them, but through rigorous analytic and mathematical procedures; to confirm theory by achieving the best possible agreement between theory and experiment; and to fulfill the expectations of an ethos of exactitude whose role in guiding his investigations was enhanced by Meyer's periodic reliance on Neumann's mathematical assistance.[12] Meyer's scientific style created a dilemma: he could not create theory as Clausius did because he did not believe he had the appropriate experimental data, yet he could not achieve the kind of exact experiment that would give him the data he wanted. His problem (as well as that of other seminar students) was not simply an overreliance on a naive empirical positivism that presumed a simple inductive approach. Theory and experiment were too closely meshed in this investigative style to give one primacy over the other. The real problem was that the demands of theory, especially of a theoretical analysis of instruments and protocol, were so tremendous that it was difficult to decide when to end an investigation. No wonder Meyer considered himself a "bad experimenter who drops not only ink but also mercury."[13] From a technical perspective, he was in fact a very good experimenter; he was just incapable of taking the kinds of shortcuts that would help him reach a definitive conclusion. Not until very late in his career, after years of experimental refinements (and frustrations), could he accept the idealization of experimental conditions as an acceptable strategy. The issue all along, however, was less the legitimacy of a particular experimental method than the determination of what conditions guaranteed the certainty of data.

Meyer's predicament was not unique. Georg Quincke's lifelong investigations of capillarity were much more than a string of ever-more-refined measurements of capillary constants. He was plagued from the start by

11. Meyer, "Ueber die innere Reibung der Gase. Dritte Abhandlung," p. 25; idem, "Ueber die innere Reibung der Gase. Vierte Abhandlung: Die Gültigkeit des Poiseuille'-schen Gesetzes für die Transpiration der Gase," *AP* 148 (1873): 1–44; idem, "Pendelbeob-achtungen," *AP* 142 (1871): 481–524, on 482; idem, "Ueber die Bestimmung der inneren Reibung nach Coulomb's Verfahren," *AP* 32 (1887): 642–59, on 645, 658.

12. As in, e.g., Meyer to Neumann, 30 March 1867, FNN 53.IIA: Briefe von Schülern.

13. O. E. Meyer to H. Wagner, 27 November 1875, H. Wagner Nachlaß, UBG-Hs.

his inability to achieve or sustain a perfect surface of rotation in his mercury (or other) sample. He was unable to compensate for this deficiency either by expressing the imperfections analytically or by making assumptions about what would happen, should a perfect surface materialize. Such experimental shortcomings generated serious misgivings about the quality of his data. Quincke's primary objective, like Meyer's, was to fashion his results on the basis of sets of data taken under different conditions for which all errors had been computed. Hence Quincke wrote in 1886 that for decades he had been trying to show how "to clarify the poor agreement between various methods of observation" so that he would be in a better position to judge the best among several different capillary theories.[14] Quite naturally he viewed his experiments collectively as individual steps in a single investigation, and he even argued that his investigations since 1858 stood "in close connection to one another and must be judged together," so one result could not be singled out for criticism without taking all others into consideration. Furthermore he believed that since he had found "a greater *number* of numerical values [of capillary constants] than anyone else," his results had to be respected.[15] Improving the condition of his data became an obsession. Quincke's desire to improve his data inhibited him from using idealizations, including any concerning the surface of his sample. His style exhibited other limitations, too. He shunned taking control measurements with the same apparatus under different conditions. Not until 1886, in response to criticism of his work, did he consider surface tension in terms of intermolecular forces.

Both Quincke's rejection of idealizations and mere estimates of errors and his emphasis on the ability to combine sets of measurements taken under different conditions meant more than establishing the epistemological soundness of data. Because he insisted on varying his experimental arrangements and combining observations from several different kinds of trials, his data, once suitably combined, transcended the material conditions of his experiments. If one did not combine sets of data from different experiments having the same purpose (e.g., measuring capillary constants), Quincke seemed to be saying, then one could be dealing with results, even laws, arising from specific experimental conditions rather than from actual properties of nature. Rigorous data analysis that paved the way for combining measurements was thus essential for separating contingent experimental conditions from the necessary conditions of theory or even of scientific truth. Data analysis of this type separated artifact from fact.

The deep concern for data analysis among Neumann's former seminar

14. G. Quincke, "Ueber die Bestimmung der Capillarconstanten von Flüssigkeiten," *AP* 27 (1886): 219–28, on 222.

15. Ibid., p. 219.

students thus created patterns of investigation that were more inclusive than any one experiment suggested. The perfection of data could become an end in itself, as it did for Ernst Dorn, who tried, in separate projects, to duplicate Königsberg's geothermal station at Breslau and to determine an absolute measure of resistance.[16] More frequently, the concern for the "proper" analysis of data meant constantly maintaining a delicate balance between theory and experiment. In their studies of Newton's rings, for instance, Sohncke and Wangerin considered it misleading to allow theory alone to guide how experimental data would be taken because, they argued, one could not always make assumptions about how theory would manifest itself in reality.[17] A theory, until it had been tested under various experimental conditions, could thus be regarded as well-formed but not as settled and stable. For instance, Heinrich Jacobson, a physician who had been a member of the seminar in 1856/57, found in his investigations of blood circulation that he had to modify hydrodynamic equations, especially Poiseuille's law, so as to use them to describe the flow of blood in the veins and the arteries, the diameters of which were larger than the tubes for which Poiseuille's law was thought to apply. Jacobson also found that Poiseuille's law might not be valid at certain temperatures or for branched tubes. Examining such conditions was necessary, "especially so that every equation does not carry the meaning of an interpolation formula but has a theoretical foundation." Like other seminar students, Jacobson was cautious about drawing theoretical conclusions from limited data, believing instead that "in order to be able to judge the course of the pressure curve for branched tubes, greater and more closely lying points than what have been given must be investigated."[18]

Investigations into the limits to the validity of certain laws always involved a consideration of errors, the presence of which affected how theory confirmation was viewed. Hence the measurements undertaken by Neumann's former students still involved several older theories—especially those of Fresnel, Poisson, Fourier, Cauchy, and Coulomb—that had not been tested under the conditions of exact experiment. About Fresnel's theory of diffraction, for instance, Voigt wrote in 1878 that even if additional observations gave "complete agreement with theory," the reliability of Fresnel's formulation could not be guaranteed because of an error that entered in consequence of Fresnel's assumptions of how ether particles operated.[19]

16. E. Dorn to Neumann, 22 February 1874, 20 December 1881, FNN 53.IIA: Briefe von Schülern.

17. L. Sohncke and A. Wangerin, "Neue Untersuchungen über die Newton'schen Ringe," *AP* 12 (1881): 1–40, 201–49, on 2.

18. Heinrich Jacobson, "Zur Einleitung in die Haemodynamik," *Archiv für Anatomie, Physiologie und wissenschaftliche Medizin* (1861): 304–28, on 319; idem, "Beiträge zur Haemodynamik," ibid. (1860): 80–113, on 113. In his investigation, Jacobson drew upon and received permission to publish Neumann's derivation of the unequal velocities of fluids in a tube, which Neumann had presented in his lectures on hydrodynamics.

19. W. Voigt, "Zur Fresnel'schen Theorie der Diffractionserscheinungen," *AP* 3 (1878): 532–68, on 533.

Conversely, it was extremely important to Neumann's former seminar students that measurements not exist without theory. Zöppritz considered his elasticity measurements of 1866 uncertain "because of a lack of *rigorous theory*" both of elasticity phenomena and of his instruments. Especially the analysis of his instruments seemed all the more necessary, he thought, because a theory had to correspond "to the superior quality of the instruments and methods used."[20] Thus his construction of theory began with an examination of instruments, experimental protocol, and errors, all considerations that Zöppritz thought made his theory more highly developed. Interested in how electrical phenomena affected geomagnetic measurements, O. E. Meyer gathered data with an apparatus designed but never used by Kirchhoff. Despite his faith in his apparatus, Meyer assumed, "The measurements probably cannot be published without a theory; without a theory they are completely worthless."[21]

Similar dialectical exchanges between theory and experiment are also found in the work of Voigt, Kirchhoff, and others.[22] Measuring constants helped that combination of theory and experiment because, as Voigt noted, determining the conditions that created variations among constants provided theory with its next task: to account theoretically for the variations and to use these results to amend the original theory.[23] So when Heinrich Weber wanted to determine how thermal conductivity varied with electrical conductivity in metals for his *Habilitationsschrift* at Göttingen (for which he drew upon his seminar exercise), he not only relied partly on Neumann's techniques, obtained several sets of data, and repeated his experiments with different metals and methods; he also was determined to use his measurements to modify the theories of Poisson and Fourier. Previous sets of data on thermal conductivity, he pointed out, showed "not inconsiderable deviations" of a "very complicated kind" because they could not be attributed to differences between metals. "The theories used to calculate the observations have only approximate validity," Weber concluded, because neither Fourier nor Poisson took into consideration thermal expansion and the change in specific heat and density with temperature. Try as he did to "rigorously fill the assumptions made in the theory," he nevertheless found that certain empirical condi-

20. K. Zöppritz, "Berechnung von Kupffer's Beobachtungen über die Elasticität schwerer Metallstäbe," *AP* 129 (1866): 219–37, on 219.

21. Meyer to Neumann, 10 May 1873, FNN 53.IIA: Briefe von Schülern.

22. See, e.g., W. Voigt, "Grundbegriffe der theoretischen Physik" [in English], WVN 8, pp. 8, 9–10; idem, *Magneto- und Electro-optik* (Leipzig: B. G. Teubner, 1908), p. iv; idem, "Phänomenologische und Atomistische Betrachtungsweisen," in *Die Kultur der Gegenwart*, ed. P. Hinneberg, Teil III, Abt. 3, Bd. 1: *Physik*, ed. E. Warburg (Leipzig: B. G. Teubner, 1915), pp. 714–31, esp. pp. 714–15; Gustav Kirchhoff, "Zur Theorie der Entladung einer Leydener Flasche" [1864], in *Ges. Abh.*, pp. 168–82, esp. pp. 169, 175; idem (with G. Hansemann), "Versuche über stehende Schwingungen des Wassers" [1880], ibid., pp. 442–54, esp. pp. 443–44.

23. W. Voigt, "Theorie der Quincke'schen Beobachtungen über totale Reflexion," *Nachrichten von der Königlichen Gesellschaft der Wissenschaften zu Göttingen* (1884): 49–68, on 49–50.

tions required by theory still could not be met, and so he subjected his apparatus, protocol, and data to a rigorous analysis of errors.[24]

From all appearances, Neumann's students were following good experimental practice by seeking rigorous agreement between theory and experiment. They might even be described as engaging in what Thomas S. Kuhn has called normal science because they were refining and determining the limits of the existing theories more than trying to create completely new conceptual frameworks.[25] But what made their work different from normal science was that normal science takes place in a community where there is a high degree of agreement over the methods and techniques considered appropriate for a scientific investigation. Neumann's former seminar students found community consensus lacking in two crucial areas of scientific practice: the role of quantification in experimentation and the role of interpolation in generalizing from experimental data, including forming theoretical conclusions. The former had been an issue when the seminar was founded but had never impeded its operation. The latter had emerged gradually during investigations based on seminar exercises. After midcentury, both became increasingly troublesome, as the controversies that erupted between Neumann's former students and others outside their circle attest.

Particularly telling are the repeated debates over the reliability of the methods and data of the French scientist Henri Victor Regnault. Much was at stake in their critical examination of his measurements. Regnault was considered one of the most accomplished French experimentalists. His measurements of specific heats in particular were regarded as authoritative, partly because he had attained an "extraordinary degree of accuracy" that far surpassed his predecessors'. Moreover, it was Regnault's careful investigative style that, according to Robert Fox, drew advanced students to his laboratory and was the basis for his national and international reputation.[26]

So it was with apparent good cause that Regnault did not react kindly to Carl Pape's scathing criticisms of his method for measuring specific heats. Regnault, who had assumed that compensating for the most prominent errors in his method was sufficient for establishing the reliability of his results, bristled at Pape's accusation that his conclusion was flawed because he had not considered *all* constant errors. Not in the habit of explaining how he viewed the practice of science, Regnault now found it necessary to articulate his philosophy of experimentation. Although he

24. Heinrich Weber to Neumann, 17 June 1869, FNN 53.IIA: Briefe von Schülern; H. Weber, "Ueber das Wärmeleitungsvermögen von Eisen und Neusilber," *AP* 146 (1872): 257–83, on 257, 258, 264, 276.

25. Thomas S. Kuhn, *The Structure of Scientific Revolutions*, 2d ed. (Chicago: University of Chicago Press, 1970).

26. Robert Fox, *The Caloric Theory of Gases from Lavoisier to Regnault* (London: Oxford University Press, 1971), p. 299.

believed that "the numerical values of these specific heats must *above all things* be compared strongly with one another," he did not consider it necessary to combine observations from different kinds of trials by analytic means (i.e., by the method of least squares), as Pape advocated. Instead Regnault thought data should be taken "with an *apparatus that as much as possible remains the same*" in the course of the trials and that the formulas used to calculate the data should remain identical as well.[27]

What bothered Regnault most, though, about Pape's discussion were the "*higher mathematical considerations* upon which [Pape] bases his method of correction." Preferring "not to follow [Pape] into this area," Regnault dismissed certain quantitative techniques as inappropriate. "Theoretical considerations of this type," he admonished, "deviate in their assumptions, one after the other, so far from reality that no experienced observer nowadays dwells on them any more." Instead, Regnault argued for a more craftsmanlike and inductive approach to experimentation, one in which the observer "draws the correction out of the experiment itself" and "uses every precaution to construct his experiment so that every [correction factor] falls out as small as possible." Yet although he thought he was eliminating extraneous considerations from experimentation by excluding excessive quantification, Regnault failed to recognize the quantitative uncertainties introduced by his own more empirical method, which was to allow the observer to find the "most probable value" of the correction "through interpolation."[28]

Regnault's method was not nearly as sound as he wanted Pape to believe. He would not, for instance, consider the "finer" points Pape brought up, for example, that some of Regnault's metal samples might have oxidized while their specific heats were being measured, a likely change that would have altered Regnault's data. Regnault did admit that his method of observation sometimes had to change from sample to sample and that when it did, he had to contend with different errors. But he did not convincingly show how he adjusted for them; nor did he substantiate either his assumption that Pape's sample differed from his or his charge that Pape's numbers were "too small." Unable to use the analytic language Pape used, Regnault tried to shift their exchange to more familiar grounds: experimental performance. Claiming Pape's method of criticism was "dictatorial" and rested on a weak foundation, he challenged Pape to meet him on his own terms: "I hope that Herr Pape, who until now has only fulfilled the function of critic, will soon give us the method of observation that makes all uncertainties, which are always present in investigations of this type, disappear. I will be the first to applaud."

27. V. Regnault, "Bemerkungen über die zur Bestimmung der specifischen Wärme fester Körper angewendeten Verfahren," *AP* 122 (1864): 257–88, on 258. Regnault was reacting to Pape's 1863 investigation; see above, Chap. 8.
28. Ibid., p. 266.

Regnault believed that Pape's experiment, if actually performed, would "probably be involved," but he was gracious enough to admit that "it will all the more serve the progress of science."[29]

Regnault's arguments in favor of the technical mastery of experimental conditions did not sway Pape, who, even before he had the opportunity to respond, emphasized the capacity of his own method (which was based on Neumann's) to account for errors quantitatively.[30] In his reply to Regnault, Pape explained that he was not trying to discredit Regnault but merely wanted "to show that his results suffered, within certain limits, from errors that can be reduced." Pape believed that Regnault was misguided in claiming "the absence of every source of error" in his trials. Although Pape acknowledged that the suitable construction of a trial could diminish an error, he still assigned priority to the quantitative determination of error because "the careful and precise execution of the trial does not alone secure the reliability of results if they are obtained according to a method that is in principle flawed." According to Pape, one had to know "in which way the various sources of error are diminished in every single trial by *corrections*" and not just by the "circumspection and skillfulness" that he claimed Regnault had used. Pape admitted that the new apparatus Regnault used reduced errors but, he warned, it did not eliminate them. Whereas Regnault wanted to ignore certain errors, Pape believed that even the "smallest error" had to be "taken into consideration with others." He cautioned Regnault from reading too much into the simple agreement between results obtained with different methods, because "an agreement of results obtained with both apparatus cannot in any case be viewed as proof of the error-free construction [of the new apparatus]."[31]

The issue of how best to perform error analysis may have dominated the exchange between Pape and Regnault, but much more was at stake in their argument than merely the role of rigorous quantification in experimentation. The correction factors Pape spoke of were not mere estimates of errors but, following the dictates of the Besselian experiment, theoretical and analytic determinations of the perturbing factors affecting instruments and protocol. Thus Pape ultimately viewed Regnault's criticisms as a challenge to the legitimacy of theoretical physics. He described Regnault's objection that mathematical theorems should not come into play in an experiment because of their alleged distance from reality as "so flashy" not only because it was "contrary to views that have been constructed about the use of mathematics in physical investigations for such a long time" but also because it appeared "to deny so entirely the great successes

29. Ibid., pp. 268, 269.

30. Carl Pape, "Ueber die specifische Wärme unterschwefligsaurer Salze," *AP* 122 (1864): 408–18.

31. Carl Pape, "Zur Kritik der Regnault'schen Versuche zur Bestimmung der specifischen Wärme fester Körper," *AP* 123 (1864): 277–98, on 280, 277–78, 281, 282, 281.

that have already been achieved on the basis of theoretical considerations in the various branches of physical investigation."[32]

The debate between Pape and Regnault demonstrated how scientific truth—obtaining it as well as reporting it—depended on one's attitude toward quantification. Regnault believed that an experiment should not contain such extensive and thorough analyses of errors that only a few persons could perform and understand it. The trustworthiness of scientific results could only be gauged, in his view, by what was essentially a subjective criterion, the craftsmanlike skill of the experimenter in eliminating errors; hence his demand that Pape perform a near-errorless experiment before Regnault would accept his criticism. For Regnault, the accessibility and communicability of an experimental investigation depended on overlooking the smaller errors that had no effect on most (he did not say all) trials, a tactic of questionable epistemological soundness in the minds of Neumann's former students. Regnault seemed to be saying that experimental results could be judged only if the investigator wrote in a clear and simple way. So, in his articles on specific heats, Regnault explained, he included only the most general parts of his experiment, editing both its protocol and results, because, he told Pape, his observations would have filled volumes and no journal would have published them completely.[33]

Pape, in contrast, believed that mathematical terms were the only ones in which problems of experimental protocol and results could be communicated and resolved. "Without mathematics," he warned, one would only be able to decide the question "with difficulty."[34] Honesty in scientific writing was thus tied to the complete disclosure and analytic portrayal of the conditions under which the data had been taken, with special attention paid to the imperfections in those conditions. When Pape reported in 1865 on Neumann's 1834 method for determining the specific heats of compound bodies and found that Neumann's results differed from Regnault's, he trusted Neumann's because the errors could be calculated and the conditions affecting the data determined.[35] Essentially, Pape wanted the experimenter to deny omnipotence but seek omniscience: the experimenter could not control all the conditions of an experiment but should try nevertheless to comprehend completely what went on in it by using the quantitative techniques of error and data analysis. Pape, like many other former seminar students, thus tied the attainment of scientific truth, as well as its degree of certainty, to the determination of error.

Precisely at this point, Pape fell into the trap that had captured so many

32. Ibid., p. 279.
33. Regnault, "Bemerkungen über die . . . Verfahren," p. 266.
34. Pape, "Zur Kritik der Regnault'schen Versuche," p. 279.
35. "Beobachtungen über die specifische Wärme verschiedener, namentlich zusammengesetzter Körper, von Prof. F. Neumann in Königsberg in Pr." (Veröffentlicht durch Dr. Carl Pape in Göttingen), *AP* 126 (1865): 123–42.

of Neumann's former students. Like Regnault, Pape believed the experimenter should try to reduce (if not eliminate) errors. But whereas Regnault could eventually become confident of his data, Pape surveyed his with perpetual skepticism. Despite the lack of certainty, Pape considered his approach advantageous because altering experimental protocol to reduce error produced new kinds of trials and hence more data, allowing the investigator to draw more informed general conclusions. In the midst of his debate with Regnault, for instance, Pape wrote to Neumann about his examination of Neumann's results on specific heats from three decades before. He mentioned his desire to create "a great observation material" that would help in the judgment of the experiment: "I intend therefore to carry out still more measurements, all the more" than initially anticipated because successive sets of measurements produce the conviction that the "perfection of the observation" is a "duty."[36]

Pape did not recognize the trap, nor did he acknowledge the extent to which his and Regnault's epistemologies of experimentation, separated by degrees of exactitude, were incommensurable even when their experiments sought the same goals. Others from the seminar (often acting on Neumann's suggestions) reworked Regnault's experiments from a theoretical perspective, always reaching the same conclusion: Regnault had not considered all errors significant enough to affect his data. Cároly Kohn-Akin, a physician who attended Neumann's seminar in 1860/61, assessed "the degree of numerical accuracy in the several formulas which have been computed by Regnault in the latter part of his celebrated memoir on Boyle's (or Mariotte's) law." Through careful experimentation, Regnault had already found that Boyle's law was only approximately valid. In reviewing Regnault's experiments, however, Kohn-Akin determined that Regnault's assumption that the pressure of the lowest layer of a mixture of compressed gases was the same as that of other layers was incorrect. Kohn-Akin concluded that Regnault's modification of Boyle's law was "inapplicable to the densities and pressures of different layers [of gases] promiscuously combined" but that the "appreciable numerical errors" in Regnault's data could be readily computed if one knew the various pressures in the gas mixture. From here Kohn-Akin found it "easy and instructive to translate the meaning of this arithmetical operation into the language of experiment." Later he also looked at Regnault's experiments on specific heats, which he judged to "have failed to yield results possessing the confidence of philosophers." He thought it "important that some new method should be devised, direct in theory and trustworthy in practice, capable of supplying the want of knowledge of the exact and real specific heats of gases in a manner that might command approbation." Like Pape, however, Kohn-Akin described but did not perform his alternative experimental method. Nevertheless he had complete confidence in it because "none of the difficulties enumerated . . .

36. Carl Pape to Neumann, 11 July 1865, FNN 53.IIA: Briefe von Schülern.

appear greater than what are habitually encountered in researches of this nature."[37]

Despite Regnault's position that "higher mathematical considerations" had no place in an experimental investigation because they were "so far from reality," he frequently deployed the one quantitative technique that Neumann's students continued to shun for exactly that reason: interpolation. To be sure, several of Neumann's students used interpolation, but always cautiously and for restricted reasons. Although they gradually accepted interpolation as a legitimate means of detecting additional errors, they rarely used it to enhance data to aid in the construction of a theoretical relation. Many shared Zöppritz's opinion that laws based on a limited range of data, particularly when the instruments used to collect the data were insufficiently understood, should be regarded with suspicion. O. E. Meyer called Poiseuille's law an "interpolation formula" when he found that it did "not hold with certainty as an expression of an empirically determined actual natural law" because the temperature range in Poiseuille's original experiments had been too small.[38] Lacking faith in both the supplemental "data" produced by interpolation and the results based on it, Neumann's students regarded interpolation as the Achilles' heel of data analysis.

So when Gustav Kirchhoff examined Regnault's measurements on the pressure (*Spannung*) of steam as a part of his own studies on the mechanical theory of heat in 1858, he took issue with parts of Regnault's analysis of data and errors, especially his use of graphical and other forms of interpolation. Although Regnault had demonstrated graphically that the curve representing the pressure of steam below 0°C was continuous with the curve representing the pressure above 0°C, Kirchhoff was not convinced. Theory, he argued, actually required that the tangents of the two curves be different at 0°C, and so the union of the two curves was "discontinuous" (i.e., not smooth) at that point. Regnault did not recognize the discontinuity, according to Kirchhoff, because he had not carried his analysis of errors far enough. Hence even though Regnault's results had the order of magnitude demanded by theory, the actual difference in the values at 0°C from the two curves was in fact "smaller than could be recognized with certainty from Regnault's trials."[39]

37. C. Kohn-Akin, "Note on the Compressibility of Gases," *Philosophical Magazine* 25 (1863): 289–300, on 289, 290, 293n, 295; idem, "On a New Method for the Direct Determination of the Specific Heat of Gases under Constant Volume," ibid. 28 (1864): 341–46, on 342, 346. So closely did Kohn-Akin's method follow Neumann's that he felt obligated to request permission to publish it; see C. Kohn-Akin to Neumann, 3 June 1862, FNN 53.IB: Briefe anderer Schüler.

38. K. Zöppritz, "Das Verhalten des Meerwassers in der Nähe des Gefrierpunktes und die Statik der Polarmeere," *AP* E5 (1871): 497–540, on 500–501; O. E. Meyer, "Beobachtungen von Adolf Rosencranz über den Einfluss der Temperatur auf die innere Reibung von Flüssigkeiten," *AP* 2 (1877): 387–404, on 395.

39. Gustav Kirchhoff, "Bemerkungen über die Spannung des Wasserdampfes bei Temperaturen, die dem Eispunkte nahe sind" [1858], in *Ges. Abh.*, pp. 482–85, on p. 484.

Later, when Kirchhoff was studying the pressure of steam produced from a mixture of water and sulphuric acid, he again looked at Regnault's trials to rigorously compare theory and experiment. This time, however, it was a matter not simply of criticizing one point in Regnault's graphical representation but of raising a far more serious issue. Regnault's graphical technique, Kirchhoff explained, was to represent the pressure of the steam mixture as a curve constructed from three measured points and then, from that curve, to interpolate a formula having three constants. Assuming that Regnault had "deemed" the value of the pressure derived from the interpolated formula "more accurate than individual observations," Kirchhoff rejected Regnault's numbers and instead worked directly from tables of measured data. In his search for errors he found that a significant amount of heat was produced when the water and the acid were mixed and hence errors could be neglected only if temperature changes were small—which was not the case, he believed, for Regnault's experiments.[40]

What is interesting about his examination of Regnault's work is that Kirchhoff, who hitherto seems to have escaped entrapment in Neumann's methods of data analysis, now found Neumann's skepticism of interpolation and graphical representation appropriate for criticizing the soundness of Regnault's reasoning in linking measured data to a generalized conclusion, a theoretical formula, that might be used as a foundation of theory. As a result, Kirchhoff, like Pape, found himself at odds with other trends in experimental practice and became engaged in controversy. The experimentalist Adolph Wüllner, who had undertaken experiments similar to Regnault's for his *Habilitationsschrift*, took issue with Kirchhoff's emphasis on error analysis in Regnault's experiment because he suspected the heat production was not as large as Kirchhoff assumed it to be and hence doubted the necessity of the corrections Kirchhoff had added to Regnault's formulas. To Wüllner, the thoroughness of error analysis was unimportant because Kirchhoff and Regnault attained experimental results whose agreement was, in Wüllner's view, "very extraordinary." Even though he agreed with Kirchhoff that several of Regnault's assumptions were "not rigorously correct," Wüllner considered Regnault's numbers to possess "extraordinary accuracy." So simply on the basis of what looked like reasonably good agreement between data sets (his own differed somewhat from Regnault's but not enough to question them), Wüllner found little to dispute in Regnault's results or even in his method.[41]

Kirchhoff found it astonishing that Wüllner considered the agreement between the two sets of results exceptional but still doubted the veracity of

40. Gustav Kirchhoff, "Ueber die Spannung des Dampfes von Mischungen aus Wasser und Schwefelsäure" [1858], in *Ges. Abh.*, pp. 485–94, on p. 486.

41. Adolph Wüllner, "Einige Bemerkungen zum Aufsatz des Hr. Kirchhoff 'Ueber die Spannungen des Dampfes von Mischungen aus Wasser und Schwefelsäure'" [26 October 1858], *AP* 105 (1858): 478–85; on 479, 485.

Kirchhoff's corrected formula. Rather than identify the exact basis of their disagreement himself (he thought it might be in the assumptions he made in his own formula, in Wüllner's measurements, or in the conclusions Wüllner drew from his measurements), Kirchhoff considered this a matter best left "to the reader," who should "judge if these differences were small enough to be considered an error of observation or not, and so to judge if my formula agreed with or contradicted Regnault's." Yet even though his methods of data and error analysis were intended to facilitate an impersonal third-party evaluation of scientific results, Kirchhoff could not in fact leave everything "to the reader" because there was no uniform view of how quantification should be used in experimentation. Moreover, unquestionably he regarded his own method as epistemologically superior because it did not rely on interpolation. As he exhorted Wüllner and the readers of the *Annalen*,

> If one wants to prove through the comparison with measurement the correctness of a theoretical formula in which unknown constants appear, then there is in my opinion no other more certain way than to look out for whether the constants can be determined so that the difference between the *direct* measurement and the corresponding value that the formula gives lies overall within the limits of the most probable error of observation. This way is the one I have followed in the proof of my formula. Herr Wüllner went another way. He did not believe in having to depend upon direct observations and tied his views to certain coefficients that Regnault had derived . . . with the help of an interpolation formula. The certainty of these coefficients cannot be evaluated without going back to the observations themselves.

Furthermore, Kirchhoff explained, one would be led "astray" by "overestimating the certainty" of Regnault's readings by relying on the regularity of the interpolation formula. But the issue was not merely one of the accuracy of interpolated values. Because error analysis demonstrated that the accuracy of Regnault's numbers was less than Regnault had claimed, Kirchhoff of course cautioned against too much confidence in the conclusions drawn from them. What Kirchhoff feared was that the uniqueness required of theoretical conclusions would be violated by interpolation because there were "abundant examples for the fact that similar observations with *the* accuracy that these have can be represented through different formulas."[42]

Like Regnault, however, Wüllner refused to assign priority to a rigorous data and error analysis in deciding the legitimacy of conclusions drawn from experimental data. He claimed to have proceeded from the view that "numbers obtained by an accurate interpolation from observa-

---

42. Gustav Kirchhoff, "Erwiderung auf die Bemerkungen des Hrn. Wüllner" [21 January 1859], *AP* 106 (1859): 322–25, on 322, 323, 324 (essay not included in the collected works).

tions must correspond just as accurately to the measurements so that one can use them instead of the numbers directly obtained." But rather than elaborate on the accuracy of his interpolated values and continue the debate in Kirchhoff's terms, Wüllner impuned the integrity of Kirchhoff's reasoning by claiming that it implied a negative judgment of Regnault's ability as an experimentalist. He interpreted Kirchhoff's opinions on the inadequacies of interpolation, especially Kirchhoff's conviction that a sequence of observations cannot be replaced by a sequence of interpolated values, as a lack of trust in the experimenter. Thus like Regnault, Wüllner believed that the proper basis for judging the reliability of data, including interpolated values, was the professional reputation of the reporting scientist. Without acknowledging that the degrees of accuracy possible or required in different kinds of investigations varied greatly, Wüllner argued that because in the present instance "the agreement of the *direct* measurements with the interpolated ones is just as accurate as in the other famous works of Herr Regnault," Regnault's conclusions and his data should be accepted.[43]

What these two debates over Regnault's results illustrate is how the commitment to certain investigative techniques translated into distinctive forms of scientific reasoning and judgment. Neither the similarity of experimental arrangements nor the attainment of comparable results provided in themselves sufficient reasons for accepting the data or the conclusions of others. Profound epistemological consequences followed. Just as the styles of experimentation represented by Pape and Regnault were incommensurable, so were the methods of data analysis practiced by Kirchhoff and Wüllner. These two debates projected a sense of smugness on the part of Neumann's former students, who believed in the technical and even the epistemological superiority of investigations combed for errors over those where errors had not been so thoroughly treated. (Indeed Karl Zöppritz even believed that the introduction of error analysis into a field of investigation made that field "*wissenschaftlich*.")[44]

To a certain degree the habits of scientific thinking fostered by the ethos of exactitude lingered among Neumann's students. The preference of Kirchhoff and other former seminar students for directly measured values over interpolated ones was especially evident in their continued reluctance to deploy the one technique that required uncompromising faith in interpolated values, the graphical analysis of data. Kirchhoff used graphs but always as he had in his maiden investigations: as visual depictions of patterns of change in the phenomena under investigation, not as

43. Adolph Wüllner, "Entgegnung auf die Erwiderung des Hrn. Kirchhoff" [15 March 1859], *AP* 106 (1859): 632–37, on 633–34.

44. See Karl Zöppritz's "Stanley's thermobarometrische Beobachtungen auf seinem Zuge durch Afrika" (*Geographische Mitteilungen* 28 [1882]: 94–98), where he differentiates "cookbook calculations" from the "scientific treatment of hypsometric materials" on the basis of whether the data had been analyzed for errors (p. 98).

resources in determining theoretical relations. Voigt, when he had only second- or even third-place accuracy, considered graphs useful only for gross qualitative conclusions or applied purposes because constants derived from them were but "raw approximation[s]" of their true values. That he still called graphs "curve tables" and used dotted lines to depict interpolated values were lingering signs of the epistemological priority he assigned to measured data. Even Carl Neumann, who belonged to the minority of Neumann's former students who did not perform exact experiment but who emphasized the mathematical side of mathematical physics, was skeptical of graphs because they represented ideal conditions and so corresponded "little to reality."[45]

The ethos of exactitude thus helped to shape a matrix of technical practices that created particular habits of scientific thinking affecting the treatment of theory as well as of experiment. Just as ideal cases and interpolation were rarely, if ever, used in experiment, analogies and hypotheses had restricted roles in discussions of theory. In the seminar Neumann had used analogies to guide physical reasoning, but for the most part such analogies were formal mathematical models based on Fourier's theory of heat, on potential theory, or elasticity theory, none of which necessarily required additional hypothetical constructions. In actual practice, analogies were regarded as approximations that had to be refined through measurement. For instance, O. E. Meyer regarded Zöppritz's attempt to carry "through a mathematical analogy" between the laws of heat conduction and those of the viscosity of fluids "very good" but considered his theory only an approximation because "the numerical observational data [were] too little guaranteed."[46] This did not mean, however, that certainty was equated merely with the ability to measure; for raw measurements were considered fundamentally unreliable and hence inadequate foundations for knowledge. Error analysis could help one determine the bounds within which knowledge based on measurements was certain.

The emphasis on error analysis was largely responsible for the paucity of philosophical speculation in the researches of several former seminar students. Although someone like Voigt considered it acceptable at times to proceed on the basis of "working hypotheses" in order to achieve "progress in the theoretical," such hypotheses were not to be confused with the underlying reality. For the most part he, like many other former seminar students, preferred a "phenomenological physics" which did not make "use of hypotheses concerning the mechanism of the process" and, more important, which did "not look for other proof than the agreement

45. W. Voigt, "Theorie der absorbirenden isotropen Medien, insbesondere Theorie der optischen Eigenschaften der Metalle," *AP* 23 (1884): 104–47, on 126; Carl Neumann, *Vorlesungen über die mechanische Theorie der Wärme* (Leipzig: B. G. Teubner, 1875), p. xi.

46. Meyer to Wagner, 4 February 1880, H. Wagner Nachlaß, UBG-Hs.

of the results of theory with the results of experiment." Hypotheses, when they were used, were subject to an "exact proof" in the sense that the laws derived from them had to be subject to observational confirmation (which included the analysis of errors).[47]

The problem with this way of thinking was that it inhibited the kinds of thinking needed to achieve theoretical results. The ethos of exactitude guided scientific reasoning beyond the mere description of a physical process, particularly as manifest under laboratory conditions, but it did not especially stimulate higher orders of explanation. Over and over again the construction of a scientific law was held up for want of more data (appropriately reduced), even in areas where additional data carried marginal meaning. When Anton Müttrich was appointed professor of physics and meteorology at the forestry academy in Eberswald in 1873 after being a secondary school teacher for fifteen years, he brought to forestry the techniques of error analysis he had learned in the seminar. He believed that meteorological observations could provide the basis for *laws* "about the connection between plants and their environment." In 1888 he tried to develop a theory of phenological observations, suggesting that periodic changes in vegetation might be understood in terms of their lawful dependency upon temperature or upon the square of temperature. But the law that he had hoped to create eluded him in the end. Although his errors were small enough to confirm that he had measured a significant effect, for want of still greater certainty, he believed he needed more observations than the several years of data he had already collected.[48]

Despite the prominent role of mathematics in the type of theoretical physics promoted by the ethos of exactitude, a purely mathematical approach to the study of physical phenomena did not strongly challenge one that relied on both mathematics and exact experiment. Among Neumann's former seminar students was a minority who had little to do with exact experiment, preferring instead to consider physical problems in the context of mathematical considerations. Alfred Clebsch was not inclined to compare mathematical results with exact experimental data rigorously—not even in the applied problems in forestry he took up in 1867. Karl Von der Mühll and especially Carl Neumann believed that progress

47. W. Voigt, "Ueber Arbeitshypothesen," *Nachrichten von der Königlichen Gesellschaft der Wissenschaften zu Göttingen*, Math.-phys. Klasse (1905): 98–116, on 102; Karl Försterling, "Woldemar Voigt zum hundertsten Geburtstage," *Die Naturwissenschaften* 38 (1951): 217–21; Voigt, "Grundbegriffe der theoretischen Physik," p. 10; idem, *Festrede im Namen der Georg-Augusts-Universität zur akademischen Preisvertheilung am 4. Juni 1894* (Göttingen: Dieterich, 1894), p. 15.

48. Anton Müttrich, "Die zu forstlichen Zwecken im Königreich Preußen und in Elsaß-Lothringen errichteten meteorologischen Stationen," *Zeitschrift für Forst- und Jagdwesen* 7 (1875): 425–33, on 425; idem, "Ueber phänologische Beobachtungen," ibid. 20 (1888): 321–47; idem, "Ueber den Einfluss des Waldes auf die periodischen Veränderungen der Lufttemperatur," ibid. 23 (1890): 385–400, 449–58, 513–26.

in physics would be achieved through purely mathematical considerations. These individuals promoted more idealized mathematical representations of physical processes, often to the point where not all expressions and terms could be linked immediately to measurement. They also had a more secure and more highly probable conception of scientific truth—one less fettered by epistemological skepticism concerning the certainty of data—than the qualified and often conditional truth that always seemed to issue from exact experiment. Among Neumann's students, however, such individuals never represented the majority, who saw dangers in using mathematics without considering exact experiment. Heinrich Martin Weber, drawing on Neumann's application of Bessel functions to the electric current, acknowledged that one may not be able to construct measuring operations from mathematical descriptions owing to certain analytic problems, such as the unwieldy convergences of some series. Voigt too observed that "involved mathematical developments" could "frequently interrupt the general physical deliberations for a long time, without leading to results of real physical significance," that is, to results amenable to measurement.[49]

Far greater threats to the integrity and usefulness of the ethos of exactitude—especially to its insistence on exact experiment and its reservations about the epistemological soundness of graphs and interpolation—came from changes in measuring practices, especially in the state of instrumentation. When Neumann created his idea of exact experiment and transferred it to physics pedagogy, where it became a model for scientific practice, precision instruments did not exist. Technical improvements by French and especially English instrument makers at the end of the eighteenth century and by German instrument makers in the first half of the nineteenth century (including the Repsolds in Hamburg, Pistor in Berlin, and Frauenhofer in Munich) had, of course, resulted in more precise data.[50] But the confirmation of several mathematical theories in physics and especially the imperative to determine which of several alternative experimental techniques were the most accurate required refined measurements that existing instruments could not provide. The problem was not merely a technical one, however, because physicists also

49. A. Clebsch, "Ueber ein Problem der Forstwissenschaft," *JRAM* 67 (1867): 248–65; Heinrich Martin Weber, "Ueber die Bessel'schen Functionen und ihre Anwendung auf die Theorie der elektrischen Ströme," *JRAM* 75 (1873): 75–105, on 105; Carl Neumann to Otto Wiener, 29 November 1902, quoted in H. Salié, "Carl Neumann," in *Bedeutende Gelehrte in Leipzig*, ed. G. Harig, 2 vols. (Leipzig: Karl Marx Universität, 1965), 2:13–23, on 14–15; W. Voigt, *Kompendium der theoretischen Physik*, 2 vols. (Leipzig: Veit, 1895–96), 1:iii.

50. J. L. Heilbron, *Electricity in the Seventeenth and Eighteenth Centuries* (Berkeley and Los Angeles: University of California Press, 1979), pp. 71–83, esp. pp. 78–83; Maurice Daumas, "Precision of Measurement and Physical and Chemical Research in the Eighteenth Century," in *Scientific Change*, ed. A. Crombie (New York: Basic Books, 1963), pp. 418–30, esp. 428; W. D. Hackmann, "The Relationship between Concept and Instrument Design in Eighteenth Century Experimental Science," *Annals of Science* 36 (1979): 205–224, esp. 219.

lacked a theoretical understanding of what and how they were measuring. Error and data analysis thus served as tools for understanding the apparatus and protocol of experiment and, until precision instruments were available later in the century, as the principal means of ascertaining the accuracy, precision, and certainty of results.

Neumann and his former seminar students did make some technical refinements in instruments and apparatus, but the principal focus of their efforts was on determining errors through the theoretical examination of experimental protocol and instruments for the purpose of increasing the accuracy and precision of data. Neumann's technical refinements were less important to him than his theoretical contributions to exact experiment, including his theory of the method of mixtures for determining specific heats, his theory of the bimetallic thermometer, and his methods for calibrating balances and thermometers and for determining the electromotive force of a Wheatstone bridge without knowing its resistance. Among his students, too, theoretical discussions of instruments and their errors far outnumbered technical improvements in instrument construction. They usually treated instruments that had been around for some time but were still imperfectly understood. Oskar Frölich showed how to compensate for the effect of geomagnetism on an electrodynamometer, noting that to solve the problem he had to use "French mathematical physics." Although Heinrich Weber thought it acceptable to make galvanometers reliable by empirical means, he considered a theoretical examination of the instrument to be a superior method for assessing reliability. His 1887 theory of the Wheatstone bridge was actually a theory of the instrument's errors. Eduard Mägis, who from St. Petersburg worked on various kinds of thermometers for Neumann, considered a "theoretical examination" of them desirable. Johann Pernet's numerous studies of thermometric calibrations and readings constituted a theory of that instrument. In 1882, Max Thiesen gave the ideal balance the theoretical treatment he thought it demanded, but he still thought it necessary to calculate correction factors based on error analysis. Thiesen's remark about the value of an instrument—that its errors not necessarily be small but be calculable—sums up the attitude toward instruments that Neumann cultivated in the seminar.[51]

This preference for the analysis of instruments, errors, and data followed the precepts of the Besselian experiment, where the mechanical

51. O. Frölich, "Das kugelförmige Electrodynamometer," *AP* 143 (1871): 643–53, on 644; H. Weber, "Zur Theorie der Galvanometer," *Carl's Repertorium* 11 (1875): 223–40; idem, "Zur Theorie der Wheatstone'schen Brücke," *AP* 30 (1887): 638–55; E. Mägis to Neumann, 14 May 1873, FNN 53.IIB: Briefe anderer Schüler; J. Pernet, "Beiträge zur Thermometrie," *Carl's Repertorium* 11 (1875): 257–309; Max Thiesen, "Zur Theorie der Waage und Wägung," *Zeitschrift für Instrumentenkunde* 2 (1882): 358–65; 3 (1883): 81–89; idem, "Eine Erweiterung der Neumann'schen Methode zur Kalibrirung von Thermometern," *Zeitschrift der österreichischen Gesellschaft für Meteorologie* 14 (1879): 426–30, on 430.

perfection of instruments and the technical improvement of experimental protocol were regarded as useful but limited in their ability to attain higher degrees of epistemological certainty. It is noteworthy that Neumann's former students regarded even self-registering instruments with suspicion. Although Heinrich Wild acknowledged the obvious advantages of self-registering instruments, he also insisted that their data be analyzed for errors in order to be useful for science. Rudolph Radau thought similarly. In a critical commentary on Bessel's personal equation (which included a discussion of a wide variety of personal errors for which Bessel had given no calculations), Radau warned that "one must not believe that the application of electricity and photography to the automatic registering of observations admits of greater accuracy" because new and different kinds of errors crop up and have to be calculated too.[52]

Technically feasible and promoted by diverse political, economic, and military interests, precision instruments were manufactured in greater numbers in Germany in the final three decades of the nineteenth century. The enhanced popularity of exploratory experimentation in physics—as well as the concomitant sharper distinction that could be drawn between theoretical and experimental physics—which surfaced at the end of the century was undoubtedly partially the result of the widespread availability of more refined instruments. Contrary to the spirit and the intent of the Besselian experiment that had guided investigations in the seminar, precision instruments offered the opportunity to reassess the function of exactitude in theory and experiment, especially to reconsider the role of error analysis. Yet in spite of this technological progress in experimentation, Neumann's former students did not completely abandon their methods of experimental practice. Georg Quincke's belief that correcting for errors was just as good if not better than using more refined instruments was one indication of the continued strength of this way of doing physics. What precision instruments did was to make arguments for the exhaustive analysis of errors and data less compelling to those outside the seminar circle because instrumentation itself now could guarantee part (if not most) of the certainty that Königsberg seminar students had been seeking in more theoretical and analytic ways. To the extent that this particular ethos of exactitude, with its origins in the Besselian experiment, survived, then, it did so largely within the seminar circle.[53]

52. H. Wild, "Die selbstregistrirenden meteorologischen Instrumente der Sternwarte in Bern," *Carl's Repertorium* 2 (1866): 161–201; Wild to Neumann, 21 November 1864, FNN 53.IIA: Briefe von Schülern; R. Radau, "Ueber die persönlichen Gleichungen bei Beobachtungen derselben Erscheinung durch verschiedene Beobachter," *Carl's Repertorium* 1 (1866): 202–18, 306–21; 2 (1866): 115–56, on 310.

53. W. Voigt, *Physikalische Forschung und Lehre in Deutschland während der letzten hundert Jahre: Festrede im Namen der Georg-August-Universität zur Jahresfeier der Universität am 5. Juni 1912* (Göttingen: Dieterich, 1912), p. 20; J. K. Rees, "German Scientific Apparatus," *Science* 12 (1900): 777–85; Georg Quincke, "Moderne Kritik der Messungen der Capillaritätsconstanten von Flüssigkeiten und die specifische Cohäsion geschmolzener Metalle," *AP* 61

Yet even within this circle, parts of the ethos weakened. Brix, Pernet, Amsler-Laffon, Thiesen, Wild, Frölich, and others applied the exacting methods from the seminar to problems from the electrical and telegraph industries, thermometry, metrology (especially the reform of weights and measures that sought to place them on a more rigorous scientific basis), and the heating power of fuels, to name just a few. In each case the practical need to reach a solution blocked the endless search for and quantification of error and forestalled an overly excessive skepticism of data. In some applications, precision could not be pursued as uncompromisingly as it had been in the formative experiences of those who had lived by it. Morever, inhibitions against using two quantitative techniques that facilitated theoretical generalizations—interpolation and graphical analysis—were ignored in the context of applied problems. Brix and others who worked on telegraphs knew that measurements of the resistance of underground cables were afflicted with disturbances not present when the same cable had been tested under laboratory conditions. One way to pinpoint the errors, Brix thought, was through graphical analysis, using the graphically determined difference between the laboratory (or ideal) and actual underground resistances as a measure of disturbing factors. Frölich deployed interpolation and graphical analysis in 1881 to understand theoretically (i.e., in terms of errors) the electric power transmission of a dynamo. He also used graphical analysis to determine the range over which instruments could be used by technicians who wanted to avoid the analytic determination of errors. Eventually even the academic physicists among Neumann's former seminar students admitted that interpolation and graphs were useful, at least for *detecting* errors from data. Voigt couched his acceptance of these techniques in terms of their *visual* advantages, maintaining that they showed at a glance and "more vividly" errors of observation that numerical tables might otherwise "hide."[54]

Although the techniques associated with the ethos of exactitude were intended to bridge theory and experiment, they also created a gap between them by not promoting more strongly certain mental habits, such as hypothesis formation and theoretical speculation, that would have allowed the investigator to extract himself from the demand of technical performance. In place of the creative freedom that is often associated

---

(1897): 267–80, on 270. For a particularly appropriate example of the persistence of the ethos see the gravitational experiments of Roland von Eötvös, especially "Untersuchungen über Gravitation mit Erdmagnetismus," *AP* 59 (1896): 354–400; "Experimenteller Nachweis der Schwereänderung . . . ," *AP* 59 (1919): 743–52; and R. von Eötvös, D. Pekár, and E. Fekete, "Beiträge zum Gesetze der Proportionalität von Trägheit und Gravität," *AP* 68 (1922): 11–67.

54. P. W. Brix, "Mittheilungen über die an unterirdischen Leitungen ausgeführten Messungen," *EZ* 2 (1881): 3–6; O. Frölich, "Versuche mit dynamoelektrischen Maschinen und elektrischer Kraftübertragung und theoretische Folgerungen aus denselben," ibid., pp. 134–40, 170–74; Voigt, "Theorie der absorbirenden isotropen Medien," pp. 129–30.

with theory formation, there were restraining compulsions and inhibitions. The postseminar investigations of Neumann's students remind one of Neumann's earlier rejection of the method of repetition in experimentation. Rather than merely *repeating* a trial (often immediately so as to guarantee identical conditions), Neumann, with Bessel's seconds pendulum investigation in mind, argued for *re*-petitioning the experiment: going back to the experiment more than one time so as to inquire about errors. In the hands of his students, however, "*re*-petitioning" often became a repetition compulsion. Neumann encouraged this *re*-petitioning of experiment, not theoretical or philosophical speculation. It is revealing that after Georg Quincke discovered diaphragm currents in 1858 and began in several letters to Neumann to speculate on the broader significance of his discovery and on the "many theoretical questions" that interested him "greatly" (including the nature of electricity), he wrote, "I know that you will probably write unhappily and that much in these considerations is very bold. But you also know how much I value your judgment and that I have always striven to follow the path which you laid out for me . . . because I was so happily your listener and your student. Unfortunately my abilities have not allowed me to improve my mathematical resources as much as I would have liked. But at least I try, as best I can, to hobble slowly along."[55] Steering his way between technical conformity and theoretical creativity, Quincke felt the tensions generated by deferring to the ways of a master while trying to follow one's own inclinations.

Exactitude in Physics Pedagogy

Exact experiment did not always figure as prominently in the institutionalization of theoretical physics outside Königsberg. The demise and eventual disappearance of exact experiment from the practice of German theoretical physics by the end of the century followed decades of ambivalence about what constituted mathematical or theoretical physics, especially in an educational setting. Although Neumann's students considered exact experiment a part of their charge, they did not always believe they could practice theoretical physics from a position in experimental physics, as almost all positions in physics were defined before the 1870s. In 1850, Kirchhoff took a position in experimental physics at Breslau, accepting the fact that he would have to "stretch out from the circle of thinking in which [he] had been used to moving." He found it difficult, however, to cultivate mathematical physics there and especially to pursue his research because he had to share the physical cabinet. Notably, when he told

55. Georg Quincke to Neumann, 29 October 1858, 27 March 1859, 24 December 1860, 14 October 1861 (quote), FNN 53.IIA: Briefe von Schülern. See Georg Quincke, "Ueber eine neue Art elektrischer Ströme," *AP* 107 (1859): 1–47; 110 (1860): 38–65.

Neumann in February 1854 that he wanted to leave Breslau, he was working on a problem taken from Bessel's continuing experiments on the pendulum—a pendulum suspended from an elastic spring—hoping that his measurements would be advantageous for determining elasticity constants. In that year he was offered a position at Heidelberg, where he had greater material resources. Wilhelm Weber regarded Kirchhoff's ability to combine theory and exact experiment highly, writing in 1854 that if Kirchhoff were to work with Robert Bunsen at Heidelberg, "the great superiority which [he] possesses in the field of theory and mathematics among the physicists of Germany (almost indeed with the sole exception of Neumann in Königsberg) combined with the talent of an exact experimenter would lead to especially great success."[56] Kirchhoff's good fortune was not shared by others, however. The ability to perform exact experiment, and hence to practice theoretical physics as they had learned it, was often compromised by the constraints of the subordinate (non-*Ordinarien*) positions and by inadequate access to funds that could provide necessary material resources.

The exact experimental side of Neumann's variety of theoretical physics was further compromised by those who emphasized the mathematical side of theoretical physics and who excluded experimentation entirely. Significantly, Voigt later called this the "left side" of theoretical physics, one that stood aloof from observations but that sought mathematical clarity and consistency in theory, in contrast to the "right side," which included experiment and used rigorous mathematical techniques to understand observations.[57] By personal predilection, some of Neumann's former students—among them, Alfred Clebsch, Carl Neumann, Karl Von der Mühll, and Eduard Gehring—did not practice experiment and taught mathematical physics in conjunction with mathematics. Yet blurred boundaries between mathematics and theoretical physics were not always of their own doing, as the case of O. E. Meyer illustrates.

While *Privatdozent* at Göttingen in February 1863, Meyer was asked to consider a position at Graz, where the faculty wanted "a mathematician who would be versed in physics." Meyer told Neumann that he could not recommend himself because he was "no mathematician," although he considered himself "capable of knowing as much mathematics" as he would have had to know to teach it. Two months later, when the physicist Gustav Wiedemann discussed Meyer's candidacy for a mathematics position at Basel, he reported that he did not know if Meyer "turned more to physics or to mathematics." Eventually Meyer did take an *außerordentlicher* position in mathematics at Breslau in 1864, but as he explained to Neumann on 16 July 1864,

56. W. Weber to R. Bunsen, 12 March 1854, GLA 235/3135.
57. Voigt, *Physikalische Forschung und Lehre*, p. 12.

I go to Breslau as the follower of [the mathematician Rudolf] Lipschitz; therefore I have to offer courses in mathematics. My position is, however, in other respects essentially different from Lipschitz's. I may consider myself at the same time and to a certain degree namely the successor to Kirchhoff and therefore I assume that I am appointed not only for mathematics but also for physics. Everything rests not on official arrangements but simply on private agreements with [the mathematician Eduard] Schröter and Olshausen [a Prussian ministry official]. . . . Both desire that mathematical physics be read again. . . . Officially they have not prescribed for me the choice of lectures or a *Fachwissenschaft*. I am pleased that the minister has left unsettled this rather delicate point. Therefore I can now entirely follow my own will and I will do that. I consider as my real task to achieve for mathematical physics the ground which it has lost since Kirchhoff's departure.

Teaching mathematics did not bother Meyer, who realized that his students would need certain mathematics courses in order to learn mathematical physics. He even confessed to Neumann that "pure mathematics exerts upon me such a significant force of attraction that earlier it tempted me to be unfaithful to physics." But months later Meyer concluded that Neumann had exaggerated his abilities when he heard that Neumann had recommended him for a full professorship in mathematics at Tübingen on the grounds that although Meyer's field was mathematical physics, he would be successful as a mathematician because of his "pure mathematical education." Yet although Meyer hoped at Breslau to "succeed in awakening an interest in mathematical physics that has been sleeping since Kirchhoff's departure," confusion persisted over what his talents as a physicist were. Tübingen considered him "as significant an observer as he is a mathematician," but when he was considered for a physics position at Bonn in 1868, Lipschitz told Neumann that Meyer was not considered an experimenter. Meyer did not succeed in his goal of tilling the ground for mathematical physics (his original *Lebensplan* based on Neumann's example, he told his mentor), but it was not because mathematics diluted the physics in his teaching and his research. He became *Ordinarius* for mathematics at Breslau in 1865, and then for physics in 1867. In subsequent years he taught primarily experimental physics, finding it a "joy to make the scientific results of rigorous research accessible to a wider audience" by putting them "in a light popular dress."[58]

58. Meyer to Neumann, 20 February 1863, FNN 53.IIA: Briefe von Schülern; G. Wiedemann to Peter Merian-Thurneysen, 29 April 1863, SAKBSt, CC27: P. Merian-Thurneysen Nachlaß; Meyer to Neumann, 16 July 1864, 18 October 1864, FNN 53.IIA: Briefe von Schülern; Tübingen Natural Sciences Faculty to the Tübingen University Senate, 9 December 1864, UAT 126/469; R. Lipschitz to Neumann, 14 June 1868, FNN 53.IIA: Briefe von Schülern; Meyer to Neumann, 10 December 1893, ibid. Additional evidence that Meyer's Breslau position was in mathematical physics is given by Wolff, "Die Rolle von Reibung," p. 173, n. 66.

Meyer's position at Breslau was unusual because it involved local bargaining that may not have been possible elsewhere. But because Meyer was smitten with an overzealous missionary spirit when it came to spreading Neumann's variety of theoretical physics and hence was at times hypersensitive to criticisms against it, he was able to exploit—at least for a while—the finer nuances of his position in order to pursue his own self-appointed goals. Indeed it was by recognizing the difference between their own theoretical physics and that of others that Neumann's former students were able to assert the role of exact experiment as well as the research values of exactitude when others simply equated theoretical physics with mathematization in general.

Meyer had earlier refused to habilitate at Breslau when Neumann advised him to do so in late 1860 or early 1861 because the faculty and the students leaned "so much in a practical direction that I may not hope as *Privatdozent* to gain ground here for theoretical physics." It would be "extraordinarily difficult," he believed, "to awaken among the students here an interest in a science of whose existence they have no idea." The mathematical abilities of the students were so wanting that two other former students from Königsberg lectured as if they were teaching secondary school, and even Meyer's brother Lothar, then teaching chemistry at Breslau, "lectured in such an elementary way that a student in *Tertia* could understand him, and still his presentation was many times too high." Moreover, Meyer reported that Breslau students seemed unable to understand the concept of constant experimental errors, which he had been accustomed to determining in part through the theory of an instrument, in any other way than as a mistake in the experiment. So when his brother Lothar tried to explain the principles of the balance and the errors that afflicted it, Meyer explained, he could not refer to the law of the lever but had to use "an intentionally false balance."[59]

More than just the deficiencies of the students bothered Meyer, however. Although the Breslau physicist Moritz Frankenheim was "extraordinarily friendly" toward Meyer and had encouraged him to lecture on theoretical physics, Meyer found that when he and Frankenheim "spoke more accurately on the matter, it unfortunately showed that by theoretical physics he understood something quite different from what I did." Frankenheim's theoretical physics, according to Meyer, boiled down in this case to "applied mathematics" and specifically to "some interpolation formulas" in which "the constants were represented as a function of temperature." Frankenheim could not, in fact, have chosen a more sensitive and, in the minds of Neumann's former students, contentious illustration of what constituted theoretical physics. Meyer's abhorrence of Frankenheim's notion of theoretical physics was thus a reaction not to its practical nature but rather to its acceptance of certain mathematical

59. Meyer to Neumann, 6 February 1861, FNN 53.IIA: Briefe von Schülern.

expressions and techniques—interpolation formulas—that Neumann and his students regarded as epistemologically suspect and as useful to experimentalists primarily when creating empirical formulas. Hence Meyer was correct in concluding that "to read something of that type [at Breslau] would be entirely superfluous because the subject of experimental physics is already represented by Frankenheim" and one of his colleagues. Meyer somewhat regretted the strident tone in which he had related this story to Neumann, but it was only, as he explained, "to say that these men have no idea of what one in Königsberg understands by theoretical physics" and hence they could have "no interest" in this "branch of science."[60]

Had he been aware of the distinction Meyer made, Frankenheim undoubtedly would have regarded it as a small point. Not so for Meyer. Even at Göttingen, where by his own admission there was "well prepared soil" for mathematical physics because it had been taught in the "proper sense" by Gustav Lejeune Dirichlet, Bernhard Riemann, and Wilhelm Weber, he nonetheless found while *Privatdozent* in 1863 that "the more intelligent [students] have for some time missed the bridge between the purely mathematical and the purely experimental representations." Meyer's sense that the Göttingen and Königsberg approaches to theoretical physics were not entirely alike was on the mark. Student notes from Weber's mathematico-physical seminar during the early 1860s reveal that Weber did not emphasize to the extent Neumann did the analytic technique that allowed precise comparisons between theoretical calculations and experimental results: the method of least squares. Hence when Meyer left Göttingen in 1864 he was glad to have heard that Carl Pape, who had also arrived as *Privatdozent* in 1862, would remain and that Neumann's former seminar students Ludwig Minnigerode and Heinrich Weber also planned to become *Privatdozenten* at Göttingen. As a result Meyer believed that Neumann's "school would continue to retain an outpost."[61] Fine distinctions such as Meyer's regarding the relation between mathematical and experimental approaches to physics were especially necessary in the second half of the century, when in addition to those who went about experiment in theoretical physics differently, there were some theoreticians, such as Rudolph Clausius, who practiced experimentation rarely or not at all.

It was often difficult, however, for Neumann's former seminar students to sustain a critical exchange between theory and exact experiment in

60. Ibid. Meyer also told Neumann that at Breslau he was called an "unpractical physicist" and that the only reason students understood the mathematician H. Schröter (also a former student from the Königsberg seminar) was because Schröter conducted his classes at a gymnasium level.

61. Meyer to Neumann, 20 February 1863, 18 October 1864, FNN 53.IIA: Briefe von Schülern; seminar notes from 1860 to 1863 by K. Hattendorff, Wilhelm Weber Nachlaß, Nr. 21 and seminar notes from 1860 to 1863 by Hermann Wagner, H. Wagner Nachlaß, Nr. 6, both in UBG-Hs.

teaching. Neumann's lecture courses on theoretical physics, "the first of their type" in Germany according to several accounts, became known more for their conceptual unification of physics than for their function in conveying the cognitive and quantitative tools essential for conducting exact experiments. Impressed by the systematic nature of Neumann's lectures, the Berlin pedagogue Karl Schellbach viewed them in 1860 as a model for school and university instruction alike, an endorsement that prompted Carl Neumann to suggest that they be edited and published. In what was known as the "Königsberg tradition" in physics instruction, some of Neumann's students imitated his lecture courses with varying degrees of success.[62] Among these students only three eventually taught from full professorial positions in theoretical physics: Kirchhoff at Berlin, Voigt at Göttingen, and Von der Mühll at Basel. The remainder taught from the subordinate positions of *Privatdozent* or *Außerordinarius*, positions that they nevertheless frequently defined as being in theoretical or mathematical physics.[63] The curricula of lecture courses taught by Neumann's former seminar students emphasized the mathematical foundation of physics in partial differential equations, organized theoretical physics conceptually, and provided a breadth of coverage that was especially necessary for educating secondary school teachers of physics. What they did not necessarily do was to promote the investigative style that had been shaped by Neumann's course on mechanics and by his seminar exercises that incorporated error and data analysis. Former seminar students continued to offer mechanics as an introductory course to theoretical physics, but it was not a mechanics that drew its inspiration from the Besselian experiment.

Differently organized and more theoretically oriented, his students' courses generally presented mechanical laws and principles without con-

62. Carl Runge, "Woldemar Voigt," *Nachrichten von der Königlichen Gesellschaft der Wissenschaften zu Göttingen* (1920): 46–52, on 49.

63. The former students who most strongly carried on the tradition of Neumann's lecture courses were, in chronological order: Kirchhoff (Berlin, 1848–50; Heidelberg, 1854–75; Berlin, 1875–87); Amsler-Laffon (Zurich, 1849–52); Wild (Bern, 1858–68); Quincke (various schools in Berlin, 1859–72; Würzburg, 1872–75; Heidelberg, 1875–1907); O. E. Meyer (Breslau, 1864–1905); Von der Mühll (Leipzig, 1868–89; Basel, 1889–1912); Sohncke (Karlsruhe, 1871–83); and Voigt (Königsberg, 1875–83; Göttingen, 1883–1919). In addition, the following students either taught lecture course sequences similar to Neumann's or tried to imitate Neumann but met with mixed success: Kirchhoff (Breslau, 1850–54); Meyer (1862–64), Pape (1862–66), H. Weber (1865–66), and Minnigerode (1866–74), all at Göttingen; Gehring (Bonn, 1863–71); Zöppritz (Giessen, 1867–80); and Sohncke (Jena, 1883–86). I have not included Roland von Eötvös (Budapest, 1871–1919) because, although he had a full professorship in theoretical physics and practiced an exact experimental physics of the type exemplified by the Besselian experiment, he was known to have had mixed feelings about Neumann and about having attended the seminar. On the university positions held by Kirchhoff, Voigt, Von der Mühll, and others see Christa Jungnickel and Russell McCormmach, *Intellectual Mastery of Nature: Theoretical Physics from Ohm to Einstein*, 2 vols. (Chicago: University of Chicago Press, 1986), 1:285–303; 2:30–58, 112–28, 144–48, 254–303.

sidering their imperfect manifestations in reality or their usefulness in analyzing experimental settings. These courses thus tended to project an image of physical knowledge as constructed of rational laws mathematically expressed, rather than as constrained by imperfections quantitatively measured. Undoubtedly these courses also created the impression that physical laws possessed a high degree of certainty (if not absolute certainty); that theoretical physics was more purely so; and that the practice of theoretical physics need not entail measurement and the quantification of error. Even where Neumann's pedagogical philosophy was allegedly explicitly adopted, considerations of measurement, instrumentation, and error analysis barely surfaced. Two mechanics courses stand out in this regard: those of Gustav Kirchhoff and Woldemar Voigt. Both demonstrate how the pedagogical reshaping of mechanics made the analytic determination of errors and the analysis of instruments less important in the curriculum for theoretical physics than they had been in Neumann's teaching.[64]

When first published in 1876, Kirchhoff's course on mechanics created a stir. His assertion that "the task of mechanics" was "to *describe* the motions occurring in nature completely and in the simplest way" was radical and phenomenological to an extreme.[65] More than a decade earlier he had viewed mechanics differently. He had argued that its object was "to determine the motion of bodies when the causes that occasion them are known." He had even espoused a Laplacian view of causality to the point of believing that "the highest goal the natural sciences must strive to obtain" is "the reduction of all natural phenomena to mechanics."[66] By 1876 he had abandoned this rigid way of viewing the intellectual purpose of physical investigation and argued instead that "mechanics used to be defined as the science of *forces*, and force as the *cause* that produces motion," but that this definition could no longer be maintained when the notion of force was itself so unclear. Eliminating force (and in his view, "darkness") from his course on mechanics meant that he was left with equations that described mechanical processes only in terms of space, time, and matter.[67]

Kirchhoff did treat two topics that Neumann had used to discuss error analysis in his course on mechanics, the pendulum and capillary theory. But Kirchhoff's discussions of both were brief and did not include descriptions of measuring operations or instructions on how to determine and compute errors rigorously. In his treatment of capillarity, he was

64. Gustav Kirchhoff, *Vorlesungen über mathematische Physik: Mechanik*, 2d ed. (Leipzig: B. G. Teubner, 1877); Woldemar Voigt, *Elementare Mechanik als Einleitung in das Studium der theoretischen Physik*, 1st ed. (Leipzig: Veit, 1889).

65. Preface to the first edition, rpt. in Kirchhoff, *Mechanik*, p. iii.

66. Robert von Helmholtz, "A Memoir of Gustav Robert Kirchhoff," *Smithsonian Institution Annual Report* (1889): 527–540, on 538 (quote from Kirchhoff's 1865 rectorial speech).

67. Kirchhoff, *Mechanik*, pp. iii, iv.

more concerned with describing capillary action accurately than he was with setting up his equations so that the capillary constant could be measured. Throughout his textbook, Kirchhoff avoided hypotheses, was mathematically elegant and clear, and ordered information systematically. The mathematical rigor of his presentation seemed to be guided by a statement he had made in his rectorial speech of 1865 in which he argued that "we have just as much right to ascribe *absolute certainty* to mechanical theorems as to geometrical." His formal presentation of mechanics, lacking as it did a more extended discussion of errors and of the qualified nature of theory, did little to diminish the attribution of mathematical truth to mechanics, even though he believed that the goal of reducing all natural phenomena to mechanical principles could at best be achieved only approximately.[68] Kirchhoff's mechanics was thus closer to the mechanics taught by Alfred Clebsch and Carl Neumann than it was to his mentor's. Kirchhoff, Clebsch, and Neumann were system builders in mechanics, describing an abstract world of material points governed by mathematical laws.[69]

Woldemar Voigt's *Elementary Mechanics as an Introduction to the Study of Theoretical Physics* (1889) was allegedly modeled directly on Neumann's course, treating mechanics "not according to its mathematical, but according to its physical, relations."[70] In the first edition of his textbook, Voigt confined his mathematical techniques to calculus and analytic geometry, making no use of the theory of series, potential functions, and elliptic functions that had cropped up in the advanced stages of Neumann's seminar exercises on mechanics. Voigt claimed to have found a model for his presentation of mechanics in Neumann's, especially "with respect to the application of mechanical equations to the theory of physical measuring instruments." He also claimed to have "emphasized such applications that have a practical significance, for example, the theory of important physical measurements," but in fact he discussed only one instrument in detail, the pendulum.[71] When he demonstrated how the pendulum could be used to determine gravity, he introduced the analysis of constant errors by following Bessel's seconds pendulum investigation, emphasizing especially Bessel's hydrodynamic correction and the method of coincidence observations. He also used the pendulum to analyze accidental errors through the method of least squares.[72] In contrast to his

68. Ibid., pp. 78–86, 136–51; Kirchhoff's Mechanik [lecture notes, 1879], C. Runge Nachlaß 53, UBG-Hs; Helmholtz, "Kirchhoff," p. 538 (emphasis added).

69. A. Clebsch, *Elementare Mechanik nach Vorträgen gehalten an der polÿtechnischen Schule, 1858–59* [lithograph of lecture notes] (Carlsruhe, n.d. [1859]); idem, *Analytische Mechanik nach Vorträgen gehalten an der polÿtechnischen Schule, 1858–59* [lithograph of lecture notes] (Carlsruhe: Geissendörfer, n.d. [1859]); Carl Neumann, "Analytische Mechanik," [Eduard Riecke's lecture notes, 1870s], Brown University Library, Manuscripts Division.

70. Voigt, *Elementare Mechanik*, 1st ed., p. iii.

71. Ibid., pp. iv, v.

72. Ibid., pp. 207–18.

treatment of the pendulum, which took up twenty pages, Voigt discussed other instruments or measuring techniques—such as bifilar suspension, the barometer, the manometer, and the balance—in only one or two pages.[73] His claims to the contrary, his discussions of instruments and measurements were outnumbered and dominated by a more abstract and mathematical approach. Voigt not only began his textbook with the mechanics of material points; he also introduced vectors and scalars into his presentation.

The contrast between Voigt's particular variety of an abstract mathematical mechanics and the remnants of a practical physics intensified in the second edition, published twelve years later, in 1901. He had in the interim firmed up the position of mechanics in physics pedagogy in his *Compendium of Theoretical Physics*, a two-volume summary of the major parts of theoretical physics. In that work, he claimed to have stressed the role of mechanical analogies and the need to keep in touch with the results of observation when constructing physical theories.[74] Later, in the second edition to his textbook, he drew attention to the alleged continuity in his approach, asserting that his "ground plan" had "remained unchanged." Voigt even went so far as to distinguish two different approaches to mechanics—the "mathematical" approach, which consisted of an "analytic or geometric investigation of the characteristics of a certain system of differential equations," and a second approach that "considered mechanics as a branch of the exact natural sciences, whose final goal is the derivation of numerical laws represented in reality"—and to identify with the latter. Mechanics in this latter sense was, he explained, "the foundation of all theoretical physics because, on the one hand, human understanding needs to explain strange phenomena through known processes and especially through the simplest and best known mechanical [ones], and because, on the other hand, almost all exact measuring instruments rest on mechanical principles." Voigt not only linked the practice of theoretical physics strongly to measurement but also seemed to make theory dependent on measurement in a positivist way.[75]

Yet what he taught his students in the pages that followed conformed minimally to what he claimed he had done. Despite the wider variety and increased precision of instruments available at the beginning of the twentieth century, Voigt still gave the most extended discussion of the one that had been a paradigm for Neumann's physics pedagogy almost seventy years earlier: the pendulum. Contrary once again to his prefatory remarks, he had in fact considerably strengthened and sharpened the mathematical foundation of his presentation by inserting a new introduc-

73. Ibid., pp. 175, 180, 206, 320.
74. Voigt, *Kompendium der theoretischen Physik* 1:v, 1.
75. Woldemar Voigt, *Elementare Mechanik als Einleitung in das Studium der theoretischen Physik*, 2d ed. (Leipzig: Veit, 1901), pp. vi, 1, 2.

tory chapter on the mathematical methods that had been laced throughout the earlier edition. The only mathematical techniques directly related
to Voigt's stated intentions, in particular to the mathematical construction
of theory suited for measuring operations, were approximation techniques, but they were discussed in the last of the eighteen sections of his
new introductory chapter. It is not even clear that his students could have
deployed approximation techniques as he wished them to because the
determination of relevant terms in a series expansion was partly dependent upon knowing the range of experimental error. Yet Voigt barely
discussed constant errors and buried the method of least squares in his
treatment of the pendulum.[76]

Voigt's new introduction concentrated on a type of mathematics suited
to identifying measurable constants but not to reducing data, especially
through error analysis: the mathematics of scalars, vectors, and now even
tensors. Although he claimed to have gathered together in the introduction to the second edition those mathematical considerations that had
been "strewn about" in the first, he had in fact introduced several new
techniques, including tensors and series expansions, that were not present earlier. The brevity of his discussions of instrumental and metrical
concerns, as well as the disparity between what he claimed to be doing and
what he actually did, made the sections of his lecture course dealing with
practical physics appear atavistic. That he included elements of practical
physics at all raises questions about why he emphasized approximation
techniques over other techniques, such as the method of least squares,
used in practical physics as he knew it. Part of the answer has to do with
Voigt's conviction that the range of problems in which approximation
techniques were useful had to be expanded because of what he perceived
as the growing inability of mechanical principles to serve all areas of
physics well. He "welcomed" approximation techniques "where rigorous
solutions are not possible or are not attainable without higher tools of
analysis." In his view, approximation techniques were necessary in the
construction of empirical equations based on measurement and in
"the application of mechanical principles to other areas of theoretical
physics."[77] The tenor of Voigt's textbook suggests that, as a guide to the
practice of physics, it would teach students to seek exactitude or a higher
degree of certainty not by trying to eradicate or quantify error but by
refining further the mathematical expression of physical relations.

Yet error analysis had not entirely disappeared from the physics curriculum under Neumann's students, even though it seemed unsupported by
reformed introductory courses on mechanics. As windows on the classroom, textbooks offer a limited view. This was especially true for the
second half of the century, when practica, formerly optional supplements

76. Ibid., pp. 262–82, 1–26, esp. 23–26.
77. Ibid., pp. vi, 2.

to physics instruction, became mandatory. Although practica by definition involved the handling of instruments, the ways in which experimental techniques were taught and the kinds of exercises assigned often differed considerably from location to location. Greater uniformity was achieved after 1870, when Friedrich Kohlrausch's *Leitfaden der praktischen Physik*, based on his and Wilhelm Weber's exercises at Göttingen, including those of the mathematico-physical seminar, was published and quickly set the standard for practica across Germany.[78] Although Kohlrausch taught the techniques of an exact experimental physics, he focused primarily on the material aspects of an experimental investigation, including the numerical computation of constant errors, the construction of instruments, and the perfection of protocol. Of less importance to the "precision" physics represented by his textbook were the calculation of accidental errors by the method of least squares, interpolation and graphical analysis of data, and especially the refinement of theory by an analytic expression that accounted for disturbing factors manifest under both laboratory and "real" conditions. The popularity of Kohlrausch's model and its concordance with other practical exercises from earlier in the century placed Neumann's practical exercises sharp relief, with their more comprehensive considerations of error and data analysis suited for evaluating and confirming theory. Even though Neumann's practical exercises did not emphasize accidental errors to the extent that Edward Pickering's did in the United States during the 1870s, the differences, both in content and in purpose, between Neumann's and Kohlrausch's exercises are enough to justify viewing them as cultivating different values, strategies, and styles in the practice of physics.[79] In spite of the popularity of Kohlrausch's textbook, several of Neumann's former seminar students either conducted or tried to teach practical exercises as Neumann had, thereby strengthening their ties to the seminar, contributing to their identification as a "school," and most important, cultivating an ethos of exactitude.

It was not always possible or desirable, though, to teach practical physics as Neumann had. Although the pressures and demands of the marketplace helped to shape practical exercises, especially in large practica that handled a broad student clientele, the local quality of mathematical instruction was often decisive in determining the level of quantitative techniques included. At Würzburg University between 1872 and 1875, Georg Quincke offered practical laboratory exercises every semester

---

78. Friedrich Kohlrausch, *Leitfaden der praktischen Physik*, 1st ed. (Leipzig: B. G. Teubner, 1870).

79. On the differences between Kohlrausch and Pickering see Kathryn M. Olesko, "Michelson and the Reform of Physics Instruction at the Naval Academy in the 1870s," in *The Michelson Era in American Science, 1870–1930*, AIP Conference Proceedings 179, ed. S. Goldberg and R. H. Stuewer (New York: American Institute of Physics, 1988), pp. 111–32, on pp. 115–17.

in conjunction with courses on both experimental and mathematical physics. He had been recommended for the position in part because he had "an unusual gift for experiment" and was considered to be among the "distinguished young physicists of the Königsberg school." But the practical exercises he taught at Würzburg were mathematically simple. Würzburg had an undistinguished mathematical seminar, and the university, like others in southern Germany, did not have strong mathematical instruction that helped students learn how to apply quantitative techniques to the sciences. Before the seminar was founded, there had been an attempt to enhance the opportunity to teach students quantitative techniques in physics by adding a physics section to the seminar, but competing proposals for sections for astronomy and mathematical geography and the absence of lecture courses on mathematical physics to prepare the students for the seminar blocked the establishment of a physical division. One reason Quincke left Würzburg for Heidelberg in 1875 was that he found the opportunity to direct Heidelberg's mathematico-physical seminar appealing.[80]

Like Würzburg, Leipzig did not have a mathematico-physical seminar, nor did it have strong practical exercises until the late 1870s. During the 1840s, Wilhelm Weber transferred to Leipzig the measuring exercises with "reliable instruments" which he had taught at Göttingen. He offered both advanced exercises on galvanic and magnetic observations and introductory exercises for teaching candidates and medical students. Later, during the 1870s and 1880s, Neumann's former students Carl Neumann and Karl Von der Mühll offered exercises in mathematical physics, for which new laboratory facilities were available after 1874. Von der Mühll's exercises especially were considerably more specialized than Weber's and closer to what Neumann had taught. Carl Neumann's physics exercises were strongly mathematically oriented; Von der Mühll's were rigorous because he could ignore elementary topics, which were taught in separate exercise sessions in physics for teaching candidates and in mathematical physics for medical students.[81]

Elsewhere, Neumann's students promoted more of the quantitative techniques of an exact experimental physics. When Von der Mühll took a position at his home university of Basel in 1889, a seminar for mathematics and the natural sciences had already been in operation there from 1866 to 1885. Largely because the students who entered it were chronically unprepared, Basel's seminar rarely fulfilled its stated purpose as an

80. Würzburg University, *Vorlesungsverzeichniss*, 1872–75; G. Kirchhoff to Würzburg University, 6 November 1866, and R. Bunsen to Würzburg University, 3 January 1867, UAW 609+, I/17/21; Acta betr. das mathematische Seminar, 1872–1905, UAW 1640, III/11/9; G. Quincke to Würzburg University Rector, 1 April 1874, UAW 713+, I/17/24.

81. J. C. W. Hoffmann, "Mathematische und naturwissenschaftliche Universitäts-seminare," *ZMNU* 5 (1874): 169–73, on 169; Otto Wiener, "Das neue physikalische Institut der Universität Leipzig und Geschichtliches," *PZ* 7 (1906): 1–14; Leipzig University, *Vorlesungsverzeichniss*, 1873–88.

institute for promoting original student investigations. Students performed so poorly in the natural sciences that the seminar stood in danger of being canceled in the mid-1880s; it survived only by becoming a seminar for mathematics alone in 1885. Just before Von der Mühll's arrival, seminar exercises in mathematical physics had resumed, but they were handled only as written problems in the application of mathematics to physics. Von der Mühll supplemented these seminar exercises with practical exercises incorporating rigorous techniques of error analysis as he had taught them at Leipzig.[82] At Bern between 1858 and 1868, Heinrich Wild offered practical exercises as his predecessor had done, but in 1861 he introduced a special course in measurement, a sign that deeper issues needed to be handled.[83] A solid (but recent) tradition in measuring exercises was already in place at Jena in 1883 when Neumann's former student Leonhard Sohncke arrived. Participants in Sohncke's practical exercises and supervised research for advanced students could draw upon Ernst Abbe's specialized courses on physical units and the method of least squares.[84]

Although seminars in physics were more enduring than exercise sessions and had a clientele more serious and committed than did practica, not all seminars promoted quantification in the same way or to the same degree. Natural science seminars, in which few former students from Neumann's seminar participated, were chronically weak in quantitative techniques, especially in using an exact experimental physics to test, refine, and confirm theory. In the physical division of Königsberg's natural sciences seminar, Ludwig Moser eschewed difficult techniques of quantification. Under Julius Plücker's guidance between 1834 and 1868, the physical division of the Bonn natural sciences seminar was a popular one. Plücker drew students into his nascent physical institute and even had them serve there as assistants. Physics students at Bonn had ample opportunity to pursue quantification and in more than one way. Plücker taught courses on mathematical physics; other faculty members offered courses on interpolation, the method of least squares, analytical mechanics, and even probability calculus applied to the natural sciences. Yet Plücker never mentioned in his seminar reports how and to what extent his practical exercises united theory and exact experiment.

Plücker's student clientele partly determined the quantitative orientation of his courses. In the mid-1850s, Eduard Heine, who had studied under Bessel, Neumann, and Jacobi before going on to teach mathematics at Bonn, found that his students showed "little interest" in mathematics and wanted to learn only what was necessary for the state teaching examination. Plücker, who did not share Heine's opinion, rejected

82. SAKBSt, Acta betr. Jahresberichte des mathematisch-naturwissenschaftlichen Seminars, 1866–1933, Rep. A12, Abt. XI, 2.8.

83. Bern Hochschule, *Vorlesungsverzeichniss*, 1858–68.

84. Jena University, *Vorlesungsverzeichniss*, 1882–87.

Heine's proposal that Plücker relinquish teaching mathematics so that an *Ordinarius* strictly for mathematics could be brought in. Yet when a mathematical seminar was considered in 1864, the Bonn University curator echoed Heine's observation by citing the poor state of mathematical instruction in the Rhine area as one reason for founding the seminar; another was the "exact" instruction that was required in physics and chemistry. At first Plücker, who insisted on codirecting Bonn's mathematical seminar with the Königsberg-trained mathematician Rudolf Lipschitz, seemed to agree. He viewed the mathematical seminar as a way to promote mathematical physics while keeping experimental physics in Bonn's natural sciences seminar. But in the first year of the seminar's operation, 1866/67, Plücker taught only topics from mathematics. He died in 1868, failing to demonstrate what he meant by mathematical physics in a seminar context. When Rudolph Clausius took over Plücker's division of the natural sciences seminar in 1869, he offered practical exercises in a more purely theoretical physics. By the time the Bonn seminar disbanded in 1887, the practical exercises, then directed by Heinrich Hertz, were for introductory students.[85]

Two other natural science seminars were established, at Halle in 1839 and at Freiburg in 1846. Both incorporated mathematics but did not achieve a strong quantitative approach to physics instruction. The only confirmed meeting of the Freiburg seminar was in the winter semester of 1846/47, when thirty students attended it. Although the Freiburg University Senate considered mathematical physics an especially important component of the seminar, the director of the physical division did not promote practical exercises with precision instruments.[86] In the longer-lived Halle seminar, physics was inconsistently promoted. Ludwig Adolf Sohncke had left Königsberg, where he had served as Jacobi's assistant, to take up a position at Halle. There he had joined the Halle meteorologist and physicist Ludwig F. Kämtz in proposing a mathematico-physical seminar for Halle in June 1837. Originating in a mathematico-physical society and tied to the state's teaching examination, their seminar was designed to promote independent investigations among students.[87]

---

85. ZStA-M, Acta betr. das naturwissenschaftliche Seminarium an der Universität zu Bonn, Rep. 76Va, Sekt. 3, Tit. X, Nr. 4, Bd. II: 1832–58; NWH, Acta betr. das naturwissenschaftliche Seminarium der Universität Bonn, Bestand NW 5, Nr. 483, Bd. III: 1859–88; Bonn University, *Vorlesungsverzeichniss*, 1848–68; Eduard Heine to [Bonn University Curator], 15 November 1854, and J. Plücker, ["Gutachten"], 4 January 1855, UAB, Personalakt betr. Dr. E. Heine; Bonn University Curator to Kultusministerium, 14 September 1864, NWH, Acta betr. den mathematischen Apparat der Universität Bonn, NW 5, Nr. 558, Bd. I: 1835–1929, fols. 17–20; Franz London and Otto Toeplitz, "Das mathematische Seminar an der Universität Bonn," in *Geschichte der Rheinischen Friedrich-Wilhelms-Universität von der Gründung bis zum Jahr 1870*, Bd. 2, *Institute und Seminare, 1818–1933* (Bonn: F. Cohen, 1933), pp. 324–34, esp. pp. 327–28.
86. J. Müller, quoted in Hoffmann, "Mathematische und naturwissenschaftliche Universitätsseminare," p. 170; D. Frommherz to Baden Ministry, 1 December 1847, GLA 235/7766.
87. L. F. Kämtz and L. A. Sohncke to Kultusministerium, 8 June 1837, ZStA-M, Acta

But their proposal had aroused the jealousy of J. S. C. Schweigger, Kämtz's former teacher, who had been directing a successful physico-chemical society at Halle for years. Schweigger had argued strongly for an integrated seminar for all the natural sciences. Convinced that the present was "an age built entirely on engineering," he had even wanted to include technical and applied physics in the seminar, especially machine construction, which he believed essential for training *Realschullehrer*. But rigorous techniques of quantification were not a part of his vision for physics instruction. Referring to the Bonn seminar which had no instruction in mathematics, he had pointed out that "mathematics plays a subordinate role" in recent experimental advances in the "nonmechanical parts of physics." Wanting to preserve peaceful relations between the natural scientists involved, the Halle faculty had considered several alternatives, including the Königsberg solution of having both a mathematico-physical and a natural sciences seminar. University authorities and the ministry had eventually agreed upon the single seminar for mathematics and the natural sciences.[88]

The Halle seminar failed to live up to the ideals expressed in its statutes. Independent study was rare; practical exercises were not common; and the three-year course of study in the seminar was infrequently followed. Yet publishable research could be funded and sent to a journal; advanced seminar students were given preference for assistantships; and dissertation support was available. Cooperation between the experimental physics and chemistry division under Schweigger and the mathematical physics division under Kämtz was not to be had. Despite his earlier success in fostering physical investigations, Schweigger now supervised primarily projects in chemistry (some of which were later published). Kämtz lacked proper instruments and found it difficult to attract students. In 1843, after eight semesters of operation, Kämtz had had a total of six students during only three of those semesters; Schweigger's section did not lack participants in any semester. Both Schweigger and Kämtz claim to have conducted demonstrations and experimental trials in their divisions. By 1848, regular practical exercises in physics were offered, but not enough instruments were available to sustain them.[89]

Reports for the Halle seminar between the late 1840s and 1876 are scarce. The physicist Hermann Knoblauch became director of the physical division in 1856 and then administrator of the entire seminar in 1876.

betr. die Errichtung eines Seminars für Mathematik und die gesammten Naturwissenschaften an der Universität zu Halle, Rep. 76Va, Sekt. 8, Tit. X, Nr. 36, Bd. I: 1837–89, fols. 3–4.

88. J. S. C. Schweigger to Kultusministerium, 8 August 1837, 21 January 1838, 30 January 1838, ibid., fols. 10, 50–51, 53–61, on fol. 50; Dean of the Halle Philosophical Faculty to Halle University Curator, 16 February 1838, ibid., fol. 20.

89. Halle University Curator to Kultusministerium, 6 March 1839, ibid., fol. 25; Halle seminar reports for 1839/40 (8 November 1840), 1842/43 (10 November 1843), 1847/48 (25 August 1848), ibid., fols. 94–97, 133–38, 178–79.

In 1875/76, eighteen students registered for the physical division but almost twice that many, thirty, in the chemical division. Throughout the 1870s the chemical division had consistently higher enrollments. Knoblauch claimed to promote the "scientific method" and to cultivate "judgment about experiment and theory," but the seminar's physics exercises and investigations were relatively unexceptional and from all appearances not rigorously quantitative. Although he purchased for the seminar more equipment and books in physics than in any other field during the 1870s and 1880s, he never identified any physical investigations undertaken in the seminar as being of publishable quality. Elementary exercises coexisted with more sophisticated projects used as the bases of doctoral dissertations. In 1878/79, Knoblauch finally purchased a copy of Kohlrausch's *Leitfaden der praktischen Physik*, but he did not say how it was used, if at all, in the seminar. At the end of the 1870s the seminar remained relevant for physics instruction largely because it had a well-stocked library. In 1884 a leading pedagogical journal reported that until recently there had been no opportunity for a young physicist to participate in a course on laboratory methods "because [at Halle] that kind of exercise was not held."[90]

Not until Ernst Dorn, Neumann's former student, arrived at Halle in 1886 did a pattern become evident in the seminar's physics exercises. Dorn's exercises were like Neumann's, and accordingly he considered his division to be for theoretical physics. He also offered a wide variety of courses on topics in mathematical physics, including one on the definition of physical constants. When Halle's physical institute opened in 1890, Dorn was appointed its director. Immediately the seminar's physics exercises were transferred to the institute, rendering the seminar less relevant for physics instruction. A year later the physical division was canceled, as were all other divisions in the natural sciences. Moritz Cantor and Neumann's former doctoral student Albert Wangerin directed the sole remaining division, for mathematics.[91]

Exact techniques of investigation were promoted more strongly in some of the eight mathematico-physical and physical seminars founded after Neumann's, at Göttingen (1850), Munich (1856), Giessen (1862), Breslau (1863), Heidelberg (1869), Tübingen (1869), Erlangen (1874) and Rostock (1880). Former Königsberg seminar students were active in establishing or directing the physical divisions of seminars at Göttingen, Breslau, Heidelberg, and Giessen. It was not the case, however, that Neumann's pedagogical strategy, especially the didactic use he made of the Besselian experiment, could be transmitted directly and without modification to new institutional settings.

90. Halle seminar reports for 1875/76 (27 August 1876), 1876/77 (5 November 1877), 1878/79 (11 August 1879), ibid., fols. 203–4, 208–9, 211–12; "Das physikalische Studium auf der Universität" (Von einem praktischen Schulmann H. G. M.), *ZMNU* 15 (1884): 638–45, on 643.

91. Halle University, *Vorlesungsverzeichniss*, 1880–94.

The mathematico-physical seminar at Göttingen incorporated the practical exercises in physics that Wilhelm Weber had offered there in the 1830s and again in the late 1840s and that J. B. Listing had taught in his absence. The Göttingen and Königsberg mathematico-physical seminars were in many respects parallel institutes. Both had mathematical and physical divisions; students could attend either. Teacher training was explicitly mentioned in the Göttingen statutes, but as at Königsberg, the seminar's physics exercises were not didactically oriented but instead were designed to go beyond lecture courses. A new physics curriculum had evolved in the Königsberg seminar; at Göttingen the seminar was organized from the start to offer a "connected systematic course of instruction."[92] Intellectually, too, the seminars were similar in that the techniques of an exact experimental physics were promoted as much as those of mathematical physics. But although the exactitude of astronomical observations was a model for Weber as it had been for Neumann, Weber (and his assistant Kohlrausch) did not translate that exactitude to physics in the same way.

The most striking difference between the Göttingen and Königsberg seminars was in the constitution and purpose of their practical exercises in physics. Whereas Neumann taught both the elementary and advanced sections of his seminar and from time to time directed original investigations, the introductory practical exercises in the Göttingen seminar were handled by a student assistant and by Listing, while Weber regularly directed the advanced exercises and student research. Göttingen's seminar statutes stipulated that instruction in theoretical physics be "bound with lectures in observation and measurement," but unlike Neumann's seminar program, which had stabilized by the 1840s, Listing had difficulty teaching elementary measuring operations from several areas of physics because, he claimed, his students entered the university with inadequate school knowledge in physics.[93]

Weber, in contrast, was far more successful with his exercises, even though they were drawn directly from his past or ongoing research in electrodynamics and geomagnetism and so covered a relatively narrow spectrum of topics. Of course from time to time he did train students in how to test fine balances; to conduct refined acoustic experiments; to use the barometer for measuring height; and to repeat a select number of key experiments, often drawn from recent literature in physics. But such exercises were offered only sporadically at first, then not at all. Weber instead concentrated more on teaching his students how to take measurements with refined electrodynamic and geomagnetic apparatus—espe-

---

92. ["Das mathematisch-physikalische Seminar"], *Göttingische Gelehrte Anzeigen* (6 March 1850): 73–79, on 73.
93. Ibid., p. 77; "Jahresbericht des Vorstandes des mathematisch-physikalischen Seminars [1850–51]," 31 March 1851, UAG 4/Vh/20, fols. 7–13, on fol. 11; "Bericht des Vorstandes des mathematisch-physikalischen Seminars zu Göttingen [1851–52]," 28 June 1852, ibid., fols. 25–28, on fol. 27.

cially ones based on the principle of bifilar suspension—so that, as his students' skills improved, he could make use of their data for his own projects. By the early 1860s, Weber's seminar focused exclusively on electrical and magnetic measuring apparatus taught from the perspective of Weber's own theories. These exercises demanded precise results, but precision was achieved primarily through the perfection of the material conditions of the instrument or apparatus and the environment of the experiment, so as to fit the conditions of Weber's theories, rather than through data analysis.[94]

When Kohlrausch was appointed Weber's assistant in the seminar in 1866, he codified the practical exercises and adjusted several for beginning students. He designed his practical exercises to serve four purposes: to aid in learning the quantitative parts of physics, to take up matters that could not be handled in lectures, to provide a foundation for scientific research, and to train teaching candidates in how to use instructional apparatus. Significantly, he chose the chemical balance, not the pendulum, as his model for precision measurement; nowhere did he even mention the pendulum. He covered instruments from all areas of physics but mostly basic ones. Under his supervision the advanced investigations that grew out of his practical exercises were similar to those begun at Königsberg: the determination of the specific heat of air at constant volume, the validity of Ohm's law for electrolytes, the experimental proof of Neumann's law for the magnetism of a rotating ellipsoid, geomagnetic measurements at Göttingen, and the influence of temperature on the elasticity of metals.[95]

When Kohlrausch's practicum became a popular model in the German universities later in the century, the Leipzig physicist Otto Wiener suggested why. But what he considered its distinguishing features—material drawn from all areas of physics, "sharp criticisms of measurements," and the "military disciplining of the observer"—also characterized Neumann's exercises, and perhaps to a greater degree.[96] What strongly differentiated Kohlrausch's practicum from Neumann's was something more fundamental: how exactitude was achieved. Kohlrausch maintained that the "content and scope of an introduction to work in physics will above all be determined by the limits of accuracy." Several aspects of his presentation, however, constrained the degree to which that accuracy could be attained. For the most part Kohlrausch used only "elementary"

94. On Weber's seminar assignments from 1850 to 1866 see UAG 4/Vh/20, esp. fols. 10, 26, 43–44, 57, 73–74, 122, 140–41, 156–57, 168, 175, 204, 219, 233, 245, 266, 277.

95. Friedrich Kohlrausch, *Leitfaden der praktischen Physik*, 4th ed. (Leipzig: B. G. Teubner, 1880), pp. iii–iv [foreword to 2d ed., 1872]; idem, "Bericht über das physikalische Institut, Abteilung für Experimentalphysik, aus den Jahren 1866 bis 1870," in *Gesammelte Abhandlungen von Friedrich Kohlrausch*, Bd. 1, ed. W. Hallwachs et al. (Leipzig: J. A. Barth, 1910), pp. 1006–8, on p. 1008.

96. Wiener, "Das physikalische Institut . . . Leipzig," p. 4.

mathematics in his textbook and in other ways adapted to what he considered to be the poor mathematical preparation of his students. Even though he believed that it was only through measurement that one could "reduce the danger" of experimental trials having "no definite goal" and becoming a "game," he thought his own presentation left "further room to play."[97]

Kohlrausch eschewed the more rigorous disciplining in the techniques of an exact experimental physics that Neumann had imposed upon his students at Königsberg. Kohlrausch began his *Leitfaden der praktischen Physik* with a discussion of how to calculate accidental errors using the method of least squares, but he did so only so that students could determine the range of experimental error overall. For Kohlrausch, accidental errors marked the numerical limit beyond which more refined determinations of constant errors need not be taken. He seemed not to acknowledge that constant and accidental errors might in any way be related to one another, and he failed to acknowledge that the determination of accidental errors could help in the evaluation of theory. Befitting the greater amount of work required to handle a larger student clientele, Kohlrausch cultivated time-consuming quantitative techniques such as the method of least squares less than did Neumann, emphasizing primarily the determination of constant errors. But even here there were limits. Because, in his view, determining constant errors exhaustively "would be very laborious," he considered it sufficient at times to approximate them.[98]

Appropriately, then, accuracy for Kohlrausch was thus less the result of rigorous mathematical calculations in data analysis than of the manual dexterity of the experimenter and the material construction of the instrument, both of which were related to the magnitude of the constant errors in the experiment. In his textbook he advocated seeking accuracy no greater than was given by "instruments customarily used and by average skill in observing." Although this conception of accuracy is what one might expect from an experimentalist, it is not from one so closely associated with precision measurement. In teaching practical physics, Kohlrausch was not, like Neumann, teaching students to ask questions of data (and hence of theory confirmation) but rather of instruments and experimental procedure. He knew that the notion of accuracy presented in his textbook was not all it could be. He admitted that he could "understandably not think of satisfying everyone's wishes," that there would be some who "will without doubt miss a more thorough treatment," whereas "to others the rigor [presented in the textbook] already appears as pedantry."[99] Kohlrausch considered the middle ground he struck sufficient for

97. Kohlrausch, *Leitfaden*, 4th ed., pp. iv, v.
98. Ibid., pp. iv, 20.
99. Ibid., pp. iv, v.

training large numbers of students from different backgrounds and with different objectives. Viewed from a pedagogical context, the Göttingen ethos of exactitude was thus less "exact" than Königsberg's.

Kohlrausch's approach to practical exercises and measurement constituted a good part of the Göttingen tradition in physics instruction when Woldemar Voigt, Neuman's last doctoral student, who had been teaching at Königsberg since 1875, was chosen to replace Listing as *Ordinarius* of "theoretical (mathematical) physics" in August 1883 and as director of the mathematico-physical institute, positions he held until 1915.[100] Undoubtedly the institutional conditions he found there were partly responsible for the tensions between two types of mathematical physics—one oriented toward measuring and theory confirmation, the other toward the construction of pure theory—in his pedagogical works, especially in his textbook on mechanics. To be sure, Voigt promoted an ethos of exactitude achieved by quantifying and eradicating error, especially in his public speeches; but he did not consistently practice the Königsberg tradition in teaching measurement at Göttingen. He considered the "theory of exact methods of observation" to be "too special" for his *Compendium on Theoretical Physics*, so he omitted them, concentrating instead on creating a comprehensive survey of what was known. Yet for the benefit of his students Voigt did include tables of weights and measures in the textbook because he found that "numerical calculation in physics is in no way comprehended as clearly as is desirable."[101]

In his mathematico-physical seminar at Göttingen he assigned problems similar to Neumann's. During his first year as director, 1883/84, he taught Laplace's theory of capillarity. Similar "classical" topics (e.g., the distribution of electricity in a plane, various elasticity theories, the theory of thermal conductivity, and interference phenomena) dominated his seminar exercises until 1890. Not until 1887/88 did he begin to introduce topics not found in Neumann's seminar exercises. At first Voigt drew new problems from his own research in crystallography, but by 1888/89 he took up the Hall effect and also the application of the mechanical theory of heat to cosmic phenomena. In 1889/90 his students discussed Maxwell's theory in the context of Heinrich Hertz's recent experiments. But for the most part, his Göttingen seminar exercises consisted of written problems; not until 1887/88 did he mention that observations were conducted weekly and then discussed in the seminar. In contrast, Voigt's Göttingen colleague in the experimental physics section, Eduard Riecke, incorporated experimental problems and demonstrations and drew more frequently upon more current topics in physics.[102]

Much suggests that Voigt's practical exercises at Göttingen lost Neu-

100. Kultusminister Falk to W. Voigt, 23 August 1883, 3 September 1883, WVN 1.

101. Voigt, *Kompendium der theoretischen Physik* 1:iv, v.

102. On Voigt's seminar exercises from 1883/84 to 1889/90 see UAG 4/Vh/24a, fols. 47, 49, 54, 57, 61, 68, 71.

mann's intense concern for error analysis in physics pedagogy. Voigt's seminar on mechanics, for instance, was much more analytic and less centered on practical examples than Neumann's had been, even though Voigt told his students that he chose and treated problems where "the mathematical difficulties are entirely unimportant so that you can devote all your attention to the physical side of the question." The pendulum played a significant role in his seminar lectures on mechanics, but in his practical exercises he merely mentioned that "for those who have the interest" he "hoped" to give them the opportunity "to test and exercise practically, that is to say through experiment and measurement, what we here will consider theoretically."[103] Voigt upheld a distinction between theoretical and experimental physics at the level of instruction because "on the one hand the topics accessible with elementary mathematical knowledge mostly permit a demonstration through experiments in lectures, and on the other hand countless fine phenomena and the entire theory of exact methods of measurement, which cannot be demonstrated in the lecture but only learned in the laboratory, require higher mathematical assistance. So lecture courses receive in fact those heterogeneous stamps, which is indicated in the names 'experimental physics' and 'mathematical or theoretical physics.'" Yet these names, he argued, "do not penetrate the deepest core of the matter"; for as he had already stated, "there is in fact only a single physics which combines the theoretical and the experimental."[104] When Göttingen's physical institute opened in 1905 and Voigt was asked to describe the difference between its divisions of experimental and mathematical physics, he merely said that experimental physics was taught in one and mathematical physics in the other but that "the goals and means of both laboratories are essentially the same."[105]

Voigt's remarks suggest more than just the opinion that pedagogical convention did not adequately represent actual practice. In denying a higher-order difference between theoretical and experimental physics, he chose at that time not to emphasize the different roles that error analysis had played in Neumann's mathematical physics and Kohlrausch's experimental physics. In 1912 when he viewed the history of practical exercises at Göttingen largely through what Kohlrausch had done, he de facto aligned himself with the Göttingen tradition. Voigt still considered practical exercises to be "first rank *Erziehungsmittel*," useful as an introduction to independent work and beneficial in acquainting teaching candidates with how their own students would learn. But the persistent lack of preparation and aptitude among his students convinced him to change his views

103. WVN 38: Seminar über Mechanik, [n.d.], fols. 7, 37–51, 82–107, on fol. 51.

104. Voigt, *Festrede . . . am 4. Juni 1894*, pp. 4–5.

105. Woldemar Voigt, "Rede," in *Die physikalischen Institute der Universität Göttingen: Festschrift im Anschlusse an der Einweihung der Neubauten am 9. Dezember 1905* (Leipzig: B. G. Teubner, 1906), pp. 37–43, on p. 38.

on the constitution and purpose of practical exercises. Sometimes ill-prepared, incompetent, and unable to follow experimental protocol, his students were a sign to him "that school instruction here and there cares too little for formal things." Voigt's complaint about his students at Göttingen is all the more remarkable because he taught the better, more serious students in physics, whereas Riecke handled the introductory practical exercises for a service clientele.[106]

At Heidelberg, Neumann's former seminar students Gustav Kirchhoff and after him Georg Quincke also had uneven success in promoting error analysis and Besselian experimentation in practical exercises. Before Kirchhoff and the mathematician Leo Koenigsberger established a mathematico-physical seminar at Heidelberg in 1869, some students had a modest opportunity to learn physics through practical exercises, primarily because Heidelberg's physical "institute" had considerably more space than fledgling institutes at other universities, including Prussian ones. In 1818, the year after the physicist Georg Muncke had arrived, Heidelberg's physical institute moved to more spacious quarters, which included an apartment for Muncke, a room for scientific research, an observatory for astronomical and meteorological observations, and four additional rooms. At first Muncke used primarily astronomical instruments to teach "the methods of the observer" and to expand upon "several of the most essential problems of higher geodesy." Soon, though, other problems attracted his attention. The accurate determination of time and the proper calibration of astronomical and pendulum clocks figured prominently in his agenda; he not only taught these methods but also used them for calibrating Heidelberg's public clocks. By 1833 he was also interested in "more accurate determinations" of meteorological observations, including temperature patterns and barometric measurements. Muncke's interest in measurement embraced the various "meters" used in new physical discoveries, including electromagnetism, thermomagnetism, and induction, and he wanted to use these instruments "in lectures as well as in [his] own observations."[107]

Among the special projects Muncke undertook were geomagnetic observations. On a trip to Göttingen in 1836, he met Gauss, who asked if Baden would contribute to the geomagnetic measurements being taken across Europe. Muncke argued to the Baden ministry in early 1837 that it would be "very useful" either "to confirm or where possible to expand" Gauss's results. Although he was initially skeptical about such an undertaking—he claimed he lacked time, funds, and assistance—Muncke considered Gauss's suggestion that he use advanced students as assistants

106. Voigt, *Physikalische Forschung und Lehre*, pp. 14, 15; E. Riecke to W. Voigt, 9, 11, and 23 January 1883, WVN 4: Briefe.

107. Georg Quincke, *Geschichte des physikalischen Instituts der Universität Heidelberg* (Heidelberg: J. Hörning, 1885), p. 15; G. W. Muncke to Heidelberg Academic Senate, 27 August 1823, UAH A587, IV3e, Nr. 52; G. W. Muncke to Universitäts- Bau- und Oekonomie-Commission, 20 January 1833, ibid.

reasonable. He later argued to the ministry that "owing to the present importance of physics for medicine, agronomy, and industry," it was no longer sufficient for "the better students" to take "a bare course of physical lectures." Between 1818 and 1841 when he began geomagnetic measurements (for which advanced students served as assistants), Muncke had begun to introduce the techniques and values of exactitude into physics instruction at Heidelberg, albeit for a select student clientele.[108]

Philipp Jolly, who had worked alongside Muncke as *außerordentlicher* professor of physics since 1839, requested additional space for a physical laboratory in 1846 for "student exercises as well as for his own research." He claimed to have engaged in "the less fruitful pure theoretical mathematical speculation in physics" because his "own restricted means [were] not enough" for undertaking "productive work in physics, independent research, and experimental discoveries." Between 1846, when he became Muncke's successor, and 1854, when he moved to Munich, Jolly expanded student use of the institute by designating one room in Muncke's former apartment for elementary student exercises and another, open all the time, for advanced student research. Jolly taught "exercises in the physical laboratory" every semester.[109]

In teaching, Jolly cultivated primarily the technical and applied aspects of physics and had limited success in promoting quantitative techniques. During his first five years at Heidelberg, from 1834 to 1839, he taught physics, technology, engineering, and applied mathematics. He claimed his students were satisfied with this curriculum. Between 1837 and 1839 he had also begun to teach differential and integral calculus and had apparently been successful enough to approach the Baden ministry about appointing a mathematician to teach science students. Yet in comparison to the kinds of mathematics and physics courses offered at Königsberg in 1839, Jolly's did not offer training in rigorous techniques of quantification. Again in 1854 he perceived the need to argue for mathematics courses that would help the natural sciences achieve "exact form." Along with Robert Bunsen and other Heidelberg natural scientists, he stressed that "mathematical knowledge is an integral part of the education of the natural scientist." Surprisingly, Jolly, Bunsen, and others, in a dissenting opinion, rejected the motion that someone distinguished in mathematical research be appointed, arguing instead that Heidelberg's need was "above all" for a teacher "who represents mathematics in its main branches and teaches in a stimulating way."[110]

When Kirchhoff arrived at Heidelberg in 1854, he strengthened quan-

108. G. W. Muncke to Staatsminister, 26 November 1836, 9 January [1837], GLA 235/3057; Quincke, *Physikalisches Institut Heidelberg*, pp. 15–16.

109. P. W. Jolly to Heidelberg University Administration, 18 June 1846, GLA 235/3135; Quincke, *Physikalisches Institut Heidelberg*, pp. 15–16; Heidelberg University, *Vorlesungsverzeichniss*, 1847–54.

110. P. W. Jolly to Staatsminister, 10 September 1839, GLA 235/3132; P. W. Jolly, R. Bunsen et al., "Votum dissensus B" [1854], ibid.

tification in physics instruction by alternating practical exercises with a course on theoretical physics, which included mechanics and other topics in Neumann's lecture cycle. His efforts to teach a rigorously quantitative physics were aided in 1857 by the appointment to Heidelberg's faculty of the mathematician Ludwig Otto Hesse, one of Neumann's and Jacobi's first seminar students. Hesse fulfilled Bunsen's desire for a mathematician with good teaching experience. Not only had he taught elementary and advanced courses at Königsberg (where he had been Richelot's assistant) and at Halle between 1845 and 1856; he also had experience teaching in Prussia's secondary schools. At Heidelberg, Hesse conducted exercise sessions with his mathematics courses. Sigmund Gundelfinger considered the exercise sessions offered by Kirchhoff and Hesse to be a "physical and mathematical seminar," which he claimed had a "decisive influence" on his course of studies. As they had done for other students, Kirchhoff and Hesse recommended that Gundelfinger continue his studies at Königsberg, which he did.[111]

But there were critical problems in Baden's educational system that made it difficult for Hesse to attract more than a small audience of mathematics students and, consequently, for Kirchhoff to use higher mathematics in physics courses: Baden's secondary schools had a weak curriculum in mathematics. The Heidelberg University Senate thought the problem could be solved by having Hesse teach "elementary mathematics as it is taught at school." The philosophical faculty, however, upheld the right of the *Ordinarius* to teach advanced subjects, in this case to teach "pure mathematics"; for it was not the obligation of the *ordentliche* faculty, they believed, to fill in the gaps left by school instruction. What needed to change, the philosophical faculty argued, was not the university but the school.[112]

In a complete misreading of the program of instruction at Königsberg, the philosophical faculty went on to argue that the high level of mathematical instruction at Prussian secondary schools began "when Bessel and especially Jacobi began to teach at Königsberg University. Both men never taught the elementary parts of mathematics, and Jacobi's lectures covered exclusively the higher levels of mathematics." The philosophical faculty did admit that at first "there was a gap between the mathematical instruction at the school and the university" but asserted that "the intellectual influence of difficult mathematical lectures," felt first in East Prussia and then over the entire monarchy, had led to improvements in teacher training and in the mathematical curriculum of secondary schools. Con-

---

111. Heidelberg University, *Vorlesungsverzeichniss*, 1854–70; "Otto Hesse's Lebenslauf," in *Ludwig Otto Hesse's Gesammelte Werke*, ed. Kgl. Bayerischen Akademie der Wissenschaften, 2d. ed. (Bronx, N.Y.: Chelsea Publishing, 1972), pp. 711–19; Sigmund Gundelfinger, Lebenslauf [1869], UAT 126/228.

112. Heidelberg Philosophical Faculty to Heidelberg University Senate, 13 March 1861, GLA 235/3132.

vinced by the arguments of the philosophical faculty, the senate argued to the Baden ministry that to train teachers properly in mathematics, Heidelberg needed a mathematical seminar but that what Jacobi and Bessel had done at Königsberg would be extravagant for Baden.[113]

Yet it was not until Baden reformed its teaching examination in 1867 and Richelot was called upon to evaluate the state of mathematical instruction at Heidelberg in 1868, when Hesse left and had to be replaced, that the state seriously considered a seminar for mathematics and physics. Richelot's sharp and frank criticisms noted that not only had Heidelberg "had no influence on the pure mathematical sciences" but also that "a deeper and more durable mathematical curriculum" was not to be found in Heidelberg, Freiburg, Tübingen, Erlangen, Würzburg, Munich, Vienna, and Prague, and not until recently in Giessen, where Neumann's former student Karl Zöppritz had just taken a position as *außerordentlicher* professor of mathematical physics. Richelot considered "the nature of south Germany" responsible for this state of affairs, especially the nature of its school and university instruction in mathematics, which he believed to have the "greatest unequivocal influence" on the mathematical sciences. The reforms he called for in Baden included a mathematical seminar "like the one established at Königsberg by Jacobi in 1833 [sic]," where teachers of mathematics could be properly trained; the complete reform of all mathematical curricula in Baden; and a replacement for Hesse, who not only had fine scholarly credentials but also an interest in school matters. Although Richelot himself was sought for the position (especially by Kirchhoff, his son-in-law), Heidelberg hired the Greifswald mathematician Leo Koenigsberger in 1869.[114]

Heidelberg's seminar thus became more than the simple institutionalization of Kirchhoff's and Hesse's exercises. The mathematico-physical seminar that Kirchhoff and Koenigsberger proposed in 1869 and that began in 1870 was designed in part to meet the needs of Baden's secondary schools. Koenigsberger ran his mathematical division as a colloquium that addressed the biweekly homework problems he assigned. Kirchhoff had seminar exercises in physics only in the summer semester, assigning both theoretical and experimental problems every week. Unlike Neumann, however, Kirchhoff included didactic exercises designed to improve the lecturing skills of teaching candidates. In other ways, too, the 1867 reform of Baden's teaching examination shaped Heidelberg's seminar and lowered the level of instruction offered in it. For instance, Kirchhoff and Koenigsberger were supposed to accommodate *all* teaching candidates, including philology students, who had to pass the mathematics and natural science portions of the examination. Adjusting to a broad-

113. Ibid.; Heidelberg University Senate to Staatsministerium, 27 March 1861, GLA 235/3132.
114. F. J. Richelot to Heidelberg Philosophical Faculty, 2 October 1868, GLA 235/3132.

based clientele proved problematic. Kirchhoff's enrollment in fact remained modest between the summer semesters of 1870 and 1874, ranging from eight to seventeen students and averaging just over twelve a semester.[115]

Yet Kirchhoff at least partially succeeded in promoting not only mathematical physics but also, for some students, the kind of exactitude cultivated at Königsberg. He alternated his seminar exercises during the summer with his course on mechanics in the winter semester; despite the fact that it was not quite the course that Neumann's had been, it nevertheless prepared students for the seminar. Kirchhoff and Koenigsberger developed such a good working relationship that Koenigsberger later remembered how he and Kirchhoff had "worked hand in hand so that occasionally in the same semester we both read mechanics, [Kirchhoff] from a physical, I from a mathematical, point of view. Every day we spoke with one another about the lecture to follow." When Koenigsberger accepted a call to the technical school in Dresden in mid-1874, both he and Kirchhoff knew the impact his departure would have on instruction in the exact sciences. Koenigsberger claimed it was his "special relation" to Kirchhoff that made it "possible to raise the study of mathematics and physics" at Heidelberg, and it was in part Koenigsberger's departure that convinced Kirchhoff to accept a call to Berlin in late 1874. He told the Baden Ministry of the Interior that with Koenigsberger gone, "it cannot be assumed that students of mathematics will assemble here," and without them, he thought, it would be impossible "to find students for the difficult lectures on mathematical physics."[116]

Initially Kirchhoff and Koenigsberger expected their exercises to provide the foundation for independent research and doctoral dissertations, but Kirchhoff rarely mentioned advanced investigations in his seminar reports. Although problems strictly in mathematical physics—requiring no experimental investigation and of the type that would issue from his lecture course on mechanics—were included among Kirchhoff's seminar exercises, most of them incorporated some techniques of an exact experimental physics. Every week he or one of his students performed an experiment in the seminar, the calculations for which were carried out at home, turned in, and then graded by Kirchhoff. Heinrich Martin Weber, who obtained his doctorate from Heidelberg in 1863 and then went on to continue his study of mathematics and physics in the Königsberg seminar, recalled of Kirchhoff's measuring exercises before the seminar was founded that they included "measurements of vibrations, the pendulum, acoustics, and the interference of light." Kirchhoff used several rooms in

115. G. Kirchhoff and L. Koenigsberger to Staatsministerium, 14 April 1869, GLA 235/3228; Oberschulrath to Staatsministerium, 15 November 1870, ibid.; Reports of the Heidelberg mathematico-physical seminar, 1870–74, ibid.

116. Leo Koenigsberger, *Mein Leben* (Heidelberg: C. Winter, 1919), p. 101; L. Koenigsberger to Staatsministerium, 12 July 1874, GLA 235/2201; G. Kirchhoff to Baden Innenministerium, 16 December 1874, GLA 76/9961.

Heidelberg's physical institute for elementary exercises. The British scientist Arthur Schuster, who attended Kirchhoff's seminar in 1871/72, later described how they were conducted: "One exercise was set aside for each week and every student—about eight altogether—had a morning or afternoon assigned to him for carrying out the experiment. There was one weekly lecture in which the results were criticized and the succeeding exercise explained."[117]

Advanced students worked differently. In 1871/72, Kirchhoff allowed Schuster and Heike Kamerlingh-Onnes to use other rooms in the physical institute for advanced investigations. Schuster, at Kirchhoff's request, tested an instrument Kirchhoff had devised for measuring metallic reflection, while Kamerlingh-Onnes demonstrated the motion of a pendulum's plane of vibration caused by the earth's motion.[118] Both advanced investigations were similar to those in Neumann's seminar in that they required the analysis of constant errors and, in the case of Schuster's project, a theoretical analysis of an instrument. Kirchhoff's aversion to "interpolation formulas" as the foundation for mathematical expressions in physics suggests that, in addition, he instructed his students in the method of least squares and would not have encouraged them to construct empirical laws, as Kohlrausch did.

The degree of exactitude Kirchhoff achieved in instruction was diminished considerably only six years after the seminar was founded. In 1875, Neumann's former student Georg Quincke came to Heidelberg as Kirchhoff's successor. Quincke's exercises were simpler and less associated with rigorous techniques in the quantification of error, the result of his self-initiated expansion of the student clientele for practical exercises. His teaching obligations were divided between a course on experimental physics held five hours weekly for a large service clientele and a smaller course on topics in mathematical physics. Upon his arrival he expanded the facilities of the physical institute to hold optional elementary practical exercises for the "beginners, chemists, and physicians" in his experimental physics course in addition to his seminar exercises. Enrollment in these optional exercises was very low at first—only three students—but by the winter semester of 1890/91 there were twenty-two, more than treble the seven in the mathematico-physical seminar. Exercises in the seminar generally dealt with classical research, usually on an experiment requiring precise measurements, which were performed first by Quincke and then by students, who worked out the calculations at home.[119]

117. "Statuten für das mathematisch-physikalische Seminar in Heidelberg," *ZMNU* 5 (1874): 397–98; Wilhelm Lorey, *Das Studium der Mathematik an den deutschen Universitäten seit Anfang des 19. Jahrhunderts* (Leipzig: B. G. Teubner, 1916), p. 73 (Weber's recollections); A. Schuster, "Biographical Byways: 4. Kirchhoff (1824–1887) and Bunsen (1811–1899)," *Nature* 115 (1924): 126–27, on 126.

118. Schuster, "Biographical Byways," p. 126.

119. [G. Quincke's courses from summer semester, 1877], Darms. Samml. Sig. Flc 1870(3); G. Quincke to Heidelberg University Senate, 27 September 1875, GLA 235/30012; Quincke to Neumann, 12 December 1893, FNN 53.IIA: Briefe von Schülern; "Nach-

In contrast, the exercises in the practicum, constrained by meager material and economic resources, utilized simple materials and instruments (often constructed by Quincke himself) that yielded, he thought, the same degree of accuracy as more complicated instruments. The practicum proved popular. By 1900, when he made it mandatory for physics students, it handled up to 120 students a semester. Space in the physical institute became "extraordinarily restricted," forcing Quincke to run the practicum in two or three sessions. His seminar, meanwhile, became less and less successful, with only four students in the summer semester of 1890. Much to his dismay, the ministry canceled the physical division of the seminar in 1900. It had operated under Quincke's direction for twenty-five years with enrollments of ten students or (more frequently) fewer per semester. One location at Heidelberg for the deeper study of physics, including the detailed analysis of errors, especially accidental errors, thus disappeared.[120]

All along, Quincke had supported instruction in theoretical physics and even tried to obtain "a smaller lecture hall for theoretical physics lectures, as is on hand in physical and chemical institutes of German universities." But with high enrollments in his experimental physics course, he preferred to have theoretical physics taught by an unsalaried *außerordentlicher* professor. Quincke subsequently devoted less time to rigorous techniques of quantification in his own physics courses. Although he favored "the inner value of an experiment" and not its "glittering" demonstration, he taught his students how to determine the range of error in an experiment largely through the calculation of constant errors. The objective of his experimental exercises, like those of Kohlrausch, was the refinement of protocol and material conditions. Further precision of the result, achieved by the calculation of accidental errors, was neither desirable (it took too much time) or necessary because students in the experimental physics course did not in general concern themselves with esoteric and detailed issues in theory confirmation.[121]

The mathematico-physical seminar at Breslau, founded in 1863, never fully cultivated critical techniques of error analysis and, by the turn of the century, evolved into an institute for mathematical physics without any connection to experiment. Like Muncke at Heidelberg, some early physicists at Breslau developed teaching styles in physics that incorporated a

weisung der Frequenz der praktischen Arbeiten im physikalischen Laboratorium" and "Nachweisung der Frequenz der Uebungen im physikalischen Seminar," 24 February 1891, GLA 235/30012.

120. G. Quincke to Staatsministerium, 18 March 1890, 9 May 1891, GLA 235/30012; Quincke to Staatsministerium, 29 May 1900, GLA 235/3228.

121. Quincke to Staatsministerium, 18 March 1890, GLA 235/30012; Dean of the Heidelberg Natural Sciences Faculty to Staatsministerium, 30 April 1894, GLA 235/3135; Quincke to Staatsministerium, 26 November 1894, GLA 235/3135; F. Braun, "Hermann G. Quincke zum 70. Geburtstag," *AP* 15 (1904): i–ix, on viii; G. Quincke, "Eine physikalische Werkstätte," *ZPCU* 5 (1892): 113–18; 7 (1893): 57–72, esp. 58–59, 61, 69, 70.

sense of exactitude, although not a very rigorous one. Before 1832, Heinrich Steffens used the three hundred taler annual budget for the physical cabinet to purchase instruments from Pistor, while Georg Friedrich Pohl taught courses in which he used astronomical and geodetic instruments and even developed apparatus for secondary school physics courses. We have already seen how Breslau's other physicist, Moritz Ludwig Frankenheim, cared so little for a rigorously mathematical approach to physics that neither Kirchhoff between 1850 and 1854 or Emil Meyer in 1860/61 could feel at home there. Exercises in mathematics and physics had been held for some time by 1862, but both Frankenheim and the mathematician Lipschitz found them wanting despite the addition to the physical cabinet of several instruments suited for quantitative measurements. Lipschitz, who proposed the seminar to the ministry, argued that at Breslau "interest in the free movement of thought, the essential nature of all scientific speculation, is very little developed" but that the spirit of research could be "enliven[ed] and strengthen[ed]" through a seminar. Frankenheim was concerned that the private exercises hitherto held in mathematics and mathematical physics had "to do without the advantages that institutes possess by state recognition and support." The mathematico-physical seminar that they founded with the mathematician Eduard Schröter, who like Lipschitz had been trained at Königsberg, was intended for cultivating independent research and was modeled on the Königsberg seminar insofar as it was to offer instruction in both mathematics and mathematical physics. It was established on 3 November 1863 with a budget of 220 taler (180 for prizes and 40 for books and materials).[122]

Under Frankenheim the physical division of the seminar seemed strong but little oriented toward mathematical physics, which Lipschitz taught. Originally the "special construction and ordering" of exercises was to be left to the students, but when the seminar opened, Frankenheim instead assigned exercises "according to the capabilities" of the twenty-seven students who had registered, some of whom produced results of "scientific value." The intellectual orientation of physics in the seminar was enhanced when Emil Meyer, who returned to teach at Breslau in 1864, directed a third section, this one devoted to teaching younger students the mathematical methods of physics, including differential calculus and equations from thermodynamics. But Meyer could not cultivate the exact

122. O. Lummer, "Physik," in *Festschrift zur Feier des hundertjährigen Bestehens der Universität Breslau*, ed. G. Kaufmann, 2 vols. (Breslau: F. Hirt, 1911), 2:440–48, on 440; R. Lipschitz to Kultusministerium [16 August 1862], ZStA-M, Acta betr. die Errichtung eines mathematisch-physikalischen Seminars bei der Königlichen Universität zu Breslau, Rep. 76Va, Sekt. 4, Tit. X, Nr. 56, Bd. I: 1862–1934, fol. 1; M. Frankenheim to Breslau University Curator, 26 March 1863, ibid., fol. 4a; cf. M. Frankenheim to Neumann, 15 November 1849, FNN 53.IA: Briefe von Collegen. On the Königsberg model see also H. Schröter and O. E. Meyer to Breslau University Curator, 21 May 1876, ZStA-M, Math.-phys. Seminar Breslau, fols. 61–62, on fol. 61.

experimental side of theoretical physics because he lacked measuring instruments and rooms for conducting practical exercises. Moreover, he was primarily responsible for teaching the general course on experimental physics and did not lecture on mathematical physics as much as he might have liked. In 1870, when he cancelled his section and took over the physical division of the seminar, he defined it as one for mathematical physics, which, he told the ministry, had been represented first by Lipschitz and "then through the undersigned present professor of mathematical physics." Five years later, when enrollments increased beyond what Meyer and Schröter could handle, they reinstituted the third seminar section. It was only for mathematical exercises, however, even though Meyer wanted to use it to lighten his burden of teaching both theoretical physics and the experimental training connected to it in the seminar's physical division.[123]

An opportunity for developing the exact experimental side of theoretical physics opened between 1873 and 1881, when Neumann's former student Ernst Dorn also taught at Breslau, but not in the seminar. At first Dorn, who had no equipment of his own, assisted in Meyer's practical exercises, but soon he reorganized physics instruction by introducing a general physics practicum. While at Breslau he was working on the calculation of absolute units for electricity and so brought to his exercises more of the Göttingen "style" in the analysis of measurements, rather than the Königsberg one. But like Meyer's courses, Dorn's on theoretical physics were classified as courses in mathematics, thus limiting the practical element he could introduce into them. The result was that the exact experimental side of theoretical physics never became strongly developed at Breslau. When Meyer asked to be released from his duties as director of the seminar in 1902, he chose Ernst Richard Neumann, Carl's nephew, as his successor, claiming that "mathematical physics, as it is represented here by Professor Neumann, is throughout mathematical."[124]

Neither of the two mathematicians then handling the mathematical division wanted mathematical physics to remain in the seminar because the intensity of work at the physical institute meant that a professor would be "less accessible" for the seminar and because they thought exercises in mathematical physics could be held independently of the seminar. But

123. "Statuten des mathematisch-physikalischen Seminars an der Breslauer Universität," ZStA-M, Math.-phys. Seminar Breslau, fol. 6; Seminar reports for 1863/64 (24 October 1864) and 1864/65 (18 January 1866), ibid., fols. 15, 23; Meyer and Schröter to Breslau University Curator, 25 January 1870, ibid., fols. 27–29 (rather than enhance the seminar exercises beyond the abilities of their students, Meyer and Schröter decided to offer seminar exercises suited for "the age, the talent, the preparation, and the inclination of the participants" [fol. 28]); Meyer and Schröter to Breslau University Curator, 29 June 1870, ibid., fols. 39–40, on fol. 39; O. E. Meyer and H. Schröter to Kultusminister Falk, 17 May 1875, ibid., fols. 59–60.

124. Lummer, "Physik," pp. 443–44; Breslau University, *Vorlesungsverzeichniss*, 1874–78; Dorn to Neumann, 22 February 1874, FNN 53.IIA: Briefe von Schülern; O. E. Meyer to Breslau University Curator, 8 May 1902, 11 June 1902, ZStA-M, Math.-phys. Seminar Breslau, fols. 100, 107–8, on fol. 107.

Neumann prevailed, arguing that mathematics and mathematical physics were closely bound to one another and that, far from abandoning the seminar for the physical institute, he considered the proper home of mathematical physics to be in the seminar, whose library was an important resource for theoretical physics. "The purchase of physical works that lean more toward the experimental direction appear to me to be entirely forbidden in the seminar's library," Neumann instructed, adding that books on experimental physics "might find a better home in the physical institute," while the literature for mathematical physics could not be transferred to the physical institute because "mathematical physics undoubtedly stands closer to mathematics than to experimental physics."[125]

In the years before World War I, Breslau's seminar exercises remained mathematically oriented but became more and more fundamental. Reduced to written work on defined problems from lecture courses, the exercises were by 1913 "not held for a limited number of students but they are open to all willing participants." To accommodate them, the seminar directors established a position for an assistant for theoretical physics who created and then corrected the seminar exercises he assigned; he also performed the "important and time-consuming calculations" needed for the physicists in the seminar and the institute. In 1920, Clemens Schaefer, theoretical physicist and director of the seminar, asked the ministry if he could transform the physical division into an "Institute for Theoretical Physics" using "the theoretical library of the deceased Göttingen physicist Woldemar Voigt"—which he hoped to purchase—as its material foundation. His request was denied. On 27 January 1928 the seminar directors told the ministry that the current seminar structure was obsolete because the student enrollment, now at ninety, was too large to use the library; because funds were insufficient for maintaining holdings for two subjects in the library; and because an assistant for theoretical physics was now required. Furthermore they thought that "the union of the two seminars stems from the time when there was no special professor for theoretical physics" (i.e., there was only an *Außerordinarius*). Now that there was an *Ordinarius* for theoretical physics, they explained, as of 1 April 1928 the seminar would operate as two, one for mathematics and one for theoretical physics. The institutional and intellectual evolution of mathematical physics at Breslau removed the discipline from the laboratory, were errors were computed and theory confirmed, and installed it in the library, where equations could be consulted and pure theory shaped.[126]

125. J. Rosanes and H. Schröter to Breslau University Curator, 31 May 1902, ZStA-M, Math.-phys. Seminar Breslau, fols. 105–6; Ernst Richard Neumann to Breslau University Curator, 10 June 1902, ibid., fols. 101–2, on fol. 101.

126. A. Kneser, E. Schmidt, and E. Pringsheim to Kultusministerium, 25 July 1913, ibid., fols. 136–37; Fritz Reiche to Breslau University Curator, 17 January 1924, ibid., fol. 239; Clemens Schaefer to Breslau University Curator, 22 June 1920, ibid., fol. 215; A. Kneser et al. to Kultusministerium, 27 January 1928, ibid., fols. 283–85.

When Neumann's former student Karl Zöppritz assumed a position as *außerordentlicher* professor of mathematical physics at Giessen University in 1867, a physical seminar directed by Heinrich Buff had been in operation for five years. Buff, who had been hired as *ordentlicher* professor of physics in 1838, had offered physics practica since the early 1840s, although at first he used the facilities at his disposal (rooms in his home, located not far from Justus Liebig's chemistry laboratory) primarily for "chemico-physical" investigations, both his own and his students'. Between 1847 and 1865, Buff, who as Liebig's doctoral student had learned the pedagogical value of precision and rigorous quantification in his mentor's laboratory, had successfully incorporated the techniques of an exact experimental physics into physics instruction. During those eighteen years, five of his students published investigations begun in either Buff's practicum or his colloquium for advanced students. All five investigations concerned precision measuring operations: the electrical resistance of liquids, the distribution of magnetism in a magnetic bar, the dependency of the electrical resistance of a fluid on temperature, the use of silver in determining units for measuring electrical resistance, and the determination of voltametric measurements.[127]

The techniques used in these investigations were like the ones taught at Königsberg, but there was little concern for theory. Students tried to achieve agreement between "a great number of observations;" to attain "the highest accuracy," and to carry out their protocol "with great care." They were extremely sensitive to the agreement that should be achieved between successive runs of an experiment, but they did not, it seems, become paralyzed by the drive for exactitude. Students determined the reliability of instruments, and they sought to reduce the deviations in measurements to below the level of the known constant errors. They recognized other irregularities in their data and used the method of least squares. Most important, they seemed not only willing to deploy graphs more readily than Neumann's students but also able to comprehend more clearly how graphs were used: one student remarked in 1850 that he "drew the most probable curve through the individual trials."[128]

127. H. Buff to the Hessian Academic Administration Commission, 27 October 1844, UAGi, Acta betr. das Lokal für die physikalischen Sammlungen der Landes-Universität, 1844–79; Wilhelm Lorey, "Die Physik an der Universität Giessen im 19. Jahrhundert," *Nachrichten der Giessener Hochschulgesellschaft* 15 (1941): 80–132, esp. 93; E. N. Horsford, "Ueber den elektrischen Leitungswiderstand der Flüssigkeiten," *AP* 70 (1847): 238–43; Ernst Becker, "Ueber die Abhängigkeit des elektrischen Leitungswiderstandes einiger Flüssigkeiten von der Temperatur," *Annalen der Chemie und Pharmacie* 73 (1850): 1–25; Wilhelm Langsdorf, "Das Silber als Einheit für die Messung des elektrischen Leitungswiderstandes," ibid. 85 (1853): 155–72; Heinrich Meidinger, "Ueber voltametrische Messungen," ibid. 88 (1853): 57–81; Georg Weihrich, "Ueber die Vertheilung des Magnetismus in weichen prismatischen Eisenstäben, die an das Ende eines Magnetstabs angelegt sind," *AP* 125 (1865): 276–92.

128. Becker, "Ueber die Abhängigkeit des elektrischen Leitungswiderstandes von der Temperatur," p. 7.

Even when Neumann's former student Alfred Clebsch established a mathematical seminar at Giessen in 1863, it did not detract from the practical orientation of the physical seminar, much as Clebsch himself had worked in a more mathematically oriented theoretical physics. Rather, ironically, it was Buff's vision for both seminars that eroded the exact experimental physics he had taught before the two seminars began. Clebsch's pedagogical abilities—which, despite an earlier poor reception among students at Karlsruhe, were considerable—were important in his appointment because Buff especially wanted someone who could teach future secondary school teachers of mathematics. Buff's early arguments for the physical seminar also stressed its role in training school teachers, especially in offering opportunities for practical exercises and in providing the kind of "direction" in learning physics that would be needed to teach it. He intended that the seminar have two divisions: one for experimental physics, where students could learn how to handle instruments, perform experiments, and become acquainted with school apparatus; another for mathematical physics, where students could deal with questions of a "physico-mechanical" nature. He considered it to be "of the greatest importance" that students have "knowledge of the meaning, the construction, and the use of countless physical instruments and apparatus, a knowledge that can only be acquired incompletely from books," and so he planned to have students in the seminar partake in practical exercises using the physical cabinet. Despite his well-intended plans, Buff did not sustain the high level of student achievement in evidence before the seminar began. His student clientele over the next sixteen years, until his death in 1878, was largely a service clientele but with few teaching candidates: students studying forestry, law, architecture, medicine, and pharmacy greatly outnumbered those from philosophy, chemistry, and mathematics.[129]

Although his students did not continue to produce original investigations, Buff made the seminar meaningful to a clientele broader than Neumann's. But Buff's abandonment of his earlier teaching style and his preoccupation with a more basic experimental physics diluted an exact direction at Giessen. After 1867 when Zöppritz tried to resurrect an exact physics and to enhance it with a greater emphasis upon theoretical considerations, he met with mixed success. In the summer semester of 1868 he began to offer a course on the method of least squares and elementary probability calculus; later he introduced a course on surveying, in which these methods were applied. These lecture courses must have gone fairly

129. UAGi, Acta betr. das mathematische Seminar, Nr. 246; Administration of the Karlsruhe Polytechnic to Baden Innenministerium, 14 August 1862, GLA 206/834; H. Buff to Giessen Philosophical Faculty, 5 January 1861, 20 November 1861, UAGi, Acta betr. Antrag Prof. Dr. Buff auf Errichtung eines mathematisch-physikalischen Seminars, 1861–62, Phil. H, Nr. 36; UAGi, Acta betr. die Eintrittsgelder der Mitglieder des physikalischen Seminars, 1863–82, Phil. H, Nr. 35.

well because by the winter semester of 1872/73 he directed his own division in Buff's seminar, one on mathematical physics, which included exercises using the methods of an exact experimental physics applied to theory. Zöpptitz's other teaching, however, lagged behind. He had offered courses like Neumann's in theoretical physics from the start—including potential theory, the mechanical theory of heat, theoretical and experimental optics, and the mathematical theory of electricity, hydrostatics, and hydrodynamics—but they were not well attended. Finally in 1873 he tried to rectify the situation by introducing a course on mathematical methods in physics, on partial differential equations. Before then, it had been the mathematical methods of an exact experimental physics, especially error analysis, that had assumed priority in his teaching. Although others later criticized Zöppritz for not taking great enough care in analyzing the data in his own work, it was Zöppritz more than Buff who tried to enhance and preserve rigorous methods of data analysis in physics pedagogy at Giessen.[130]

After Buff's death in 1878 and Zöppritz's departure in 1880, Giessen's practical exercises in physics met a mixed fate. Wilhelm Röntgen created a practicum in experimental physics that "essentially operated on the basis of Kohlrausch's practical physics," as one student remembered it. In place of Buff's seminar, now considered superficial by the ministry, Röntgen created a colloquium. Carl Fromme, Zöppritz's successor for theoretical physics, at first held practical exercises in measurement for theoretical physics, but these were soon absorbed by the practicum. The rigorous quantitative methods of measurement cultivated in the Giessen seminar survived elsewhere. Karl Noack, who had attended the seminar before Buff's death in 1878 and had worked on the relation between capillarity and temperature for his seminar investigation, made the pedagogical role of measurement a central concern in his own seminar for physics teachers, established at Giessen in 1888. Noack emphasized the accuracy and reliability of instruments, their amenability to error analysis, and the critical comparison of similar instruments and their results. But he also asked teaching candidates to consider "the value of measuring experiments for the upper forms" and the didactic purpose served by requiring students to attain accuracy.[131]

The four other mathematico-physical seminars—at Munich, Tübin-

---

130. Giessen University, *Vorlesungsverzeichniss*, 1868–73; K. Zöppritz to H. Wagner, 7 April 1879, 25 January 1880, H. Wagner Nachlaß, UBG-Hs; Meyer to Wagner, 4 February 1880, ibid. One of Zöppritz's biographers attributed his low enrollments to the poor regard in which mathematics education was held in Giessen (S. Günther, "Karl J. Zöppritz," *Allgemeine Deutsche Biographie* 45: 434–37), but O. E. Meyer thought it equally likely that Zöppritz himself was responsible (Meyer to Neumann, 10 May 1873, FNN 53.IIA: Briefe von Schülern).

131. Lorey, "Die Physik an der Universität Giessen," pp. 108 (quote), 109, 114; Karl Noack, "Ueber die Vorbildung von Lehrern des physikalischen Unterrichts am pädagogischen Seminar in Giessen," *ZPCU* 3 (1889/90): 103–4, on 104.

gen, Erlangen, and Rostock—did not always offer exercises in an exact experimental physics and so differed substantially from their counterparts. Teacher training dominated Rostock's seminar, which existed primarily to give students an opportunity to improve their lecturing skills.[132] Tübingen's offered more involved activities, in part because it was preceded by practical exercises that had accompanied theoretical lectures; such exercises were considered professional training in that they were viewed as corresponding to the practical training received by a theologian or physician. Unlike Tübingen's exercises in experimental physics, which involved critical techniques in measurement and the analysis of instruments, its seminar exercises in theoretical physics concerned abstract, specialized problems stemming from lecture courses and generally did not require the use of instruments. Even had these seminar exercises included a rigorous consideration of constant and accidental errors, problems in relating the numbers of experiment to the mathematics of theory similar to those that had arisen at Königsberg would probably have occurred. After all, the mathematical division of Tübingen's seminar was for a while directed by the former Königsberg seminar student Sigmund Gundelfinger, who had been brought to Tübingen to teach elementary mathematics, especially analytic geometry.[133]

At the Bavarian universities of Erlangen and Munich, the relatively late reform of the state teaching examination in 1853 as well as the fact that physics was not made an obligatory subject in Bavarian gymnasiums until 1854 meant that the state's science policy for secondary school teaching tended to crowd out other objectives in the mathematico-physical seminars at these two universities. Documents on the Erlangen mathematico-physical seminar's operation are sketchy, but those there are suggest that its most important offerings were its library and the practical experience it offered teaching candidates, especially in the handling of school apparatus.[134] More propitious circumstances were present at Munich. After the reform of Bavaria's science policy in 1847, Bessel's student Karl August Steinheil and Jacobi's student Ludwig Seidel proposed a mathematico-physical seminar in 1848; both would have taught a practical physics like Neumann's. By the time the seminar was founded in 1856, after a second wave of secondary school educational reforms, Philipp Jolly had replaced Steinheil as director of the proposed physical division; he had already begun offering practical exercises in 1854. Jolly managed to improve the

132. G. Becherer, "Die Geschichte der Entwicklung des physikalischen Instituts der Universität Rostock," *Wissenschaftliche Zeitschrift der Universität Rostock*, Math.-naturwiss. Reihe, 16 (1967): 825–37, on 827; G. Kelbg and W. D. Kraeft, "Die Entwicklung der theoretischen Physik in Rostock," ibid., 839–47, on 839.

133. Sigmund Gundelfinger to Tübingen University Rector, 16 February 1872, 10 February 1873, UAT 126/228; A. Brill to Tübingen University Rector, 28 November 1885, 11 March 1887, UAT 117/882.

134. UAE I/20V/9, especially documents dated 9 and 14 December 1874, 23 April 1875, and 25 February 1902.

physical cabinet by getting a new laboratory, and after the seminar was founded, he instituted an assistant's position intended for advanced seminar students. Even though his own investigations centered on the analysis of instruments and the determination of physical constants, the suggestion to introduce the mathematics necessary for an exact experimental physics came not from Jolly but from Seidel, who viewed the function of mathematical instruction not just as providing a "general *Bildungsmittel*" for students in the philosophical faculty but also as having a "value for the exact sciences." In 1859 and then again in 1869, Seidel wanted to offer a course on the elements of probability calculus and "the theory of the combination of observations belonging to it," but he could not without having someone else take over the lectures on analytic geometry. Munich's exercises were never outstanding, although they were suited for training secondary school teachers. By the 1880s, after Eugen Lommel had taken over Jolly's position, the material resources for practical instruction were so old, deteriorated, and scarce that exercises were hardly possible.[135]

Practical exercises were also offered where seminars did not exist, and by 1893, physics courses at all German universities had optional or mandatory exercise sessions.[136] The mathematical rigor of these exercises, the regularity with which they were offered, and the degree to which they promoted exactitude varied. Greifswald, not known for the quality of its physics instruction, nonetheless had exercises that were well-enough attended, sufficiently quantitative, and regularly supported by the state for there to be a mathematico-physical institute at the university from around the 1840s, out of which emerged a seminar for theoretical physics sometime after 1899.[137] In contrast, practical exercises at Berlin, one of the undisputed centers of German physics in the nineteenth century, varied over the course of the century. With the opening of his private laboratory in the 1840s, Magnus offered practical exercises for select students. Even after he opened a university laboratory in 1862/63, only a limited number of students (between six and eight) were admitted, a pattern that continued for a while after Helmholtz's arrival in 1870. Between the 1840s and 1870, the handling of data in Magnus's laboratory became less similar to that in Neumann's seminar; the critical perspective evident in the investigations of Magnus's students during the 1840s was replaced by one more exploratory and strictly experimental. Neither Helmholtz nor

---

135. Karl Neuerer, *Das höhere Lehramt in Bayern im 19. Jahrhundert* (Berlin: Duncker & Humblot, 1978), pp. 103, 110–11, 118–19, 129; Ludwig Seidel to Munich Philosophical Faculty, 10 May 1869 (quote), UAM 208; P. Jolly to Munich Academic Senate, 11 December 1876, UAM 289; E. Lommel to Munich Philosophical Faculty, 10 April 1886, UAM 289.

136. For a listing of practica see *Die deutschen Universitäten*, ed. W. Lexis, 2 vols. (Berlin: Asher, 1893), 2:164–65. On facilities built for practica see David Cahan, "The Institutional Revolution in German Physics, 1865–1914," *HSPS* 15 (1985): 1–66.

137. ZStA-M, Acta betr. das mathematisch-physikalische Institut der Königlichen Universität zu Greifswald, Rep. 76Va, Sekt. 7, Tit. X, Nr. 25, Bd. I: 1829–62; Bd. II: 1863–99.

Kirchhoff, who arrived at Berlin in 1875, conducted practical laboratory work in the same way Neumann had at Königsberg. Admission to the physical laboratory was restricted to eighteen students, who had to pay eighty marks a semester and apply one semester in advance. In 1888 the experimentalist August Kundt reorganized Berlin's practicum on the basis of Kohlrausch's textbook, offering problems "from all areas of physics in a systematic sequence," but apparently the changes he introduced were inadequate for meeting the needs of all students (including teaching candidates and introductory students) and even for promoting a "quantitative, mathematical treatment of physical problems."[138]

This survey of German practical exercises in physics illustrates the uneven promotion of exactitude, places Neumann's exercises with their more involved determination of error and more rigorous idea of exactitude into relief against the background of practices at other German universities, and provides some sense of how his former students brought his exercises elsewhere but failed to sustain them. It also suggests that although some practical exercises were rigorously exact elsewhere (as they were at Göttingen), they were directed at different objectives; that outside Königsberg, exactitude had a pedagogical as well as a research function; and, recalling the investigations that issued from practical exercises at Giessen, Göttingen, and Berlin, that how and to what degree one sought exactitude had epistemological consequences in the construction of scientific knowledge. Insisting upon the computation of accidental errors and eschewing interpolation, as they did at Königsberg, made achieving scientific truth technically more difficult. And yet despite variation in practical exercises, the differences among them were ones of degree, not kind. Precise measurements and the calculation of at least constant errors were essential components of practical exercises at most German universities.

There is no easy explanation for why exactitude and precision were sought in different ways and to different degrees, especially in the final decades of the century. The decline of rigorous precision in practical exercises certainly correlated with the pressure of the marketplace—the growth in enrollments, the diversity of student clienteles, and the expansion of practica—but it is unlikely that demographics were totally responsible for simplifying practica. What the expansion of practica signified was the acceptance of the educational ideology of the seminar, which assumed that lectures were incomplete without exercises that applied what had been learned. The type and quantity of material resources also set the boundaries of what was possible in practical exercises, but much could be

138. "Das physikalische Studium auf der Universität," p. 643; H. Rubens, "Das physikalische Institut," in *Geschichte der Königlichen Friedrich-Wilhelms-Universität zu Berlin*, ed. M. Lenz, 4 vols. (Halle: Waisenhaus, 1910–18), 3:278–96, on 289; R. W. Pohl, "Von den Studien- und Assistentenjahren James Francks: Erinnerungen an das physikalische Institut der Berliner Universität," *Physikalische Blätter* 28 (1972): 542–44, on 544.

accomplished even with meager means. Of all the "external" conditions that shaped the nature of practical exercises and the degree to which they promoted exactitude and precision, two regional characteristics appear to have been the strongest: mathematical instruction at the secondary and higher levels, and the various state teaching examinations. Richelot was not alone in his negative assessment of mathematical instruction at universities south of the Main River.[139] Without question, then, the possibility of cultivating a rigorous and exact experimental physics was severely constrained without advanced mathematics. This was true even for locations in Prussia where mathematics was not promoted, as in some of the seminars for the natural sciences. State teaching examinations were often crucial in promoting quantification in the sciences. Prussia's consistently required more mathematics than any other and hence created an institutional reinforcement for attempts to teach a rigorously quantitative physics, either theoretical or experimental.[140]

The support the pedagogues Karl Schellbach in Berlin and Karl Noack at Giessen gave to assigning measuring exercises in secondary schools reinforced the notion that learning physics meant knowing how to achieve precision in measurement. Some of Neumann's students, we have seen, took the ethos of exactitude into secondary schools. Ludwig Otto Hesse, one of Neumann's and Jacobi's first seminar students, was instrumental in having practical expertise with instruments made a part of the Bavarian teaching examination after the 1860s.[141] But by the late 1870s, support for precision measuring exercises waned as teacher training in all German states, especially Prussia, began to emphasize pedagogy more than the learning of specialized disciplinary knowledge. Prompted by the "observation that teachers of physics coming out of the university with adequate theoretical training frequently do not have the required acquaintance with teaching apparatus in physics and with desirable exercises in experimentation," the Prussian *Kultusministerium* issued a regulation in 1876 that made clear the shortcomings of existing seminars and practica in training *Physiklehrer*. The state considered independent inves-

139. For like-minded assessments see, e.g., Carl Neumann to R. Radau, 29 April 1869, Darms. Samml. Sig. Fle(3) 1868; H. Weber, "Ueber die Stellung der Elementarmathematik in den mathematischen Wissenschaften," *JDMV* 12 (1903): 398–401, on 400; Eduard Reusch to F. Neumann, 3 October 1868, FNN 53.IA: Briefe von Collegen.

140. Braunschweig's examination, for instance, covered only elementary mathematics, and Prussia did not recognize the examinations of Baden and Württemburg as equivalent to its own (Wilhelm Lorey, *Staatsprüfung und praktische Ausbildung der Mathematiker an den höheren Schulen in Preußen and einigen norddeutschen Staaten* [Leipzig: B. G. Teubner, 1911], pp. 90, 97).

141. O. Volk, "Mathematik, Astronomie, und Physik in der Vergangenheit der Universität Würzburg," in *Vierhundert Jahre Universität Würzburg: Eine Festschrift* (Neustadt: Degener, 1982), pp. 751–85, esp. pp. 769–70; Karl Schellbach, "Die wissenschaftlichen Seminarien auf der Universität," 30 June 1860, ZStA-M, Acta betr. die Errichtung eines mathematischen Seminars bei der Königl. Universität zu Berlin, Rep. 76Va, Sekt. 2, Tit. X, Nr. 77, [Bd. I:] 1860–1916, fols. 4–5.

tigations undertaken in physical institutes with "several delicate apparatus" to be "inadequate" for teaching candidates and so recommended that special courses be created where teachers could become acquainted with *school* apparatus, instruments, and exercises. On the 1890 state teaching examination, experimental physics was given priority over theoretical or mathematical physics, especially for candidates wishing to teach the upper forms. All teaching candidates in physics had to demonstrate the ability to handle school apparatus and create exercises.[142] Journals for science pedagogy, just beginning to appear in the 1870s and 1880s, echoed the sentiments of the Prussian ministry, which in a reversal of its earlier views now stressed the pedagogical element in teacher training.[143] By the turn of the century, Bavaria had improved its teacher training in physics to the point where its institutions of higher learning offered teaching candidates the opportunity to gain experience with school apparatus in specially designed courses.[144]

Physiklehrer and physicists alike, among them some of Neumann's former seminar students, questioned the pedagogical usefulness of precision. The problem, one teacher pointed out, was that a teaching candidate did not learn at the university what he needed to teach secondary school. The apparatus was too complicated; the lectures on theoretical physics were too mathematical and specialized; and in general, physical laboratories were not large enough to handle teaching candidates in addition to physics majors. But even where practical experience was possible, there were problems because

> the trials conducted by the young practicant are all the so-called precision trials and therefore take a lot of time. The consequence is that if one wants to profit from them, one has to devote almost all of one's time exclusively to the laboratory and everything else must be put aside. . . . We are not entirely against measuring trials as such but against the exclusive preoccupation of the practicant with precision measuring trials that have meaning only for the physicist who later has a well-equipped laboratory at his disposal. . . . Precision measurements, which are thought proper in these laboratories, are of little significance for the prospective school man.[145]

The complaints against precision measuring exercises cut deeper, into the epistemological issues that Neumann's own seminar students had strug-

142. "Königl. preußische Ministerial-Verordnung, betreffend die Anleitung künftiger Lehrer der Physik zur Bekanntschaft mit den physikalischen Lehrmitteln und im Experimentiren während der Studienzeit" [20 June 1876], *ZMNU* 8 (1877): 186; *Ordnung der Prüfung für das Lehramt an höheren Schulen in Preußen vom 12. September 1898 und Ordnung der praktischen Ausbildung der Kandidaten für das Lehramt an höheren Schulen in Preußen vom 15. März 1890* (Berlin: Hertz, 1898), p. 14.

143. See, e.g., ["Review of *Zeitschrift zur Förderung des physikalischen Unterrichts*"], *ZMNU* 16 (1885): 217–20, on 217–18.

144. Staatsministerium to Erlangen University, 25 February 1902, UAE I/20V/9.

145. "Das physikalische Studium auf der Universität," pp. 643, 644, 645.

gled with. Former student Friedrich Pietzker pointed out that when emphasis was placed on measurement, "formal techniques" dominated over "the execution of fundamental thinking" and that even "the construction of the law lying at the basis of the experiment" was hardly considered.[146] Teachers alluded to the dangers arising when students were led to believe that accuracy in measurement could be equated with the attainment of scientific truth. For one teacher, "absolute accuracy" was "an empty word without content" in the natural sciences because, in contrast to mathematical values, "natural scientific values are themselves never defined with unlimited accuracy."[147] The evolution of seminars in which teaching candidates had hitherto been trained—especially their turn to a mathematical physics devoid of experiment, as at Breslau under Ernst Richard Neumann—contributed to the sense that precision measuring exercises were no longer relevant to teacher training.

Important as such pedagogical factors were by the end of the century, they were not the only reasons why precision in physics attracted less attention in educational settings. Changes in the curricular role of mechanics, in the form of practical exercises, and in the function of seminars were but signs of a more fundamental transformation taking place. As the postseminar investigations of Neumann's students, as well as the later investigations by the students of Kohlrausch and Magnus suggest, the character of experiment itself was evolving. Although Neumann's type of experiment was never practiced everywhere, precision had nonetheless played a dominant role in both theoretical and experimental physics from the late 1820s or early 1830s to sometime in the 1870s or 1880s. But by the final years of the century, precision experimentation had become but one experimental style, identified either with the refined measurements associated with theoretical physics or with the official economic, commercial, and scientific concerns of such institutes as the Physikalisch-Technische Reichsanstalt in Berlin. Even so, however, experiment had loosened its ties to several of the techniques that Neumann had borrowed from the Besselian experiment. The type of experiment emerging in the final decades of the century was one oriented less toward the exact determination of physical constants or the rigorous expression and confirmation of theory than toward the discovery of something new. Although the historical lineage of the exploratory experiment in German physics is at present sketchy, it runs through the work of the students of Magnus and Kohlrausch. Referring to Magnus's student August Kundt, Woldemar Voigt observed in 1899 that "the Kundtian pure experimental school

146. F. Pietzker, "Der Göttinger Entwurf eines Studienplanes für die Kandidaten des höheren Lehramts in Mathematik und Physik," *ZMNU* 24 (1893): 470–74, on 473.

147. H. Burkhardt, "Mathematisches und naturwissenschaftliches Denken," *JDMV* 11 (1902): 49–57, on 52; see also A. Postelmann, "Zur Reform des mathematischen und naturwissenschaftlichen Unterrichts auf dem Gymnasium," *Programm*, Kgl. Gymnasium zu Bartenstein, 1909–10, p. 9.

direction with its more striking successes is the standard in Germany." The hegemony of an exploratory experimental physics had consequences for theory, especially for the type of theoretical physics promoted in Neumann's seminar, which did not escape Voigt. "Exact theory," he continued, had become less meaningful in an age when there were "very gifted experimenters" but only a few who showed an interest and inclination for theory and when those who did, neglected precise observation.[148]

The techniques of the Besselian experiment did not entirely disappear from sciences other than astronomy, however. Bessel's former student Karl August Steinheil, for example, tried to preserve them at Munich, but after he died in 1870, his Physico-metronomical Institute passed into the hands of Philipp Jolly, who did not use Steinheil's instruments for precision exercises.[149] Of greater significance in the preservation of the rigorous mathematical techniques used in error and data analysis was the seminar for scientific calculation, established at Berlin in 1879. Most of the seminar's exercises concerned problems in astronomy, but they were intended as exemplars in the art of precision measurement, where the techniques of error and data analysis were most highly developed. Students learned numerical integrations, interpolation, the method of least squares, approximation methods, how to handle tables, and other methods for the reduction and testing of measurements.[150] The third location where not only the techniques but also the model of the Besselian experiment survived was in physics instruction at Königsberg.

Exactitude at Königsberg

Whereas by the end of the century, practical exercises in physics at several German universities either ceased to emphasize error analysis so strongly or concentrated primarily (if not exclusively) upon the analysis of constant errors, at Königsberg the analysis of constant and accidental errors, especially the pedagogical role of the Besselian experiment, remained important parts of instruction in theoretical physics. Exactitude persisted at Königsberg largely because the values, intellectual and pedagogical, that had given it meaning proved strong among Neumann's successors, who had either been trained under him or had developed strong ties with his way of teaching physics. Without any compelling disciplinary reason to promote error analysis as Neumann had done in

148. W. Voigt to L. Koenigsberger, 6 June 1899, Darms. Samml. Sig. Fla(4) 1891.
149. J. von Müller, "Das physikalisch-metronomische Institut," in *Die wissenschaftlichen Anstalten der Ludwig-Maximilians-Universität zu München*, ed. K. A. von Müller (Munich: Oldenbourg, Wolf & Sohn, 1926), pp. 278–79.
150. "Reglement für das Seminar zur Ausbildung von Studirenden im wissenschaftlichen Rechnen an der Königlichen Universität zu Berlin," *Centralblatt für die gesammte Unterrichtsverwaltung in Preußen* 21 (1879): 165–67.

the past (theoretical physics was securely institutionalized at the end of the century and relied less on the techniques of an exact experimental physics), Neumann's successors sometimes made the meaning of exactitude deeply personal and subjective.

Neumann's last doctoral student, Woldemar Voigt, was in the process of revising his doctoral dissertation on the elasticity constants of rock salt in 1875 when he was called as *außerordentlicher* professor of theoretical physics to Königsberg. He had just completed a semester of secondary school teaching at Leipzig's Nicolai Gymnasium in 1874/75 and had received the *venia legendi* at Leipzig University. The financial arrangements of his position were at first complicated; for he was expected to make a living on student fees alone. As he had explained in September 1875, although he looked forward to Neumann's company and guidance once again, he found it troublesome that he would "probably be working for many years without a salary." It was his hope that Königsberg would establish a third salaried professorship in physics—a position that seemed likely, he thought, because there were three professorships in chemistry. But a full professorship was not to materialize, leading Voigt to accept an offer from Göttingen in 1883. When he was first called to Königsberg, however, he thought there was a chance he could remain there. Since Neumann had retired, someone else was needed to teach theoretical physics. Voigt also noted that Moser "only offer[ed] popular lecture courses in physics for medical students" and was, at the time of Voigt's appointment, very ill. Upon receiving his appointment, Voigt pointed out too that "during the three years [June 1871 to March 1874] I studied at Königsberg, no physical-exact lecture course was held for two summer semesters straight." Perhaps because he was hoping for a more permanent and financially secure appointment, he had in fact exaggerated the gaps in Königsberg's physics curriculum, forgetting that although Neumann had not taught during the summer semester of 1872, he had during the summer semester of 1873, when he had offered his "chapters in mathematical physics."[151]

Voigt's position at Königsberg turned out to be more satisfying than he had at first anticipated but still less than he had hoped it would be. In October 1875 he was awarded a salary of twenty-eight hundred marks, which he soon found he had to draw upon to supply his students with the necessary apparatus for practical exercises.[152] He later remembered that one of the first lectures he prepared for a physics class at Königsberg concerned the history of pendulum experiments. Significantly, in that lecture he focused on the techniques and corrections Huygens, Borda,

---

151. Kultusminister Falk to Woldemar Voigt, 2 October 1875, WVN 1; Voigt to [?], 4 September 1875, Darms. Samml. Sig. Fla 1891(4); "Anmeldungsbuch [Königsberg Universität]," WVN 1.

152. Falk to Voigt, 2 October 1875, WVN 1.

and especially Bessel made to increase the accuracy of their measurements.[153] So even before he became director of the seminar in the summer semester of 1876, Voigt had made the Besselian experiment a model for his pedagogical physics.

As he prepared for the seminar, Voigt's emphasis on the techniques used to close the gap between theoretical and experimental results, especially error analysis, intensified. Away in Leipzig in Sepember 1876, he wrote Neumann about his continuing preparations for the seminar. Although he had known firsthand of Neumann's pedagogical style and especially of his seminar exercises, Voigt found the task of creating seminar exercises suitable for his students to be nonetheless considerable. In the few weeks before the winter semester began, he worked in earnest on problems he could assign, occupying himself "with the capillary theories of Gauss and Laplace as well as with the construction of seminar exercises based on them." But he was finding the discrepancies between theory and experiment difficult to reconcile. He had "found so many problems" in capillary observations (especially Georg Quincke's) that he was determined to achieve a satisfying agreement between theory and experiment himself. Voigt told Neumann that he considered many of Quincke's results in further need of proof; he meant that the cause of the deviations produced by Quincke's method of observation had to be more precisely ascertained through error analysis. Whereas Neumann had often left the task of comparing observational data with theory to his students, Voigt wanted to compare certain calculations from capillary theory with observations himself before he "recommend[ed] them to the seminar."[154] His review of the data thus simplified the seminar exercises but also made them less original. The urgency he felt to rectify the data himself detracted from the potential impact seminar exercises would have on training students in the investigative techniques of physics and on inculcating in them the same values Voigt himself had learned from Neumann.

Preparing for the seminar also involved mastering classical literature in physics. Mentioning Poisson's work in particular, Voigt asked Neumann if they could get together when he arrived in Königsberg in order to talk about it, believing he could profit more by spending an hour with Neumann than by working on it alone by himself for a week. He also planned to review other classical investigations, including the work of Ohm and Ampère, in preparation for taking over the seminar.[155] Neumann's years of experience and the canonical state of his seminar exercises had not made Voigt's task that much easier. But that Voigt believed it necessary to consider these theories in terms of their experimental confirmation, even

---

153. Woldemar Voigt, "Der Kampf um die Dezimale in der Physik," *Deutsche Revue* 34.3 (July–September 1909): 71–85, on 75–77.
154. Voigt to Neumann, 9 September 1876, FNN 53.IIA: Briefe von Schülern.
155. Ibid.

after other seminar students such as Kirchhoff had done so in an exemplary fashion, is a testament to the powerful pedagogical role of the Besselian experiment, as well as to the seemingly endless tasks created by the ethos of exactitude it entailed.

Shortly after he had arrived at Königsberg but before he had assumed the directorship of the seminar, Voigt conducted practical exercises in private for a few students. Following Neumann's earlier practice, he put aside two rooms in his own living quarters for practical exercises and purchased the necessary equipment from his own salary. The arrangement he had thus made was at first a special one, designed for advanced students from the mathematico-physical seminar. He could accommodate the four students he was then instructing quite easily, but should that number increase by only two, he would be unable to conduct these exercises any longer. Refusing to let his students use his own, more delicate instruments, Voigt asked the ministry for a thousand marks to purchase equipment for his exercises, which he received in March 1876 before he took over the seminar.[156]

Much as Neumann had done, Voigt argued for the construction of a laboratory for mathematical physics, drawing the ministry's attention to the fact that in Germany, Neumann was considered "master" of this "new science" and to the fact that it would become increasingly difficult to sustain a tradition in mathematical physics at Königsberg without one. Significantly, the institute he had in mind was one where error analysis would play a central role; for he wanted the laboratory principally so that he and his students could test the "correctness" of what had been discovered elsewhere. But now Voigt believed that the model for the kind of laboratory he wanted was to be found in the *Technische Hochschulen* of Germany. In July 1876 circumstances looked favorable for breaking ground on a new institute because a suitable location had been found and the university curator was ready to put aside one hundred thousand marks to outfit it, but no firm steps were taken.[157]

By January 1877, Voigt was handling twelve to seventeen students per semester in his practical exercises. Not only were his instruments and rooms insufficient; he was also finding that he just did not have the time to prepare adequately for the exercises or for the seminar, so he asked for a salaried assistant. At the end of 1877 he finally admitted that it was becoming impossible for him to handle practical exercises every semester without an assistant; eventually he managed to convince an older student to assist him without pay. Despite the technical sophistication that ex-

156. Königsberg University Curator to Kultusministerium, 13 January 1876, ZStA-M, Acta betr. den mathematisch-physikalischen Apparat und das physikalische Institut der Universität zu Königsberg in Preußen, Rep. 76Va, Sekt. 11, Tit. X, Nr. 15, Bd. II: 1876–86 (unpaginated when seen); Kultusministerium to Königsberg University Curator, 3 March 1876, ibid.

157. Königsberg University Curator to Kultusministerium, 7 July 1876, ibid.

ercises in experimental physics had by then attained, especially through the widespread use of Kohlrausch's textbook, Voigt was of the opinion that the instruments used in experimental physics did not suit his purpose and that exercises for mathematical and experimental physics were sufficiently different to justify separate facilities for conducting them. He consistently viewed his practical exercises as ways to achieve *"scientifically exact* measurements and investigations."[158]

That Voigt was able to maintain a distinction between the practical laboratory exercises of theoretical physics and those of a more strictly experimental physics was due in part to the evolution of experimental physics at Königsberg, which since the 1830s had been unusual (but not unique) among the German universities because the university had two full physics professorships. Ironically, after Moser retired in 1876, it was one of Neumann's former seminar students, Carl Pape, who received his chair. By the time he returned to Königsberg in 1878, however, Pape had a different perspective on physics instruction than he had had a decade earlier. Although he had not planned well for a career, he became in 1862 *Privatdozent* for mathematical physics at Göttingen, where he taught until 1866. Still trying to find a niche for himself in academics, he used several of his courses on mathematical physics as vehicles for teaching physical chemistry, in which, as it turned out, he included several topics he would have covered in mathematical physics anyway, including the theory of specific heats. When his enrollments fell off slightly in 1865, he asked for a laboratory and for a permanent appointment as assistant in it so that he could teach students how to "completely evaluate" laws in physical chemistry. The exercises Pape proposed to teach were those of Neumann's seminar: the theory of the balance and of weighing; the determination of specific heats; the theory of the apparatus used to calculate specific heats, including the calibration of scales; and thermometry. In other ways he proposed to bring topics from mathematical physics into physical chemistry, including what the mechanical theory of heat meant for this subordinate and still emerging subdiscipline of chemistry.[159] Pape did not obtain a permanent position at Göttingen, but he and other former seminar students from Königsberg contributed to the breadth of courses in mathematical physics offered there. During the summer semester of 1866, for example, Pape, Ludwig Minnigerode, and Heinrich Weber taught five of the eight courses in physics, including mathematical physics, the mechanical theory of heat, the theory of attraction, magnetic measurements, and physical demonstrations.[160]

Pape's ability to teach and to practice a critically oriented mathematical

158. Ibid.; Voigt to Königsberg University Curator, 4 January 1877, 2 December 1877, 26 March 1878 (quote), ibid.

159. Carl Pape to Friedrich Wöhler, 23 June 1865, UAG 4/Vc/105.

160. Göttingen University, *Vorlesungsverzeichniss*, summer semester 1866.

physics waned considerably after he took an appointment at the agricultural academy in Proskau in 1866. The nature of his position there demanded that he accommodate his lectures to students not always well prepared to study physics and that he develop a talent for demonstrations and experiments that would attract and keep students in physics courses. Intellectually isolated, unproductive, and unable to offer advanced courses in physics, he nonetheless considered himself qualified for a university position and tried but failed to obtain a joint position at Bonn University and the Poppelsdorf agricultural academy in 1869. Still at Proskau in 1876 when Moser retired, he was one of the leading candidates for the professorship in experimental physics at Königsberg, primarily because of his research experiences in experimental physics and because he had taught "the entire area of experimental physics."[161]

Two conditions stood in the way of Pape ever returning to the type of critical techniques he had taught earlier at Göttingen. He could no longer offer any help in teaching mathematical physics, Emil Meyer told Neumann, because "he has not forgotten what he learned with you, but he did not remain active in it and appears to have little inclination for mathematical lectures because although he was appointed in mathematics as well, he has always taught only physics." Even if Pape had the motivation to deploy more sophisticated mathematics in teaching, it would not have been possible for him to do so at Königsberg, where his position was designed for the benefit of students who needed broad survey courses in experimental physics. It was quite clear from the start not only that he would be teaching in ways different from Voigt but also that he would not be able to draw upon a common pool of resources for physics teaching at Königsberg, where the apparatus for theoretical and experimental physics remained strictly separated.[162]

The physical laboratory that both Neumann and Voigt considered essential for the execution of critical and evaluative techniques in mathematical physics and that had been promised and partially funded in 1876 did not materialize while Voigt was at Königsberg. Voigt continued to draw upon a budget of about one thousand marks per year for exercises in mathematical physics. Until the early 1880s, other funds designated for physical instruments were used to cover deficits elsewhere in the university (the curator promised Neumann that such reappropriations would not affect the new institute because the cost of outfitting the new institute would be included in its construction costs[163]). When Voigt was called as *ordentlicher* professor of theoretical physics to Göttingen in 1883 and the

161. Pape to Neumann, 20 April 1869, FNN 53.IIA: Briefe von Schülern; Meyer to Neumann, 28 November 1876 (quote), FNN 61.9: Berufungssachen.

162. Meyer to Neumann, 28 November 1876 (quote), FNN 61.9: Berufungssachen; Pape to Neumann, 24 December 1877, ibid.

163. Königsberg University Curator to Neumann, 26 April 1879, FNN 62: Kassensache; see also 18 November 1880, 17 July 1881, and 4 November 1881, ibid.

Königsberg faculty had voted about the same time to promote him to *Ordinarius* as well, it was Neumann who pointed out that even if Voigt were to remain at Königsberg, the faculty stood a good chance of losing him because Voigt's research and teaching needs had not been met by the construction of a physical laboratory. Neumann viewed Voigt's departure as a loss for the physical division of the seminar, creating "a gap difficult to fill."[164]

Ironically, in 1884, a year after Voigt's departure, construction on Königsberg's physical institute finally began. When completed, it more firmly institutionalized the difference between the critical practical techniques of theoretical physics and the more exploratory ones of experimental physics. The symmetrical, U-shaped building completed in 1888 was divided almost perfectly in half, with experimental physics on one side and mathematical physics on the other. The rooms for the divisions were duplicated almost one for one; each division had its own rooms for galvanic experiments, balances, manometers, and optics (Figure 30). But the two divisions did not simply define two separate realms for conducting similar measurements. Pape's experimental division offered primarily elementary practical exercises, and although it was equipped for advanced student investigations, few were actually conducted there. Between 1889 and 1902, Pape reported only one independent investigation: a study of an air thermometer undertaken by a mathematics student, who used his investigation as the basis of his doctoral dissertation. Although Pape employed an assistant and had at his disposal other resources that would have helped him to conduct and direct more extensive and sustained investigations, he seems not to have utilized effectively what was available to him. During his directorship, for example, students used the assistant's position in his laboratory as a steppingstone to secondary school teaching, not to academic positions.[165]

Under Paul Volkmann, Voigt's former student and successor, the laboratory for mathematical physics had, in contrast, a more vigorous activity, but it was an activity that brought Neumann's critical seminar exercises based on the mathematically and technically difficult quantification of error to a high baroque stage of expression. Volkmann, who began his studies at Königsberg in 1875, entered the mathematico-physical seminar in the summer semester of 1876, when Voigt became director. He attended the seminar for nine semesters straight until the summer semester

164. Neumann to [Königsberg University Curator?], n.d. [1883] (draft), FNN 61.9: Berufungssachen.

165. Königsberg University, *Chronik*, 1889–1902, esp. 1896–97, p. 36; 1889–90, p. 14; 1900–1901, p. 35. On the physical laboratory and institute at Königsberg see ZStA-M, Physikalisches Institut Königsberg; "Königsberg: Das physikalische Institut," in *Anstalten und Einrichtungen des öffentlichen Gesundheitswesen in Preußen: Festschrift zum X. internationalen medizinischen Kongress Berlin 1890*, ed. M. Pistor (Berlin: J. Springer, 1890), pp. 338–42; "Neubau des physikalischen Instituts in Königsberg i. Pr.," *Zentralblatt der Bauverwaltung* 7 (1887): 13–14; and Cahan, "Institutional Revolution."

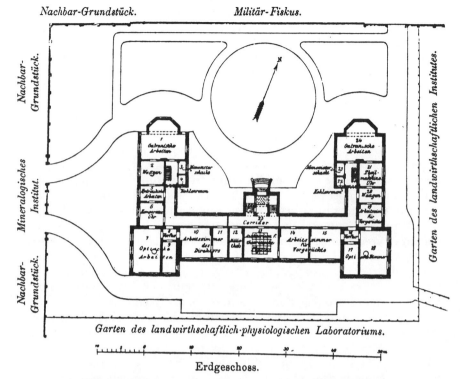

Erdgeschoss.

Mathematisch-physikalische Abtheilung.

1. Galvanische Arbeiten.
2. Waagen.
3. Manometerschacht.
4. Kohlenraum.
5. Hydraulische Arbeiten.
6. Komparator, Uhr.
7 u. 9. Optische Arbeiten.
8. Vorflur.
10 u. 11. Arbeitszimmer des Direktors.
12. Bibliothek.

Experimentell-physikalische Abtheilung.

13. Auditorium für theoretische Physik.
14, 15 u. 19. Arbeitszimmer für Vorgerückte.
16. Vorflur.
17 u, 18. Optische Zimmer.
20. Waagen.
21. Theilmaschinen, Uhr.
22. Kohlenraum.
23. Manometerschacht.
24. Galvanische Arbeiten.

25. Korridor.

Figure 30. Ground floor plan of the Königsberg Physical Institute, with rooms used for (1) galvanic work, (2) weighing, (3) manometer pit, (4) coal room, (5) hydraulic work, (6) comparator and clock, (7 and 9) optical work, (8) vestibule, (10 and 11) directors' workrooms, (12) library, (13) auditorium for theoretical physics, (14, 15, and 19) workrooms for preparations, (16) vestibule, (17 and 18) optical rooms, (20) weighing, (21) threading machine and clock, (22) coal room, (23) manometer pit, (24) galvanic work, and (25) corridor. From "Königsberg: Das physikalische Institut," in *Anstalten und Einrichtungen des öffentlichen Gesundheitswesen in Preußen*, p. 339.

of 1880. Although he had taken courses taught by Moser and Neumann, he considered himself more strictly a student of Voigt, under whom he completed a doctoral dissertation on capillarity in 1880. Dedicated to Voigt, Volkmann's dissertation began as a seminar exercise; incorporated the methods of error analysis associated with the seminar, including the method of least squares; and was based on observations conducted in what Volkmann called Voigt's "physical institute." It was Voigt who, as Volkmann explained, pressed Volkmann to recalculate capillary constants by using a less accurate but more direct method and who persuaded Volkmann to make the method more rigorous by giving it a theoretical foundation. Both the topic and the style of his dissertation exhibited the penchant for "error over truth" that had characterized earlier dissertations at Königsberg, including that of his advisor Voigt. Neumann, as one of the readers of Volkmann's dissertation, thought that Volkmann had convincingly questioned several older sets of capillary observations and had conducted enough better ones of his own to cast doubt on Laplace's capillary theory.[166]

In 1877/78 and intermittently thereafter, Volkmann had been Voigt's unpaid assistant in the physical practicum. He became Voigt's salaried assistant in 1880 and held that position until 1882, when he became *Privatdozent* at Königsberg, and then *außerordentlicher* and *ordentlicher* professor of theoretical physics there in, respectively, 1886 and 1894. In 1886 he assumed the directorship of both the physical division of the seminar and the laboratory for mathematical physics. Along with Voigt he became a major spokesman for the tradition that had emerged under Neumann. His admiration for the past, however, frequently resulted in uncritical, if not romantic, views of the current state of physics. At work assembling his lectures on theoretical optics for publication in 1888, he wondered if he should follow electromagnetic theory or elasticity theory, which "a pupil of F. E. Neumann would prefer." He eventually chose both representations, considering it the "obligation" of a university teacher to present alternative theories so that he could help shape in his students "the ability to make scientific judgments."[167]

Soon after Volkmann acquired the physical laboratory that Neumann had for so long considered essential for instruction in physics, the appeal and activity of the physical division of the mathematico-physical seminar at Königsberg waned considerably. Between 1883 and 1890 the number

166. Paul Volkmann, *Ueber den Einfluss der Krümmerung der Wand auf der Constanten der Capillarität bei benetzenden Flüssigkeiten*, Inaugural-Dissertation (Leipzig: Metzger & Wittig, 1880); Neumann, c. 1880 (draft), FNN 61.5: Gutachten.

167. Volkmann, *Ueber den Einfluss der Krümmerung der Wand*, p. 42 (Lebenslauf); Voigt to Königsberg University Curator, 26 March 1878, ZStA-M, Physikalisches Institut Königsberg; Paul Volkmann to Hermann von Helmholtz, 23 May 1888, DM 1976–30/A; Paul Volkmann, *Vorlesungen über die Theorie des Lichtes unter Rücksicht auf die elastische und die elektromagnetische Anschauung* (Leipzig: B. G. Teubner, 1891), p. iii.

of seminar students ranged between 3 and 23, but after 1887 there was never more than 10 students and between 1892 and 1902, no more than 5, and most often only 2 or 3. Volkmann attributed the decline in the activity, not only of his seminar but also of others, to the ministry's cancellation of seminar premiums for students in 1884, a step taken to cut costs. But in fact the seminar's purpose eroded once a physical laboratory opened. When the physical laboratory was completed, practical exercises moved out of the seminar and into the laboratory, leaving the seminar with the function of assigning, reviewing, and grading written homework problems drawn from ongoing lecture courses. At times Volkmann considered these written problems to be so elementary that he openly acknowledged that students could not go "into depth" in them. Although his introductory seminar exercises were similar to Neumann's in that they were drawn from mechanics, they were more theoretical than practical; Volkmann claimed that through them, "theoretical knowledge" could be enhanced through "theoretical ability."[168]

But although "theoretical ability" continued to mean the ability to express and manipulate physical theory mathematically, the mathematical techniques in seminar exercises no longer included those used to analyze data or constant and accidental errors, which had been so essential in Neumann's exercises. Volkmann's seminar exercises, in contrast, concerned ideal conditions, not the constraints of an experiment. Ironically Volkmann spoke of his exercises in terms reminiscent of the ideology of *Bildung*: that at the start, a student did not know where he was going, but after working on the exercises he soon became introduced to "the individuality of scientific activity" through the guidance of his teacher, who would help him transform "reproductive" learning into "productive" learning. Through seminar exercises, Volkmann thought, a student would be able to organize "all knowledge in an organic way according to his own experience" and so "to awaken" the drive to seek science for science itself. Solving written problems being its central activity, the most important material resource the seminar offered was its library, always at the disposal of students for their consultation. As problems in the seminar became more book oriented (in the sense of "looking up the answer"), Volkmann advocated the study of classics in science— Bessel, Helmholtz, and others—as a way to fill in the gaps left behind by lectures.[169] Whereas Neumann had used mathematics and other analytic

---

168. Königsberg University, *Chronik*, 1889–1902, esp. 1889–90, p. 14; 1899–1900, p. 33. For the course of study in the seminar, see Ferdinand Lindemann and Paul Volkmann, *Rathschläge für die Studirenden der reinen und angewandten Mathematik an der Königl. Albertus-Universität zu Königsberg i. Pr.* (Königsberg: R. Leupold, 1887), pp. 5–12.

169. Königsberg University, *Chronik*, 1901–02, p. 35; Paul Volkmann, *Rathschläge für die Studierenden der theoretischen Physik an der Universität Königsberg i. Pr.* (n.p., June 1901), pp. [5–6]; Lindemann and Volkmann, *Rathschläge für die Studirenden der . . . Mathematik*, p. 5; Paul Volkmann to Königsberg University Curator, 26 February 1890, Darms. Samml. Sig. Fla 1880(3).

means to knit together what his students were learning in physics, Volkmann used history.

Paralleling the decline in the seminar was the growth of the physical practicum in the new institute. Volkmann had recognized that it was the small number of students in Neumann's seminar that made it possible to weave "an intimate bond between teacher and student." The growth in the size of his own physical practicum, made possible by the mathematicophysical laboratory, essentially prevented such intimacy from developing again. In the three year period from 1899 to 1902, for example, the number of students in his practicum grew from twelve to twenty-five.[170] But it was not only a growth in numbers that distanced Volkmann from his students. The practical exercises incorporating error analysis which Neumann had taught in the seminar became, under Volkmann, the principal component of laboratory training in his institute. Initially, Neumann had constructed his practical exercises so as to maximize creative student involvement in all parts of them, especially in the attainment of precision and the assessment of accuracy. Later, Voigt introduced an element of artificiality into them by working through all data before his students did. Now Volkmann strengthened the supervisory element in them even further.

Volkmann claimed that the work of the physical laboratory "consists in the solution of a series of physico-practical problems" chosen to promote the study of physics in what he called "material" and "formal" ways. The "material" elements of physics instruction were physical facts and instruments, whereas the "formal" elements were the art and method of observation, including the calculational techniques necessary to analyze observations. Volkmann frequently spoke of observation as an art form, learned like a craft in a guild under the supervision of a master. "Working through and putting the finishing touches [on observations] has to be reviewed [by the instructor] and in many cases be verified [by him] in order to instruct the student in the artistically sound execution of calculations," he wrote about how students learned to analyze errors and data competently and accurately. But he also created some exercises in which his judgment and discipline were not necessarily required; some of his problems incorporated an "inner control" of the correctness of their solutions; if the "control" failed, then the student knew promptly and automatically to begin solving the problem all over again.[171]

Although Volkmann referred frequently to Kohlrausch's textbook

170. Paul Volkmann, "Franz Neumann als Experimentator," *Verhandlungen der Deutschen Physikalischen Gesellschaft* 12 (1910): 776–87, on 787; Volkmann to Königsberg University Curator, 14 July 1902, Darms. Samml. Sig. Fla 1880(3).

171. Paul Volkmann, *Vorbemerkungen für die praktischen Uebungs-Arbeiten und ihre Behandlung im mathematisch-physikalischen Laboratorium der Universität Königsberg i. Pr.* (n.p., June 1900), p. [1]; Volkmann to Königsberg University Curator, 14 July 1902, Darms. Samml. Sig. Fla 1880(3).

when instructing his students in the art of practical observations, the practical physics that he taught in his laboratory was not what Kohlrausch advocated. Practical exercises at the Königsberg institute, in contrast to its seminar, were still distinguished from Kohlrausch's and those of others by a greater emphasis on accidental errors and by a more sophisticated mathematical calculation of errors in general. Volkmann called the data analysis undertaken by his students "critical": the investigator, understanding the conditions under which the observations had been taken, calculated the limits to which the data were accurate. Volkmann viewed himself as continuing the tradition that Neumann and Jacobi had begun in the seminar, and he even drew upon Neumann's pool of exemplars, including Bessel's seconds pendulum experiments. The pedagogical strategy in Volkmann's practical exercises was designed less to train students in the art of taking measurements for computing physical constants than to analyze errors for understanding the imperfections of experimental conditions and the limits to laws mathematically expressed, just as the model Besselian experiment implied. Although students began their exercises by adjusting their instruments, they moved quickly to taking "preliminary measurements," but not with the intention of solving the problem at hand. Instead, "the preliminary measurements made known the order of magnitude and limits to the accuracy" of the data and how small changes in the parameters of the experiment affected the precision of the result. These control observations they used to determine the magnitude of the corrections that had to be made. The final stage, completed by the students at home, was the determination of accidental errors (every result had to have "±" affixed to it, Volkmann instructed his students) including a discussion of the possible reasons for the remaining uncertainty.[172]

"In the analytic completion and perfection" of data, Volkmann explained, "is found one of the most important sources of self-education in physics, which represents the inner goal of every university course in physics." Thereby what Humboldt and others had earlier believed could be achieved through *Bildung*, Volkmann thought best attained through rigorous procedures in data and error analysis. Every student in the practicum thus handled "a special problem from the realm of precision measurement," first by learning how to handle instruments in the laboratory and then by taking data home "in order to work through the calculational and mathematical side." Referring to both aspects of his practical exercises—the instrumental and the calculational—Volkmann explained

172. Paul Volkmann, *Anleitung zur Aufzeichnung physikalischer Beobachtungen, zur Anlegung von Beobachtungsprotokollen und ihrer Verwertung für die Praktikanten des mathemat[isch]-physikalischen Laboratoriums der Universität Königsberg i. Pr.* (n.p., January 1902), pp. [1–2]; idem, *Aufgaben für die Praktikanten des mathematisch-physikalischen Laboratoriums der Universität Königsberg i. Pr.* (n.p., January 1901); idem, *Vorbemerkungen für die praktischen Uebungs-Arbeiten*, pp. [1–4], esp. p. [3].

that "this activity of the practicum is grounded in the nature of the subject [*Fach*] of mathematical physics."[173]

Whether a student would lean toward the theoretical or the practical side of physics, Volkmann thought, would depend on his talents, desires, inclination, and "individuality." But the choice was neither completely voluntary nor entirely inner directed. Volkmann considered it unthinkable to take up physics without also studying pure mathematics, or even to study pure mathematics without also taking courses in theoretical physics. *Physiklehrer* in particular were obligated to study theoretical physics because it endowed "the teacher with the capability to deepen physics instruction, to structure it clearly, and to concentrate on the essential." The practicum, too, Volkmann considered meaningful for teaching candidates because "in no lecture course did the sources of knowledge step so far into the foreground as in the practicum."[174]

But by the turn of the century, the practical exercises that Neumann had initially developed to achieve certainty in mathematical physics and that fit the professional ideology embodied in the state's teaching examination were viewed by Volkmann as having a greater relevance to modern life itself. He considered it a *"Lebensfrage"*—a vital question—for the physics student to decide early to gain practical experience because there were now so many occupations that demanded it.[175] In a state that had industrialized so quickly as to surpass all other nations economically and that had institutionalized in its bureaucracies, businesses, weights and measures, and factories the practical benefits and ideological components of precision, practical exercises in measurement and data analysis were a way to achieve certainty and exactitude not only in physics but also in a now more technically oriented and rationally directed life.

Volkmann's pedagogy embraced an epistemology built on the notion that human apprehension of nature was partly subjective and flawed with errors that could be eliminated or reduced by systematizing physical knowledge and quantifying error. Hence although he believed a theorist was supposed to unify and order physical theories, he also believed that "reality is not comprehended through theoretical speculation alone but above all is studied through observation and measurement." Knowing a theory, according to Volkmann, thus meant having "knowledge of the conditions under which it is valid and applicable." Accordingly, as "the

173. Volkmann, *Vorbemerkungen für die praktischen Uebungs-Arbeiten*, pp. [1–4] on p. [4]; Volkmann to Königsberg University Curator, 14 July 1902, Darms. Samml. Sig. Fla 1880(3).

174. Volkmann, *Rathschläge für die Studierenden der theoretischen Physik*, p. [2]; Lindemann and Volkmann, *Rathschläge für die Studirenden der Mathematik*, pp. 13–14; Paul Volkmann, *Fragen des physikalischen Schulunterrichts* (Leipzig: B. G. Teubner, 1913), p. 29; idem, *Anleitung zur Aufzeichnung physikalischer Beobachtungen*, p. [2].

175. Volkmann, *Anleitung zur Aufzeichnung physikalischer Beobachtungen*, pp. [1–2]; idem, *Rathschläge für Studierenden der theoretischen Physik*, p. [3].

main task of the laboratory" Volkmann cited "the testing of physical theory," and only after that did he mention that, "hand in hand with that," went "the determination of fundamental constants." He furthermore claimed that the precision measurements that deepened and strengthened what was known "correspond[ed] to the critical standpoint of research" and had to be distinguished from experimental investigations that widened the scope of knowledge but "correspond[ed] to the naive standpoint of research."[176]

In Volkmann's view, precision measurements were difficult to achieve because one had to know how "to interpret and analyze observations" and "to set up frequently equivalent [experimental] conditions under which one can repeat results" so "that one strives with all energy to attain under equivalent conditions exactly equivalent results." He considered it "the first requirement of an exact precision measurement, indeed of every experimental investigation," to examine and diminish as much as possible every source of error. "Not until after such a study," he stressed, "may one step over into the essential subject of the determination of a physical constant; not until then will one be capable of conceiving a definitive experimental arrangement in which the error, if not diminished, has its influence suppressed to a minimum." In emphasizing so strongly the view that the reliability of physical knowledge was determined by the quantification of error, Volkmann knew that he was preserving from the past an element of physical investigation that was not valued so highly in the present. Not only had theoreticians turned their attention to matters more theoretical in nature, but, as he noted about the physics community of which he was a part, "satisfying calculations" of the sources of error were "not always carried out today by well-known experimental physicists," an admission that experimental procedure was itself changing.[177]

Yet Volkmann believed that the diminution or eradication of error could not lead to the certainty believed to characterize mathematical theorems. Because "no measurement" could be "absolutely accurate," the task of practical physics was less to exceed a certain accuracy "than to prove the rigor and sharpness of fundamental theorems and natural laws." Assuming the position of the pious and the chosen, Volkmann insisted that it would be "deceitful" in physics "to erase" the difference between the accuracy that was attainable in practice and the "unlimited accuracy" that theory seemed to demand. Hence, pure mathematics could not, in his view, be taken as "the ideal of an exact science" because

176. Paul Volkmann, *Einführung in das Studium der theoretischen Physik, insbesondere in das der analytischen Mechanik* . . . (Leipzig: B. G. Teubner, 1900); idem, *Vorlesungen über die Theorie des Lichtes*, p. viii; idem, *Erkenntnistheoretische Grundzüge der Naturwissenschaften und ihre Beziehungen zum Geistesleben der Gegenwart* (Leipzig: B. G. Teubner, 1910), pp. 172–75, on pp. 172, 175, 174.

177. Volkmann, *Erkenntnistheoretische Grundzüge der Naturwissenschaften*, pp. 289, 292, 179; see also pp. 292–96.

quantifying error had nothing to do with pure mathematics. Instead, one model of the type of investigation he had in mind was Bessel's on the seconds pendulum. Volkmann considered Bessel's experiment, especially its error analysis, a corrective for pride and arrogance, a check on the "overestimation of our intellectual capabilities that so easily pretend to be able to view all relations a priori."[178]

So although Volkmann clung to an ethos of exactitude, it was in the sense that, as he put it, "we remain in the area of the exact if we can keep the value of the error in view." Not only did he seem to realize that his views were atavistic; he also knew that present institutional conditions did not necessarily promote them. "The sense for accuracy is little developed in contemporary education," he explained, because often numbers have ascribed to them an accuracy greater than what was allowed by experimental conditions and because—if his own secondary schooling were representative of contemporary practices—the determination of accuracy in numerical calculations is considered "an exaggerated game wasteful of time and energy" having "no relation to external reality."[179]

The cordial relations between the mathematical and physical divisions of the seminar became somewhat strained under Volkmann after the turn of the century. In the last decades of the century, the mathematical division had offered students instruction in both elementary and advanced mathematics, as it always had. The topics covered were drawn from analytical mechanics, analytic geometry, differential calculus, elementary problems concerning plane curves, second-order linear differential equations, variational calculus, elliptic functions, Hadamard's theory of functions, some of Hilbert's theories, and other subjects. Introductory students completed weekly homework problems, while advanced students discussed recent mathematical literature with the director. Not only did Volkmann acknowledge the importance of advanced mathematical methods that the professional theoretical physicist needed to analyze errors and make approximations; he also considered it essential that *Physiklehrer* know mathematics as well. But he vehemently opposed theoretical physics becoming a "mathematical discipline," or an "appendix" to mathematics, or even a "physical mathematics." As he wrote to Arnold Sommerfeld in 1899, theoretical physics was an "independent discipline," which had much to be thankful for in mathematics, but it did not need to be propped up by mathematics.[180]

Other conditions exacerbated the increasingly strained relation between mathematics and physics at Königsberg. Positions in mathematics were not readily filled, and representatives of the exact sciences not only disapproved of how Volkmann handled the practicum in mathematical

178. Ibid., pp. 179–80, 175, 177, 181, 176, 190.
179. Ibid., p. 181.
180. Königsberg University, *Chronik*, 1883–1902; Volkmann, *Fragen*, p. 29; Paul Volkmann to Arnold Sommerfeld, 30 October 1899, DM 1977.28/A, 348.

physics but also considered his "introduction to the study of theoretical physics" to be "an inadequate foundation" for other theoretically oriented lectures.[181] Along with the transfer of practical exercises from the seminar to the laboratory for mathematical physics and the subsequent internal changes in the operation of the seminar itself, especially Volkmann's turn to historical literature, the tensions between mathematics and physics further weakened the operation of the physical division.

Almost all the eight mathematico-physical seminars founded throughout the German states in the nineteenth century changed significantly in the first decades of the twentieth. With the possible exception of the mathematico-physical seminars founded at Göttingen in 1850 and Tübingen in 1869, which persisted at least nominally into the twentieth century, the remaining six seminars met similar fates. The physical division of the Heidelberg seminar was absorbed into the university's physical institute in 1900, leaving behind a seminar for mathematics. The same occurred at Erlangen in 1914/15 and at Munich in 1922/23. Separate seminars for mathematics and for physics replaced the mathematico-physical seminars at Rostock in 1922 and at Breslau in 1928. In July 1937, the director of the Königsberg mathematico-physical seminar finally admitted to the minister of education that

> the name "mathematico-physical seminar" does not correspond to actual conditions. Both physical institutes have for a long time been completely separated from the mathematico-physical seminar materially and in terms of space. The rooms and books of the mathematico-physical seminar are equipped only for mathematical purposes, and they serve only such. A physical division does not exist. If it still did exist, then it probably would have been united a long time ago with the second existing physical institute, the present institute for theoretical physics. In fact there is no difference between the mathematico-physical seminar here and the mathematical institutes at other universities that would justify a different name. The existing name has already often caused confusions and mistakes.

The seminar that Neumann and Jacobi had established in 1834 for mathematics and physics thus became, over a century later, one for mathematics alone.[182]

The End of the Ethos

In March 1909, Woldemar Voigt spoke before an academic audience in Amsterdam on a theme that he thought required no special knowledge of physics: the "struggle for the decimal in physics." The increasingly accu-

181. Paul Volkmann to David Hilbert, 20 January 1897, 26 May 1905, David Hilbert Nachlaß 4.16, UBG-Hs.

182. Seminar director to Kultusministerium, 27 July 1937, GSPK, Acta betr. das mathematisch-physikalische Seminar, Rep. 76, Nr. 906 [Königsberg University], Abt. X, Nr. 25, Bd. III: vom Januar 1935, fol. 13.

rate determination of physical constants was, he believed, "in no way to be attained by the rise in the accuracy of measuring instruments alone" because "the accuracy of a measuring instrument [was] something entirely different from the accuracy of a number determined with its help." The problem was that a physical constant was rarely measurable directly but had to be computed from one or more values that could be measured. Hence, in his view, "it [was] the task of the experimenter to choose the order of the measurements" so that disturbances were minimized and mathematical calculations facilitated. For Voigt, attaining the next decimal place was not merely a matter of being more accurate; it was also a matter of quantifying error. Dealing with each source of error was, he believed, "in no way a scientific sport" but, rather, a "scientific necessity," which if handled properly would lead to greater degrees of accuracy— that is, to certainty in the next decimal place. To illustrate what he meant, he drew upon the history of pendulum experiments from Huygens to Bessel, detailing the errors that had been calculated by each investigator in order to attain an accuracy greater by one-tenth. He attributed the more accurate measurements to greater mastery of the theory of the pendulum's motion in real conditions, a mastery exemplified best, he thought, by Bessel.[183]

Voigt ended his speech with an allegory. He spoke of a fairy tale in which a prince struggled to reach and capture a king's daughter ensconced in a magical castle. To reach the captive princess, the prince had to work his way through doors in the many walls that surrounded the castle. As he opened each door, another adversary confronted him. Not until he was victorious in his struggle could he move on to the next wall; not until he had overcome all adversaries did he reach his prize. "So it also goes for the scientific investigator in the struggle for knowledge," Voigt claimed, including "the struggle for the decimal. The goal of absolute truth lies for the most part unreachable in the infinite. The investigator will never attain it. But happiness and satisfaction are found in struggling and pushing forward. And [happiness and satisfaction] are so much greater for the investigator when he feels himself in this struggle as a member of a large organism, the totality of scientific efforts. The organism stretches over all civilized nations and gives them, in spite of material oppositions, an ideal bond." Although earlier in his speech Voigt had spoken of "the scientific necessity of more accurate observations" and, when discussing Bessel, had linked the need for accuracy to the reform of weights and measures and hence to the commercial interests of a century earlier, neither the ideal nor the economic justification for greater accuracy seemed appropriate in 1909. Perhaps to neutralize momentarily the growing international tensions in the years before World War I, Voigt alluded instead to the leavening effect of scientific internationalism.[184]

183. Voigt, "Kampf um die Dezimale," pp. 73, 74, 75–83, 81.
184. Ibid., pp. 85, 78.

One of the interesting features of Voigt's fairy tale and story of physics, of the fulfillment of the ethos of exactitude as the conquering of errors, is that he told it in masculine metaphors. The eradication of error was not, as he said, a sport; but that only confirmed that it was not an enterprise where victory could be attained. Cast in terms of a fairy tale, conquering error was a masculine undertaking nonetheless: a prince rescuing his princess. Voigt was not unique in representing the activity of practical physics as masculine. In 1881, Emil Warburg, then director of Freiburg's physical institute, spoke of practical exercises and the achievement of accuracy in similar terms: "There may scarcely be found a better means for training an earnest, masculine scientific character than these exercises."[185]

What Voigt's fairy tale meant, however, was that the struggle could in principle never end. The ethos of exactitude demanded the attainment of accuracy and precision through error analysis, but it was a goal that could never be completely realized. Absolute certainty lay somewhere in the infinite, not only because it was unattainable in practice but also because if it were achieved, the investigative enterprise in physics would itself collapse. In terms of the practice of physics, the ethos emphasized performance over product, means endlessly pursued over ends decisively achieved. There might have been greater truth in each decimal place, but it was not and could not ever be either absolute or mathematical truth, as Bessel himself had admitted. The kind of physics taught and practiced genuinely in the tradition of the Königsberg mathematico-physical seminar could only achieve a truth qualified by the quantitative determination of error, as the Besselian experiment demanded.

The ethos of exactitude characterizing some of the research and teaching of former Königsberg seminar students over the years had become an ethic. It guided professional actions and decisions by providing the ways and means of separating right from wrong, truth from error, and even the called from the damned. It helped to define professional identities, structure investigative strategies, and identify significant problems. It shaped the interaction between exact experiment and theory, showing one way to create the certain epistemological base that both experimental data and theoretical results required. But the ethos of exactitude also limited the range of choices in an investigation by blocking alternative means of solving problems, thus narrowing both the field of potential investigative techniques and the range of possible solutions. In the end, the ethos proved impossible to sustain. Its eventual demise, notably coinciding with the growing popularity of more exploratory experimentation and of a more purely "theoretical" physics, contributed to the redefinition of the character and role of exact experiment in physics.

185. Emil Warburg, "Das physikalische Institut," in *Die Universität Freiburg seit dem Regierungsantritt seiner königlichen Hoheit des Grossherzogs Friedrich von Baden* (Freiburg: J. C. B. Mohr, 1881), pp. 91–96, on p. 93.

Epilogue

The BEGINNING OF the nineteenth century in Germany marked a moment in the evolution of physical experimentation when the tools for analyzing and calculating constant errors sharpened considerably, paving the way for precision measurement, and when a consideration of accidental errors, computed by the method of least squares, was of very recent origin. A considerably more rigorous exact experiment, incorporating both kinds of errors, had a profound effect on the evolution of theoretical physics and on the assessment of scientific truth within it.

What a fine-grained look at scientific practice as it was taught in Neumann's seminar revealed was that theoretical physics involved much more than merely expressing physical relations in mathematical form. Mathematization was, to be sure, enlightening, but it also had its dark side insofar as there was no uniform agreement about the meaning of equations and data, especially when it came to determining what degree of certainty was achieved and how that certainty was to be judged, as the debates and controversies between Neumann's former students and others demonstrated. What illuminated the path to truth in theoretical physics was not mathematization alone but mathematization in conjunction with a type of precision measurement strongly qualified by the results of a quantitative analysis of errors. In this sense, exact experiment performed a crucial epistemological function in legitimizing mathematization, especially in separating interpolation or empirical formulas from legitimate theoretical ones and in curbing the use of questionable quantitative techniques of generalization, such as the interpolation of data, especially in graphs.

In German settings between roughly 1800 and the late 1830s, theory in physics was still relatively new, not well understood, but arriving in prodigious quantities from across the Rhine. Neumann's creative adaptation of the Besselian experiment—one variety of experiment among several—provided a much needed bridge between theory and experiment. It also seemed to supply, through its techniques of error analysis, the means of ascertaining the certainty of both, thus helping to bring French mathematical physics into the foreground of German physical research. Neumann's oft expressed argument that physics instruction would fall entirely into the hands of technical institutes if universities were not given physical laboratories was a plea to keep exact experiment where it was most needed: where pure theory was evaluated.[1] The skeptical attitude with which his students learned to survey their data was intended to achieve a certainty qualified by the quantitative measure, through error analysis, of what was not known. For criticizing and sometimes even substantiating theoretical reasoning and results, their investigative strategy worked. But it was not always needed. Higher theoretical formulations often eluded them. Not all former students learned to accept the partial certainty that pragmatic action in scientific investigation, both theoretical and experimental, demanded. It is not that their desires for greater certainty were wrong but rather that they were at times pursued with a fervor that became increasingly out of place in the German physics community when a more purely theoretical physics was formulated without the help of experimental results worked over by rigorous techniques of error and data analysis and when experimental physics loosened its ties to, but did not entirely abandon, those same techniques.

It is often argued that the deterministic and causal nature of nineteenth-century physics—whose mechanical foundation enabled it at times to lay claim to absolute truth—meant that experimental errors could only be viewed as uncertainties in knowledge about the world. Some of Neumann's former seminar students certainly viewed errors as uncertainties representing future areas of investigation that could refine known laws. But they also acknowledged that the universal persistence of error in an experimental investigation was an indication that absolute truth could not be fully achieved. Error circumscribed the degree of certainty inherent in any mathematical theory of natural phenomena. Hence what emerged in this group was what might be called a probabilistic notion of scientific truth, albeit one apparently devoid of notions of chance. In an investigation one was supposed to achieve a small enough measure of the most probable error so as to claim a high degree of probability of the result, or of the theory (if one were seeking to determine the limits of confirmation). Objectification was thus not coextensive with mathematization or even with precision, because objectivity (in the form of the mathematical expression of natural laws) was itself tied to the assessment of results on the basis of the method of least squares.

1. Cited in S. Flügge, "In Memoriam: Zur fünfzigsten Wiederkehr des Todestages Franz Neumann," *Zeitschrift für Naturforschung* 1 (1946): 46–48, on 47.

Although Neumann's course on mechanics as an introduction to theoretical physics incorporated a strong practical element and de-emphasized causal considerations, investigations based on its methods could not entirely escape some of the expectations built into a mechanical physics. Investigations originating in the seminar incorporated a fundamental tension between a deterministic conception of absolute truth (or certainty) and the only truth that seemed possible with exact experiment, probabilistic or approximate truth. Neumann's former seminar students frequently groped in the area separating probable truth from absolute truth. Even though they knew that absolute truth was not attainable and that scientific truth was not mathematical truth, they nevertheless persisted in their grasp for certainty. Holding up the ideal of mathematical truth at the philosophical or epistemological level created expectations that scientific practice could not fulfill. Neumann and his students did not address in sustained ways the questions that far more incisive minds (such as James Clerk Maxwell and Ludwig Boltzmann) did: whether nature itself *behaves* in statistical ways. Yet as a result of their use of the method of least squares, they began to appreciate that knowledge about nature may only have limited certainty.

This epistemic function of probability calculus in physics was evident, for the most part, only in the investigative practices of some former seminar students, few of whom addressed the issue in more philosophical or reflective ways. It was out of the ordinary, then, when Heinrich Martin Weber explicitly addressed the issue of probable truth in a speech on causality in the natural sciences, given at the end of his prorectorship at Königsberg University in 1881. He acknowledged that all human thought suffers from "the imperfection that it never leads to knowledge of truth with absolute certainty. All our knowledge rests only on a more or less great probability, and it is the task of scientific research to increase the probabilities or to ascertain the greater of two contrary probabilities." For Weber, probability calculus was supposed to guide scientific thinking by setting up "firm standards" for how to make decisions, standards that, in his view, were not devoid of moral content. He explained that the type of calculation used in probability calculus "makes use of no other methods and principles than those used by a man *who is healthy and judges rightly from the start in all of his decisions*." Weber added that the methods of probability calculus were "related to the utterances of healthy human understanding" in the same way that "measurement with a compass or a ruler [was] to the estimation of size by visual reckoning." Since none of these methods, including measurement, could produce absolute certainty, lawful regularities were just matters of probabilities. Significantly, the examples he used to illustrate his points were not drawn from the kinetic theory of gases or the emerging science of statistical mechanics but rather from astronomy, especially the calculation of orbits, where error and data analysis were crucial to framing the result. And so it was in the "collective measuring sciences," Weber concluded, where "the balancing of errors of observation" was achieved by the application of probability calculus.[2]

During the first stage in the evolution of exact experiment, in the late eighteenth century, error analysis was confined to constant errors, and mathematization consisted in the construction of empirical formulas. In the second stage, at the beginning of the nineteenth century, the introduction of accidental errors, computed by the method of least squares, not only added quantitative rigor to exact experiment but also created the possibility (not the necessity) of going beyond the material conditions of the experiment to consider epistemological issues linked to the quality of data. It is this second stage in the history of exact experiment that seems to have been so crucial to the beginnings of theoretical physics in Germany. A considerably enhanced exact experiment supplied the methods for examining French mathematical physics and other mathematical expressions of theory; for achieving consensus on their veracity; and for training students in physics, especially theoretical physics.

The procedures of exact experiment became the practical foundation of physics instruction almost everywhere in the German states. In the preceding chapters, examples drawn from the investigations of students of Heinrich Buff at Giessen, Gustav Magnus at Berlin, and Friedrich Kohlrausch at Göttingen suggested that the way in which those procedures were used varied from location to location. At two universities the combination of exact experiment with mathematization was especially pronounced: Königsberg and Göttingen. At Königsberg, exact experiment with an emphasis on data analysis was inseparable from theoretical considerations. The same cannot be said for Göttingen. Why?

It seems as if the crucial difference, especially before 1850, was the way in which Gauss and Bessel influenced the practice of exact experiment at their home institutions. Of note for the development of exact experiment at Göttingen was Gauss's contention in 1831 that even though "physical apparatus [were] a great deal less complicated than astronomical ones," he believed that the skills of practical astronomy were transferable to physical investigations.[3] In the years ahead, the construction of refined apparatus capable of extremely precise measurements were central to the projects he undertook with Weber, especially in geomagnetism. The model of how astronomical measurements were taken also became the template for physics instruction at Göttingen, where students before 1870 were trained in geomagnetic measuring techniques more for the didactic goal of acquiring practical skills in precision measurement than for the purpose of obtaining scientifically interesting results. In many respects Königsberg was like Göttingen, but precision instruments played a less

---

2. Heinrich Martin Weber, *Ueber Causalität in den Naturwissenschaften* (Leipzig: W. Engelmann, 1881), pp. 15, 16 (emphasis added), 27.

3. C. F. Gauss to Cabinetsrath, 27 February 1831, UAG 4/Vb/95a.

prominent role in physics instruction at Königsberg, where there was a greater concern for the density and epistemological certainty of data. Bessel and Gauss had comparable views regarding the certainty of observations—that they were always in some sense approximate, never having absolute accuracy—but it was Bessel who, through applications involving a denser series of data than Gauss often took, developed the principles of data analysis for which Gauss had laid the foundations, especially the method of least squares.[4] Bessel's concern for the epistemological status of data had a deep and lasting influence on physics instruction at Königsberg.

The most significant difference between Gauss and Bessel, however, concerned the truth of scientific laws constructed on the basis of increasingly precise results. It will be recalled that Bessel, who was ever suspicious of "too many instruments" in an investigation despite his ability to analyze thoroughly the constant errors of an experiment, believed that results based on observations "may never be found with the certainty that mathematical truth claims by right of law."[5] Gauss, in contrast, placed great weight on what could be done with improved instrumentation, especially the ability to work with absolute units and to produce more precise observations. He went so far as to argue that, as a result of more precise measurements of geomagnetism, "one may consider the theory [of geomagnetism] as coming very close to the truth" and even suggested that mathematical accuracy might be ascribed to geomagnetic theory.[6] That is not to imply that Gauss and Weber took no interest in the method of least squares or the condition of their data—they certainly did—but that deeper epistemological issues did not trouble them and their students as much as they concerned Neumann in the seminar. So although Weber, like Neumann, distinguished *"empirical interpolation formulas"* from *"true* laws of nature," he seems not to have been so obsessed with the difference as to allow it to influence significantly his teaching of theory and exact experiment.[7] Hence a more pragmatic program of practical

4. Stephen M. Stigler, *The History of Statistics: The Measurement of Uncertainty before 1900* (Cambridge, Mass.: Harvard University Press, 1986), p. 203.

5. F. W. Bessel to C. F. Gauss, 10 March 1811, GN; F. W. Bessel to Alexander von Humboldt, 24 January 1838, A. v. Humboldt Nachlaß, SBPK-Hs.

6. C. F. Gauss, "Vorrede," in *Atlas der Erdmagnetismus nach der Elemente der Theorie entworfen*, ed. C. F. Gauss and W. Weber (Leipzig: Weidmann, 1840), pp. iii–iv, on p. iii. Alas, whereas Gauss was overjoyed with the agreement between his geomagnetic measurements and his theory (ibid., p. iii), Bessel considered it a "pity that the observations *still* fall short of the theory and the averages fail to give proof of the value *V* [the magnetic potential] with the completeness and certainty that is desirable." F. W. Bessel to A. v. Humboldt, 9 August 1839, A. v. Humboldt Nachlaß, SBPK-Hs. Bessel was here reacting to the first two or three volumes of results upon which Gauss's *Atlas* was based: *Resultate aus den Beobachtungen des magnetischen Vereins im Jahre 1836–41*, 6 vols., ed. C. F. Gauss and W. Weber (Göttingen: Dieterich, 1837–38; Leipzig: Weidmann, 1839–43).

7. Wilhelm Weber, "Versuche mit Zugenpfeifen" [1829], rpt. in *Wilhelm Weber's Werke*, ed. Kgl. Gesellschaft der Wissenschaften zu Göttingen, 6 vols. (Berlin: J. Springer, 1892–

instruction in physics evolved at Göttingen, one that became embodied in Kohlrausch's textbook and remained relatively immune to the excessive skepticism about data so often apparent at Königsberg.

That excessive skepticism was the principal weakness of Neumann's seminar exercises and of the investigations based on them. It was a consequence of the role Neumann assigned to exact experiment in theoretical physics, especially of its implications for how one had to assess scientific truth. In 1902, E. H. Stevens, with reference to Georg Quincke's experience at Königsberg, argued that "Neumann allowed his pupils too little scope for originality."[8] It cannot be denied that seminar exercises and investigations generated a tension between theoretical creativity and technical certainty. It also cannot be denied that Neumann instilled in his students an incessant compulsion to measure. Johann Pernet recalled that Neumann gave his students completed experimental projects "whose correctness had already been proven by a series of trials," but because the "experimental foundation" of these projects "did not appear comprehensive and accurate enough to him," students had to carry out still more trials with improved means, which they often did over vacations.[9] Accepting Neumann's gifts entailed fulfilling an expectation sometimes difficult to meet: that one surpass the accuracy of Neumann's own data.

Robert Fox has written that symptoms of the decline of French physics in the first decades of the nineteenth century can be seen in "the growing disenchantment with theory and a consequent preoccupation with the accumulation of data and with mathematical treatments in which the discussion of causes was eliminated or made irrelevant."[10] It is ironic that at about the same time, German physicists, at Königsberg and elsewhere, were beginning to consider how more accurate data could be used to establish the legitimacy of mathematical methods in physics; that French theoretical investigations played a prominent role in their inquiry; and that their efforts helped to secure their hegemony in theoretical physics, a discipline that the French had done so much to inaugurate. Although the investigative style practiced in Neumann's seminar proved incapable of dealing with higher theoretical issues and, like French physics at the beginning of the century, was distanced from a discussion of causes, it nonetheless contributed in positive ways to shaping the discipline of theoretical physics in its early years in Germany. The style of experimen-

---

94), 1:276–91, on 288–89. I develop the differences between the Königsberg and Göttingen styles in physics instruction in my "Exact Experiment and the Formation of Schools of Physics at Berlin and Göttingen."

   8. E. H. Stevens, "The Heidelberg Physical Laboratory," *Nature* 65 (1902): 587–90, on 587.

   9. Quoted in J. Hann, "Nekrolog [Franz Neumann]," *Almanach der Kaiserlichen Akademie der Wissenschaften zu Wien* 64 (1896): 271–80, on 277.

   10. Robert Fox, *The Caloric Theory of Gases from Lavoisier to Regnault* (London: Oxford University Press, 1971), p. 317.

tal practice cultivated in Neumann's seminar, built on the Besselian experiment and surpassing what the French had done, helped to integrate a critical epistemological function into theoretical physics, making physical knowledge as well as nature subjects of an investigation; enhanced the authority of theoretical investigations by integrating into them a means for discussing the certainty of results; and helped to create a social hierarchy in physics instruction. The greater mathematical abilities required for courses in theoretical physics and practica in exact experiment contributed to the view that theoretical physics was intrinsically more difficult, more scholarly, and more advanced than experimental physics. Hence, in contrast to experimental physics, which drew the rank-and-file student who did not plan to major in physics, theoretical physics came to be identified with more serious and often gifted students.[11]

Education disseminates the standards and values of scientific practice. The institutionalization of exact experiment in physics practica promoted the acceptance of the mathematical methods appearing in physical theory. The rich story of Neumann's seminar, and of nineteenth-century science education in general, suggests that the history of laboratory science may profit in two ways from a closer examination of how the role of experiment in science education evolved. The first is by expanding the range of contexts in which experiment is considered. Most historians of experiment argue vociferously for examining "actual practice, not reconstructed practice," yet the location where experimental skills are first learned, the educational laboratory, has only been acknowledged in passing and given little sustained scrutiny.[12] Social constructionists such as Steven Shapin have argued that "'experimental design' cannot be divorced from the commitments of the communities that frame and evaluate experiments," but few have examined the educational institutions where such values and judgments are first shaped.[13] Part of the reason for the neglect stems from lingering doubts about the historical value of an influence study, which any analysis of science education inevitably entails; "Teaching," warns Shapin, "does not *determine* future career-choices or judgments."[14] Certainly the skills and values of experimental practice as

11. On the association of theoretical physics with a clientele of gifted students see, e.g., Lothar Meyer to Neumann, 21 October 1870, FNN 53.IB: Anfragen wegen Berufungen.

12. *The Uses of Experiment: Studies in the Natural Sciences*, ed. D. Gooding, T. Pinch, and S. Schaffer (Cambridge: Cambridge University Press, 1989), p. xv. In his article "The Rhetoric of Experiment" in this volume (pp. 159–80), Geoffrey Cantor does mention the importance of the pedagogical context in understanding how scientists perform experiments. Also Peter Galison mentions a "pedagogical continuity" in twentieth-century high energy physics (*How Experiments End* [Chicago: University of Chicago Press, 1987], p. 249). Neither Cantor nor Galison, however, elaborates on how the pedagogical context actually influences experimental practice.

13. Steven Shapin, "History of Science and Its Sociological Reconstructions," *HS* 20 (1982): 157–211, on 162.

14. Ibid., p. 168.

first learned do not necessarily persist; but neither are they of no effect on subsequent experimental practice. If we are to continue in our efforts to write a history of experiment, building on the base already created and offering a much-needed complement to the history of scientific theory and an equally needed supplement to the history of scientific institutions, then we must examine more seriously what, since the first half of the nineteenth century, has been the setting in which a scientific practitioner was first exposed to the art of experimentation: practica and laboratory courses.

The second way in which a consideration of science education may add to the history of experimentation is by providing a more complete foundation for discussing changes over the long term, since the beginning of the nineteenth century. Some historians have found the conceptions of historical time developed by Fernand Braudel and partially applied in his magisterial study of the Mediterranean world during the age of Philip II useful for synthesizing the history of experimentation and formulating the new periodizations that mark its course. Peter Galison has shown how the Braudelian notions of geographic time, social time, and individual time lend themselves to analyzing the long-term, middle-term, and short-term constraints characterizing experimental practice in twentieth-century microphysics. Frederic L. Holmes has used with profit Braudel's notion of the *longue durée* (Braudel's abstract generalization of the notion of geographic time) in describing the continuities in chemical laboratories, and hence in the investigative practices of chemistry, in the seventeenth and eighteenth centuries.[15] What this study of Neumann's seminar and related institutes suggests is that the Annales paradigm can be taken one step further in carving out the long-term constraints that affect experimental practice (and also scientific practice in general).

Central to Annales-inspired historiography is the investigation of the mental and psychological dimensions of life. Often grouped together under the now-indeterminate label "history of mentalities," such studies are not strictly examples of intellectual history, although they intersect with it; do not deal with ideas but rather with the tools and patterns of thinking; are concerned not only with what is possible to think but also with the constraints upon thinking. I have in this book employed two terms borrowed from this tradition: mental equipment (strictly the tools of thinking, but in this case embracing various techniques used in experimentation) and mental habits (loosely speaking, the psychology of thinking, especially the way in which individual categories of thought are

15. Fernand Braudel, *The Mediterranean and the Mediterranean World in the Age of Philip II*, trans. Siân Reynolds, 2 vols. (New York: Harper & Row, 1972–73), esp. 1:16, 21; idem, "History and the Social Sciences: The *Longue Durée*," in F. Braudel, *On History* (Chicago: University of Chicago Press, 1980), pp. 25–54; Galison, *How Experiments End*, pp. 246–55; F. L. Holmes, *Eighteenth-Century Chemistry as an Investigative Enterprise* (Berkeley: Office for History of Science and Technology, University of California at Berkeley, 1989), pp. 19, 124.

combined and assigned value).[16] What is useful for the historical study of experimental practice is that mental equipment and habits persist over the long term—and hence are partitioned temporally by the *longue durée*—and that educational institutions are considered the principal repositories and disseminators of both.[17] Hence practica and laboratory courses—specifically the exercises assigned to or projects undertaken by the students in them—can be studied for the mental equipment and habits of experimentation that form the foundation of experimental practice for the individual investigator as well as for groups or generations of students over a long period of time.

The introduction of error and data analysis into physics instruction created a long-term constraint on experimental practice that has persisted in various forms and in different degrees to the present day. At Königsberg, one of the first locations where error and data analysis were firmly institutionalized, they formed a constraint on theoretical practice as well. They were constant reminders of how imperfectly nature actually fits precise mathematical laws and overarching theories. By making error and data analysis a central part of his seminar exercises, Neumann tried to instill caution into the search for truth by teaching his students how to hear the noise surrounding the edge of theory. Error and data analysis were part of the mental equipment conveyed in the seminar along with higher mathematics, especially partial differential equations and approximation techniques; theoretical commitments, such as a phenomenological approach to physical phenomena; technical skills in the material aspects of experimentation, such as methods of thermometric calibrations; and standards of experimental protocol, such as the rejection of the method of repetition. But at Königsberg it was largely error and data analysis, especially the latter, which determined how all these tools worked together and which shaped the habits of thinking that set the standards of scientific judgment as well as the terms of epistemological certainty and scientific truth. Practitioners at other locations may have shared the same tools but not necessarily the same habits of thinking, as the comparison between Königsberg and Göttingen illustrates. The persistence of these habits of thinking, expressed in the ethos of exactitude, became long-term cognitive and psychological commitments for several of the students in the seminar.

In the seminar, it was exact experiment that made the notion of a "calling" ambiguous, possessing both modern and traditional elements. Neumann, like others of his generation, helped to shape the modern

16. See above, Introduction and Roger Chartier, "Intellectual History of Sociocultural History? Two French Trajectories," in *Modern European Intellectual History: Reappraisals and New Perspectives*, ed. D. LaCapra and S. L. Kaplan (Ithaca, N.Y.: Cornell University Press, 1980), pp. 13–46, esp. pp. 16–22.

17. According to Braudel, "mental frameworks too can form prisons of the *longue durée*" ("History and the Social Sciences," p. 31).

notion of a calling by basing it upon the possession of an esoteric body of knowledge and upon the ability to work with specific, often technical, skills acquired through training. To the extent that Neumann emphasized means over ends and made scientific labor an end in itself, he contributed to the ascetic rationalism that Max Weber later believed characteristic of all callings in the modern world.[18] But Weber's excessively pessimistic conclusion—that rationalization was accompanied by a loss of meaning, purpose, and a sense of higher values—overlooks the subjective context in which such rational thinking often took place. The rational discipline demanded by seminar exercises in exact experiment and the investigations that stemmed from them also cultivated inner psychological commitments, both epistemological and moral, similar to the kinds of commitments characterizing earlier notions of "calling." The ethos of exactitude required a genuine belief in practicing theoretical physics according to the dictates of exact experiment; also an inner faith that greater certainty but not absolute truth could be achieved in this way; and finally, an admission that although precision and accuracy were technical goals based on skill, they were also outward signs of personal and professional integrity.

18. Max Weber, *The Protestant Ethic and the Spirit of Capitalism*, trans. T. Parsons (New York: Scribner's, 1958 [orig. pub. 1904–5]), pp. 180, 181, 182.

Provisional Statutes of the Seminar for Mathematics and Physics at the University of Königsberg

Divisions and Directors of the Seminar

§1. The seminar for mathematics and physics is composed of the division for pure and applied mathematics (mechanics, physical astronomy) and the division for mathematical physics. {The mathematical division is directed by Professor Jacobi and Doctor Sohncke; the physical division is directed by Professor Neumann.} No ordinary member may participate in only one division. With the exception of the condition described in §7, members of one division are at the same time members of the other.

Conditions of Admission

§2. To qualify for admission, the directors may find it necessary to require a written assignment or to administer an oral examination. For the mathe-

---

Translation of "Vorläufige Statuten des mathematisch-physikalischen Seminars an der Königsberger Universität," from ZStA-M, Acta betr. das mathematisch-physikalische Seminar an der Universität zu Königsberg, Rep. 76Va, Sekt. 11, Tit. X, Nr. 25, Bd. I: 1834–61, fols. 4–5.

Author's note: Material in braces was later deleted. The journals referred to in §6 were the *Journal für die reine und angewandte Mathematik, Astronomische Nachrichten*, and the *Annalen der Physik und Chemie* edited respectively by August Crelle, Heinrich Schumacher, and Johann Poggendorff. New statutes for the seminar were never written. In 1839 the following changes were made: the section for mathematical exercises under Sohncke was canceled; provisions were made for the mathematician Friedrich Julius Richelot to assist Jacobi and, in Jacobi's absence, to substitute for him; and on 16 March 1839, funding for the seminar (§6) was increased from 150 to 350 taler (see ibid., fols. 55, 63–65). The journal noted in §9 seems not to have been kept; only annual reports exist.

matical division, knowledge of differential calculus and the foundations of integral calculus is required. For the physical division, knowledge of the main features of the subjects treated in [Ernst Gottfried] Fischer's textbook is required.

Membership

§3. Members can be students of mathematics and physics at this university. They can remain members after graduating, until they have found a permanent position.

Work in the Mathematical Division

§4. The work of the mathematical division is of two kinds: (1) Connected lectures, given alternately by various members, on a designated part of pure or applied mathematics, for which one or more books serve as a foundation. The director and those who listen to these lectures are allowed to raise objections and make other remarks. {Dr. Sohncke will direct these lectures.} (2) a) Small problems in pure and applied mathematics, which are to be solved by all of the members; (b) Larger compositions prepared on a definite theme from pure or applied mathematics. Themes for this work either are independently chosen or are assigned by the director. The solutions to the small problems as well as the larger compositions will be discussed in a session set aside especially for that purpose. The larger compositions will be circulated beforehand to the students, one of whom will evaluate each one. {Professor Jacobi will direct this work.}

Work in the Physical Division

§5. The work of the physical division consists likewise in (1) connected lectures, which will be held by the members on a definite branch of mathematical physics; (2) independent works in mathematical physics, that either are purely theoretical or require specific observations and measurements on the basis of a mathematical theory. {Professor Neumann will direct this work.}

Remuneration of Members

§6. For the larger compositions, the directors can request from the ministry through the local curatorium a remuneration of at most twenty taler. In addition, for the physical work, where it is found necessary, a reimbursement can be granted for the costs incurred in undertaking observations. In the proposal for the remuneration, the student's relative participation in the other work of the seminar will be considered, as well as his year of study and the foundations required for his investigation. In the case of compositions, which have scientific value, if they are subsequently worked up into a form, in which they can be submitted for publication, then an extraordinary remuneration of ten taler will be awarded. — These works will be recommended for publication in a scholarly journal. {Those works in mathematics or astronomy will be recommended to Crelle or Schumacher, and those in physics to Poggendorff, for publication in their scholarly journals. The seminar has a yearly budget of 150 taler, which may not be overspent. Savings from one year may, however, be carried over to the next in order to defray greater needs and to make possible extended physical investigations.}

Connection of the Seminar for Mathematics and Physics to That for Natural History

§7. The members of the seminar for natural history can take part in the physical division of the seminar for mathematics and physics without being members of the mathematical division. Members of the seminar for mathematics and physics can take part in the section for experimental physics in the seminar for natural history without taking part in the remaining work of the natural history seminar.

Auditors

§8. Students not yet qualified for membership, as well as employed teachers, can, after application to the directors, take part as free auditors in all the exercises of the seminar, but these auditors are not eligible for the remuneration.

Annual Report and Journal of the Seminar

§9. Each of the directors keeps a journal, in which the day-by-day work of the seminar is noted and mention is made of the larger works of the seminar members. The original of this journal will be submitted to the ministry every year, along with a summary annual report and the work of the seminar published during the year.

{This draft shall be in effect for one year. After that year, the statutes will be reformulated on the basis of experience.
   F. E. Neumann   C. G. J. Jacobi   L. Sohncke}

Neumann's Cycle of Lecture Courses

WS	26/27	*Crystallography* (?;3)
WS	27/28	*Physics of the Earth* (2;25)
		Mineralogy (4;5)
SS	28	Crystallography (2;3)
		Mineralogy (?;4)
WS	28/29	*Physical Characteristics of Minerals* (2;2)
		Mineralogy (4;14)
WS	29/30	*Experimental Physics* (5;8)
SS	30	*Theory of Light* (3;12)
		Mineralogy (4;6)
WS	30/31	Crystallography (2;2)
		Physics of the Earth (4;6)
SS	31	*Analytical Theory of Heat* (2;10)
WS	31/32	Theory of Light (3;4)
		Mineralogy (4;4)
SS	32	Topics in Mathematical Physics (2;10)
WS	32/33	Topics in Mathematical Physics (2;4)
SS	33	Theory of Light (2;4)
		General Physics (4;12)
WS	33/34	Mineralogy (4;10)

Source: Paul Volkmann, *Franz Neumann: Ein Beitrag zur Geschichte Deutscher Wissenschaft* (Leipzig: B. G. Teubner, 1896), pp. 56–58. Volkmann constructed his list from university treasury records (indicating the students who actually paid to attend the course) rather than from the university's lecture catalog. WS, winter semester; SS, summer semester.

Author's note: New courses are in italic. When Neumann offered two lecture courses, the first was public, the second, private. The first number in the parenthesis indicates hours per week; the second, enrollment. In 1864, class hours were reduced.

SS	34	[No courses offered]
WS	34/35	Theory of Light (4;3)
SS	35	Crystallography (2;6)
WS	35/36	Theory of Heat (2;6)
		Mineralogy (4;8)
SS	36	Theory of Light (4;12)
WS	36/37	Topics in Mathematical Physics (2;6)
		Mineralogy (4;5)
SS	37	Theory of Heat (2;6)
WS	37/38	Crystallography (2;4)
		Theory of Light (4;8)
SS	38	Topics in Mathematical Physics (2;10)
		Mineralogy (4;2)
WS	38/39	*Capillarity* (2;8)
		Theoretical Physics (4;10)
SS	39	Crystallography (2;15)
		Theory of Light (4;12)
WS	39/40	*Elasticity Theory* (2;5)
		Mineralogy (4;8)
SS	40	Theoretical Physics (4;15)
WS	40/41	Topics in Mathematical Physics (?;6)
		Theory of Light (4;6)
SS	41	Theory of Heat (2;8)
		Mineralogy (4;17)
WS	41/42	Topics in Mathematical Physics (2;9)
		Theoretical Physics (4;14)
SS	42	Topics in Mathematical Physics (2;14)
		Theory of Light (4;7)
WS	42/43	Topics in Mathematical Physics (2;12)
		Mineralogy (4;7)
SS	43	Topics in Mathematical Physics (2;12)
		Theoretical Physics (4;12)
WS	43/44	*Theory of Magnetism* (2;12)
		Theory of Light (4;9)
SS	44	*Electrodynamics* (2;14)
		Mineralogy (4;8)
WS	44/45	Topics in Mathematical Physics (2;12)
		Theoretical Physics (4;7)
SS	45	Topics in Mathematical Physics (2;6)
		Theory of Light (4;5)
WS	45/46	Topics in Mathematical Physics (2;5)
		Mineralogy (4;4)
SS	46	Topics in Mathematical Physics (2;5)
		Theoretical Physics (4;8)
WS	46/47	Topics in Mathematical Physics (2;4)
		Theory of Light (4;7)

SS	47	Electrodynamics (2;10)
		Mineralogy (4;7)
WS	47/48	*Hydrodynamics* (2;10)
		Theoretical Physics (4;7)
SS	48	Theory of Light (4;7)
WS	48/49	Topics in Mathematical Physics (2;8)
SS	49	Topics in Mathematical Physics (2;8)
WS	49/50	*Theory of the Electric Current* (2;4)
		Theoretical Physics (4;6)
SS	50	Topics in Mathematical Physics (2;4)
		Mineralogy (4;4)
WS	50/51	Theoretical Physics (4;5)
SS	51	Topics in Mathematical Physics (2;4)
		Theory of Light (4;3)
WS	51/52	Theory of the Electric Current (2;5)
		Mineralogy (4;7)
SS	52	Topics in Mathematical Physics (2;2)
		Theoretical Physics (4;6)
WS	52/53	*Potential Theory* (2;8)
		Theory of Light (4;6)
SS	53	Topics in Mathematical Physics (2;6)
		Mineralogy (4;7)
WS	53/54	Topics in Mathematical Physics (2;7)
		Theoretical Physics (4;4)
SS	54	Topics in Mathematical Physics (2;8)
		Theory of Light (4;9)
WS	54/55	Topics in Mathematical Physics (2;7)
		Theoretical Physics (4;11)
SS	55	Topics in Mathematical Physics (2;6)
		Mineralogy (4;10)
WS	55/56	Topics in Mathematical Physics (2;8)
		Theory of Light (4;4)
SS	56	Topics in Mathematical Physics (2;8)
		Theoretical Physics (4;7)
WS	56/57	Topics in Mathematical Physics (2;9)
		Potential Theory (4;5)
SS	57	Theory of Magnetism (2;9)
		Theoretical Physics (4;6)
WS	57/58	Elasticity Theory (2;8)
		Theory of Light (4;4)
SS	58	Topics in Mathematical Physics (2;8)
		Mineralogy (4;13)
WS	58/59	Topics in Mathematical Physics (2;10)
		Theoretical Physics (4;10)
SS	59	Topics in Mathematical Physics (2;8)
		Theory of Light (4;11)

WS	59/60	Elasticity Theory (2;10)
SS	60	Theory of Heat (2;20)
		Theoretical Physics (4;18)
WS	60/61	Potential Theory (2;12)
		Theory of Light (4;11)
SS	61	Theory of Light (continued) (2;8)
		Theory of the Electric Current (4;12)
WS	61/62	*Induced Electric Current* (2;8)
		Theoretical Physics (4;10)
SS	62	Elasticity Theory (4;10)
WS	62/63	Potential Theory (2;8)
		Theory of Light (4;15)
SS	63	Theory of Light (continued) (2;12)
		Mineralogy (4;22)
WS	63/64	Elasticity Theory (2;12)
		Theoretical Physics (4;14)
SS	64	Elasticity Theory (continued) (2;12)
WS	64/65	Theory of the Electric Current (3;18)
SS	65	Mineralogy (3;19)
WS	65/66	Theory of Heat (1–2;15)
		Theoretical Physics (3;21)
SS	66	Potential Theory (1;15)
		Theory of Light (3;27)
WS	66/67	Theory of Light (continued) (1;15)
		Elasticity Theory (3;18)
SS	67	Topics in Mathematical Physics (1;15)
		Mineralogy (3;29)
WS	67/68	Theory of the Electric Current (3;18)
SS	68	[No courses offered]
WS	68/69	Potential Theory (1;12)
		Theoretical Physics (3;25)
SS	69	Electrostatics (1;16)
		Theory of Light (3;23)
WS	69/70	Topics in Mathematical Physics (1;15)
		Elasticity Theory (3;27)
SS	70	Elasticity Theory (1;11)
		Theory of the Electric Current (3;20)
WS	70/71	Topics in Mathematical Physics (1;5)
		Theoretical Physics (3;16)
SS	71	Topics in Mathematical Physics (1;12)
		Mineralogy (3;21)
WS	71/72	Topics in Mathematical Physics (1;7)
		Theory of Light (3;23)
SS	72	[No courses offered]
WS	72/73	Topics in Mathematical Physics (1;17)
		Theory of the Electric Current (3;24)

SS   73   Topics in Mathematical Physics (1;14)
              Induced Electric Current (3;15)
WS   73/74   Topics in Mathematical Physics (1;4)
              Elasticity Theory (3;17)
SS   74   [No courses offered]
WS   74/75   Theoretical Physics (3;21)
SS   75   [No courses offered]
WS   75/76   Theory of Light (3;18)
SS   76   [No courses offered]
WS   76/77   Potential Theory (2;18) [course not completed]

Directory of
Identified Seminar Students

Albrecht, F. Hermann (1814–?)
Allemann, F. J.
Amsler-Laffon, Jakob (1823–1912)
Arnold, G.
Aronhold, Siegfried Heinrich (1819–84)
Baechler, J.
Bardey, Ernst (?–1897)
Baumgarten, Gustav Louis (1846–?)
Behr, G. S. Hermann von (?–1896)
Besch, Adolph Carl Wilhelm (1849–?)
Bessel, Karl Wilhelm (1814–40)
Bock, L. O.
Boehm, Carl Friedrich
Borchardt, Carl Wilhelm (1817–80)
Böttcher, Johannes Eduard (1847–1919)
Brandeis, S.
Brandis, Ferdinand C. T.
Brix, Philipp Wilhelm (1817–89)
Busolt, Carl Albert (1810–?)
Bylda, E.
Clebsch, Rudolph Friedrich Alfred (1833–72)

Note: Italicized names are those of Neumann's doctoral students. All names were compiled from ZStA-M, Acta betreffend das mathematisch-physikalische Seminar an der Universität zu Königsberg, Rep. 76Va, Sekt. 11, Tit. X, Nr. 25, Bd. I: 1834–61, and FNN 48: Seminar Angelegenheiten.

Cohn (first name unknown)
Czwalina, Julius Eduard (1810–96)
Dorn, Friedrich Ernst (1848–1916)
Durège, Jakob Heinrich Karl (1821–93)
Ebel, E. F. J. Theodor (?–1851)
Ebel, P. Wilhelm G. E. (?–1884)
Ellinger, Franz Julius Gustav (?–1882)
Eötvös, Roland von (1848–1919)
Feyerabendt, E. W.
Fleischer, G. F. F.
Friedrich, J. A.
Fritsch, Hugo J.
Frölich, Oskar (1843–1909)
Fuhrmann, Wilhelm Ferdinand (1833–1904)
Gehring, Franz Eduard (1838–?)
Gundelfinger, Sigmund (1846–1910)
Hagen, Otto Albert (1835–62)
Hagen, Robert Hermann Heinrich (1815–58)
Haub, Eduard
Haveland, A.
Heideprim, W. P.
Heinemann, J. A. E. (1846–?)
Hermes, Johann H. Gustav (1846–1912)
Hesse, Ludwig Otto (Anton) (1811–74)
Heusler, G.
Hoffmann, H. O.
Höslin, G. von (?–1842)
Hossenfelder, H. A. Emil
Hübner, Emil Louis Hermann Ferdinand (1850–?)
Hübner, O. H. J. Eduard (1849–?)
Hutt, Eduard Johann (1843–?)
Jacobson, Heinrich (1826–90)
Jänsch, Ernst Robert
Just, Friedrich Gustav Adolph (1836–63)
Kade, Gustav Heinrich (1812–?)
Kaul, E. F. H.
Kayser, F. Ernst (1830–1907)
Kiessling, Karl Johann Hermann (1839–1905)
Kirchhoff, Gustav Robert (1824–87)
Kleiber, C. G. H.
Kohn-Akin, Cároly (1830–93)
Koss, E. L. von
Kostka, Carl G. Franz Albert (1846–1921)
Krakow, F. J.
Kratz, Heinrich (?–1863)

Krause, C. G.
Kreyssig, G. R. E.
Krueger, H. A. W.
Lampe, C. J. H.
Lange, C.
Lautsch, C. G.
Lorek, Emil Franz (1842–71)
Luther, F.
Mägis, Eduard
Matthias, G. O. R.
Mecklenberg, B.
Meyer, J. J. H. T.
Meyer, Julius Lothar (1830–95)
Meyer, Karl Otto (1815–93)
Meyer, Oskar Emil (1834–1909)
Michelis, A.
Milinowski, C. G. Alfons (1837–88)
Minnigerode, Ludwig Bernhard (1837–96)
Mischpeter, Carl Emil (1847–?)
Möller, P.
Momber, Albert
Mörstein, August E. F. von
Münch, A. J. W.
Müttrich, Gottlieb Anton (1833–1904)
Nagel, H. A.
Neumann, Carl Gottfried (1832–1925)
Nicolai, Otto C. R.
Noske, R. (1853–?)
Oelsnitz, A. W. F. G. C. von der
Pahlen, Louis Eduard Carl von (1810–?)
Pape, Carl Johannes Wilhelm Theodor (1836–1906)
Pauley, M.
Pebal, Leopold von (1826–87)
Pernet, Johann (1845–1902)
Peters, Johann Paul Gotthilf (1850–1933)
Peters, R. H.
Pietzker, Wilhelm Friedrich Christian (1844–1916)
Powel, F. Adalbart
Prophet, C.
Quincke, Georg Hermann (1834–1924)
Radau, Jean Charles Rudolph (1835–1911)
Radicke, Eduard Albert (1845–99)
Rahts, Johannes (1854–1922)
Rathke, Heinrich Bernhard (1840–1923)
Reichel, Otto E. J.

Reuter, F. V.
Rumler, G. O. E.
Ruppel, F. Oskar
Saalschütz, Louis (1835–1913)
Sanio, G. Theodor (1847–?)
Sanio, Paul
Schäwen, J. A. Hermann von (1844–?)
Schäwen, P. C. A. von (1846–1918)
Scheeffer, E.
Schindler, Ernst G. H. (1835–?)
Schinz, A. Emil (1817–87)
Schlicht, Paul Conrad (1848–?)
Schoch, W.
Schönemann, Theodor (1812–68)
Schroeder, F. W. K. Ernst (1841–1902)
Schroeder, L. F. H.
Schumann, Johannes Hermann Eduard (1844–1914)
Schumann, Julius Heinrich Carl Eduard (1810–?)
Schwarz, C. G. E. F.
Siebeck, Friedrich Hermann (1819–?)
Sohncke, Leonhard (1842–97)
Steiner, A. (?–1865)
Sucker, F. W.
Thalmann, J. Ernst G. (1847–?)
Thiesen, Max Ferdinand (1849–1936)
Tischler, Friedrich Carl Adalbart (1844–70)
Tischler, Otto Emil Friedrich (1843–91)
Voigt, Woldemar (1850–1919)
Von der Mühll, Karl (1841–1912)
Wangerin, Friedrich Heinrich Albert (1844–1933)
Weber, Heinrich (1839–1928)
Weber, Heinrich Martin (1842–1913)
Wegner, August Carl Robert (1845–?)
Westphalen, Hermann L. (1822–46)
Wiesing, Otto Hermann (1840–?)
Wild, Heinrich (1833–1902)
Wittrien, Otto (1851–1936)
Zimmermann, M. R.
Zöppritz, Karl Jacob (1838–85)

Glossary of
German Terms

ABITUR: State-regulated examination for leaving secondary school.

ALLGEMEINE BILDUNG: Liberal, all-round, or well-rounded education or cultivation with an emphasis on cultural elements.

ALLGEMEINE VOLKSBILDUNG: General popular education.

ALLTAGSGESCHICHTE: The history of everyday life.

AUSBILDUNG: Training, usually through a structured, goal-directed curriculum.

AUßERORDENTLICHER PROFESSOR: Extraordinary professor, roughly equivalent to associate professor.

AUßERORDINARIUS: Extraordinary professor (*see* AUßERORDENTLICHER PROFESSOR).

BERUF: Calling (older meaning); profession (modern meaning).

BILDUNG: Generally, a cultured education; in the nineteenth century, a self-cultivation believed best achieved through a neohumanist curriculum based on the classics and leading to the refinement of one's personality and the development of one's talents.

BILDUNGSMITTEL: Educational or curricular resources.

BROTSTUDIEN: Literally, "bread studies"; studies taken up only to achieve professional goals.

BÜRGERSCHULE: Early nineteenth-century secondary schools with a nonclassical curriculum, including the natural sciences and sometimes applied subjects.

ERZIEHUNGSMITTEL: Educational resources (*Erziehung* means education in the sense of upbringing or rearing, rather than in the sense of cultivation, as is implied by BILDUNG).

FACH: In academics, subject or specialty.

FUSS: The Prussian measure of length, equal in 1835 to 0.313854 meters. The Fuss is divided in 12 ZOLL or 144 LINIE.

GYMNASIUM: Secondary school with a classical curriculum emphasizing the study of Latin and Greek.

[473]

HABILITATION: Process of becoming a PRIVATDOZENT at a university.

HABILITATIONSSCHRIFT: A major research essay presented in fulfillment of part of the requirements for becoming PRIVATDOZENT.

HÖHERE BÜRGERSCHULE: *See* BÜRGERSCHULE.

HÖHERE GEWERBESCHULE: A trade school, sometimes dedicated to a specific area of study such as agriculture.

INAUGURAL-DISSERTATION: The doctoral dissertation.

INNERER BERUF: An "inner calling."

KULTUSMINISTER: The minister of culture and education in a German state.

KULTUSMINISTERIUM: The ministry of culture and education in a German state.

LEBENSBERUF: Professional life.

LEBENSPLAN: Literally, a plan for one's life.

LEHRBUCH: Textbook.

LEHRERBERUF: The profession of secondary school teaching.

LEHRERPRÜFUNG: The state examination for secondary school teachers.

LEHRFREIHEIT: Literally, the freedom of professors to teach what they want to; also academic freedom.

LERNFREIHEIT: Literally, the freedom of students to take whatever courses they want to.

LINIE: 1/144 of a Prussian FUSS; *see* FUSS.

MATHEMATIKLEHRER: A secondary school mathematics teacher.

MATHEMATIKLEHRERBERUF: The profession of teaching mathematics at a secondary school.

NATURLEHRE: Natural philosophy; sometimes considered synonomous with physics.

OBERLEHRER: Secondary school teachers in the upper forms of PRIMA and SECUNDA.

OBERLEHRERARBEIT: An essay completed in fulfillment of the written portion of the state teaching examination.

ORDENTLICHER LEHRER: An ordinary, or full-time, secondary school teacher.

ORDENTLICHER PROFESSOR: An ordinary, or full, professor.

ORDINARIUS: An ordinary, or full, professor.

PÄDAGOGIUM: A Swiss secondary school.

PHYSIKLEHRER: A secondary school physics teacher.

PHYSIKLEHRERBERUF: The profession of teaching physics at a secondary school.

PRIMA: The uppermost form of the gymnasium, usually spanning two years of study.

PRIVATDOZENT: University lecturer who has obtained the right to teach (and hence to collect course fees from students). The position is generally otherwise unsalaried.

REALSCHULE: A semiclassical secondary school offering Latin but not Greek and placing instructional emphasis on "real" subjects, such as mathematics and the natural sciences.

REALSCHULLEHRER: A teacher at a semiclassical secondary school.

SECUNDA: The second-to-last form in the gymnasium, usually spanning two years of study.

TECHNISCHE HOCHSCHULE: Technical "high" school emphasizing the natural sci-

ences and applied subjects. As parallel institutions to the university, the TECH-
NISCHE HOCHSCHULEN were given the right to award doctorates in 1899.

TERTIA: The third-to-last form in the gymnasium, usually spanning one year of
study.

VENIA LEGENDI: The right to teach, awarded to PRIVATDOZENTEN.

WISSENSCHAFT: Systematic knowledge and its pursuit; considered the means to
BILDUNG.

WISSENSCHAFTLICH: Scholarly or scientific.

WISSENSCHAFTLICHE BILDUNG: Scholarly education or cultivation.

WISSENSCHAFTLICHER HILFSLEHRER: An assistant teacher in a secondary school.

ZOLL: 1/12 of a Prussian FUSS; *see* FUSS.

Index

Abbe, Ernst, 405
Abitur, 4, 323–24
Accuracy. *See* Error analysis; Precision measurement
Albrecht, F. Hermann, 469
Allemann, F. J., 237, 469
Altenstein, Karl Freiherr vom Stein zum, 30
Ampère, André-Marie, 94, 172, 175, 435
Amsler-Laffon, Jakob, 179, 201–2, 224, 392, 469; polarplanimeter of, 339, 344; research, 343–44, 347–48; seminar investigation, 188–89
Analytical mechanics, 103, 108, 124, 125, 143, 145, 155–56, 157, 158, 219. *See also* Mechanics, didactic role in theoretical physics
Anger, Carl Theodor, 69
Ångström, A. J., 306–7
Arago, Dominique-François, 58, 69
Arnold, G., 237, 469
Aronhold, Siegfried Heinrich, 178, 469
Ausbildung, 14, 114, 126, 140, 217, 311, 312

Baechler, J., 469
Bardey, Ernst, 217, 363, 469
Basel University: seminar for mathematics and the natural sciences, 404–5
Baumgarten, Gustav Louis, 289, 291, 294, 322, 363, 469; research, 340–41, 347; school teaching, 326–27
Behr, G. S. Hermann von, 112, 117, 134, 136, 139, 144, 469
Berlin Physical Society, 282, 332, 363, 364
Berlin polytechnic institute, 87–88
Berlin University, 21, 25, 33, 34, 241; physi-

cal laboratory, 261, 428–29; seminar for scientific calculation, 433
Bern University: practical exercises at, 405
Beruf. See Calling
Besch, Adolph Carl Wilhelm, 469
Bessel, Friedrich Wilhelm, 43, 56–57, 82, 121, 140–41, 178, 204, 341, 421; and colleagues of, 36, 81, 56–57, 82; early career, 26–27; on epistemology of measurement, 72–73, 455n; on error distribution, 67, 68; on experimental errors, 66–73, 76, 149, 159–60, 162, 203, 260, 391; on Laplace's treatment of errors, 67, 162; neohumanist curriculum, criticism of, 53; personal equation, 68; physics instruction, influence on, 16, 93, 98, 110, 146, 153, 159, 163, 296, 433, 435–36, 444, 447, 449, 454–55; political views of, 57, 116, 194–95; on precision and accuracy, 66–73; science and mathematics education, views on, 29–31, 37, 59; seconds pendulum investigation, 68–72, 81, 155, 168, 283, 289, 371, 372, 373, 393, 394, 400, 435–36, 444, 447, 449; teaching, 29, 31, 37–38, 40, 42, 43, 45, 416–17. *See also* Besselian experiment; Instruments; Königsberg University: observatory; Method of least squares
Bessel, Karl Wilhelm, 106, 122, 320, 469
Besselian experiment, 261, 275, 367, 380, 432, 433, 450, 452, 457; defined, 169–70; in physics pedagogy, 224, 313, 408, 414, 435, 436, 444, 457; and precision instruments, 232, 275, 391–92
Bildung, 13–14, 22, 44, 118, 123, 140, 327

Biot, Jean-Baptiste: measurement accuracy of, 58, 162
Bischoff, Theodor, 49, 50
Bock, L. O., 272, 469
Boehm, Carl Friedrich, 178, 210, 469
Boltzmann, Ludwig, 453
Bonn University natural sciences seminar, 2; curriculum, 47–49, 51; mathematics, exclusion of, 51–52; as model, 43, 216; operation, 50–51; origins, 47, 88; practical exercises, 49–50, 405–6; student research in, 54; teacher training in, 49–50; teaching examination privilege, 52, 268
Borchardt, Carl Wilhelm, 31, 133, 134, 469
Böttcher, Johannes Eduard, 289, 322, 363, 364, 469; research, 337, 338; school teaching, 326, 330–31
Bradley, James: Bessel on fixed-star catalog of, 67
Brandeis, S., 178, 469
Brandes, H. W., 94, 155
Brandis, Ferdinand C. T., 134, 469
Braudel, Fernand: on historical time, 458
Breslau University mathematico-physical seminar, 2, 3n, 408, 448; practical exercises in, 420–23
Brettner, H. A.: Leitfaden für Unterricht in Physik, 327–28
Brewster, David, 58, 131, 276
Brix, Philipp Wilhelm, 133, 134, 140, 144, 168, 170, 181, 182, 188, 255, 282, 363, 392, 469; dissertation and seminar investigation, 137–39
Brunner, C., 262
Buch, Leopold von, 34
Buchwald, Jed Z.: on French experimental style, 162
Buff, Heinrich, 2, 454; practical exercises at Giessen, 424–26
Bunsen, Robert: on mathematical instruction at Heidelberg, 415–16; seminar students study with, 230, 297, 300, 322; teaching at Heidelberg, 241–42
Burdach, Karl Friedrich, 31, 39, 40, 191, 198
Burschenschaften, 33, 40, 47
Busolt, Carl Albert, 103, 106, 469
Bylda, E., 469

Calling, 13–14; meaning in seminar, 313, 459–60; and theoretical physics, 311–12
Cantor, Moritz, 408
Carnot, Sadi, 278
Cauchy, Augustin-Louis, 164
Causality, 145, 150, 211, 453, 456
Cavendish, Henry, 104
Clausius, Rudolph: practical exercises of, 406; scientific style, 279, 397; thermodynamics and kinetic theory, 274, 278, 284, 306, 374
Clebsch, Rudolph Friedrich Alfred, 213, 214, 216, 246, 388, 394, 400, 425, 469
Coulomb, C. A., 281, 282–83, 284, 285
Crelle, August Leopold, 88; evaluations of seminar, 115–16, 135–36, 183, 199–200
Czwalina, Julius Eduard, 103, 153, 470

Data analysis. See Error analysis; Exact experiment; Graphical analysis; Interpolation; Interpolation formulas; Method of least squares
Dirichlet, Gustav Lejeune, 87, 397
Discipline: defined, 14–15; and error analysis, 161; formation, teaching and, 6; and mechanics, 154
Dorn, Friedrich Ernst, 470; and data perfection, 376; geothermal studies, 349, 350–52, 357; at Halle, 408; practical exercises at Breslau, 422
Dove, Heinrich Wilhelm, 42, 83, 192; teaching, 32–33, 36, 95
Du Bois-Reymond, Emil, 219, 220
Du Bois-Reymond, Paul, 320, 321n; education, 219–20; on interpolation, 255–56; on surface tension, 220–22; teaching, 362
Dulk, Friedrich Philipp, 32, 46, 92, 192
Dumas, Wilhelm, 213
Durège, Jakob Heinrich Karl, 201–2, 470

Ebel, E. F. J. Theodor, 470
Ebel, P. Wilhelm G. E., 134, 135–36, 470
Educational reform: in clinical instruction, 39; in German universities, 21–23, 115–16, 197–99; at Königsberg University, 25, 31–32, 40–43, 45–46, 59–60; in secondary schools, 23–25
Eichhorn, Johann Albrecht Friedrich, 182; on exercise sessions, 197–99
Einstein, Albert: and teachers of, 18
Electricity, studies of, 105, 231–32; current, 242–46; curves of equal tension, 180–83, 243–46; induced currents, 175–78; induction constant, 184–86
Ellinger, Franz Julius Gustav, 470; school teaching, 325
Empirical formulas. See Interpolation formulas
Eötvös, Roland von, 398n, 470
Erlangen University: mathematico-physical seminar, 2, 427, 448
Error analysis, 4, 38, 179, 255, 302; accidental errors, 12, 62, 68, 70, 75, 78, 130, 135, 147–48, 215, 369, 444, 451, 454; constant errors, 12, 66–68, 70–72, 74,

75, 77–78, 130, 135, 139, 147–49, 157, 168, 187, 215, 226–27, 283–84, 411, 451, 454; and craftsmanlike skills, 378–82; didactic function of, 131, 154, 161, 167–68, 207, 215, 223, 233–34, 261; epistemological function of, 139, 233–34, 373, 385, 387, 445–47; and hypotheses, 387–88; as impediment, 373, 374, 375, 388; as instrument of persuasion, 347, 335, 351–52; Magnus's and Kohlrausch's students' practice, 261–64; metaphorical expression of, 449–50; Neumann's students' practice, 137–39, 180–82, 184, 187–88, 221–22, 226–28, 242, 243, 282–85, 286–87, 288, 290–92, 294–96, 297–98, 347; before nineteenth century, 232–33; and pendulum, 147–49, 154; personal equation, 68, 161, 391; in science teaching, 38, 396, 424; in seminar, 12, 13, 17, 135–36, 150, 151–52, 206–7, 223, 233, 279–80, 302–4; Stichweh on, 13n; and theoretical physics, 10, 12, 13, 130–31, 135, 161–67, 174, 233–34, 312–13, 367, 369–70, 376, 377–78. *See also* Bessel, Friedrich Wilhelm; Exact experiment; Königsberg University mathematico-physical seminar, physical division; Method of least squares; Neumann, Franz Ernst; Physics instruction, in secondary schools; Practical exercises; Secondary school teachers; *names of individual science seminars*

Exact experiment: and data analysis, 375–76; defense of, 378–86; demise in physics teaching, 429–33; in eighteenth century, 64, 451, 454; epistemological function of, 13, 136, 165–66, 382, 451–52, 454, 457; and hypotheses, 371–74, 375; and ideal cases, 370–75; incommensurability of types, 378–86; in Königsberg seminar, 12, 13, 133–40, 144–54, 456; Königsberg vs. Göttingen styles, 403, 454–57; and material conditions of experiment, 375; moral qualities of, 381, 382, 450, 460; and pendulum, 146–49; in postseminar investigations, 368–78, 380–86; and scientific truth, 72–73; and theory, 81, 161–67, 369–74, 377–78, 451–52; transformation of, 18, 64, 347–48, 366, 371, 397, 398–402, 432–33, 454. *See also* Bessel, Friedrich Wilhelm; Besselian experiment; Error analysis; Exactitude, ethos of; Kohlrausch, Friedrich; Mechanics, didactic role in theoretical physics; Method of least squares; Practical exercises; Precision measurement; *names of individual science seminars*

Exactitude, ethos of: and analogies, 387; defined, 367–68; demise of, 392–93, 446, 447, 448–50; as ethical standard, 450, 460; as impediment, 367, 388, 392–93; and mathematical physics, 388–89; persistence at Königsberg, 374, 386–87, 433–34, 435–36, 443–47; and precision instruments, 389–91; transmission of, 17

Faraday, Michael, 173, 175, 177
Febvre, Lucien: on mental equipment, 5
Feyerabendt, E. W., 470
Fischer, Ernst Gottfried, 23, 35; *Lehrbuch der mechanischen Naturlehre*, 99, 100, 122, 156–57, 327, 328
Fischer, Johann Karl, 155
Fleck, Ludwik: on science teaching, 18; on trained person, 12
Fleischer, G. F. F., 470
Foucault, Michel, 14
Fourier, Joseph, 7, 35, 64, 65, 78, 79, 81, 94, 162, 208, 278, 348, 349, 377; *Théorie analytique de la chaleur*, 34, 58, 62–63
Fox, Robert: on decline of French physics, 456; on Regnault, 378
Frankenheim, Moritz, 396–97, 421
Freiburg University: seminar for mathematics and the natural sciences, 2, 406
French mathematical physics, 12, 105, 121, 312–13, 376, 454, 456; certainty of, 63n, 162; classified as mathematics, 35, 93–94, 156; data reliability in, 78; didactic function of, 163–66; error treatment in, 162–63, 262; Magnus's students' treatment, 261–63; as model for mathematization, 63–64; Neumann's early reading of, 62–64. *See also names of individual scientists*
Fresnel, Augustin Jean, 7, 58, 80, 81, 128–29, 376; error treatment of, 162–63
Friedrich, J. A., 470
Fritsch, Hugo J., 337, 470
Frölich, Oskar, 392, 470; education, 286; on instruments, 390; on interpolation, 256; thermometric studies, 286–87, 349, 352–57
Fromme, Carl, 426
Fuhrmann, Wilhelm Ferdinand, 217, 236, 299, 303, 470

Galison, Peter: use of Braudelian time, 458
Gauss, Carl Friedrich, 27, 30, 31, 43, 72, 121, 134, 172–73, 175, 238, 273, 275, 283, 285, 414; and experimental errors, 66, 67, 75, 76, 160; physics instruction, influence on, 454–55
Gay-Lussac, Joseph, 68, 104
Gehring, Franz Eduard, 235, 237, 394, 470

Geison, Gerald L.: on science teachers, 18; on scientific schools, 8

Geometry: analytic, 124, 141, 213, 214, 252–53; descriptive, 252, 254–55; didactic function of, 217, 252–53

Giessen University: chemistry laboratory, 95, 203; pedagogical seminar (for physics and mathematics teachers), 2, 426; physical seminar, 2, 408, 424–26

Goldstein, Jan: on discipline, 14–15

Gordan, Paul, 253

Göttingen University, 47; geomagnetic station, 72, 121; philology seminar, 44, 327; physics instruction at, 454–56

Göttingen University mathematico-physical seminar, 2, 3n, 4, 408, 448; method of least squares in, 397; practical exercises in, 409–14

Grafton, Anthony, 302n, 364

Graphical analysis: acceptance of, 264–65; and applied problems, 392; barriers to use, 13, 17, 250–55, 260–61, 367, 386–87; Buff's students' treatment of, 424; in eighteenth century, 251–52; and error determination, 392; limitations of, 234, 250–52, 254–57, 260–61, 263–65; Magnus's and Kohlrausch's students' treatment of, 263; and precision, 228–30, 251, 252; Regnault's treatment of, 383–84; seminar students' treatment of, 180–81, 188, 228–30, 239–40, 243–46, 255, 257–59, 292–93, 295–96, 341–42, 349–50; and theory, 256–59; truth value of, 256; visual function of, 254, 257–59

Greifswald University, 21, 25; chemical society, 88; mathematical society, 86–87; mathematico-physical institute, 428

Gundelfinger, Sigmund, 416, 427, 470

Hagen, Karl Gottfried, 25–26, 29, 32, 39, 82–83

Hagen, Otto Albert, 235, 250, 256, 470; on optical absorption, 239–41

Hagen, Robert Hermann Heinrich, 112, 337, 470

Halle University, 21; mathematical society, 86; mathematico-physical society, 88; philology seminar, 44; physico-chemical society, 88; seminar for mathematics and the natural sciences, 2, 406–8

Haub, Eduard, 470

Haüy, René-Just, 65; error treatment of, 162

Haveland, A., 134, 470

Heat, studies of: geotemperatures, 120–21, 286–87, 348–60; latent heats, 137–39; specific heats, 61–65, 77–80, 297–98; thermometer calibration, 68, 104, 130

Heidelberg University, 241–42; mathematico-physical seminar, 2, 404, 408, 414, 417–20, 448; physical institute, 414–17

Heideprim, W. P., 470

Heine, Eduard, 140, 342, 405–6

Heinemann, J. A. E., 470

Helmholtz, Hermann von, 203, 219, 220, 320–21, 428

Herbart, Johann Friedrich, 37, 38, 42, 43, 45, 47, 86, 92

Hermes, Johann H. Gustav, 470

Hertz, Heinrich, 406, 412

Hesse, Ludwig Otto, 103, 110, 111, 112, 113, 122, 129, 141, 172, 430, 470; at Heidelberg, 300, 322, 416–17; optical studies of, 104, 107–8

Heusler, G., 470

Heussi, J.: Die Experimental-physik, methodisch dargestellt, 328

Heyne, Christian Gottlob, 44, 327

Hoffmann, H. O., 470

Holmes, Frederic L.: on technical creativity, 308; use of Braudelian time, 458

Höslin, G. von, 470

Hossenfelder, H. A. Emil, 470

Hübner, Emil Louis Hermann Ferdinand, 352, 470; school teaching, 324, 329

Hübner, O. H. J. Eduard, 326, 470

Humboldt, Alexander von, 30, 73, 182

Humboldt, Wilhelm von, 26; reforms of, 21–22, 23, 25, 44

Hünefeld, D. C., 88

Hutt, Eduard Johann, 335–36, 363, 470

Huygens, Christian, 58

Hypotheses: Magnus's and Kohlrausch's students' use, 263–64; in seminar, 304

Instruments, 12, 377; Bessel's reformed treatment, 66–68, 72, 81; as didactic examplars, 135, 147–51, 161, 174, 218, 223, 273, 275, 310; ministry's comments on, 120, 132–33, 430–31; precision, 225–228, 231–32, 330, 344, 389–91, 454–55; scientific, 150; self-registering, 391

Interpolation: and applied problems, 392; barriers to use, 17, 221, 255–57, 260–62, 358, 360; and error detection, 383, 392; and graphical analysis, 265; Kohlrausch's students' use, 263–64; Pape's criticism of Regnault's use, 298; Regnault's use, 379, 383–84; seminar students' use, 211, 228–230, 243–48, 250–51, 255–57, 263–65, 267, 295, 350–52, 403, 451

Interpolation formulas, 455; analytic, 247–48, 250, 256, 358; certainty of, 256–57, 372, 376; defined as theoretical physics, 396–97; and ethos of exactitude, 367; Kirchhoff's criticism of Regnault's use, 383–85; Neumann's sense of, 131; Regnault's use defended by Wüllner, 385–86; seminar students' use, 340, 342; theoretical, 243–46, 248, 250, 358; W. Weber on, 455

Jacobi, Carl Gustav Jacob, 3, 37, 43, 56–57, 59, 84, 176, 177, 182–83; early career at Königsberg, 32, 81–82; on measurement, 73; mechanics course, 158; and method of least squares, 160; move to Berlin, 140–41, 178; political views, 37, 116; and students' research, 108; teaching, 86, 89, 92, 103, 114, 124–25, 416–17. *See also* Königsberg University mathematico-physical seminar: mathematical exercises
Jacobi, Moritz, 73, 108
Jacobson, Heinrich, 230, 346, 376, 470
Jänsch, Ernst Robert, 470
Jarausch, Konrad H.: on historiography of education, 5
Jena University: practical exercises, 405
Joachimsthal, Ferdinand, 133
Jochmann, E.: *Grundriß der Experimentalphysik*, 328
Jolly, Philipp: at Heidelberg, 415; at Munich, 427–28, 433
Joule, James Prescott, 278, 306
Just, Friedrich Gustav Adolph (Fritz), 237, 238, 272, 470; seminar investigation, 270–71, 299–300

Kade, Gustav Heinrich, 103, 109, 111, 335, 470
Kamerlingh-Onnes, Heike, 419
Kämtz, L. F., 406–7
Kant, Immanuel, 25, 53
Karlsbad Decrees, 40, 41
Kaul, E. F. H., 234, 236, 253, 470
Kayser, F. Ernst, 470
Kiessling, Karl Johann Hermann, 470; education, 331–32; research, 339–40; school teaching, 332–34, 364
Kirchhoff, Gustav Robert, 176, 178, 230, 241–43, 255, 377, 429, 470; at Breslau, 393–95; criticism of Regnault, 383–86; education, 179–80, 187–88; on error in Poisson's elasticity theory, 370; first seminar investigations, 180–83; and graphical analysis, 386–87; at Heidelberg, 241, 300, 322, 369, 395, 414–19; mechanics

course, 399–400; prize question and doctoral dissertation, 184–86; Quincke's comment on electric current studies of, 242–43; and reputation of seminar, 5, 16, 202; and tricentennial celebration, 191
Kleiber, C. G. H., 470
Klein, Felix: and descriptive geometry, 254–55
Knoblauch, Hermann, 407–8
Koenigsberger, Leo: at Heidelberg, 414, 417–18
Kohlrausch, Friedrich, 261, 454; *Leitfaden der praktischen Physik*, 166, 403, 408, 411–12, 426, 429, 443–44, 456; practical exercises at Göttingen, 411–12; students' treatment of errors, 263–64
Kohlrausch, Rudolph, 173
Kohlrausch, Wilhelm, 264
Kohn-Akin, Cároly, 470; criticisms of Regnault, 382–83
Königsberg Physico-economic Society: and geothermal station, 350, 354; viewed as agent of political and economic change, 57, 194
Königsberg school of theoretical (or mathematical) physics: identifications of, 13, 264, 300, 397, 403, 404; problems in defining, 8–9
Königsberg University: anatomical institute, 39, 40; botanical garden, 39, 40, 41; chemical institute, 41; clinical institutes, 39, 40; declining enrollment in natural sciences, 269–70; experimental physics at, 437–39; geothermal station, 66n, 120–21, 206, 238, 348–60, 376; history seminar, 45, 92; juridical seminar, 90; language seminars, 45; mineralogical collection, 39, 40, 135; natural history museum, 56; observatory, 26, 27–29, 38, 41, 67, 121; pedagogical seminar, 37, 47, 84, 86, 92; philology seminar, 45, 92; tricentennial celebration, 16, 178, 190–92; zoological museum, 39
Königsberg University mathematico-physical seminar: auditors, 196, 270; design of, 89; enrollees, diminished quality of, 112–13; founding, 59, 94–95; funding, 110, 112, 132; initial operation, 103–14; mathematical exercises, 103–4, 106, 108–9, 113, 133, 140–42, 196–97, 212–14, 218–19, 237, 252–54, 272–73; mature activity, 215–16; ministerial evaluation, 114–15; as model, 3, 421; novelties, 2–3; organizational growth, 195; original research in, 91–92, 100; statutes, 99–100, 103, 461–63; teacher training in,

Königsberg University mathematico-physical seminar (*cont.*)
89, 103. *See also* Crelle, August Leopold; Jacobi, Carl Gustav Jacob; Königsberg University mathematico-physical seminar, physical division; Mechanics, didactic role in theoretical physics; Neumann, Franz Ernst
——, physical division: cancellation of, 447–48; curriculum, evolution of, 104–14, 123–26; documents on, 9–11; enrollment in, 10; identified enrollees, 469–72; inauguration of introductory and advanced sections, 136; mental equipment conveyed, 459; operation, 4–5; research, problem in sustaining, 172; routinization of exercises, 273; seminar-based investigations, 104, 107–8, 137–39, 143–44, 180–83, 188–89, 216, 225–30, 231–32, 239–41, 246–50, 270–71, 280–87, 290–93, 294–96, 297–98, 299–301, 335, 345; seminar exercises, 11, 17, 89, 100, 104–14, 134–40, 151–54, 161, 164–65, 167–68, 172–73, 178–90, 196–97, 201–2, 206–8, 212, 214–15, 217–18, 222–25, 230–31, 235–39, 271–80, 302–4, 307–10; student memories, 4, 266–67, 275, 476; students' professions, 13–14, 318–19, 368; theorist's perspective cultivated in, 261; Voigt's exercises in, 435–37; Volkmann's exercises in, 441–43
Königsberg University natural sciences seminar, 2, 100; absence of mathematical instruction, 119; cancellation, 208–9; clientele, 117, 119; curriculum, 54; enrollment, 117, 120; founding, 43, 46–47, 52–53, 94; operation, 55–56, 117–20; practical exercises in, 46, 117, 118, 120, 405; student research in, 54–55, 91–92, 118; teacher training, 46, 54, 55, 200
Königsberg University physical laboratory, 4, 16; construction of physical institute, 438–41; disputes over use of equipment, 84–85; equipment funding, 82–85, 144, 195, 196–97, 269; Neumann on necessity of, 310–11; Neumann's requests for, 41, 142–44, 192–93, 195–96; Neumann's residential laboratory, 196, 205–7; physical cabinet, 29, 40, 284; Voigt's request for, 436; Volkmann's exercises in, 443–47
Koss, E. L. von, 470
Kostka, Carl G. Franz Albert, 337, 470
Krakow, F. J., 470
Kratz, Heinrich, 470
Kraus, Christian Jakob, 25, 53
Krause, C. G., 178, 471
Kreyssig, G. R. E., 213, 471

Kries, F.: *Lehrbuch der Naturlehre für Anfänger*, 327
Krueger, H. A. W., 471
Kuhn, Thomas S.: on measurement, 12–13; normal science, 378
Kundt, August: practical exercises at Berlin, 429; and transformation in experimental physics, 18n, 432–33

Labor, scientific, 15; and error analysis, 38, 168; seminar's image of, 304
Lambert, Johann Heinrich: on graphs, 251
Lampe, C. J. H., 230, 236, 471; investigation of Danzig water lines, 344–48; school teaching, 325, 334
Langberg, Christian: on confirmation of French mathematical physics, 262
Lange, C., 471
Laplace, Pierre Simon, 104; Bessel's criticism of, 67, 162; capillary theory of, 248, 276; and method of least squares, 76
Laplacian program, 63
Lautsch, C. G., 471
Legendre, Adrien Marie: and method of least squares, 75
Leibig, Justus von, 95, 143, 203
Leipzig University, 288–89; practical exercises at, 404
Lenz, Emil, 173, 175, 177
Leyst, Ernst: and geothermal station prize competition, 356–57
Lindemann, Ferdinand: and geothermal station, 355–56
Lipschitz, Rudolf: at Bonn, 406; at Breslau, 395, 421; school teaching, 361
Listing, J. B.: and Göttingen mathematico-physical seminar, 409
Lommel, Eugen, 428
Lorek, Emil Franz, 471
Lorey, Wilhelm: on Königsberg seminar, 3
Ludwig, Carl, 236, 319
Luther, Eduard: courses on least squares at Königsberg, 290, 300; elementary mathematics lectures, 252
Luther, F., 471
Lyotard, Jean-François: on performativity in modern science, 308

Mägis, Eduard, 390, 471
Magnetism, studies of: at Göttingen, 134, 172, 173, 174, 454, 455n; in Königsberg seminar, 134, 136–37, 189
Magnus, Gustav, 33, 192, 246, 454; colloquium and practicum, 4; private laboratory, 203; students' treatment of errors, 221–22, 261–63; teaching, 428

Mathematical instruction: south German weakness in, 404, 406, 416–17, 430
Mathematical physics: contribution of Neumann's lecture courses to, 5–7; distinctions made at century's end, 279–80; institutional classifications of, 35, 155–56, 157; institutional constraints upon, 393–97; institutional reinforcement at Königsberg, 85; promotion on 1866 teaching examination, 268–69; separation from exact experimental physics, 286; teaching of, 35; Voigt's definition, 413. *See also* French mathematical physics; Theoretical physics
Matthias, G. O. R., 471
Mayer, Robert, 306
Maxwell, James Clerk: electrodynamics, 412; kinetic theory of gases, 274, 371, 372, 373; and meaning of statistics, 453
Mechanics, didactic role in theoretical physics: analytical mechanics in, 145; in E. G. Fischer's textbook, 156–57; institutional classification of, 126, 156; in Neumann's lecture course and seminar, 125–26, 133–35, 144–61, 165–71, 215; and pedagogical definition of theoretical physics, 154–56; pedagogical reshaping under Kirchhoff and Voigt, 398–402; pendulum in, 146–49, 154; as response to student needs, 10, 125–26, 142, 153; stability of seminar due to, 126. *See also* Besselian experiment; Causality; Error analysis; Exact experiment; Königsberg University mathematico-physical seminar, physical division
Mechanics, studies of: capillarity, 143, 220–22, 246–50; elasticity, 109; viscosity, 280–85
Mecklenberg, B., 470
Method of least squares, 12, 139, 156, 163, 211, 228, 233, 242, 254, 260, 379, 451, 452; Bessel on, 66, 68, 70, 73, 76, 160–61; and Besselian experiment, 170; epistemic function of, 453; and ethos of exactitude, 367; French unpopularity, 163n; Gauss on, 66, 75, 160; introduction into physics, 76–77, 95; Jacobi on, 160; Königsberg interest in, 159–60; Neumann on, 62, 64, 77, 78, 160–61; pedagogical function of, 167–68, 403, 405, 411, 419, 424, 433, 444; and pendulum, 146, 147–48; in seminar, 135, 199, 237; students' use of, 35, 187–88, 283, 295, 346; theoretical physics, role in, 64, 130–31, 135, 164, 215, 260, 288; in Voigt's mechanics course, 400, 402; Wüllner's objections to, 278–79. *See also* Error anal-
ysis; Mechanics, didactic role in theoretical physics
Meyer, Ernst Heinrich Friedrich, 32, 46, 117–18, 209
Meyer, J. J. H. T., 471
Meyer, Julius Lothar, 230, 236, 471; memories of seminar, 266
Meyer, Karl Otto, 112, 134, 136, 144, 471
Meyer, Oskar Emil, 230, 236, 237, 238, 266, 303, 345, 361, 377, 471; on analogy, 387; dissertation on viscosity, 280–85; on interpolation and graphical analysis, 256–58, 383; positions in mathematical physics, 394–97; practical exercises at Breslau, 421–22; research, 371–74; as Von der Mühll's teacher, 285
Michelis, A., 471
Milinowski, C. G. Alfons, 471
Minnigerode, Ludwig Bernhard, 236, 237, 272, 397, 437, 471; seminar investigation, 271
Mischpeter, Carl Emil, 355, 471; geothermal studies, 349, 353–54
Mitscherlich, Eilhard: objection to interpolation, 247–48, 250, 256; objection to Laplace's capillary theory, 248, 256
Möller, P., 471
Momber, Albert, 471; research, 338; school teaching, 324
Mörstein, August E. F. von, 471; school teaching, 329
Moser, Ludwig: gymnasium teaching, 122; at Königsberg, 83–85, 434; lecture courses, 125, 306; and natural sciences seminar, 92, 93, 117, 118–20, 200; physics teaching, 209–10, 405; political views, 194; private physical laboratory, 269; quantification in physics teaching, views on, 85–86, 92, 94; student clientele, 306
Müller, Johannes: on exercise sessions and practica, 198
Münch, A. J. W., 471
Muncke, Georg: on method of least squares, 76; on physics, definition of, 155–56; teaching at Heidelberg, 414–15
Munich University mathematico-physical seminar, 2, 408, 426, 448; practical exercises in, 427–28
Müttrich, Gottlieb Anton, 217, 219, 223–34, 257, 471; dissertation on optical constants, 257, 259, 287; forestry studies, 388; school teaching, 323–24, 361

Nagel, H. A., 471
Naturphilosophie, 33, 47, 156
Navier, Claude, 80, 81

Neohumanist curriculum: criticisms of, 30, 53, 59
Neumann, Carl Gottfried, 210, 216, 217, 336, 394, 400, 471; on graphs, 387; as mathematical physicist, 388–89; as teacher of Franz Neumann's seminar students, 289, 322, 342; teaching, 362, 404
Neumann, Ernst Richard: practical exercises in mathematical physics at Breslau, 422–23
Neumann, Franz Ernst: Bessel's experimental style, adaptation of, 66, 73–75, 77–81, 130; dissertations mentored: 137–39 (Brix), 135–36 (Ebel), 286–87 (Frölich), 299–300 (Just), 184–86 (Kirchhoff); 256–58, 281–85 (Meyer), 287 (Müttrich), 256, 286–87 (Saalschütz), 144 (Schinz), 290–93 (Voigt), 287–88 (Wangerin), 368; early physics teaching, 57–59; education, 33–34, 96–97; and error analysis, 74–75, 130–31; French mathematical physics, reading of, 62–64; French mathematical physics, reclassification of, 93–94, 98; geothermal measurements, 120–21, 349–50, 357; on hypotheses, 65; on induced currents, 175–78, 188; instruments, modifications of, 390; Königsberg appointment, 32, 34–35; lecture courses, 6–8, 10, 85–86, 109, 124–26, 134, 144–54, 157–61, 163–64, 165–71, 172–75, 210–11, 304–7, 464–68; mathematical physics, early interest in, 35–36; mechanical foundation of research, 157; on mechanical theory of heat, 306; and method of least squares, 62, 64, 75, 77, 160–61; on method of repetition, 65, 74–75, 77, 393; and natural sciences seminar, 46, 104; on optics, 80–81, 128–31; pedagogical strategy in seminar, 5, 10, 13, 16–17, 105–6, 183, 190, 196, 200–204, 206–7, 212, 272, 308; personal life, 126–27, 131; political views, 116, 194; on precision and accuracy, 61, 63–64, 65, 73–75, 77, 82, 130–31, 159, 174; on quantification, 61–64, 130–31; role in establishing mathematical physics, 5–7, 17–18, 81, 95–96, 143; seminar, missed opportunities in, 202–4; on specific heats, 61, 77–80; on thermal conductivity, 306–8; thermometer calibration, 138. *See also* Königsberg University mathematico-physical seminar, physical division; Königsberg University physical laboratory; Mechanics, didactic role in theoretical physics
Neumann, Luise, 10

Newton, Isaac, 58, 283
Nicolai, Otto C. R., 471
Noack, Karl, 426, 430
Nordmann, Gottlob, 96
Noske, R., 471

Oelsnitz, A. W. F. G. C. von der, 471
Ohm, Georg Simon, 173, 175, 435; factors contributing to acceptance of galvanic theory, 12, 165
Optics, studies of, 102–3, 104, 107–8, 109, 128–31; double refraction, 80–81; Newton's rings, 287–88; optical absorption, 239–41; optical constants, 287; photometry, 225–30; rainbow, 299–300; reflection and refraction, 285–86

Pahlen, Louis Eduard Carl von, 103, 471
Panofsky, Erwin: on mental habits, 5
Pape, Carl Johannes Wilhelm Theodor, 237, 238, 355, 397, 471; criticism of Regnault's specific heats, 297–98, 378–82; early career, 296–97; education, 96; Göttingen position, 437; gunpowder investigation, 297; Königsberg position in experimental physics, 437–38, 439; on school teaching, 321; seminar investigation, 271
Paucker, Magnus Georg: on method of least squares in physics, 76
Pauley, M., 471
Pebal, Leopold von, 236, 471
Pernet, Johann, 392, 471; memories of seminar, 266–67, 456; on teacher training and theoretical physics, 322; thermometric studies, 390
Pestalozzi, Johann Heinrich, 115; pedagogical methods in science instruction, 37–38, 59
Peters, Johann Paul Gotthilf, 352, 471; school teaching, 329; seminar notes of, 150–52, 275–78
Peters, R. H., 471
Physics instruction, in secondary schools, 121–23, 216, 325–26; courses and curricula, 100–101, 327, 330, 331; error analysis in, 322, 329–30, 333, 334–35; instruments in, 329, 330–31, 334; physical cabinets and laboratories in, 327, 329–31, 332, 333; practical exercises, 332–33, 334–35; precision measurement, 330, 333; problem sets, 328; quantification in, 328–29; textbooks, 327–28. *See also names of individual teachers*
Pickering, Edward, 403

Pietzker, Wilhelm Friedrich Christian, 471; criticism of precision measurement, 432

Pistor, Karl Philipp Heinrich, 83

Plücker, Julius: teaching, 405–6

Poggendorff, Johann Christian, 33, 128, 192, 247; on Neumann's investigation on induced currents, 176; reaction to Kirchhoff's investigation, 182

Pohl, Georg, 421

Poiseuille, Jean Léon Marie: criticisms of his theory, 166–67, 383; fluid motion in tubes, 166–67, 372, 373, 376

Poisson, Siméon-Denis, 35, 63–64, 81, 94, 130, 173, 377, 435; absorption of solar heat, Frölich's criticism of, 286; elasticity equations, Kirchhoff's criticisms of, 370; measurement in work of, 162

Powel, F. Adalbart, 471; teaching examination, 336

Practical exercises, 2, 402–5; early at Königsberg, 25, 32, 38, 39, 40, 41, 42–43, 59–60; educational reform, part of, 25; exactitude in, 402–33; in law, 90; in mathematics, 86–88; ministerial recommendations on, 115–16, 197–99; in the natural sciences, 38, 40, 55, 88–89, 117–19. *See also* Precision measurement; *names of individual science seminars*

Practice: defined, 15; and science curricula, 14; seminar theme, 17

Precision measurement, 70, 72, 77, 78, 139, 247, 255, 281–82, 283, 284, 288, 290–92, 294; in Britain, 14n; and corrected results, 226; culture of, 348; economic value of, 344, 348; Göttingen vs. Königsberg styles, 121, 454–56; Kohlrausch's students, as performed by, 264; pedagogical function of, 224, 308, 426, 429–33, 444, 446; and pendulum, 147–49; and theory, 377. *See also* Bessel, Friedrich Wilhelm; Neumann, Franz Ernst; Physics instruction, in secondary schools; Practical exercises; *names of individual science seminars*

Probability calculus, 210, 405, 425, 428; and reasoning, 453–54. *See also* Method of least squares

Prophet, C., 471

Pure mathematics, 445, 446–47. *See also* Truth: mathematical

Pyenson, Lewis: on science teaching, 18

Quantification: controversies over, 85–86, 92–96, 378–86. *See also* Error analysis; Exact experiment; French mathematical physics; Graphical analysis; Interpolation; Interpolation formulas; Mathematical physics; Method of least squares; Physics instruction, in secondary schools; Precision measurement; Theoretical physics

Quincke, Georg Hermann, 222, 230, 235, 236, 253, 256, 336, 363, 456, 471; affection for Neumann, 241; education, 241–42; practical exercises at Heidelberg, 414, 419–20; practical exercises at Würzburg, 403–4; preference for error analysis over the material improvement of an instrument, 391; research, 242–50, 374–75, 393, 435

Radau, Jean Charles Rudolph, 219, 306–7, 391, 471

Radicke, Eduard Albert, 471

Rahts, Johannes, 471

Rathke, Heinrich Bernhard, 272, 471

Ravetz, Jerome: on craft knowledge and skills, 360, 366; on judgments of value, 261; on scientific schools, 8

Regnault, Henri Victor, 278; measurements criticized by seminar students, 297–98, 378–86; objections to analytic methods of exact experiment, 378–80

Reichel, Otto E. J., 235, 237, 471

Remer, Wilhelm, 39, 40, 42

Repsold, Johann Georg, 69, 81

Research: ethos, 22; and teaching, 86, 91–98, 158–71, 360–65

Reuter, F. V., 472

Richelot, Friedrich Julius, 3, 108, 141, 310; contribution to 1866 Prussian teaching examination, 267–68; evaluation of Heidelberg mathematical instruction, 417; and Königsberg seminar, 132, 133, 141–42, 178; lectures, Heinrich Martin Weber on, 310; on pedagogical role of geometry, 196–97, 252–54; pedagogical strategy in seminar, 196–97, 212–14, 216–17. *See also* Königsberg University mathematico-physical seminar: mathematical exercises

Riecke, Eduard: experimental style, 264; practical exercises at Göttingen, 412, 414

Riemann, Bernhard, 397

Röntgen, Wilhelm, 426

Rose, Gustav, 34

Rosenberger, August: as Bessel's student, 320

Rosenhain, Johann Georg, 112

Rostock University: mathematico-physical seminar, 2, 408, 427, 448

Rowe, David E.: on teaching of descriptive geometry, 255
Rumler, G. O. E., 472; research based on Neumann's data, 336–37
Ruppel, F. Oskar, 472; and teaching examination, 336

Saalschütz, Louis, 217, 236, 256, 272, 349, 472; dissertation, 286–87; on geotemperatures, 322–23; on Neumann's geothermal data, 287; seminar-based investigation, 271
Sanio, G. Theodor, 472
Sanio, Paul, 472
Schaefer, Clemens: as theoretical physicist at Breslau, 423
Schäwen, J. A. Hermann von, 472; research, 336; school teaching, 334
Schäwen, P. C. A. von, 337, 472
Scheeffer, E., 352, 472
Schellbach, Karl, 2, 430
Scherk, Heinrich Ferdinand: mathematical exercises at Halle, 86, 88
Schindler, Ernst G. H., 235, 237, 272, 472; school teaching, 325–26
Schinz, A. Emil, 143, 363, 472; professional discontent, 361, 362; research, 144, 338–39
Schlicht, Paul Conrad, 472
Schmidt, Adolf: response to geothermal station prize question, 357–59
Schoch, W., 349, 350, 472; professional discontent, 361–62
Schön, Theodor von, 190–91, 194–95; support of natural sciences at Königsberg, 37, 53
Schönemann, Theodor, 103, 472; research on weighbridges, 330, 333, 334, 338, 364
Schroeder, F. W. K. Ernst, 322, 362, 472
Schroeder, L. F. H., 472
Schröter, Heinrich Eduard, 213, 361, 395, 421, 422
Schubert, Friedrich Wilhelm: Königsberg history seminar, 45
Schubring, Gert, 47, 87–88
Schumacher, Heinrich, 69
Schumann, Johannes Hermann Eduard, 323, 472
Schumann, Julius Heinrich Carl Eduard, 103, 105, 107, 109, 111, 113, 153, 172, 192, 216, 472; research, 339, 348–49, 350
Schuster, Arthur: on Heidelberg mathematico-physical seminar, 419; on Königsberg school of theoretical physics, 13

Schwarz, C. G. E. F., 472
Schweigger, August Friedrich, 31
Schweigger, J. S. C.: and Halle seminar for mathematics and the natural sciences, 407; pedagogical methods of, 88
Schweins, Franz Ferdinand: on mathematics seminar for Heidelberg, 87
Science education, historiographic considerations: documents for *Alltagsgeschichte*, 9–12; educational laboratories and the Annales paradigm, 457–59; effect on practice and value judgments, 260–61; history of mentalities and practica, 458–59; instruction and the structure of disciplinary knowledge, 169; nonelites, study of, 364; problem defined, 5; textbooks, limitations of, 327–28; treatment of lecture courses, 6–8; treatment of science teachers, 17–18
Secondary school teachers, 17, 317, 323; and error analysis, 322, 330, 335, 339–40, 342, 344, 346–48, 349, 354, 364–65; at Königsberg schools, 323–24; negative opinions of profession, 320–21; perceived security of profession, 320; percentage identifying with physics, 323; permanent teachers, 317, 317n–19n; on practical exercises, 338–39, 343–48; and precision measurement, 330, 336–337, 339, 342, 346, 348; professional identity of, 322, 335, 347–48, 360, 363–65; professional tensions, 326, 327, 336, 337, 360–65; and research, 324, 326, 327, 330, 331, 334–54; Richelot's influence on work of, 337–38; seminar's importance to, 322, 333–37, 360, 363, 364; teaching responsibilities, 319–20, 324, 326, 361–62; temporary, 317, 321–22. See also names of individual school teachers
Seconds pendulum, 68–72; in Britain, 159n; exact experiment, model for, 146–49; in physics pedagogy, 73, 104, 146–49, 153, 159. See also Bessel, Friedrich Wilhelm; Mechanics, didactic role in theoretical physics
Seebeck, August, 33; Neumann on optical data of, 129–30, 276
Seebeck, Thomas, 174
Seidel, Ludwig von, 140, 158; at Munich, 427–28
Seminars: evolution of, 1–2; forerunners of, 42; in the humanities, 44–46; ideology of learning in, 44–45; in the natural sciences, 2; restrictions on membership, 41. See also names of individual science seminars

Senff, Carl Eduard: Bessel's comments on, 31; on Neumann's mathematical physics, 6; polarization experiments, 101–3
Shapin, Steven: on experiment, 457; on teaching, 457
Siebeck, Friedrich Hermann, 472
Social construction: neglect of educational laboratories, 457
Sohncke, Leonhard, 291, 336, 472; research, 341–42, 347, 369, 370–71, 376; teaching, 405–6
Sohncke, Ludwig Adolf, 108, 341; exercise session in Königsberg seminar, 59, 100, 103–4, 132
Stefan, Josef: criticism of O. E. Meyer, 371
Steffens, Heinrich, 421
Steiner, A., 472
Steinheil, Karl August, 427, 433
Stewart, Balfour, 373
Stichweh, Rudolf: on error analysis, 13n
Student research, 22–23, 88, 91–92, 172; at Berlin, 246–50, 261–63, 271n; at Bonn, 50, 54; at Giessen, 424; at Heidelberg, 242–46; under F. Kohlrausch, 263–64; at Königsberg, early, 37, 40; at Königsberg physical institute, 439; objections to, 91; by private arrangement at Königsberg, 101–3, 220–22. *See also* Königsberg University mathematico-physical seminar, physical division; *names of individual science seminars*
Sucker, F. W., 472

Tait, P. G., 373
Teacher training, 24, 37, 43, 199–201; in humanistic seminars, 44, 45; in mathematics, 86–88; in the natural sciences, 46, 55; and original research, 91–92. *See also names of individual science seminars*
Teaching examination: Baden's, 417; Bavaria's, 427, 431; and Bonn seminar, 52; Neumann's assistance on, 336; Neumann's criticisms of, 106, 111–12; Prussia's, 3, 24, 96, 132–33, 267–69, 430–31
Thalmann, J. Ernst G., 472
Theoretical physics: definitions of, 155, 156, 312, 313, 447; elite clientele for, 311–12; hypotheses in, 284–85; and interpolation formulas, 396–97; Königsberg vs. Göttingen styles in, 397; pedagogical definitions of, 126, 144–56; role of seminar in shaping, 303, 312, 456–57; students' lecture courses in, 397–98. *See also* Graphical analysis; Inter-

polation; Interpolation formulas; Mathematical physics; Mechanics, didactic role in theoretical physics
Thiesen, Max Ferdinand, 390, 392, 472
Tilling, Laura: on graphical analysis, 251
Tischler, Friedrich Carl Adalbart, 472
Tischler, Otto Emil Friedrich, 472
Tollinger, Johann: on error analysis, 263
Truth: mathematical, 72, 73, 94, 156, 170–71, 367, 400, 450, 453, 455; probable, 367, 452–54; scientific, 13, 72, 73, 85–86, 94, 139, 170–71, 313, 367, 381, 389, 432, 453, 455, 456
Tübingen University mathematico-physical seminar, 2, 408, 448; practical exercises in, 426–27

Voigt, Woldemar, 6, 11, 163, 354, 423, 472; allegory on error and accuracy, 448–50; and Bessel's pendulum, 289, 296, 435–36, 449; dissertation and its reworking, 290–96; education, 288–90; on experimental physics, changes in, 432–33; on graphs, 387; on hypotheses, 387–88; in Königsberg seminar, 266–67; lecture course on mechanics, 400–402; on mathematical physics, varieties of, 394; on mathematics and measurement, 389; position at Königsberg, 434–39; practical exercises at Göttingen, 412–14; school teaching, 327; seminar work, 266–67, 271; theory confirmation, 376, 377; as Volkmann's mentor, 441
Volkmann, Paul, 10, 11; and Bessel's pendulum, 447; on geothermal station, 354–55, 357; physics teaching at Königsberg, 439, 441–47
Volta, Allessandro, 173, 174
Von Baer, Karl Ernst, 192; departure from Königsberg, 56; and Königsberg natural sciences seminar, 43, 46, 52, 54, 92, 93; and Königsberg Physico-economic Society, 57; on science education, 31–32, 39, 40, 41, 42, 45
Von der Mühll, Karl, 298, 322, 336, 340, 388–89, 394, 472; dissertation on reflection and refraction, 285–86; teaching, 404–5

Wangerin, Friedrich Heinrich Albert, 6, 10, 11, 303, 472; education, 287; at Halle, 408; on interpolation, 257; memories of seminar, 275; professional tensions, 363, 364; research, 287, 370, 376
Warburg, Emil: on practical exercises, 450

Warner, John Harley, 327
Weber, Heinrich, 377–78, 390, 397, 437, 472
Weber, Heinrich Martin, 310, 418, 472; on causality and probability calculus, 453–54; on mathematics and measurement, 389
Weber, Max: on calling, 460; on ideal types, 367; on rationalization, 308, 312; on tensions in teaching and research, 324; on workaday world, 365
Weber, Wilhelm, 3, 72, 88, 116, 121, 191, 273, 275, 455; criticisms of instruments of, 149, 231–32, 343; in Göttingen mathematico-physical seminar, 409–10; on interpolation formulas, 455; on Kirchhoff and Neumann, 394; research, 173–74, 185–86, 188, 279; teaching, 397, 404
Wegner, August Carl Robert, 472
Weights and measures reform, 68, 69, 160, 168, 392, 449; and seconds pendulum, 159n
Weiss, Christian Samuel, 33, 34, 61, 62, 77, 127
Westphalen, Hermann L., 178–79, 472

Whewell, William: on truth value of graphs, 256
Wiener, Otto, 410
Wiesing, Otto Hermann, 472; school teaching, 334
Wild, Heinrich, 234, 256, 286, 322, 362, 391, 392, 405, 472; research, 225–32, 369
Wissenschaft, 13, 22, 44
Wittrien, Otto, 472
Wolf, Friedrich August, 44
Wrede, E. F., 82, 86
Wüllner, Adolph: and error analysis, 278–79; on experimenter's trustworthiness, 385–86; Lehrbuch der Experimentalphysik, 278; and measurement, 278–79; on Regnault's accuracy, 384
Würzburg University, 31, 403; mathematical seminar, 404

Zimmermann, M. R., 472
Zöppritz, Karl Jacob, 237, 238, 272, 387, 472; on data, 383; education and early career, 300–301; on error analysis, 386; mother's worries, 300; practical exercises at Giessen, 424, 425–26; research, 300–301, 377; seminar investigation, 271

Library of Congress Cataloging-in-Publication Data

Olesko, Kathryn Mary.
   Physics as a calling : discipline and practice in the Königsberg seminar for physics /
Kathryn M. Olesko.
      p.   cm.—(Cornell history of science series)
   Includes bibliographical references and index.
   ISBN 0-8014-2248-5 (alk. paper)
      1. Physics—Study and teaching (Higher)—Germany—History—19th century.
2. Physics—Germany—History—19th century.   I. Title.   II. Series.
QC47.G3043   1991
530'.71147'47—dc20   90-55717